Hybrid Electric Power Train Engineering and Technology:

Modeling, Control, and Simulation

Antoni Szumanowski
Warsaw University of Technology, Poland

Managing Director: Lindsay Johnston
Editorial Director: Joel Gamon
Book Production Manager: Jennifer Yoder
Publishing Systems Analyst: Adrienne Freeland
Development Editor: Austin DeMarco
Assistant Acquisitions Editor: Kayla Wolfe
Typesetter: Lisandro Gonzalez
Cover Design: Jason Mull

Published in the United States of America by
Engineering Science Reference (an imprint of IGI Global)
701 E. Chocolate Avenue
Hershey PA 17033
Tel: 717-533-8845
Fax: 717-533-8661
E-mail: cust@igi-global.com
Web site: http://www.igi-global.com

Copyright © 2013 by IGI Global. All rights reserved. No part of this publication may be reproduced, stored or distributed in any form or by any means, electronic or mechanical, including photocopying, without written permission from the publisher.
Product or company names used in this set are for identification purposes only. Inclusion of the names of the products or companies does not indicate a claim of ownership by IGI Global of the trademark or registered trademark.

Library of Congress Cataloging-in-Publication Data

Szumanowski, Antoni.
 Hybrid electric power train engineering and technology : modeling, control, and simulation / by Antoni Szumanowski.
    pages cm
 Includes bibliographical references and index.
 Summary: "This book provides readers with an academic investigation into HEV power train design using mathematical modeling and simulation of various hybrid electric motors and control systems"-- Provided by publisher.
 ISBN 978-1-4666-4042-9 (hardcover) -- ISBN 978-1-4666-4043-6 (ebook) -- ISBN 978-1-4666-4044-3 (print & perpetual access) 1. Hybrid electric vehicles--Power trains. I. Title.
 TL260.S98 2013
 625.26'3--dc23
                                2013001607

British Cataloguing in Publication Data
A Cataloguing in Publication record for this book is available from the British Library.

All work contributed to this book is new, previously-unpublished material. The views expressed in this book are those of the authors, but not necessarily of the publisher.

Table of Contents

Preface vi

Acknowledgment xxiii

Chapter 1
Introduction to The Hybrid Power Train Architecture Evolution 1
    Introduction 2
    Hybrid Power Trains Architectures Engineering Development Review 9

Chapter 2
The Energy–Power Requirements for HEV Power Train Modeling
and Control 23
    Introduction 24
    Energy and Power Distribution Dynamic Modeling 25
    The Modeling of the Hybrid Power Train Energy Flow 39
    An Approach to the Control of Hybrid Power Trains 42
    The Method of Determination of the Discharging Accumulator Factor
        (SOC): Minimal Internal Losses of Energy 46

Chapter 3
Electric Machines in Hybrid Power Train Employed Dynamic
Modeling Backgrounds 52
    Introduction 53
    AC Asynchronous Induction Motor Modeling 53
    PM Synchronous Motor Modeling 77

Chapter 4
Generic Models of Electric Machine Applications in Hybrid Electric
Vehicles Power Train Simulations 95
    Approach to a Power Simulation Model of a Driving System with an AC
        Induction Motor 96
    PM Permanent Magnet Motors Modeling 120
    Approach to a Power Simulation Model of a Drive System with a PM
        Synchronous Motor 123

Chapter 5
Nonlinear Dynamic Traction Battery Modeling 151
    Introduction 152
    Main Features of Most Common Batteries Applied in HEV and EV
        Power Trains 155
    Fundamental Theory of Battery Modeling 157
    The Basic Battery Dynamic Modeling 160
    Nonlinear Dynamics Traction Battery Modeling 175

Chapter 6
Basic Design Requirements of an Energy Storage Unit Equipped
with Battery 194
    Introduction 195
    Battery Management System Design Requirements 195
    Battery and Ultra Capacitor Set in a Hybrid Power Train 206
    Influence of Temperature on Battery and Super Capacitor's Voltage
        Equalization 221
    Voltage Equalization 222

Chapter 7
Basic Hybrid Power Trains Modeling and Simulation 229
    Introduction 230
    The Internal Combustion Engine as a Primary Energy Source:
        Dynamic Modeling 230
    Series Hybrid Drive 238
    Drive Architecture Equipped with an Automatic
        (Robotized) Transmission 253
    Split Sectional Drive 260

Chapter 8
Fundamentals of Hybrid Power Trains Equipped with
Planetary Transmission 267
    Introduction...268
    Planetary Gear Power Modeling...268
    Design of the Planetary Gear with Two Degrees of Freedom Applied
        to the Two-Source Hybrid Electric Drive Systems......................271
    Planetary Gears Possible for Application in Hybrid Power Trains............273
    Compact Hybrid Planetary Transmission Drive (CHPTD) 283
    Design of Power-Summing Electromechanical Converters 309

Chapter 9
Basic Simulation Study during the Process of Designing the Hybrid Power
Train Equipped with Planetary Transmission 322
    Introduction...323
    Simulation Studies of the Hybrid Electric Power Train Based on an
        Urban Bus 325
    Simulation Studies of the Hybrid Electric Power Train Based on the
        Shuttle Service Bus..336
    Analysis of Vehicle Performance Sensitivity to Mechanical Ratio 342

Chapter 10
Plug-In Hybrid Power Train Engineering, Modeling, and Simulation 357
    Introduction...358
    Pure Battery Mode Power Train Operation 359
    Mechanical Transmission Concept Proposals 361
    Exemplary Plug-In Hybrid Power Train Analysis 380

Appendix..406

Compilation of References 408

About the Author 420

Index...421

Preface

Not many books have been published so far devoted to the designing of hybrid electric vehicle power trains. The existing publications delineate mainly the other types of power trains or introduce only basic knowledge concerning the operations of Hybrid Electric Vehicle (HEV) power trains.

However, many articles and presentations are produced during such symposiums as the EVS (Electric Vehicle Symposium), which do indeed focus on EV and HEV power train construction. These publications, though, are usually based on particular solutions and research (mostly simulations) and lack the synthetic approach to the HEV power train construction, while their authors often use simple linear mathematical models, even resorting to such academic basics as the ADVISOR formula. From the formal point of view, such an approach is unacceptable as far as HEV power trains are concerned. HEVs are characterized by the multiplicity of components operated according to a complex control function, which renders them highly nonlinear. Thus, mathematical modeling has to take into consideration this particular dynamic and the non-linear capacity of HEV power trains. Factors that are especially demanding while modeling are such power train components as the electro-mechanical battery, combustion engine, electric machine, and their control systems. Hence, the preparation of the adequate mathematical model is crucially responsible for the outcomes of simulation studies.

Fortunately, we have at our disposal the results of the laboratory studies concerning every component of the HEV power train. Relying on these results, we can construct the "maps" as the static functions of, for example, the torque and angular velocity, which can be later applied in the form of tables in engineering simulation studies. Yet, from the academic point of view, this approach is only supplementary.

Car producers eagerly employ in the emerging hybrid solution components that are predominantly applicable to the classic internal combustion engines and mechanical transmission production. Such an attitude is understandable, and economical, due to the marginality of the current market of electric vehicles. However, Toyota, who started to mass-produce self-designed hybrid cars in 1997, is a creditable exception to this mainstream routine. The challenge that is addressed in the contemporary automotive industry is integrated thinking. If we want to design an efficient and economical HEV power train, the elements have to be adjusted individually, at each stage to the given power train, which concerns the adjustment of the battery.

If we assume that the development of hybrid vehicles and their usage are essential, it appears compulsory to design a construction that would be highly effective electrically, inexpensive, and thus generally available. The designing of HEV power trains fundamentally differs from the designing of the classic Internal Combustion Engine Vehicles (ICEVs), for example in terms of mechanical transmission, which does not have to be that expensive, as the existing Automatic Manual Transmission (AMT) or Dual Clutch is constructed only for the conventional power train.

In general, the designing of hybrid power trains takes little notice of the mechanical parts of the power train, which is not proper practice, as the correctly designed mechanical components can enhance the efficiency of the power train by about 10%, especially during regenerative braking, and duly lower the cost of the hybrid system. These gains cannot be disregarded in the optimization of the mechanical parts in EVs and HEVs.

Hence, the key to HEV power train designing is the nonlinear dynamic modeling applied in simulation studies. Of course, the other crucial element in such designing is the original power train structure concept, as well as the construction, and a suitable choice of its components on the basis of simulation-designing results.

This book presents such a HEV power train designing process, and its aim is to introduce the application of nonlinear dynamic modeling to the general design of the vehicle, as well as its power train and control strategy.

This book is an attempt to apply a holistic approach on an academic level to HEV power trains design by mathematical modeling and simulations based on the long-enduring research experience of the author. Emphasis is put on the importance of the energetic analysis of the power train, which allows for the minimization of energy parameters in both primary and secondary sources of energy, and stands as the first step in proper design.

This publication also discusses the electric machines modeling and their control systems, drawing on their synthetic model for the motor and generator operation with emphasizing field vector-oriented control modeling, as well as for asynchronous induction motor (AC) and permanent magnet motor (PM).

My work also refers to the Internal Combustion Engine (ICE) applied in hybrid power trains, using the solution based on experimental research data.

Special attention has been paid to traction batteries, both of High Power (HP) and the High Energy (HE) types, which are essential in power train operation, yet are the most costly, and thus require re-designing.

Finally, the book focuses on the designing of the mechanical transmission. The planetary transmission with two degrees of freedom controlled by electric machine and additionally equipped with constant ratio transmissions controlled by the proper electromagnetic clutches has been considered, too. Moreover, the design of simple automatic transmissions, exemplified by the ball or belt transmissions, has been submitted. Regarding the internal combustion and electric motors, the application of their "maps" in simulation studies has been simultaneously discussed, with a comparison of the results of these studies and the studies employing nonlinear models.

The fundamental goal of the book is to present the methodology of designing, allowing for the construction of the most energy-efficient power train. Such high efficiency entails minimization, respecting the limiting conditions, of both mechanical and electric energy consumption, and in practice, involves the minimization of fuel consumption in ICE (with regard to the limitations, referring to the change of the battery's state of charge).

The properly-designed HEV has to consider the following features:

1. Vehicle start is in pure electric mode (generally vehicle acceleration must be supported by motor-battery set).
2. ICE operation has to start referring to momentary mechanical transmission ratio, when the engine working area corresponds to its best efficiency (hybrid mode of power train operation).
3. Regenerative vehicle braking has to exist as well during hybrid mode power train operation, decreasing vehicle speed by motor-generator braking, or by pure electric mode (ICE switch off) effectively for as long as possible. Increasing the entire hybrid power train efficiency and the elongation of the regenerative braking range during vehicle deceleration are two of the most important problems to be discussed in this book.

This publication, on the one hand, shows the changing of the mechanical transmission ratios influence on electricity and fuel consumption and, on the other hand, on the total efficiency of the hybrid power train. This has significant meaning during the battery parameters design. The decreasing internal power losses of freeing from the battery electric motor (caused by its load current decreasing) is the result of motor co-operation with proper design specific automatic transmission. It significantly influences the entire power train's efficiency (total internal loss reduction) causing the minimization of energy consumption. The properly adjusted momentary mechanical ratios (automatic gear transmission) is not only very important during vehicle acceleration, but above all during its regenerative braking. The time extension of vehicle regenerative braking, according to its acceptable minimal linear speed, can be obtained by proper automatic transmission ratio adjustment. This problem is duly considered in this book.

The part devoted to electrochemical energy sources refers also to the ultra capacitors, especially to their cooperation with the battery.

Modeling and simulation studies presented are applicable to the designing of both types of power trains—full hybrid and plug-in hybrid.

The simulation and experimental studies carried out by the author and the group of his PhD students and researchers constituted the base of the enclosed analyses. As a result of this research, some construction proposals are presented. This is important because it allows for usage of the mathematical models verified in the laboratory, ensuring the originality of the presented solutions.

Other similar published solutions are also referred to, but are found in mainly insignificant symposium papers, which, according to the author's opinion, are currently the most essential source of knowledge concerning hybrid electric vehicle power trains.

In obtaining maximum efficiency from the EV & HEV power trains, the design of the entire power train, as well as its components, should be made by proper modeling and simulation. On the other hand, the power train component parameters and its operational control are defined by simulation studies. These studies, when applied to standard statistical vehicle driving cycles, should not be carried out. The static vehicle driving cycles are mostly only useful for different power trains' architecture and control comparisons. The research tests of power train operation for required real driving conditions are necessary.

This book emphasizes the role of mechanical components applied in HEV power trains. It shows mechanical and electrical device integration, as it is the best source of knowledge for both mechanical and electrical engineers.

Specific engineering problems concerning real machine construction in this book are neglected. The enduring experiences in, for example, ICE, motors, batteries, or mechanical transmissions, permit the adjustment of its construction to EV & HEV requirements. For instance, thermal management during power train component operation should be individually designed, with, respectively, its dynamic load or overload, and time at its most lasting.

The minimization of these components, mass and volume is the construction goal, which is especially important for its implementation in future ultra-light vehicle bodies.

An accurate and adequate non-linear dynamic modeling and simulation method suitable for everyone's power train energetic optimization done by computer computation, which is presented here, is a "sine qua non condition" for EV& HEV power train design.

The hybrid power train control strategy should be based on hybrid power train minimization internal losses, which means obtaining minimal energy consumption, in relation to the entire power train, as well as to everyone's considered drive component, respectively to its input-output torque and speed distribution. Control strategy, especially defined in this way, is the background to local controller construction.

The main criteria of control strategy are:

- The internal combustion engine fuel consumption should be as low as possible. This could be resolved by the IC engine operating in the area of minimal specific fuel consumption;
- The battery electricity consumption should be as low as possible.

The balance of battery energy has to be equal to zero (only for full hybrids), while the k-factor (State of Charge) should be the same at the beginning and at the end of a static vehicle driving cycle. In the case of PHEV, pure electric and hybrid power train operation has to be strictly defined to obtain the most efficient area of battery operation, which means its state of charge factor value alteration must be correctly controlled in relation to the maximal range of vehicle driving. The good condition of electricity transfer by battery must be defined in choosing power train architecture and real driving requirements:

- Alterations to a battery's internal resistance should be limited (keeping it in range of the minimal value of analyzing a battery's internal resistance)
- The battery's current should not exceed the determined limited value
- Electric motors must operate in the area of highest efficiency during a vehicle driving cycle

The above-mentioned target can be obtained using only proper modeling, as well as all drive components as a whole power train and powered by vehicle. It means the preferred nonlinear dynamic models provide more accurate simulations as the best tools for HEV and PHEV design.

This book consists of ten chapters and the scope of all of them is in the details depicted below:

Chapter 1 includes presentation of the most important hybrid power train architectures, as well as their function and construction evolution. The main hybrid vehicle power trains, selected according to their function, are considered here in detail. Generally, there are two main hybrid drive types that are possible to define.

The first is the "full hybrid" drive, which is a power train equipped with a relatively low capacity battery that is not rechargeable from an external current source and whose energy balance—State Of Charge (SOC)—is obtained by regenerative braking and proper Internal Combustion Engine (ICE) operation during the entire time of driving the vehicle.

The second one is the "plug in hybrid," which means the necessity of recharging the battery by plugging into the grid when the final State Of Charge (SOC) of the battery is not acceptable. In this case, one would recommend the low current overnight charge.

Both of these hybrid drive types are not mainly differentiated by their power train architecture. The "plug in hybrid" is suited to the larger capacity high-energy battery (HE), supported only by the internal combustion engine. The solution consists of downsizing the internal combustion engine, contemporarily called the "drive range extender," which is closer to the pure electric power train. The "plug in" version connects the pure electric and the "full hybrid" drive's features, and the assumed further battery development is very promising.

The hybrid power train is a complex system. It consists of mechanical and electrical components, and they are all important. The evolution of the Hybrid Electric Vehicle (HEV) and the electric battery-powered vehicle (EV) power trains is presented here from a historical point of view.

The fuel consumption difference between the pure Internal Combustion Engine (ICE) drive and the hybrid drive is especially emphasized. The chapter contains the comparison of "maps" of the internal combustion engine operation, for example the ICE's static characteristics of its shaft output torque versus its rotary speed referring to the selected vehicle driving cycle.

This is not only important for fuel economy, but also for the emission of the internal combustion engine. It has to be particularly stressed that CO_2 emission is a derivative factor induced by the energy utilization. Thus, the political and economic incentives refer directly to the reduced use of energy. The energy cost is defined by its amount ratio (MJ/l) and the CO_2 emission by its g/km factor. The lowering of the cost of energy is related to the general drive system efficiency and the vehicle's weight. The suitable drive cycles are taken into account to compare the new evaluation of energy and emission costs. The fuel "calorific value" is an important element in relation to the energy density. However, the energy density's stored energy in the contemporarily best lithium–ion battery is about 30 times lower, and in terms of the space taken 8 to 10 times higher than in the case of liquid fuels. Fortunately, the average efficiency of an exemplary pure electric drive system is 3 times higher in comparison with the internal combustion engine drive. Hybrid drives, which improve the combustion efficiency of the enumerated fuels in thermal engines (in our case the internal combustion engine) are able to be used comprehensively.

The hybrid power train's power efficiency also depends on the type of the Internal Combustion Engine (ICE), equipped with a drive. This chapter presents different engines applied in the hybrid power trains, whose entire efficiency is compared to the conventional drives with gasoline or diesel internal combustion engines. Additionally, the chapter shows the Fuel Cell (FC) power train efficiency, which is also a hybrid drive, because the internal combustion engine is replaced by the fuel cell cooperating with the battery, which makes it typical from the energy flow point of view for a series hybrid power train.

This chapter discusses the review of the hybrid power trains architectural engineering. It includes development of the hybrid vehicle power trains construction from the simple series and parallel drives to the planetary gear hybrid power trains.

Finally, the chapter focuses on the fuel cell series hybrid power train, which is only shown because its operation and design are beyond the scope of this book.

Chapter 2: The first step in the hybrid vehicle power train design, of course, after choosing the drive architecture, is analysis of the power distribution and energy flow between the Internal Combustion Engine (ICE) (considered in this book only as the primary source of energy – PS) and the energy accumulator (called the source of power, or a secondary energy source – SS). The role of the Primary Source (PS) is to deliver to the system the basic energy, while the Secondary Source (SS) feeds the hybrid power train during its peak power loads and first of all stores the vehicle's kinetic energy during regenerative braking. The flow of energy among the vehicle's traction (road) wheels, the storage unit, in the case of the battery discussed in this book (SS), and the primary energy source (ICE) is the crucial problem at the beginning of the hybrid power train design process. The target of these considerations is to search for the minimal necessary power of the Primary Source (PS) and the minimal energy capacity of the Secondary Source (SS). Certainly, this computation requires the proper energy flow model and the basic vehicle driving cycle, in the role of which the static driving cycle is recommended. The main aim of this chapter is the depiction of the above problem, as well as the finding of its solution.

In general, in this chapter, the two hybrid power train systems, as the background to other more developed and advanced propulsion systems are taken under modeling and mathematical analysis. These are the series and parallel hybrid power train. They follow the basic computation equations.

An essential problem encountered in the designing of power trains for vehicles with hybrid or pure electric drives is to make a precise assessment of their energetic effectiveness. Such driving structures differ from each other, for example in the types of batteries (primary sources in purely electric drives, and secondary ones in those of the hybrid nature), electric motors, CVT (Continuously Variable Transmission) assemblies, and the like. The determination of the internal efficiency of each of these components separately does not allow for making an overall assessment of each particular drive, enabling us to compare all available drives and to choose the most appropriate from them, which is of crucial importance in all designing processes. The background to the energetic evaluation of the hybrid drive structure is the dynamic determination of the internal watt efficiency of each of the propulsion system components, according to the momentary external load depicted as the vehicle's required power alteration, which is reduced on its road wheel. It is necessary to emphasize that all the energy and power calculations should be addressed to only one point of the hybrid power train system. The vehicle's traction (road) wheels (all torques and speeds of components of the hybrid power train have to be reduced to this point) are strongly suggested.

The complex construction of the hybrid drives requires an appropriate control strategy from its designers. In order to achieve this aim, numerical optimization methods of nonlinear programming (by decomposing the dynamic optimization problems into the nonlinear programming problem) can be applied. The control functions that provide the realization of the required vehicle speed distributions by the output shaft torque of the analyzed power train have to minimize the assumed criterion of quality. The quality of the hybrid vehicle operation in a significant way depends on the battery or, alternatively, on the entire energy storage unit. The method of determination of discharging the accumulator factor "k" as the indication of the battery's state of charge is discussed in this chapter.

Chapter 3 depicts the advanced modeling of motors, which is also suggested as a solution for hybrid power trains modeling and simulations.

At the present time, there are only two types of motors that can be applied. These are: the alternative current induction asynchronous (AC) motor and the permanent magnet synchronous (PM), or the brushless permanent magnet direct current (BLDC) motor, which are, in fact, types of the permanent magnet (PM) synchronous machines. This chapter presents the fundamental theory as a necessary background to the mentioned motors' generic, dynamic, nonlinear model determination.

The differential equations based on the phase quantities as the complete system of equations describing the transients should include the equations of winding voltages and the equations of motion for the rotating parts of the machine. Here, the phase quantities in terms of the resultant phasors as the basis of dynamic modeling are taken into consideration. Introducing a complex (α, β) plane, stationary and relative to the stator of a two-pole model equations set is carried out, including transformation from the α- and β- axis components of the stator quantities to the d- and q- axis components of rotor quantities. In addition, the magnetic field, flux linkage phasor, in terms of motor current phasors, is considered in this chapter. There is a definition of the voltage equation in terms of the α- and β- axis components. The voltage equation is written for the particular stator phase, by an equation determined in terms of the resultant phasor functions. The voltage equations are presented in terms of d- and q- axis components, and in terms of components, along axes rotating at an arbitrary velocity. The electromagnetic torque is expressed in terms of the resultant current along with the flux linkage phasors and their components.

The fundamental electric machine theory necessary for dynamic equations as the mathematical model, is addressed via the alternative current induction asynchronous motor (AC). This basic theory can be viewed also as a basis for the permanent magnet, as well as for the alternative current synchronous motor. In this last case, there is an explanation of its operating principles and construction evolution. A strong emphasis is put on the permanent magnet, synchronous, motor dynamic modeling, because this type of electric machine is in modern times, the most popularly applied in hybrid vehicle power trains. The reason for this wide application is that this kind of electric machine is the most efficient among universally known motor constructions.

To make the discussion simpler, the following assumptions are made for an ideal model of a synchronous machine:

1. The discussed machine is a symmetrical one (its armature windings are identical and the phase resistances are distributed symmetrically);
2. Only the fundamental harmonic of ampere-turns and magnetic flux density are taken into account, which means that the higher harmonics of the magnetic field distribution in space (due to the discrete form of the windings and the magnetic circuit geometry) occurring in the air-gap are neglected.

This chapter is a background source of the advanced knowledge concerning the principles of electric machine modeling. It may well be useful for mechanical engineers engaged in the hybrid vehicle power trains design process, but also for electrical engineers, especially those attending Masters and Doctoral courses.

Chapter 4 presents the approach to obtaining the power simulation model of electric machines that would be practically useful in hybrid power trains simulation studies. The models presented in this chapter, namely the alternative, current induction, motor (AC) model and the permanent, magnet motor (PM) mathematical, dynamic model, are based on the necessary and fundamental knowledge conveyed in the previous chapter. These generic models are here adapted to the hybrid power train's requirements, while the mechanical characteristics of the vehicle's driving system are relegated to the background. These characteristics have two zones. In the first zone, the value of the torque is constant and the value of the motion's power linearity reinforces the rotary speed. In the second zone, the motion's power is constant and the torque hyperbolically decreases with the reinforcing rotary speed. In the first zone, the rotary speed is small (from 0 to ω_b), and at this stage, vehicles usually speed up as the driving system has to overcome the resistance of inertia. In the second zone (between ω_b to ω_{max}) the motion is more uniform—there are not any big accelerations—so, the torque can be smaller and can be adequate only for driving. From the motors' control point of view, the most common method of an electric machine's torque-rotational speed regulation is the Pulse-Width-Modulation (PWM), which is taken into consideration in this chapter.

The relationships between the motor performance and the motor design parameters, and a description of the inverter/motor control strategy is presented later in this chapter. This description is a fundamental basis for most considerations involved in the selection and design of the motor, for particular application of the motor in the traction drive system.

The induction motor (AC) and the permanent, magnet motor (PM) as commonly used in hybrid electric vehicle power trains are analyzed. The permanent, magnet motors are currently rapidly gaining in popularity. There are two types of these motors that are especially common: the Permanent, Magnet Synchronous (PMS) and the permanent, magnet Brushless Direct Current (BLDC). The approach to the dynamic modeling of these motor construction types is the same and the synchronous, permanent, magnet machine is the foundation of both. The vector field-oriented control of induction AC and permanent magnet motors is applied in the conducted mathematical modeling. The influence of the controlled voltage frequency is discussed as well. In the case of permanent, magnet motors, the adjusted method of the magnetic field weakening is very important during the pulse modulation (PWM) control. The chapter presents the model of synchronous, permanent, magnet motor magnetic field weakening.

The basic simulation studies' results, dedicated especially to the previously mentioned upper electric motors, are attached. One of the targets of these simulations is the determination of these electric machines static characteristics as the function: output mechanical torque versus the motors' shaft rotational speed.

This feature is indicated as the map of electric machines connected with its efficiency in a four quarterly operation (4Q), which basically means the operation of the motor/generator mode in two directions of the shaft rotational speed, which appears very useful in practice.

Chapter 5 presents the method of determining the Electromotive Force (EMF) and the battery internal resistance as time functions, which are depicted as the functions of the State Of Charge (SOC). The model is based on the battery's discharge and charge characteristics under different constant currents, which are tested in a laboratory experiment.

Moreover, the method of determining the battery SOC, according to the battery modeling result, is considered. The influence of the temperature on the battery's performance is analyzed according to the laboratory-tested data and obtaining the theoretical background for calculating the SOC. The algorithm of the battery State Of Charge (SOC) indication is depicted in detail. The algorithm of battery State Of Charge (SOC) "online" indication, considering the influence of temperature, can be easily used in practice. The nickel metal hydride (NiMH) and lithium ion (Li-ion) batteries are taken into consideration and thoroughly analyzed. In fact, the method can also be used for different types of contemporary batteries, if the required test data is available.

The hybrid electric (HEVs) and electric (EVs) vehicles are remarkable solutions for the worldwide environmental and energy problems caused by automobiles. The research and the development of various technologies in hybrid electric vehicles (HEVs) are being actively conducted. The role of the battery as the source of power in Hybrid Electric Vehicles (HEVs) is basic and significant. The dynamic nonlinear modeling and simulations are the only tools for the optimal adjustment of the battery's parameters, according to the analyzed driving cycles. The battery's capacity, voltage, and mass should be minimized, considering its over-load currents. This is the way to obtain the minimal cost of the battery, according to the demands of its performance, robustness, and operating time.

The process of battery adjustment and its management is crucial during the hybrid and electric drives design. The approach to battery modeling, based on the linear assumption (such as the Thevenin model) and then adopted to the data obtained in experimental tests, is ignored here (see paragraph 5.3 of Chapter 5). The generic model of the electrochemical accumulator, which can be used in every type of battery, is introduced in its place. This model is based on the physical and mathematical modeling of the fundamental electrical impacts during energy conservation by the battery.

The model is oriented toward the calculation of the parameters of the Electromotive Force (EMF) and internal resistance. It is easy to find direct relations between the State Of Charge (SOC) and these two parameters. If the Electromotive Force (EMF) is defined, and the function versus the battery state of charge is known, it is easy to depict the discharge/charge state of the battery.

The model is actually nonlinear because the correlation parameters of the equations are the functions of time, or the functions of the battery State Of Charge (SOC), during battery operation. The modeling method presented in this chapter should use the laboratory data (for instance, voltage for different constant currents or internal resistance versus the battery SOC), which are expressed in a static form. These types of data have to be obtained in discharging and charging tests. The considered generic model is easily adapted to the different types of battery data and is expressed in a dynamic way using approximation and iteration methods.

The Hybrid Electric Vehicle (HEV) operation puts unique demands on the battery when it operates as the auxiliary power source. To optimize its operating life, the battery must spend minimal time in overcharge, or over the discharge. The battery must be capable of furnishing or absorbing large currents, almost instantaneously, while operating from a partial state of charge baseline of roughly 50%. For this reason, knowledge about the battery internal loss (efficiency) is significant, because it influences the battery state of charge (SOC).

Chapter 6 presents the basic requirements in energy storage unit design. Basically, the storage unit is understood as the battery, and this is practically true in the majority of cases. However, another type of electrochemical energy storage unit can be considered, which is the capacitor.

Electrochemical capacitors applied in hybrid power trains are commonly called "super" or "ultra" capacitors. The application of ultra capacitors in Hybrid Electric Vehicle (HEV) power trains does not seem to be a strong alternative to the batteries. Anyhow, the exemplary yet complex solution of the parallel connection of the battery and the capacitor as a possibility of increasing the cell's lifetime and decreasing its load currents is also discussed in this chapter.

Most important is, of course, the battery, and emphasis is put on the battery's thermal behavior, its State Of Charge (SOC) indication, and monitoring as the mainstay of the Battery Management System (BMS) design. The chapter discusses also the original algorithmic base of the nonlinear dynamic traction battery modeling. This algorithmic base includes the battery temperature impact factor. The battery State Of Charge (SOC) co-efficient presented in this chapter has to be determined in terms of its maximum accuracy. This is very important for the control of the entire hybrid power train. The battery State Of Charge (SOC) signal is the basic feedback in power train on-line control in every operation mode: pure electric, pure engine, or, in the majority of cases, the hybrid drive operation.

The battery pack modeling and design mentioned above are equally useful in terms of the High Power (HP) or High Energy (HE) batteries.

In full Hybrid Electric Vehicle (HEV) power trains, the applied battery is the High Power (HP) one. In plug-in hybrid vehicles (PHEV), the battery type is designed closer to the high energy (HE), and it is similar to this in pure electric vehicles. The differences between these two batteries are discussed earlier in Chapter 1.

In the part of this chapter devoted to the ultra capacitor, its generic dynamic model is proposed. Most important attention is paid to the lithium ion battery cooperation with the mentioned capacitor.

The modeling and simulation presented above, show some general advantages of the coupled battery/ultra capacitor storage unit (the battery plays the role of the energy source, the ultra capacitor is the power source), which is caused by the system's inertia and *RC* time-constant. The main advantages of this parallel connected set are determined by the current reduction (especially of the battery), the soothing battery voltage, and the dropping of its average value, as well as the connection in this storage set of high energy with high power density, etc. Certainly, there are some disadvantages, such as the higher cost and the set weight and volume. The analyzed hit distribution shows the highest temperature increasing on the battery's terminals.

Both energy storage devices need voltage equalization: for the ultra capacitor, this is necessary. The newest lithium ion batteries are characterized by the highest quality, which means that every cell has the same parameters as others, which permits the avoidance of costly and complex, electronic, cell voltage balance devices.

As was mentioned earlier, the contemporary super power lithium-titanate battery (see Chapter 5) is the real substitute for ultra capacitors. Anyway, the nonlinear dynamic modeling method presented in this chapter can be used successfully, also, in the case of ultra capacitors.

Chapter 7 is devoted to what is basic and existent in present vehicles, power train modeling, and simulation. There are generally a series of parallel hybrid power trains. In both cases, the role of the Internal Combustion Engine (ICE) and its dynamic modeling is significant. The two aspects of the internal combustion modeling should be considered. One devoted to energy distribution modeling, the second to local internal combustion engine control. Both issues are discussed in this chapter.

At first, the Internal Combustion Engine (ICE) as the primary energy source, and the one possible approach to the dynamic modeling method, is proposed.

Modeling of the ICE is complicated. The best solution for the hybrid drive design is to use the engine's operating maps, which are possible to obtain after special laboratory bench tests. The control of the ICE is based on its torque creation dependent on fuel injection. In this case, generally, the output response is the angular velocity of the engine's shaft, depending on the external load conditions.

When the map of the internal combustion engine has been well determined, it means that the proper dynamic model, based on real laboratory bench test permits, indicate static engine characteristics as an internal combustion engine output torque versus its rotational speed. The map can be also used for energy flow analysis.

Chapter 1 presented the power distribution process in the series hybrid drive. The power generated by the Internal Combustion Engine (ICE), theoretically, can be permanently constant or interrupted (see Figure 1 of Chapter 2). In the practical application, it is necessary to consider different control strategies of the Internal Combustion Engine (ICE) operation. The most important are the "constant torque" and the "constant speed" engine functions.

The theoretical analyses mentioned above cannot be strictly realized in practice. Firstly, the Internal Combustion Engine (ICE) generator unit is a nonlinear object. Secondly, the problem is connected to the accuracy of the controls. It means that in real time control, there is the hesitation of the Internal Combustion Engine (ICE) operating points. This impact is shown in the attached simulation results.

The other important problem is the modeling of the permanent, magnet generator connected by the shaft with the internal combustion engine. The vector graph analysis of the Permanent Magnet (PM) synchronous generator with the construction of buried magnets (see Chapter 4) is included in the chapter.

The modeling and simulation results of the series hybrid power train are discussed. The fifteen-ton mass urban bus equipped with a series power train is used as an example for the discussion.

As for the common parallel, hybrid power train, two of its types are in dynamic modeling tested by simulation. One of them is the hybrid power train equipped with an automatic (robotized) transmission. Generally, one can state that this transmission can be used as the Automatic Manual Transmission (AMT) or the Dual Clutch.

The second one is the split, sectional, hybrid power train seen as the most simple solution. The Hybrid Split Sectional Drive (HSSD) applied in an urban bus is presented in this book. Taking into consideration only the energy analyses of this bus with the hybrid drive, it is easy to note that the Hybrid Split Sectional Drive (HSSD) has typical features of the regular parallel system.

Chapter 8 describes the most advanced hybrid power trains, generally depicted in Chapter 1. The presented figures consist of the planetary two degrees of freedom planetary gears. It seems to be the best system of energy split between the Internal Combustion Engine (ICE), the battery, and the electric motor, but unfortunately, it is also the most costly solution for the manufacturing world. Nevertheless, this type of hybrid power train should be preferred as the best drive architecture composition from the technical point of view. For this reason, this chapter, in a detailed way, describes the features and the modeling approach to the planetary hybrid power train. Certainly, most attention is paid to the planetary two degrees of freedom gear, not only to this one. Cooperating with this planetary gear, additional and necessary clutches and mechanical reducers are considered as well.

The planetary gear with two degrees of freedom changes the angular velocity of the output shaft with the constant ratio of input and output torques. Therefore, it is not a classic torque and velocity continuous ratio transmission (Continuous Variable Transmission – CVT). To get the Continuous Variable Transmission (CVT) function, the planetary gear has to be torque-controlled. The best torque converter for controlling the planetary gear is the electric motor operating in a four quarterly operation (4Q), which means controlled in 4 quarters of the coordinate system. It also means that the power train system can be applied only when two of three shafts of planetary two degrees of freedom gear are connected with the Internal Combustion Engine (ICE), and, for instance, the Permanent Magnet (PM) motor fed from the electrochemical battery. Thus, the internal combustion engine, and the battery, are the two sources of energy supply. Certainly, the third shaft via the reducer and main differential gear is connected to the vehicle's traction wheels. For this reason, the function of two degrees of freedom in the hybrid power train has to be strictly defined. Its operation must be possible and effective, as well as having to provide the vehicle's acceleration—its steady speed—especially during regenerative braking. Three modes of the hybrid planetary power train are possible: the pure electric operation, partly while the vehicle is accelerating and only when the vehicle has braked, and also sometimes during the vehicle's steady speed drive, when its speed is low; the pure engine when the battery State Of Charge (SOC) factor value is too low; and finally, in hybrid operation.

The exemplary hybrid power train equipped with the mentioned planetary gear considered as the propulsion system, presented here, is the system called the Compact Hybrid Planetary Transmission Drive (CHPTD). This drive architecture is characterized by the following shaft connections: the ICE, via the mechanical reducer and the clutch-brake system is linked with the planetary sun wheel. The electric machine is connected to the crown wheel. The planetary yoke wheel transmits the sum (with a positive or negative sign, depending on the drive-operating mode) of power generated by the engine and the motor through the main and the differential gear set of traction wheels.

The role and modeling auxiliary drive components, such as the automatic clutch-brake device and mechanical reducers are discussed in this chapter, which also contains the control strategy discussion and the analysis of the vehicle's pure electric and pure engine start.

Moreover, the other possible combination of the two degrees of freedom planetary gear and the power-summing electromechanical converters is also taken into consideration.

The design of electromechanical drives related to the planetary gear of two degrees of freedom controlled by the electric motor can be transformed to the purely electromagnetic solution. An example of the mentioned gear is given in the chapter. It is a complicated construction, with the rotating stator of a complex electrical machine requiring multiple electronic controllers. The increasing output torque of the electromechanical converter and its connection with the mechanical two degrees of freedom planetary gear are depicted as well.

Chapter 9 is devoted to the simulation research showing the influence of changes in the power train parameters and control strategy on the vehicle's energy consumption, depending on different driving conditions.

The control strategy role is to manage how much energy, more specifically, how much of the torque-speed relations referring to the power alteration, is flowing to or from each component. In this way, the components of the hybrid power train have to be integrated with a control strategy, and, of course, with its energetic parameters, in order to achieve the optimal design for a given set of constraints. The hybrid power train is very complex and non-linear, respectively, with its every component. One effective method of system optimization is numerical computation, the simulation, as in the case of a multivalent suboptimal procedure regarding the number of an electrical mechanical drive's elements whose simultaneous operation is connected with the proper energy flow control. The minimization of power train internal losses is the target. The quality factor is the minimal energy, as well as the minimal fuel and electricity consumption.

The fuel consumption by the hybrid power train has to be considered in relation to the conventional propelled vehicle, where one of the sources of energy is the internal combustion engine. The Compact Planetary Transmission Drive (referred to in Chapters 1 and 8) and its improved solutions, adapted to the real power distribution requirements, as the structure of the analyzed hybrid power train was tested. Two types of the vehicle were considered: the urban fifteen-ton bus and the five-ton shuttle service bus.

After the assumption of the exemplary selected basic vehicle and its developed drive architectures, the analyzed power trains were tested according to the suggested, and the obtained simulation procedure regarding many considered cases, which is considered necessary to the proper hybrid power train design. All of them concern the power train's internal losses decreasing and its influence on fuel and electricity consumption. First of all, the commonly chosen statistic driving cycles should be taken into consideration. Unfortunately, this is not enough. Additional tests for the vehicle's climbing, acceleration, power train behavior referring to real driving situations are strongly recommended during the drive design process. For this reason, the chapter contains multiple simulation results causing its greater capacity.

Chapter 10 presents the principles of the plug-in hybrid power train (PHHV) operation. The power trains of the battery-powered vehicle (BEV – pure electric) are close to the plug-in hybrid drives. For this reason, the pure electric mode of operation of the plug-in hybrid power train is very important. The vehicle range of driving autonomy must be extended. It means the design process has to be focused on energy economy, emphasizing the electricity consumption especially carefully. Simultaneously, the increasing battery capacity, means its mass and volume is not recommended. After many tests, one can observe the strong dependence between the proper multiple gear speed, the proper mechanical transmission adjustment and the vehicle's driving range, which, in the case of the plug-in hybrid power train, means long distance of the drive using mainly battery energy. The mechanical ratio's proper adjustment and its influence on the vehicle's driving range autonomy is discussed in the chapter.

The reduction of the time frequency of the electric motor operation in the zones of its lowest efficiency area and the decreasing motor torques means that their currents cause, among other factors, decreasing energy consumption, yet increasing the total power train's efficiency.

The application of the range extender unit consists of a strongly downsized engine, as well as the Wankel engine, or the free piston engine connected with rotating or linear components. In this case, the possible solution of the electronic control system consists of a generator controller and the charger designed as a one-unit device. However, the main emphasis is put on the concept of different types of automatic transmissions. Three types of automatic mechanical transmission are depicted: the tooth gear (ball); the belt's continuously variable transmission and the planetary transmission system called the Compact Hybrid Planetary Transmission Drive (CHPTD – see Chapters 1 and 9) equipped additionally with teeth gear reducers connected or disconnected by the specially constructed electromagnetic clutches. The number of mechanical ratios (gear speed) depends on the vehicle size, mass, and function, which, in the majority of cases, means the maximal speed value. Strictly for the city's ultralight cars, two to three transmission speeds are enough. In case of multifunctional cars powered, for example, by the Compact Hybrid Planetary Transmission, the number of mechanical transmission ratios has to be higher. The influence of the automatic changes in the mechanical transmission ratio, according to the properly adjusted control algorithm, during vehicle regenerative braking is considered. The received result of such a solution is the increasing power train's efficiency. Only during the vehicle's braking, the growth of efficiency of recuperated energy stored in the battery is about eight percent.

Each chapter of the book begins with an adequate abstract, introducing the main theme of the chapter to the reader and guiding him/her through the discussed issues.

This book is dedicated to students (at both MS and Ph.D. level), engineers, and researchers who are interested in hybrid electric vehicle propulsion systems. The book concerns the designing of hybrid electric power trains and the proper adjustment of their parameters with the use of nonlinear dynamic modeling and simulation. The book generally focuses on hybrid vehicle power trains' design, including engineering, modeling, and control strategy. These three elements are especially emphasized, and the main target of the book is to present the possible searching tools for the analyzed drive structure, for the determination of maximum effectiveness of propulsion systems, including entering power trains, duty cycles, and the traction characteristics of vehicles. The tools mentioned above are the only possible basic approach to the design of hybrid electric vehicle power trains.

Antoni Szumanowski
Warsaw University of Technology, Poland

Acknowledgment

I would like to express special thanks to my present and former PhD students whose assistance during the seemingly eternal common research I appreciated very much. Their efforts put into simulation studies were certainly necessary.

These special thanks have to be directed to Dr. Yuhua Chang, Dr. Arkadiusz Hajduga, Dr. Piotr Piórkowski, Zhiyin Liu, MSc., Paweł Krawczyk, MSc., and Michał Sekrecki, MSc.

Additional thanks must be addressed to Agnieszka Kłys, MSc., Artur Kopczyński, MSc., and Paweł Roszczyk, PhD.

I would like to also pay tribute to my proofreader, Peter Whiley, who corrected all my linguistic and general mistakes in the writing of this book.

Finally, I would like to express my gratitude to everybody who believed in the positive results of my research activities, especially my family.

Antoni Szumanowski
Warsaw University of Technology, Poland

Chapter 1
Introduction to the Hybrid Power Train Architecture Evolution

ABSTRACT

The hybrid power train is a complex system. It consists of mechanical and electrical components, and each of them is important. The evolution of the Hybrid Electric Vehicle (HEV) power trains is presented from the historical point of view. This chapter discusses the selected review of the hybrid power train's architectural engineering. It includes the development of the hybrid vehicle power train's construction from the simple series and parallel drives to the planetary gear hybrid power trains. The fuel consumption difference between the pure Internal Combustion Engine (ICE) drive and the hybrid drive is especially emphasized. Generally, there are two main hybrid drive types that are possible to define. Both these hybrid drive types are not mainly differentiated by their power train architecture. The first is the "full hybrid" drive, which is a power train equipped with a relatively low capacity battery that is not rechargeable from an external current source, and whose battery energy balance—its State-Of-Charge (SOC)—has to be obtained. The second one is the "plug in hybrid," which means the necessity of recharging the battery by plugging into the grid when the final State-Of-Charge (SOC) of the battery is not acceptable. Additionally, the chapter focuses on the fuel cell series hybrid power train, which is only shown because its operation and design are beyond the scope of this book.

DOI: 10.4018/978-1-4666-4042-9.ch001

INTRODUCTION

The initial image of the Hybrid Electric Vehicle (HEV) and the Electric Vehicle (EV), corresponding simply to the fuel efficient and environmentally friendly car, is now evolving in response to wider customer expectations. New power train architectures, and the augmented functionalities are required to provide increased performance, improved drive ability and comfort, as well as even the 'fun-to-drive' so strongly demanded from the new generations of vehicles.

Indeed, the main initial fuel–efficiency appeal of the hybrid electric vehicle has also been found to depend heavily on the conditions of its operation, which is good in city driving, but relatively poor on highways, or during uphill manoeuvres. Correspondingly, the new generation of HEVs follows a new trend, in which low fuel consumption is marketed together with the best overall performance.

This chapter first presents analysis of the possible hybrid power train architectures, and then follows an overview of the recent constructions of hybrid vehicles.

Since the year 2000, when one of my books, 'the Fundamentals of Hybrid Vehicle Drives' was published, the price of oil has doubled more than two times, while air pollution has increased enormously as well. In particular, the increasing emissions of CO_2 have had a negative influence on the composition of the earth's atmosphere, and this has presumably contributed to the recent climate changes. Now, about 25% of total industrial CO_2 emissions are produced by vehicles. The world's reservoirs of oil have been heavily exploited. The oil exploitation peak occurred about one year ago. The recent prediction assumes that oil, as we know it today, will probably be exhausted within the next forty years. These are important reasons for showing the necessity of road transport electrification. The growing market success of hybrid vehicles will be consolidated as an important part of the green electric transportation system. In 2020, about 15% of the total number of cars on the world's roads is expected to rely on hybrid and electric vehicles. This means that the permanent improvement and development of hybrid engineering and technology must by necessity start now.

In this situation, an increasingly strong tendency to introduce hybrid or vehicle power trains is satisfied, and HEVs give us hope for achieving ecological balance in the future. Hope based on a simple rule which states that the less natural fuels are combusted, the less exhaust gases are emitted. Furthermore, the more stable combustion processes result in easier control over the combustion of the emitted gases, and the hybrid drive provides all these advantages.

Introduction to the Hybrid Power Train Architecture Evolution

Of course, if we aim to achieve the traction characteristics similar to classic vehicles, we should put a similar driving energy onto the tyres. As in hybrid drives, the primary energy should be lower (e.g. obtained through the combustion of primary fuels in a thermal engine – in the majority of cases, in the internal combustion engine), so how can we get the required energy balance?

The missing part of the energy is supplied by accumulators, as secondary power sources, also called the 'power sources', which are usually of the electrochemical battery type. Their purpose is to accumulate the kinetic energy of the vehicle during its regenerative braking, and also to support the internal combustion engine during the temporary overloads of the drive. The proper energy power distribution, according to the minimum fuel and electricity consumption, is a significant problem with the optimization of the hybrid power trains.

Generally, there are two main hybrid drive types. One of them, the full hybrid, relies on the power train equipped with a relatively low capacity battery, which cannot be recharged by use of an external current source, and whose energy balance – state of charge (SOC), in the entire time the vehicle is being driven, is obtained by regenerative braking and proper operation of the internal combustion engine.

The second type is the plug-in hybrid. The operation of this drive means the necessity of battery recharging by plugging into the grid, when the battery's SOC is not acceptable. Hence, a low current charge during the night is recommended.

Both hybrid drive types can be designed by employing the same power train architecture. The plug-in hybrid type is fitted with a larger capacity high energy battery, the HE, supported only by the Internal Combustion Engine (ICE), nowadays called the 'drive range extender.' The plug-in hybrid connects, in itself, the features of the pure electric and the full hybrid power train. Moreover, the assumed further battery development is very promising due to the Internal Combustion Engine's (ICE) fuel consumption and CO2 emissions being very low—for present and future applications.

This chapter includes the presentation of the most important hybrid power train architectures and the method of proper modeling, the energy-power distribution, by considering the system of power efficiency for minimum primary energy consumption, and secondarily, a power- sourced energy balance.

The hybrid power train is a complex system. It consists of mechanical and electrical components, both of which are important. The proper drive design has to involve the modeling and optimal adjustment of each of them. The advantage of the hybrid drive, in comparison with the pure engine, is the operation of the ICE limited to its minimum fuel consumption, emissions, and the possibility of charging the battery—the source of power—by the process of regenerative braking.

Figure 1 presents the evolution of the HEV & EV power trains from the historical point of view.

Introduction to the Hybrid Power Train Architecture Evolution

Figure 1. The evolution of the contemporary HEV power train

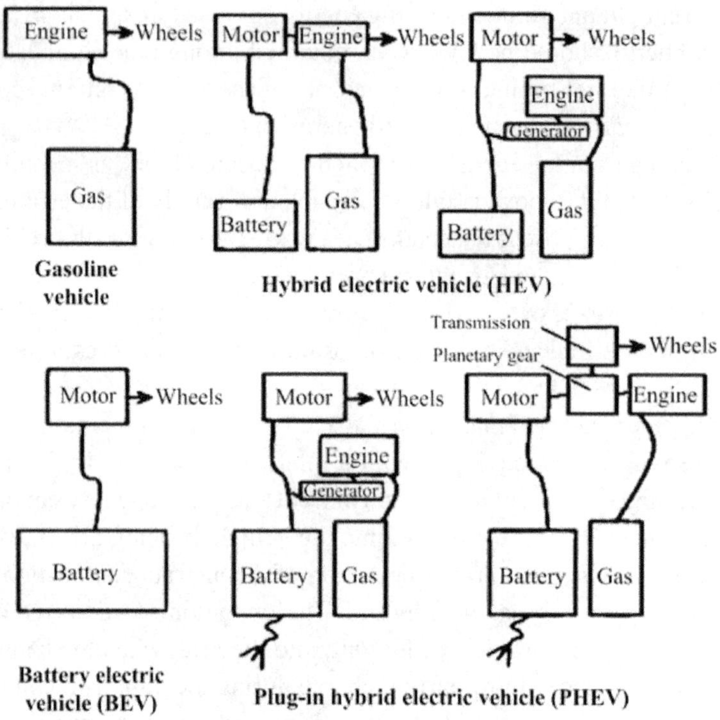

The most important aim is to obtain a smart and effective Plug-In Hybrid Electric Vehicle (PHEV) power train, equipped with a new generation combustion engine as a range extender. This solution is a bridge between the pure electric Battery Electric Vehicle (BEV) drive and the full Hybrid Electric Vehicle (HEV) power train. It does not mean that the classic series or parallel hybrid power trains are not improved or developed.

Anyway, for modeling purposes, simulation and propulsion system architecture, which are all mentioned above, the HEV drives need the same approach design process, which is discussed later in this chapter.

Figures 2 and 3 show the maps (internal combustion engine torque versus its angular velocity with curves of equal fuel consumption indication). One of them is for the pure ICE drive, the second for the hybrid planetary drive (compact hybrid planetary transmission drive—CHPTD (see in Figure 16), equipped with the same components as in the first case of the diesel ICE). The points of engine operation for a 12-ton urban bus, driving in the same city cycle, are strictly connected with the fuel consumption of the internal combustion engine. It is thus easy to note that the area of engine operation in a hybrid drive is limited to a higher ICE efficiency than in the case of the pure engine drive.

Introduction to the Hybrid Power Train Architecture Evolution

Figures 2 and 3 present the Internal Combustion Engine's (ICE) operating points in the shape of static maps. The repetition frequency of the points indicated in these figures can be shown only in a dynamic way for a comparison based on assuming the same driving cycle. In the next chapters, there are dynamic analyses of the hybrid power train operation. Certainly, these maps were obtained in the process and by the procedure of dynamic modeling and simulation, permitting the computation of the total ICE fuel consumption for both power trains under consideration: the pure internal combustion engine and the hybrid one.

The fuel consumption difference between the pure classic diesel combustion engine ICE drive and the Compact Hybrid Planetary Transmission Drive (CHPTD, see Figures 13 and 16) drive is shown in Figure 4.

It has to be stressed that the emission of CO_2 is a derivative factor induced by energy utilization. So, the political and economic incentives refer directly to the reduced use of energy. The energy cost is defined by its amount ratio (MJ/l) and a CO2 emission divided by its g/km factor. The lowering of energy cost is related to the general drive system's efficiency and the vehicle's weight. To compare the new evaluation of energy and the emission costs, only the suitable drive cycles are taken into account. The fuel 'calorific value' is an important element in reference to the energy density.

Figure 2. Example of an ICE map with operation points indicated for the pure engine classic drive. Note: The curves of equal fuel consumption are defined as g/kWh

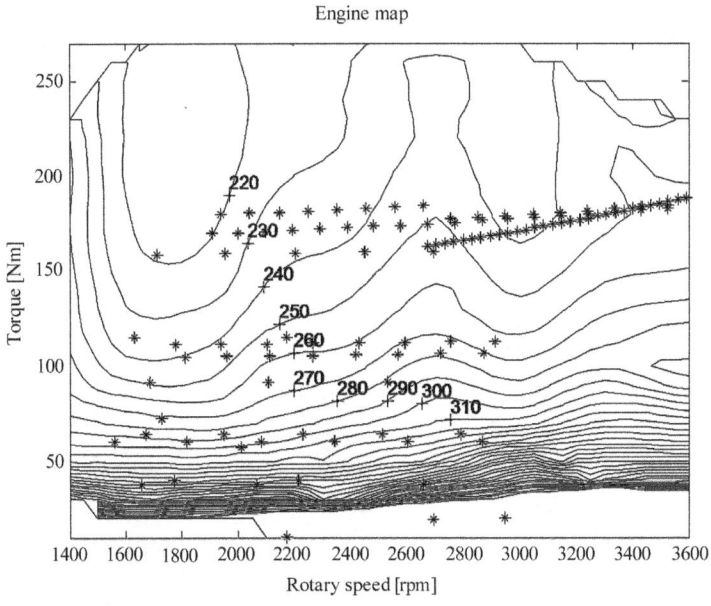

Figure 3. The operation points indicated for the compact hybrid planetary transmission drive (CHPTD) equipped with the same engine whose map is shown in Chapters 7 and 8. Note: The curves of equal fuel consumption are defined as g/kWh

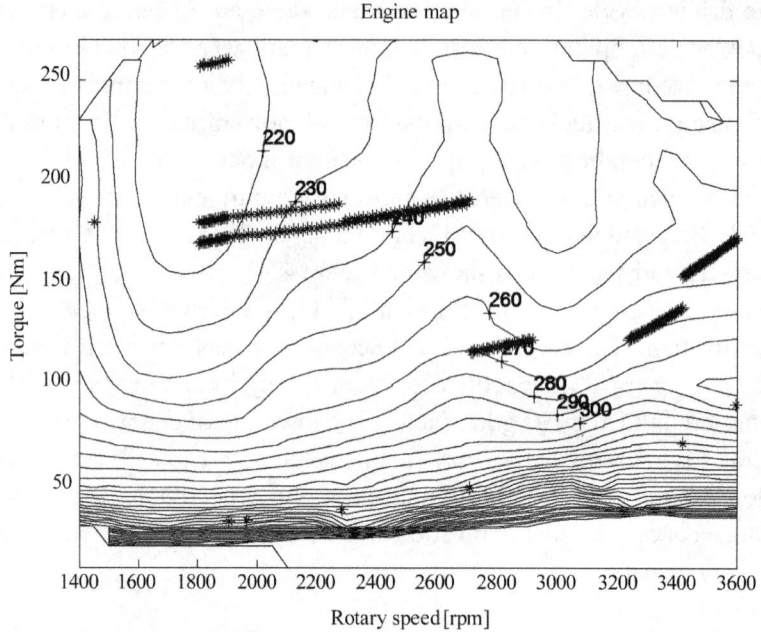

Figure 4. Fuel consumption compared to the classic and the hybrid CHPTD power trains. Note: A five-ton vehicle was tested.

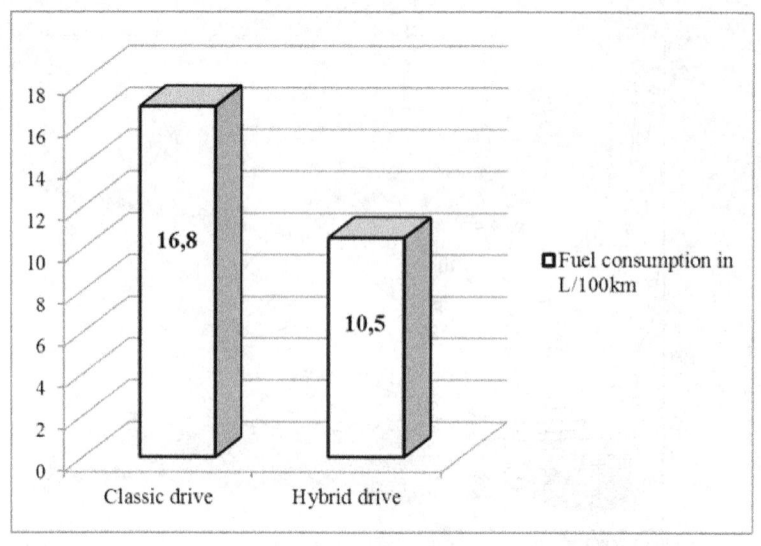

Introduction to the Hybrid Power Train Architecture Evolution

Figure 5. The bulk energy density comparison: different sources

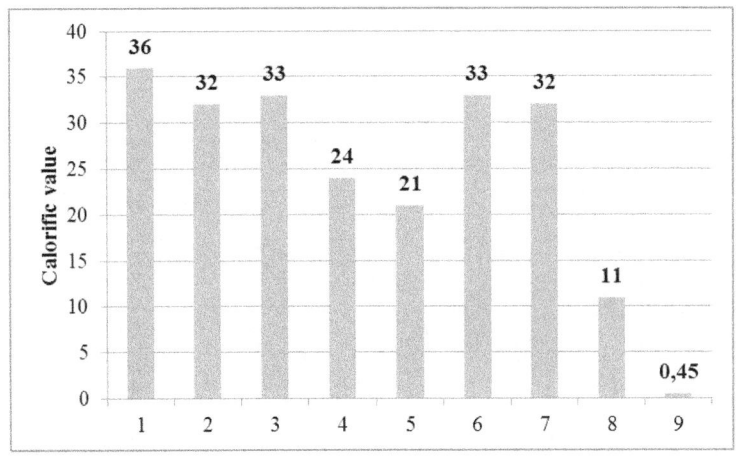

Energetic Values
1. Gas oil (MJ/l); 2.Petrol (MJ/l) ; 3.Natural gas / biogas (MJ/l);
4. Liquefied petroleum gas (MJ/l); 5.Ethanol (MJ/l); 6.Biodiesel (MJ/l);
7. Emulsive fuel (MJ/l); 8.Hydrogen (MJ/Nm3); 9.Li–ion battery (MJ/kg).

The bulk density (Figure 5) when compared to the mass of the best li-ion electrochemical batteries is ca. 30 times lower and as to the space taken—8 to 10 times higher than in the case of liquid fuels. Fortunately enough, the average efficiency of a pure electric drive system is 3 times higher, compared to the combustion engine drive. However, we have to admit that the vehicle's one-charge range is a few times lower than the traditional car's parameters. Thus, it is reasonable to apply the pure battery Electric Vehicles (EV) for city or local traffic at the time when the batteries will be efficient enough. Hybrid drives, which improve the combustion efficiency of the enumerated fuels in thermal engines, as well as in the Internal Combustion Engine (ICE), may be used comprehensively.

A vehicle's weight plays a substantial role in energy and emission costs. At present, research and development centers try to lower the vehicle's mass of existing cars, even to as much as 50% less. Yet, the new compounds are not the only fundamental factor: the production technology matters as well.

The efficiency of the hybrid power train power depends also on the type of the Internal Combustion Engine (ICE) equipped with a drive. Figure 6 presents the efficiency of different engines applied in hybrid power trains compared to the conventional drives with gasoline or Diesel ICEs. Additionally, Figure 6 shows the Fuel Cell (FC) power train efficiency, which is also a type of hybrid drive, because the Internal Combustion Engine (ICE) is replaced by the FC cooperating with the battery. The New European Driving Cycle (NEDC) is taken into consideration for the sake of this comparison.

Figure 6. Efficiency comparison of different hybrid power trains in the NEDC driving cycle

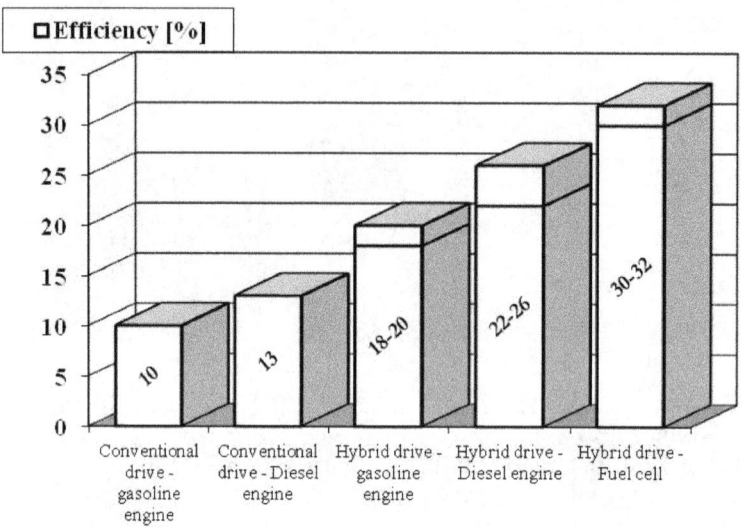

It is easy to note what significant advantage can be obtained, using the hybrid power train instead of a conventional drive, for the same mass of the car.

Special effort has to be made for the proper adjustment of the hybrid drive architecture. This means obtaining an optimal compromise between the mechanical and the electrical drive's components, as well as a way of evaluating their connection and the control design, which are necessary. This is practically important because of the role of the hybrid ('full hybrid' or 'plug-in hybrid') and electric vehicles, especially for city traffic, which is rising systematically. The price of oil, as mentioned before, is rapidly increasing. The problem of the world's oil reservoirs' rapid decline (the common prediction that oil will be fully depleted after 40 years) also should be taken into consideration. In common understanding, the hybrid drive with a thermal (combustion) engine is a bridge between the present technology of HEV and EV power trains and the hybrid fuel cell vehicle with its infrastructure that is expected in the future. At present, the replacement of the thermal engine by the fuel cell is very costly and requires a very expensive infrastructure such as hydrogen production, transportation, and a great number of safety filling stations.

In city traffic, the problem of air pollution is presently the most important. The hybrid drive offers the possibility of decreasing this pollution significantly, to a much greater extent than the classical drive. Special emphasis is put here on CO_2 emissions, which can be lowered. This result can be obtained by the lower fuel consumption of the internal combustion engine, and the maintenance of its operating points in the optimal range, using the highest possible operating efficiency of the engine.

Introduction to the Hybrid Power Train Architecture Evolution

The main feature of the hybrid drive is the balance of the battery. It means that the state of charge (SOC) of the battery should be at the same level at the beginning, and at the end of the driving cycle for the 'full hybrid' power trains, or SOC alterations have to be limited up to 0.3 as a minimum for the 'plug-in hybrid' power trains. In the latter case, battery charging overnight is necessary. The electrical energy consumption also should be as low as possible. The minimization of the consumption of fuel and electricity by the proper adjustment of the hybrid drive architecture and its control strategy results in minimal energy losses or maximal drive efficiency. It is the main goal in the process of designing the hybrid electric vehicle. As for motors, in order to obtain their highest efficiency, induction AC and PM, and for similar reasons, NiMH and Li–ion batteries are taken under consideration. Highly efficient mechanical transmissions, including automatic speed fitting – or CVTs – controlled electrically, e.g. by step motors and electromagnetic, low-energy consumption clutches, selecting the modes of hybrid drive operation, have to be simultaneously included in the design process. This book is aimed to advance such a complex approach.

1. HYBRID POWER TRAINS ARCHITECTURES ENGINEERING DEVELOPMENT REVIEW

There are two common architectures of the hybrid drive that have the chance of obtaining the optimal driving performance: the series and the parallel one. Both are shown in Figure 7 and 8.

There are three types of parallel hybrid drive: the single shaft, double shafts and the split–sectional drives. The 'double shafts' parallel drive is characterized by the double shafts, one shaft connected with the engine, and the second shaft connected with the electric machine. Also, both shafts are connected with the transmission.

If the two degrees of freedom planetary gear is used as a transmission, the hybrid drive can be equipped with one or two electric machines. The hybrid drive used in the Toyota Prius consists of two electric machines connected with two independent shafts of a planetary gear. The ICE engine is connected with the third one (see Figure 11). In this case, the operation of the Internal Combustion Engine (ICE) is more stable than in the case of a typical parallel drive and it is limited to the proper area of its map. For this reason, this drive is sometimes called the 'Series-Parallel'. The same result can be obtained by the use of only one electric machine. The decrease of the number of components can reduce the cost of the drive. Hence, the problem that should be resolved is to develop a proper control strategy, which will be more complicated, in comparison with the two electric machines with double shafts, series

Figure 7. Series power train: a) the general scheme; b) the energy flow. Note: Pure internal combustion engine (ICE) driving is not possible for the series power train. It is possible, only in the case of the parallel power train (see Figure 8).

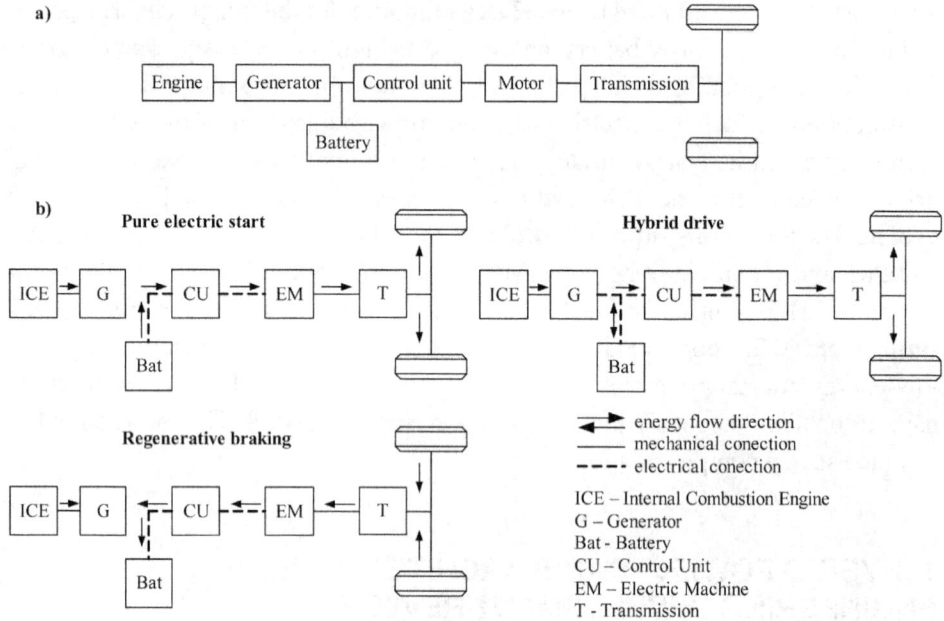

or parallel split–sectional hybrid drive (or the parallel double drive, or the parallel TTR–Through The Road) architectures.

The high efficiency of the hybrid drive (which results in minimal energy consumption) presented in this book is the planetary drive system with one electric machine (see Chapters 7, 8, 9, and 10), which has been invented by Antoni Szumanowski, and developed by his co–workers.

The role of this book is to present the method of designing the hybrid electric drive, especially for urban traffic application. The following hybrid drive architectures are used for analysis and comparison: series, split–sectional parallel and planetary transmission with two degrees of freedom equipped with one electric machine. The following figures show the architectures of the above–mentioned hybrid drives (see Figures 8, 9, & 10).

The application of the planetary gear in hybrid power trains is very attractive, firstly because of its small volume and mass proportionate to the shaft loads. Secondly, because of the use of the two degrees of freedom transmission, it means that its three rotating shafts can be connected: one for the internal combustion engine, one for the motor, and the third one for the traction wheels, with an additional gear. The manufacturing technology of the planetary gear production is highly developed

Introduction to the Hybrid Power Train Architecture Evolution

Figure 8. Parallel power train: a) double-shaft drive; b) single-shaft drive; c) energy flow

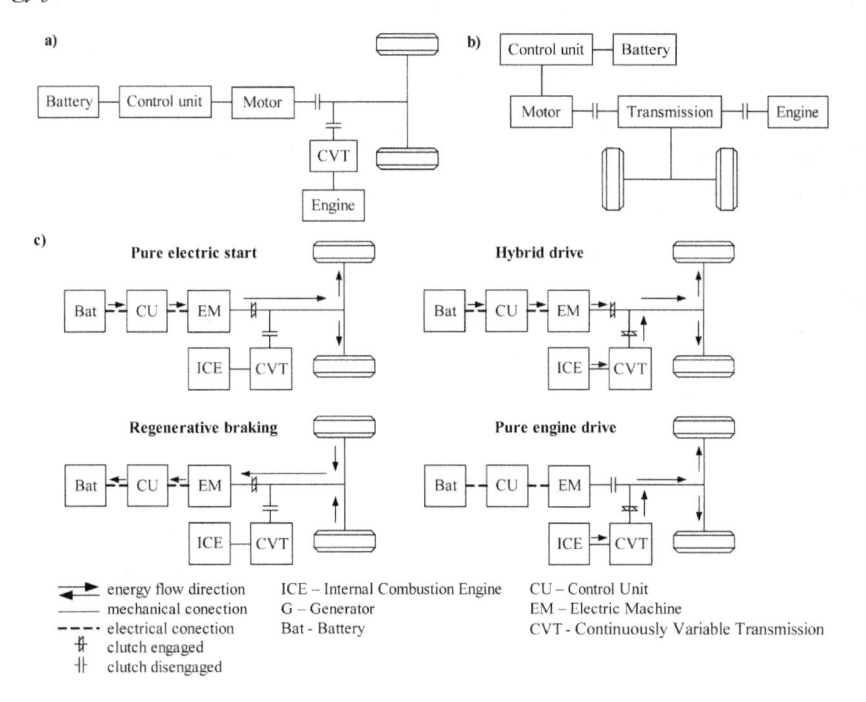

Figure 9. The split-sectional hybrid drive

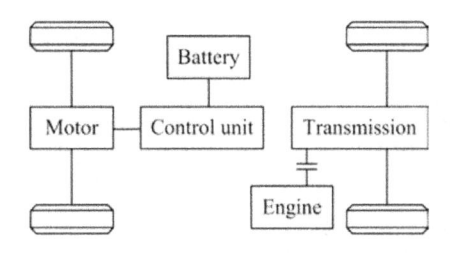

Figure 10. Fiat research center (CRF's) ecodriver parallel hybrid power train scheme (Szumanowski, 2000)

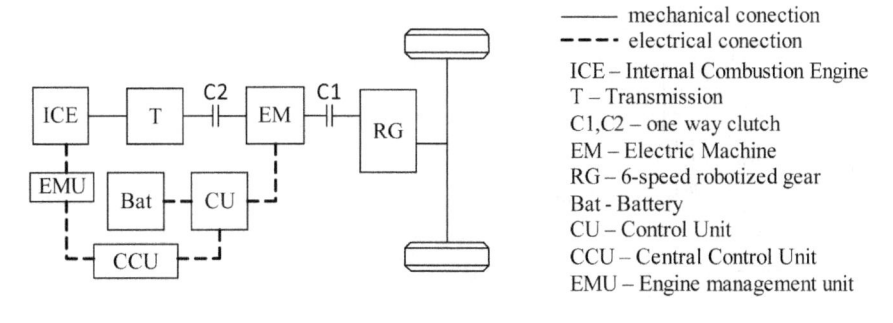

Introduction to the Hybrid Power Train Architecture Evolution

for the conventional vehicle's application. In hybrid power trains, the role of the planetary transmission is different than in conventional drives. Apart from the planetary two degrees of freedom gear system, it has to consist of clutch–brake devices operating according to the optimal, (generated before the design by modeling and simulations), control functions.

Figure 11 shows the typical power train architectures, where the role of energy–power distribution is played by the two degrees of freedom planetary gear.

Figure 11. Equos Research Co. Ltd. Toyota Hybrid Drive system[1] – HV (this hybrid architecture is also called the split or parallel-series hybrid drive): a) drive scheme; b) energy flow during the operation of this hybrid power train (Yamagouchi, 1996; Yamagouchi, Miyaishi, & Kawamoto, 1996)

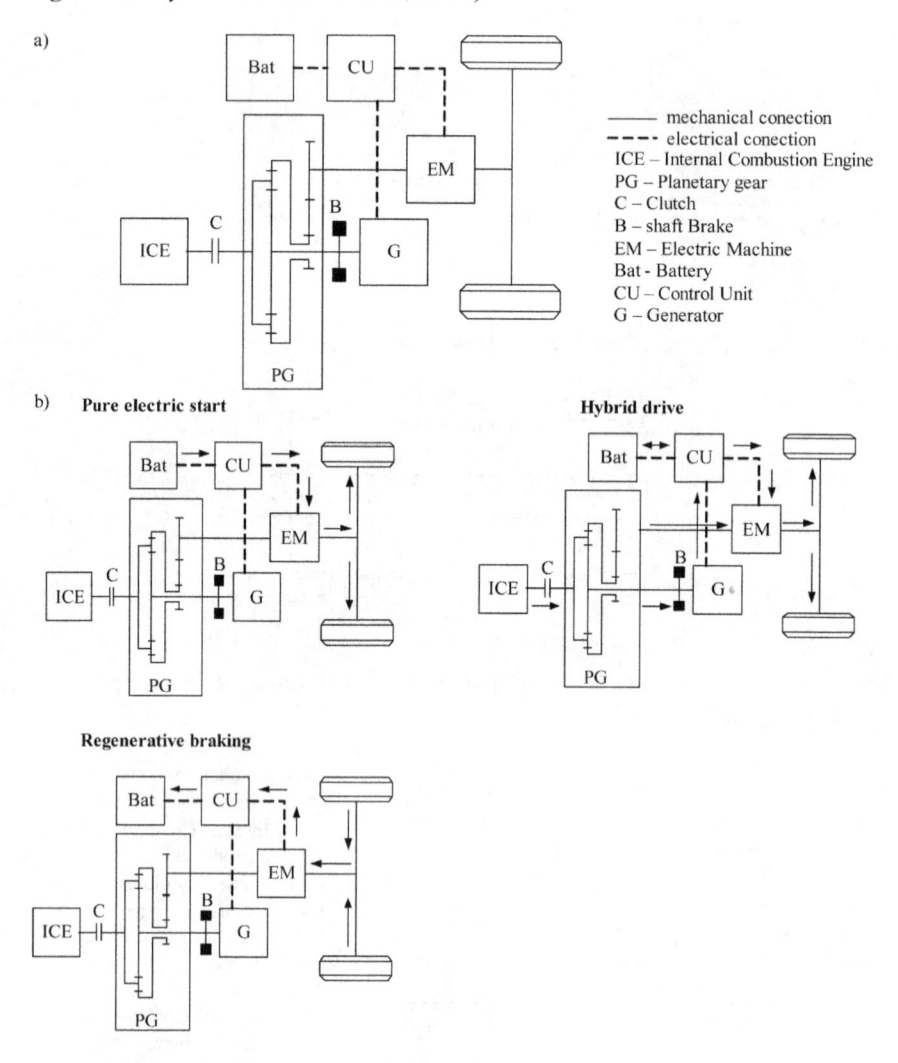

Introduction to the Hybrid Power Train Architecture Evolution

Figure 12. Toyota plug-in hybrid (PHEV) power train system based on the previous full hybrid (HV) shown in Figure 15: a) the power train layout, b) the electric scheme. Note: The difference between the HV and PHEV power trains is in: bigger capacity and battery type, additional equipment such as the battery charger connected to the grid and charger relay (Takaoka & Komatsu, 2010)

―――― mechanical conection
- - - - electrical conection
I – Inverter
BC – Boost Converter
ChR – Charger Relay

ICE – Internal Combustion Engine
PG – Planetary gear
C – Clutch
B – shaft Brake
BCh – Battery Charger

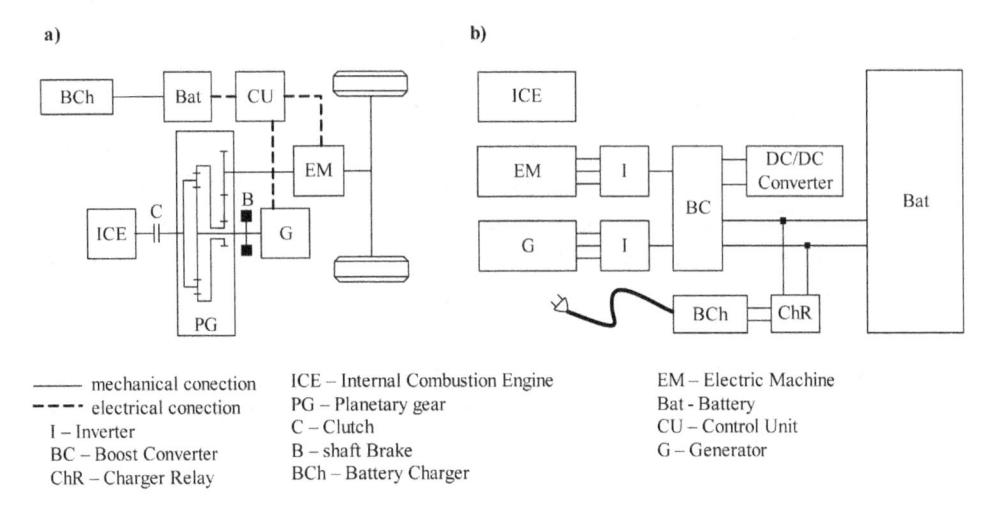

EM – Electric Machine
Bat - Battery
CU – Control Unit
G – Generator

Figure 12 depicts the system configuration of the newly-developed plug-in hybrid power train (PHEV) by Toyota. In comparison with that of the base of the full hybrid power train (HV) (see Figure 11), the main differences are: the High Power (HP) Lithium-ion battery, the 100V/200V recharging system, and finally, the wide pure electric driving range, which means higher vehicle available power and the maximum speed of a car powered only by the battery.

Adopting the same basic hybrid system from the full hybrid power train (HV), the plug-in hybrid power train (PHEV) has an increased battery power/energy and battery recharging capability from an external power source. For these reasons, the compact and lightweight Li–ion battery was newly developed for the PHEV. This battery has higher energy density compared with the one applied in the previous full hybrid (HV) nickel metal hydride (Ni-MH) battery. The data of the nickel metal hydride battery (Ni-MH) and the lithium–ion battery (Li-ion) are as follows:

- **The Full Hybrid Power Train (HV):** Battery Ni-MH; 202V; 1.3 kWh; 42 kg;
- **The Plug-in Hybrid Power Train (PHEV):** Battery Li-ion; 345.6V; 5.2 kWh; 160 kg, as nominal values.

One of the big issues for Li-ion batteries is to obtain a constant pressure structure, along with the upgrading of the cooling performance and preventing.

Introduction to the Hybrid Power Train Architecture Evolution

The planetary two degrees of freedom gear controlled by only one electric machine is depicted in the next diagram. In comparison to the solution shown in Figure 12, the clutch/brake units control the hybrid drive operation modes (see Chapters 7, 8, 9, and 10). There are three operation modes: if the clutch (CB1) is disengaged and the shaft connected, the ICE with the planetary gear is put in brake mode, and the motor, fed on the battery, propels the planetary crown wheel—ensuring the pure electric drive is obtained.

When the brake (CB1) is released and the clutch is engaged, the ICE propels on wheels and the motor drive's crown wheel. In this situation, the hybrid operation takes place. If the voltage (from the inverter) at the motor terminals is switched off and the brake B2 is active—the pure engine drive is received. This power train and its more complex options, equipped with special electromagnetic controlled clutch and brake devices (see Chapters 7, 8, 9 and 10), whose characteristic features are the small volume and very efficient operation, is called the Compact Hybrid Planetary Transmission Drive (CHPTD) and is shown in Figure 13.

The other planetary hybrid power train called Two Mode was developed by Allison GM (2000–2001). Its main feature is a strong output torque reinforcement and for this reason this power train is applied in luxury passenger cars, as well as in SUVs—e.g. Daimler (Figure 14)—and, for instance, in buses. This is a complex and expansive power train.

The new transmission concept developed by FEV Motorentechnik GmbH (Aachen) is based on three simple planetary gear sets with not more than three clutches and two brakes. Four forward gears can be used by the Electric Motor (EM) of the hybrid system, which, together with the first planetary gear and the first lockup clutch (C1), are installed in the transmission's bell housing. The three members of the first planetary gear set (PGS1) are connected as follows:

- **Sun Wheel:** Combustion engine (ICE).
- **Crown (Ring) Wheel:** PGS2.
- **Yoke (Carrier) Wheel:** PGS3 and electric motor.

Figure 15 shows the layout of the new concept of the hybrid powertrain.

PGS2 and PGS3 serve as two-speed-transmissions, each with one brake and one clutch each (B1/C2 and C3/B2, respectively). PGS1 has two different functions: with the clutch C1 closed, the internal combustion engine and the electric motor are locked together and can use four direct gears which are selected by engaging one of the shift elements B1, C2, C3, or B2. With the clutch C1 open, PGS1 acts as a mechanical power–split device distributing the combustion engine torque to PGS2 and PGS3, where one shift element has to be closed (combinations B1/C3, B1/B2, C2/C3 and C2/B2). This gives four power–split gears.

Introduction to the Hybrid Power Train Architecture Evolution

Figure 13. The basic compact hybrid planetary transmission drive (CHPTD; developed by A. Szumanowski in 1994) with the two degrees of freedom planetary transmission (see Chapter 8)

─── mechanical conection
─ ─ ─ electrical conection
ICE – Internal Combustion Engine
PG – Planetary gear
C – Clutch
B1, B2 – shafts Brakes
EM – Electric Machine
Bat – Battery
CU – Control Unit
1 – sun wheel
2 – crown wheel
3 – yoke
4 – planet wheel

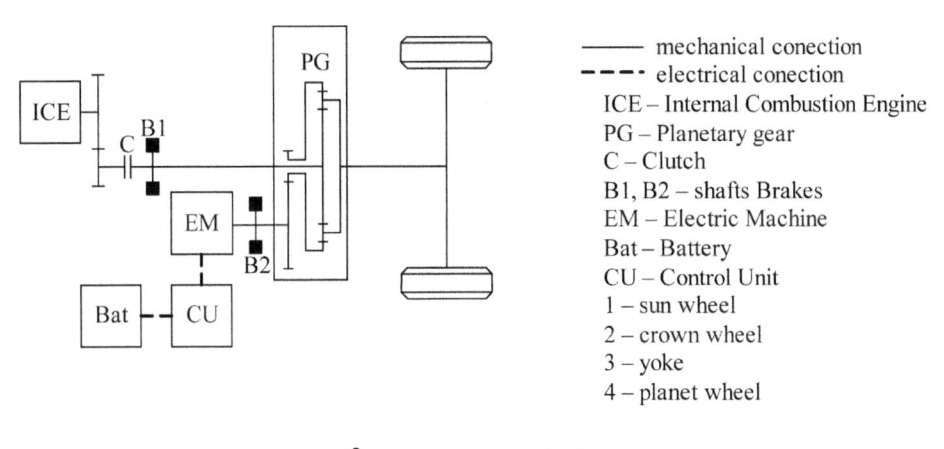

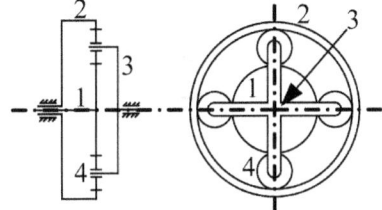

Figure 14. The exemplary hybrid power train with two degrees of freedom planetary gears corresponding to a two mode system

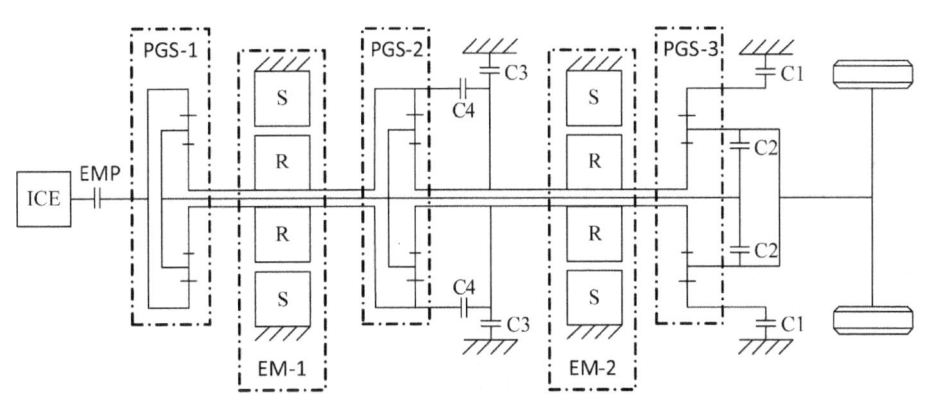

EMP – electromagnetic input clutch plate
PGS-1, PGS-2 – two degree of freedom planetary gears
C (1-4) – clutch/brake systems
R – electric machine's Rotor

ICE – Internal Combustion Engine
PGS-3 – planetary output reducer
EM – Electric Machines (motors/generators)
S – electric machine's Stator

Introduction to the Hybrid Power Train Architecture Evolution

Figure 15. The hybrid power train scheme based on FEV's electric power-shift transmission (2010) (Hellenbrouch, Lefgen, Janssen, & Rosenburg, 2010)

ICE – Internal Combustion Engine
PGS – Planetary gear section
C – Clutch
B – shaft Brake
EM – Electric Machine

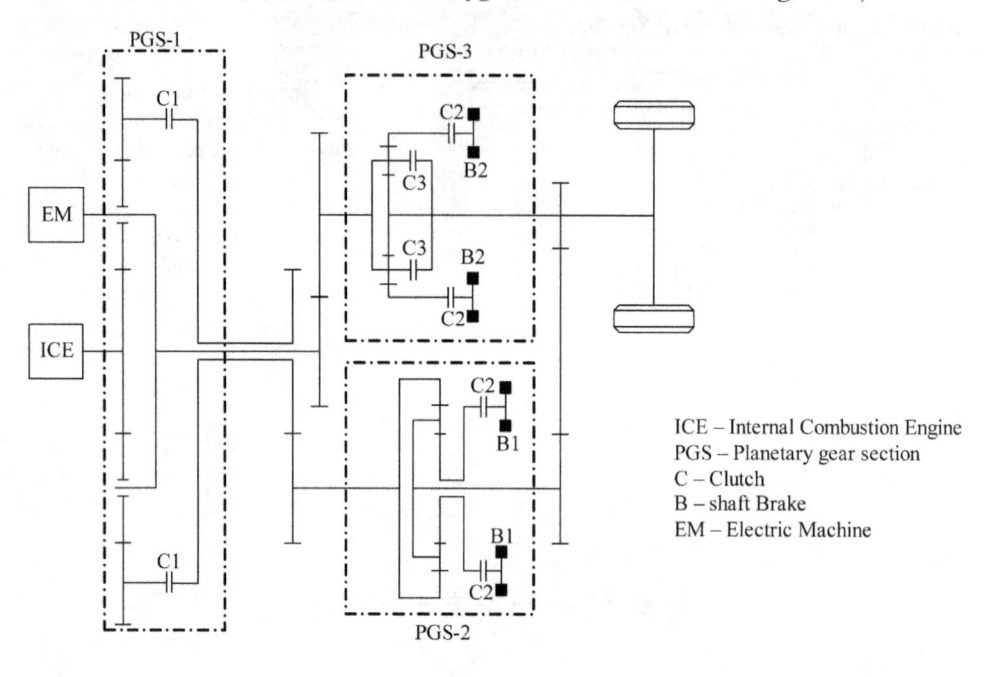

Figure 16. One of the options of a compact hybrid planetary transmission drive (CHTPD; A. Szumanowski) power train, equipped with three planetary gears (see Chapter 9)

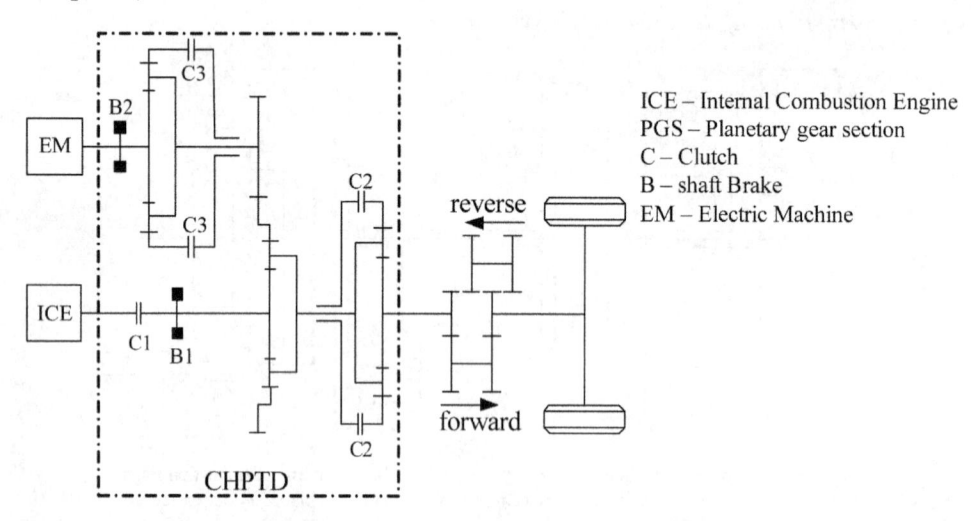

ICE – Internal Combustion Engine
PGS – Planetary gear section
C – Clutch
B – shaft Brake
EM – Electric Machine

Introduction to the Hybrid Power Train Architecture Evolution

A similar concept to the one of the CHPTD power train option is depicted in Figure 16. It offers four speeds, including the basic ratio of the connected with the ICE main planetary two degrees of freedom gear. Two speeds are available to its crown wheel and two to the carrier (yoke) wheel.

Due to the high repetitiveness of the city driving cycles, the lowest energy consumption can be obtained by using the hybrid drive. It is obvious that the hybrid drive is more expensive than the classic drive. However, this additional cost is less important for the city bus, but for the future of the passenger car, this cost can be significantly decreased and will be very competitive in comparison with the advantages of the hybrid drive, which they are possibly obtained by the proper design. These advantages are: the significant decrease of fuel consumption, air pollution, and the noise of driving. These features are very important to downtown and urban traffic.[2]

This book presents the approach of designing the hybrid drive based upon modeling and simulation.

All mathematical models used in this book are verified at the laboratory bench–test stands. For the thermal engine (ICE), the map (torque versus angular velocity) of the selected engines is used, because it is very difficult for ICE mathematical modeling. In the case of the 'black box' model, the difficulty also exists, because of the lack of availability of the adequate bench–test data (see Chapter 6).

As for the secondary power source, the battery modeling (Ni-MH and Li-ion) and ultra capacitor are taken here into consideration.

The Fuel Cell (FC) power train is also a hybrid which cannot be left unmentioned. The fuel cell cooperation with the battery is the same, as in the example of the series hybrid power train. However, the Internal Combustion Engine (ICE) is replaced by the fuel cell as shown in Figure 17.

The internal combustion engine operation is limited and made stable in the series hybrid power train (see Chapter 7). Similarly, obtaining the best fuel cell effectiveness and additionally the longest lifetime, its operation stabilization and limitation are also necessary. For this reason, the battery support has to be demanded. Unfortunately, the content of this book does not include FC power train modeling and simulation.

ACKNOWLEDGMENT

Special thanks to the companies mentioned in this chapter, whose presentations (e.g. during the summit world conference in the field of hybrid and electric vehicle design – Electric Vehicle Symposium [EVS] – EVS 20–25, EVS 25 and EVS 26) are permitted to depict their great technical achievements.

Figure 17. A series hybrid power train equipped with FC – Fuel Cell-PEM (Proton Exchange Membrane): a) the FC operation principle; b) the power train energy flow; c) the exemplary 5 kW powered FC PEM (author's HEV and EV laboratory)

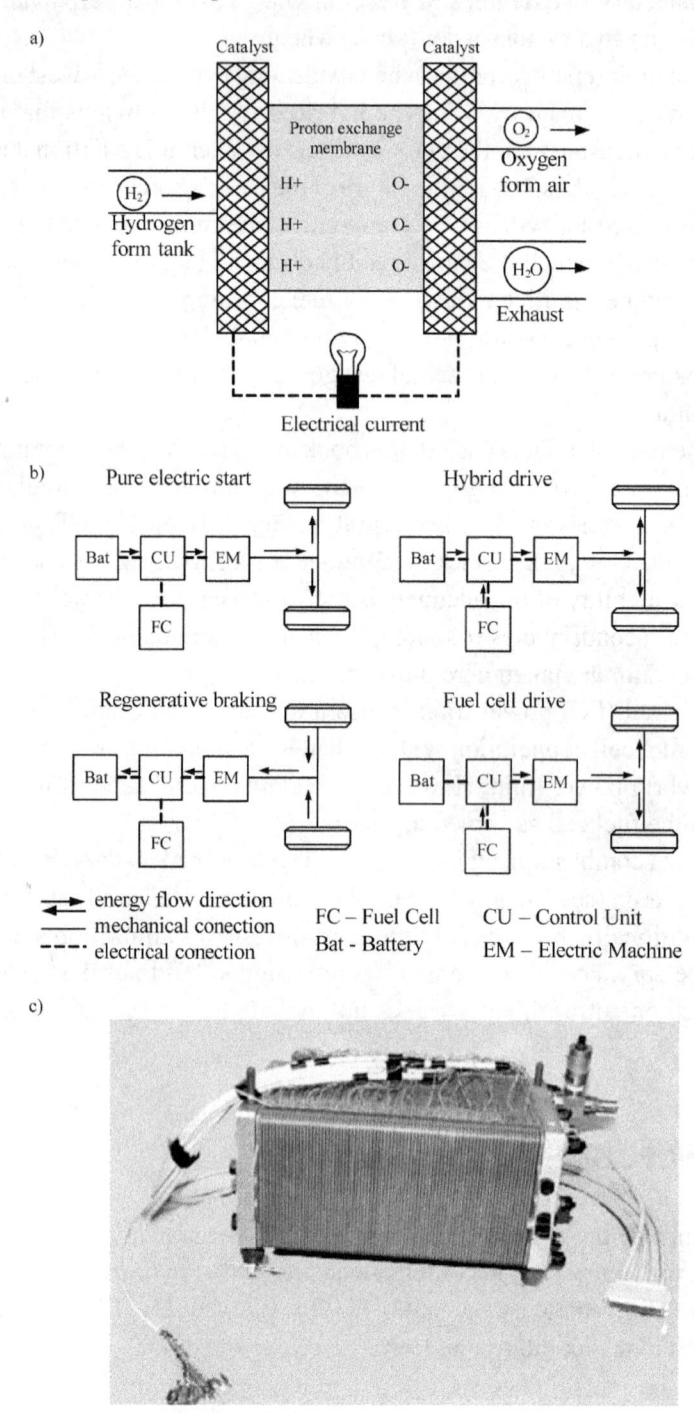

Figure 18. The proton exchange membrane fuel cell (FC PEM) voltage and power versus current density basic functions

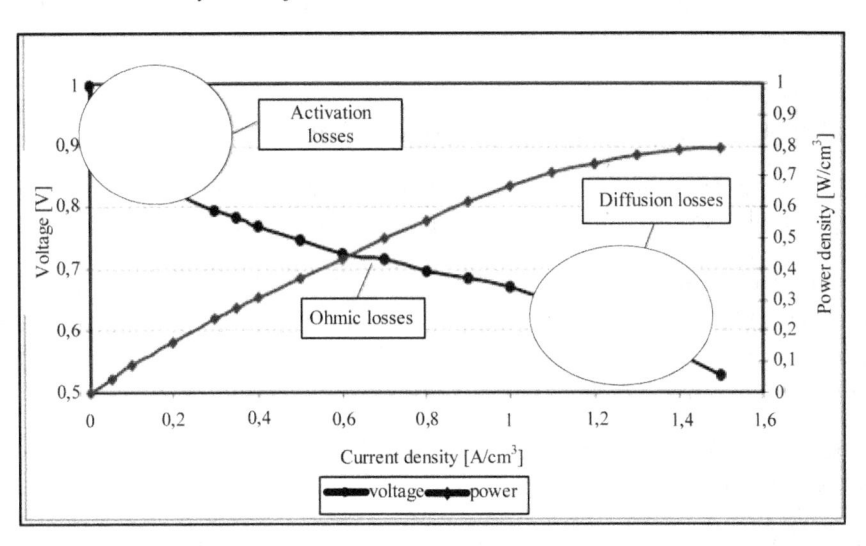

REFERENCES

Cackette, T., & Evaoshenk, T. (1995). *A new look at HEV in meeting California's clean air goals*. Paper presented at the EPRI North American EV & Infrastructure Conference. Atlanta, GA.

Dietrich, P., Ender, M., & Wittmer, C. (1996). *Hybrid III powertrain update*. Electric & Hybrid Vehicle Technology.

Ehsani, M., Gao, Y., Gay, L., & Emadi, A. (2004). *Modern electronic, hybrid electric and fuel cell vehicles – Fundamentals, theory and design*. Boca Raton, FL: CRC Press. doi:10.1201/9781420037739.

Hayasaki, K., Kiyota, S., & Abe, T. (2009). The potential of parallel hybrid system and nissan's approach. In *Proceedings of Aachen Colloquium*. Aachen Colloquium.

He, X., Parten, M., & Maxwell, T. (2005). Energy management strategies for HEV. In *Proceedings of IEEE Vehicle Power and Propulsion Conference*. Chicago, IL: Illinois Institute of Technology.

Hellenbrouch, G., Lefgen, W., Janssen, P., & Rosenburg, V. (2010). New planetary based hybrid automatic transmission with electric torque converter on demand actuation. In *Proceedings of Aachen Colloquium*. Aachen Colloquium.

Hofman, T., & Van Druten, R. (2004). Research overview – Design specification for HV. In *Proceedings of ELE European Drive Transportation Conference*. Estorial, Portugal: ELE.

Jozefowitz, W., & Kohle, S. (1992). Volkswagen golf hybrid – Vehicle result and test result. In *Proceedings of EVS 11*. Florence, Italy: EVS.

Keith, H. (2007). Doe plug in hybrid electric vehicles R&D plan. In *Proceedings of European ELE – Drive Transportation Conference*. Brussels: ELE.

Martines, J. E., Pires, V. F., & Gomes, L. (2009). Plug-in electric vehicles' integration with renewable energy building facilities – Building vehicle interface. In *Proceedings of Powereng IEEE 2nd International Conference on Power Engineering, Energy and Electrical Drives*. Lisbon, Portugal: IEEE.

Moore, T. (1996). Ultralight hybrid vehicle principles and design. In *Proceedings of EVS 13*. Osaka, Japan: EVS.

Neuman, A. (2004). Hybrid electric power train. In *Proceedings of ELE European Drive Transportation Conference*. Estorial, Portugal: ELE.

Overman, B. (1993). Environmental legislation may initiate the EV and HV industry will economics sustain. In *Proceedings of ISATA*. Aachen, Germany: ISATA.

Portmann, D., & Guist, A. (2010). Electric and hybrid drive developed by Mercedes-Benz vans and technical challenges to achieve a successful market position. In *Proceedings of Aachen Colloquium*. Aachen Colloquium.

Ruschmayer, R. Shussier, & Biermann, J.W. (2006). Detailed aspects of HV. In *Proceedings of Aachen Colloquium*. Aachen Colloquium.

Sporckman, B. (1992). Comparison of emissions from combustion engines and European EV. In *Proceedings of EVS 11*. Florence, Italy: EVS.

Sporckman, B. (1995). *Electricity supply for electric vehicles in Germany*. Paper presented at EPRI North American EV & Infrastructure Conference. Atlanta, GA.

Szumanowski, A. (1996). Generic method of comparative energetic analysis of HEV drive. In *Proceedings of EVS 13*. Osaka, Japan: EVS.

Szumanowski, A. (2000). *Fundamentals of hybrid vehicle drives*. Warsaw, Poland: ITE Press.

Szumanowski, A. (2006). *Hybrid electric vehicle drives design*. Warsaw, Poland: ITE Press.

Szumanowski, A., & Hajduga, A. (1998). Energy management in hybrid vehicles drive. In *Proceedings of Advanced Propulsion Systems GPC*. Detroit, MI: GPC.

Szumanowski, A., & Piórkowski, P. (2004). Ultralight small hybrid vehicles why not? In *Proceedings of ELE European Drive Transportation Conference*. Estorial, Portugal: ELE.

Szumanowski, A., Piórkowski, P., Hajduga, A., & Ngueyen, K. (2000). The approach to proper control of hybrid drive. In *Proceedings of EVS 17*. Montreal, Canada: EVS.

Takaoka, T., & Komatsu, M. (2010). Newly-developed Toyota plug-in hybrid system and its vehicle performance. In *Proceedings of Aachen Colloquium*. Aachen Colloquium.

Trackenbrodt, A., & Nitz, L. (2006). Two–mode hybrids = adoption power of intelligent system. In *Proceedings of Aachen Colloquium*. Aachen Colloquium.

Vaccaro, A., & Villaci, D. (2004). Prototyping a fussy based energy manager for parallel HEV. In *Proceedings of ELE European Drive Transportation Conference*. Estorial, Portugal: ELE.

Yamagouchi, K. (1996). *Advancing the hybrid system*. Electric & Hybrid Vehicle Technology.

Yamagouchi, K., Miyaishi, Y., & Kawamoto, M. (1996). Dual system – New developed hybrid system. In *Proceedings of EVS 13*. Osaka, Japan: EVS.

ENDNOTES

[1] A similar hybrid drive with a planetary gear was partly presented by Szumanowski (patented in 1994) (see Figure 13) during the EVS-Electric Vehicle Symposiums – No.12 in Los Angeles and EVS13 in Osaka, Japan, from Equos Research Co. Ltd. in Osaka (EVS13). This Japanese Hybrid Drive was applied in the Toyota passenger car, the PRIUS, in 1997. The main difference in the design of the mentioned hybrid drives is the number of the used electric machines. In the first case, only one electric machine is used, and in the second case, the generator and traction motor are separate. Certainly, in the first case, the system operation is also different, which means the electric motor should change its operating mode from generating to motoring during vehicle acceleration, simultaneously changing the direction of its angular velocity – passing by 0 point of power and rotation. If we use two machines, the problem will not exist. The two–source series parallel hybrid drive with one motor is more compact and can be more efficient.

Introduction to the Hybrid Power Train Architecture Evolution

2   The other solution of the hybrid drive architecture is based on the floating stator motor. This means that both parts of the electric traction machine (generally the stato-exciting circuit and the rotor-armature) are rotating. The floating stator is connected through the clutch system to the ICE and the rotor to the differential gear. This series–parallel unit gives the possibility of summing the power from the ICE and the battery, changing the output rotational speed and torque. It is possible to be applied as a mild hybrid electric drive with a limited range of operation. A similar hybrid electric drive, equipped with three rotating elements of the special electric machine, with a common axis called EMCVT (Electromagnetic Mixed Hybrid) was tested on the bench–stand by CRFiat. This drive operation is similar to the operation of the two degrees of freedom planetary transmission (see Chapter 8). Anyway, it is much more costly and complicated to control.

Chapter 2
The Energy–Power Requirements for HEV Power Train Modeling and Control

ABSTRACT

The first step in the hybrid vehicle power train design, of course, after choosing the drive architecture, is the analysis of the power distribution and energy flow between the Internal Combustion Engine (ICE) (considered in this book only as the primary source of energy – PS) and the energy accumulator (called the source of power, or a secondary energy source – SS). The role of the Primary Source (PS) is to deliver to the system the basic energy, while the Secondary Source (SS) feeds the hybrid power train during its peak power loads and stores the vehicle's kinetic energy during the regenerative braking. The target of these considerations is to search for the minimal necessary power of the Primary Source (PS) and the minimal energy capacity of the Secondary Source (SS). Certainly, this computation requires the proper energy flow model and the basic vehicle driving cycle, in the role of which the statistic driving cycle is recommended. The main aim of this chapter is the depiction of the above problem, as well as the finding of its solution. The background of the energetic evaluation of the hybrid drive structure is the dynamic determination of the internal watt efficiency of each of the propulsion system's components. The complex construction of the hybrid drives requires an appropriate control strategy from its designers. In order to achieve this aim numerical optimization methods of nonlinear programming can be applied.

DOI: 10.4018/978-1-4666-4042-9.ch002

INTRODUCTION

The main aim of the proper adjustment of the hybrid vehicle drive is to determine the power distribution and the energy flow between the internal combustion engine – ICE (called the primary source of energy – PS) and the energy accumulator (called the source of power or the secondary energy source – SS). The role of the primary source (PS) is to deliver the basic energy, regarding the assumed control strategy – in a continuous or interrupted way (e.g. see Figure 3). The secondary source (SS) has to supply the energy for power peaks of the external load, while the vehicle accelerates, or recuperate kinetic energy when the regenerative braking takes place. This source of power operation involves the internal combustion engine (ICE), which is the most common element used as a primary energy source, with a stable working condition, significantly increasing this engine's (PS) efficiency, which also means the decrease of its fuel consumption. Next to the vehicle's kinetic energy recuperation, this is a fundamental feature of the hybrid power trains. The hybrid drive is complex and consists of several components. The power train's total efficiency directly influences fuel and electricity consumption. The first step in the hybrid power train design is the energy/power dynamic distribution modeling presented in this chapter.

The process of power distribution strongly depends on the selection of the drive structure and the transitional efficiency of its components, especially at the Primary Source (PS) (in this book, the Internal Combustion Engine (ICE) is basically considered) and on the secondary power source (SS), as well as the nickel metal hydride (Ni–MH), and lithium ion (Li–ion) batteries.

There are two well-known hybrid structures: the series and the parallel, whose standard simple compilations have too many bad features of the practical application. It is necessary to look for a new concept of the hybrid drive structure, which can connect the positive properties of the above-mentioned hybrid systems in one propulsion structure. The target of this searching procedure is to keep the engine at its highest possible operational efficiency level, which entails the minimal fuel consumption and atmospheric pollution. This design target can be obtained by entire power train nonlinear dynamic modeling and the internal combustion engine – as the basic energy source (PS) – a mathematical model is directly connected with the energy/power distribution.

Modeling and simulation methodology is the only way to acquire such a Formulated goal. The most important problem in modeling is the determination of the efficiency of the energy split as the function of the driving time. The solution to this problem is the approach of the power distribution determination on the generalized hybrid drive system.

The Energy–Power Requirements for HEV Power Train Modeling and Control

An essential problem encountered in designing the structure for vehicles with hybrid or pure electric drives is to make a precise assessment of their energetic effectiveness. Such driving structures differ from each other in many aspects, e.g. in the types of batteries (primary sources in purely electric drives and secondary ones in those of the hybrid nature), the electric motors, the CVT (Continuously Variable Transmission) assemblies, and the like. The determination of the internal efficiency of each of these components separately, does not allow us to make an overall assessment of each particular drive, enabling us to compare all available drives and to choose the most appropriate for them, which is of crucial importance in every designing process.

The continuous power addition of two or more energy sources takes place during each hybrid drive work. However, in each case, the hybrid drive may be reduced to one source – e.g. the purely electric or combustion one.

The background of energetic evaluation of the hybrid drive structure is the dynamic determination of the internal watt efficiency of each of the propulsion system components, according to the momentary external load ($N_{i=3}(t)$) alterations referring to the analyzed time of the power alterations (e.g. according to the driving cycle), which is expressed by the Equation 7.

1. ENERGY AND POWER DISTRIBUTION DYNAMIC MODELING

Figure 1 illustrates the block diagram of the two-source hybrid drive.

Figure 1. Block diagram strictly connected with the series hybrid power train

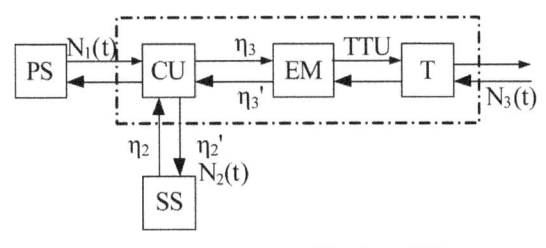

PS – Primary Source
SS – Secondary Source
EM – Electric Machine
N_1 – output PS power
N_3 – power on road wheels
CU – Control Unit
T – mechanical Transmission connected with road wheels
TTU – Torque Transmission Unit
N_2 – SS power

η_2 – energy discharging watt efficiency of SS
η_2' – energy charging watt efficiency of SS
η_3 – watt efficiency of torque transmission unit for vehicle acceleration and steady speed
η_3' – watt efficiency of torque transmission unit during vehicle regenerative braking (recuperation)

Figure 2. The substitute scheme of driving the vehicle expressed by its elementary components' moments of inertia and its angular velocities reduced on the one selected propulsion shaft of the vehicle power train

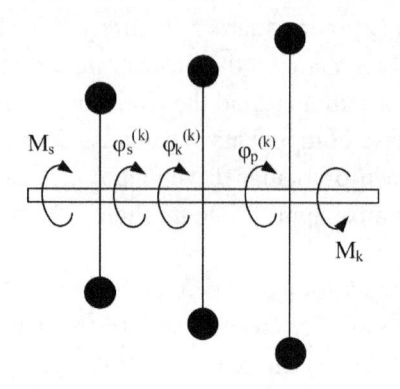

The cycles in the form of $N_k(t)$[1] and $M_k(t)$ for the considered vehicle can be resolved on the basis of the speed relation $V(t)$ (N_k – the power reduced on vehicle traction wheels, M_k – the torque reduced on traction wheels, t – time). For this purpose, vital relations are considered to carry out the energy calculations. Figure 2 is accepted as a sufficiently precise model of that.

For infinitely big angle rigidity and the driving engine, as well as the main gear damper equal to zero, the vehicle's movement can be described by an equations system in the D'Alembert form:

$$\left(J_s + J_k + J_p\right)\frac{d^2\phi^{(k)}}{dt^2} = M_s - M'_k$$
$$\phi_s^{(k)} = \phi_k^{(k)} = \phi_P^{(k)} = \phi^{(k)} \tag{1}$$
$$M'_k = b + c\frac{d\phi^{(k)}}{dt}$$

where:

- J_s: The moment of inertia of the driving engine rotor;
- J_k: The moment of inertia of vehicle wheels, reduced on the driving engine shaft;
- J_p: The moment of inertia corresponds to the entire vehicle mass, reduced on the driving engine shaft;
- $\varphi_s^{(k)}$: The rotation angle of the driving engine shaft;

- $\varphi_k^{(k)}$: The rotation angle of vehicle wheels, reduced on the driving engine shaft;
- $\varphi_p^{(k)}$: The displacement of the vehicle referred to the driving engine shaft;
- M_s: The engine's driving torque;
- M'_k: The friction torque, reduced on the driving engine shaft;
- **b, c:** The equations parameters dependent on the relevant vehicle data and driving conditions, reduced on the driving engine shaft.

The reduction of the mechanical values in the mathematical model is carried out by the methods of balance of the kinetic, potential energy and the power of the dissipation energy in the dissipative elements.

In practice, the easiest method is to utilize the equations describing the driving friction torque M_k at the vehicle's traction wheels:

$$M_k = \left(mgf_t + 0{,}047 A c_x V^2 + m\delta_b a\right) r_d^* \qquad (2)$$

$$V[km/h] \rightarrow M_k[Nm]$$

where:

- **m:** The vehicle's total mass (kg);
- **g:** The acceleration of gravity (m/s²);
- f_t: The rolling friction resistance;
- **A:** The frontal area (m²);
- c_x: The aerodynamic coefficient;
- δ_b: The rotation elements factor;
- r_d: The vehicle dynamic radius (m);
- **a:** The vehicle acceleration (m/s²).

As for:

$$\delta_b = 1 + \frac{J_s j^2 \eta_m}{m r_d^2} + \frac{\sum J_k}{m r_d^2} \qquad (3)$$

where:

- **j:** The amount ratio between the driving engine shaft and the vehicle wheels;
- η_m: The torque transmission efficiency;

The total power Nk at the vehicle wheels is described by the following equations:

$$N_k = F_k V = \frac{M_k}{r_d} V = \frac{1}{3600}\left(mgf_t V + 0,047 A c_x V^3 + m\delta_b a V\right) \quad (4)$$

$M_k[Nm] \rightarrow N_k[kW]$

$V[km/h]$

The amount of energy that the vehicle is driven by the power N_k is:

$$E_k = \int N_k dt \quad (5)$$

Energy analysis of the vehicle's movement, according to a given cycle, is done as the function of time. Hereafter, it is comfortable to present the Equation (5) as $E_k = f(a,t)$. In order to do this, the following expression should be taken into the Equation (4) Formula:

$$V[km/h] = 3,6 \cdot a[m/s^2] \cdot t[s] \quad (6)$$

Then, the value of E_k is expressed in kilojoules (kJ).

In general, the vehicle's energetic analysis is carried out for horizontal roads. In particular cases, gradability force should be taken into account in the Equations (3) and (5).

Figure 3 illustrates a simplified driving cycle presented as V – the vehicle's speed and N – the power distribution. It shows a part of the real power requirement, reduced at the vehicle's traction wheel.

The constant power operation of an internal combustion engine as a primary energy source is indicated in Figure 3a. This form of primary energy source operation is assumed as basic for energetic analysis of a hybrid power train. Of course, in practice, an internal combustion engine constant power operation can be replaced by interrupting the ICE work, as is shown in the afore-mentioned figure.

The primary energy source PS, equipped with the ICE's (Internal Combustion Engine) possible propulsion strategy in the HE (Hybrid-Electric) series drive, is also the engine's constant speed (angular rotational velocity) operation. In this case, the engine's torque (M) versus its angular velocity ω and its power (N_1) – time (t) characteristics are shown in Figure 3b).

The Energy–Power Requirements for HEV Power Train Modeling and Control

Figure 3. The simplified driving cycle and the control strategies of the PS operation: a) 'continuous' or 'interrupted' power N1 (in the discussed case – the ICE power reduced on the traction wheel) – power N3 respectively equals N_k on the traction wheels; b) 'constant speed'; c) 'constant torque'

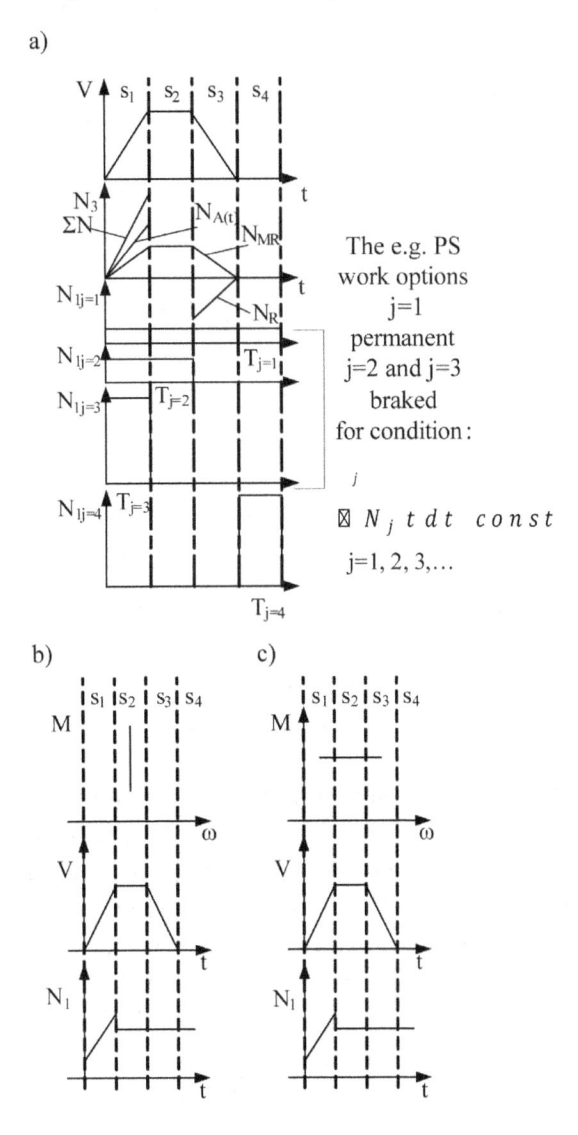

A similar shape of power $N_1 = f(t)$ can be obtained from $M(\omega) = $ const., which is called constant torque operation of primary energy source (see Figure 3c). The simulation studies of these three different control strategies of operation of an internal combustion engine as the primary energy source in a series hybrid power train are included in Chapter 7.

Figure 4. The real power requirement referring to the vehicle's driving cycle reduced on its traction (road) wheel

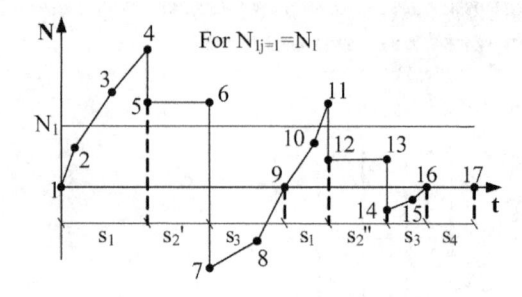

The power of the traction wheel (road wheel) is defined as follows:

- N_3: The traction wheel power:

$$\begin{cases} -N_3(t) = N_{MR}(t) + N_A(t) \forall sign N_3(t) = 1 \\ -N_3(t) = N_R(t) = N_A(t) - N_{MR}(t) \forall sign N_3(t) = 1 \end{cases}$$

- $N_{MR}=N_k$: The resistance power of traction wheels;
- N_A: The acceleration or deceleration power;
- N_R: The regenerative braking power;
- S_d: The time of each driving cycle phase; where: $d = 1,2,3...$

The result of this analysis is the determination of minimal energetic parameters of the hybrid drive as a power stream distributed by the drive components, specially controlled, and by the minimal level of energy storage in the accumulator (SS) used as the power supplier, a supported unit, according to the propulsion structure and the considered drive cycle. The most important problem in the proper (optimal) adjustment of the hybrid drive energy transmission is the determination of the dynamic efficiency of power distribution, according to the hybrid drive structure. The amount of losses strongly depends on the drive structure. Using an additional variable ratio gearbox between the electric traction motor and the road wheels, for example, makes it possible to limit heat losses in the motor copper windings, the momentary power of the battery, and to increase the total drive efficiency. Certainly, for simulation, it is necessary to use accurate and realistic drive component modeling.

The solution of the Equation (1) means that the minimal value determination of the output ICE (PS) power N_1 should be reached by iterative methods. The background of the dynamic analysis of the power distribution is described by the following methodology. The function of loads $N_3(t)$ has a strong non–linear form and has a repetitive cycle. An example of the computer straight–line approximation of the function is shown in Figure 4. The most favorable working point of the primary source is the point of stable power. Using energy accumulation and suitable automatic regulation, one can (on the diagram of power) assume that the primary source of power has been designated by the N_1 line (Figure 4).

The $N_3(t)$ function refers to changes in the power of the system loading by the active receiver. Its dependent variables may be positive or negative. If N_1 denotes the primary source of power, then the three energetic states of the energy accumulator can be generally described, according to the vehicle's acceleration, steady speed, and deceleration.

While determining the load cycle of particular elements of a multi–source system, one cannot leave out the definition of the respective functions of efficiency, referring them to the definite cycle phases. One can introduce the notion of averaged efficiency functions on the grounds of the extended mean value theorem.

$$\sum_j \int_0^T N_j(t) \cdot \eta_j^a(t) dt = \eta_j^a \cdot \sum_j \int_0^T N_j(t) dt \qquad (7)$$

while:

$$\eta_j^a = \overline{\eta}_j^a(\zeta); \qquad \zeta \in <0, T>$$

where: $\overline{\eta}_j^a$ – averaged efficiency function for the j– cycle segment, and defined as the internal watt efficiency of each of the propulsion system's components according to the momentary of its external load alterations:

$$\sum_j \int_0^T N_j(t) \eta_j^a(t) dt = \overline{\eta}_j^a \int_0^T N_j(t) dt$$

where:

- **T:** The analyzed time of power alterations (e.g. according to the driving cycle);
- **a = ±1:** Depends on the direction of the power flow (a=−1 during the kinetic energy recuperation of the vehicle);
- **j:** 1, 2, 3, ... j∈S_d;

- **T:** The cycle duration time, $T = \sum_{d=1}^{n} S_d$, where S_d – the cycle phase composed of j– segments.

The solution that is currently being sought is the determination of the lowest value of N_1 for a given load cycle and a definite structure of a multi–source system.

1.1. The Energy Balance of the Power Train

If one assumes the following notations:

- **E1:** The amount of energy required for the realization of the cycle in the form of $N_3(t)$ for:

$$\frac{dN_3(t)}{dt} \neq 0$$

and $N_3(t) \geq 0$;

- **E2:** The amount of energy taken from the primary source when:

$$\frac{dN_3(t)}{dt} \neq 0 \text{ and } N_3(t) \geq 0;$$

- **E3:** The amount of energy produced by the active receiver and transmitted to the accumulator defined by the field below $N_3(t)$ for:

$$\frac{dN_3(t)}{dt} \neq 0 \text{ and } N_3(t) \leq 0;$$

- **E4, E5:** The amount of energy accumulated, when the primary source momentary power is larger than the load power for $N_3(t) = \text{const}$:

$$\frac{dN_3(t)}{dt} = 0;$$

$N_3(t) \in <0, K>$, when K means the greatest value of the load power in the definition phase of the $N_3(t)$ cycle and for $N_3(t) <0$ (during the process of recuperation), then in $\eta = 1$ the energy balance of the two – source system has the following Formula:

$$E_1 - E_2 = E_3 + E_4 + E_5 \qquad (8)$$

The subtraction on the left side of the Equation (8) refers to the energy expended from the accumulator for $N_3(t) \geq 0$,

$$\frac{dN_3(t)}{dt} \neq 0,$$

and the sum on the right side of the equation signifies the accumulated energy. When the cycle phases: $S_t \in T$ and $j \in S_d$ are attributed to the respective changes of $N_3(t)$, the Equation (7) can be presented for $\eta \neq 1$ and for the starting point from the primary source in the form shown in the following form:

$$\bar{\eta}_s^{-1} \sum_{j \in S_1} \int_{t_j}^{t_{j+1}} N_3(t)\, dt - N_1' \cdot \bar{\eta}_3 \sum_{j \in S_1} (t_{j+1} - t_j) =$$

$$= \bar{\eta}_r \sum_{j' \in S_3} \int_{t_{j+1}}^{t_j} N_3(t)\, dt + \bar{\eta}_{ak} N_1' \sum_{j' \in S_3 \in S_4} (t_{j+1} - t_j) + \qquad (9)$$

$$+ \bar{\eta}_{ak} \sum_{j \in S_2} \left(N_1' - \bar{\eta}_3^{-1} N_3(t) \right) \cdot \left(t_{j+1} - t_j \right)$$

where:

- N'_1: The first approximation of the primary source of power;
- $\bar{\eta}_s = \bar{\eta}_2 \bar{\eta}_3$: The efficiency (according to Figure 1) of the active receiver of the considered control;
- $\bar{\eta}_r = \bar{\eta}_2' \bar{\eta}_3'$: The recuperation efficiency;
- $\bar{\eta}_{ak} = \bar{\eta}_2'$: The accumulator efficiency;
- S_i: The cycle phase for:

$$\frac{dN_3(t)}{dt} \neq 0, \quad N_3(t) > 0;$$

- S_4: The cycle phase for:

$$N_3(t) = 0;$$

- S_3: The cycle phase for:

$$\frac{dN_3(t)}{dt} \neq 0, \quad N_3(t) < 0;$$

S_2: The cycle phase for:

$$N_3(t) = \text{const} \neq 0;$$

- t'_j: The beginning of the j' – cycle segment.

The Equation 8 includes two approximations:

1. The respective fields are not separated below the curve $N_3(t)$ and the line N'_1 for $j \in S_{jd}$,
2. Lack of ay determination of the quantitative relation.

$$N'_1 \text{ and } N_3(t) = \text{ const. for } j \in S_2;$$

$$N'_1 \geq N_3(t) = \text{ const. for } j \in S''_2;$$

$$N'_1 \leq N_3(t) = \text{ const. for } j \in S'_2.$$

The second approximation can be eliminated after determining N'_1, and more closely investigating the relationship between N'_1 and $N_3(t) \in S$ The first approximation is concerned with the analysis of the intersection of the line N'_1 with the curve $N_3(t)$ as a set of points (Figure 5).

The point of intersection of the line N_{z1} and the curve $N(t)$ can be determined by using the Lagrange interpolation multinomial, as well as the Eitken interpolation and iteration procedure. The objective is to determine the time t_j, then compare with Figure 5:

$$E_a = \sum_{j \in S_1} \left(\int_{t'_j}^{t'_{j+n}} N_3(t)\,dt - E'_n - E''_n \right) \tag{10}$$

where $n = 1,2,3,...$

Figure 5. The illustration of E_a and E_{ak} definitions, referring to part of the vehicle's driving during its acceleration

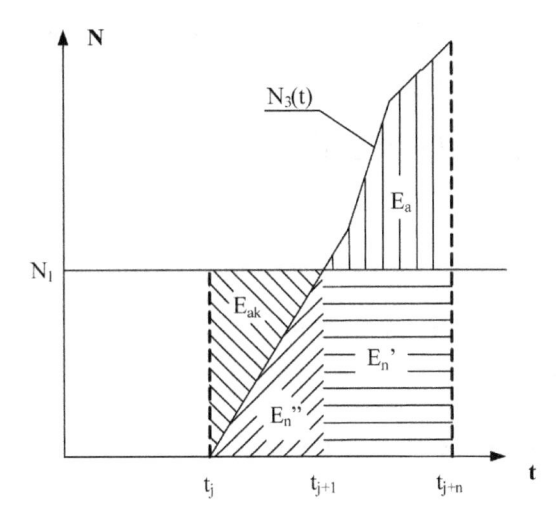

The energy E_a is expended from the accumulator to S_j:

$$E_a = \int_{t_{j'}}^{t_{j'+n}} N_3(t)\,dt - \bar{\eta}_3 N'{}_n\left(t_{j'+n} - t_{j'+1}\right) - \int_{t_j}^{t_{j+1}} N_3(t)\,dt \tag{11}$$

The subsequent form of the Equation (11) is as follows:

$$\bar{\eta}_s^{-1}\left[\sum_{j\in S_1} E_a + \sum_{j\in S_4}\left(N_3(t) - N''{}_1\bar{\eta}_3\right)\left(t_{j'+1} - t_{j'}\right)\right] =$$

$$= \bar{\eta}_r\left[\sum_{j\in S}\int_{t_{j'+1}}^{t_{j'}} N_3(t)\,dt + \bar{\eta}_{ak}\sum_{j\in S_2''}\left(N''{}_1 - \bar{\eta}_3^{-1}N_3(t)\right)\right] \tag{12}$$

$$\left(t_{j'+1} - t_{j'}\right) + \bar{\eta}_{ak}N''{}_1\sum_{j'\in S_4 S_3}\left(t_{j'+1} - t_{j'}\right) + \bar{\eta}_{ak}\sum_{j\in S_1} E_{ak}$$

where:

$$E_{ak} = \bar{\eta}_3 N'{}_1\left(t_{j'+1} - t_{j'}\right) - \int_{t_j}^{t_{j+1}} N_3(t)\,dt$$

- **N''₁**: The second iterative approximation of power N_1.

The solution to the Equation (12) should be reached by the application of iterative methods:

$$F\left(N_1^{(h+1)}\right) = N_1^{(h+1)} A - B\left(N_1^{(h)}\right) = 0 \qquad (13)$$

where:

- **A**: The numerical index of the equation;
- **B**: The equation parameter is dependent on the *h*–iterative value of $N_1^{(h)}$.

This function fulfills the Lipschitz condition, which gives the Newton algorithm the convergence:

$$N_1^{(h+1)} = N_1^{(h)} - \frac{N_1^{(h)} A - B\left(N_1^{(h)}\right)}{A} \; ; \left|N_1^{(h+1)} - F(N_1^{(h+1)})\right| < C \qquad (14)$$

where: *C*<1 – the constant determining the accuracy of the result after *h*+1 iterations. Another method of solving the Equation (12) is the bisection, which is slowly convergent.

In cases when the power N_1 of the primary source of power (PS) reduced at the traction wheel for the direct comparison to power N_3 is the variable function of time, the energies E_{ak} and E_a (see Figure 6) are defined, based on the Equation (12), as follows:

$$E_{ak} = \int_{t_j}^{t_{j+1}} N_1(t)dt - \int_{t_j}^{t_{j+1}} N_3(t)dt \Rightarrow E_{ak} = \int_{t_{j+n}}^{t_{j+(n+1)}} N_1(t)dt - \int_{t_{j+n}}^{t_{j+(n+1)}} N_3(t)dt$$

$$E_a = \int_{t_{j+1}}^{t_{j+2}} N_3(t)dt - \int_{t_{j+1}}^{t_{j+2}} N_1(t)dt \Rightarrow E_a = \int_{t_{j+(n+1)}}^{t_{j+(n+2)}} N_3(t)dt - \int_{t_{j+(n+1)}}^{t_{j+(n+2)}} N_1(t)dt \qquad (15)$$

$$j = 1,2,3..., n = 1,2,3...$$

This case is illustrated in Figure 6.

Figure 6. The variable during the vehicle's driving cycle, the power of the primary source of energy (PS) versus the time function

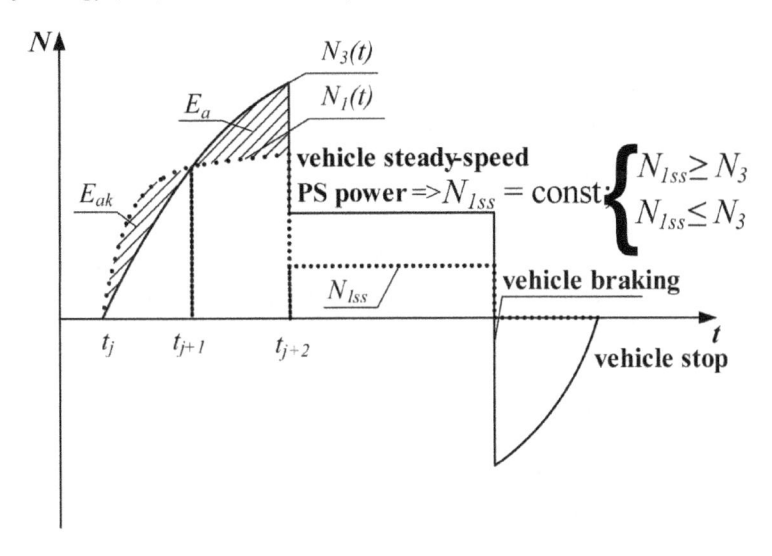

($N_1(t)$ – is the known function approximated by a proper polynomial – it is often assumed as $N_{1SS} \cong N_{10}$ – e.g. idle state of the engine's operation, as the well-known internal combustion engine parameter).

The above-mentioned case is typical for the engine's (PS) operation in the parallel hybrid drive structure. During the vehicle's steady speed, braking and stopping time, the value of the internal combustion engine power is practically constant, but different (see Figure 6). For this reason, in the Equation (12) no change is necessary, but it is worth remembering that each value of N_1 for the vehicle's steady speed:

$$(t_j \in S_2' \quad \text{and} \quad t_j \in S_2'')$$

is different from the value during vehicle's braking ($t_j \in S_3$) and stopping ($t_j \in S_4$). For the creation of the new form of the Equation (18), there should be defined (assumed) the proper values of N_1 power, according to the vehicle's steady speed, braking and stopping driving cycle phases.

In the series structure, the engine most often works on constant or interrupted constant power mode, as illustrated in Figure 3.

The block diagram of the two–source hybrid drive connected with the parallel structure is shown in Figure 7.

Coming back to the extended and typical hybrid drive structures, it is necessary to emphasize the possibility of the creation of the structure consisting of two ac-

Figure 7. Block diagram of the parallel power train structure (depictions as in Figure 1)

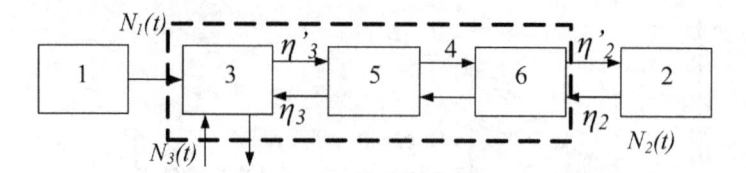

Figure 8. The illustration of the method of determination of the minimal secondary source energy capacity E_{SSmin} and the initial value E_{SSin} as the starting point for the analyzed vehicle driving cycle

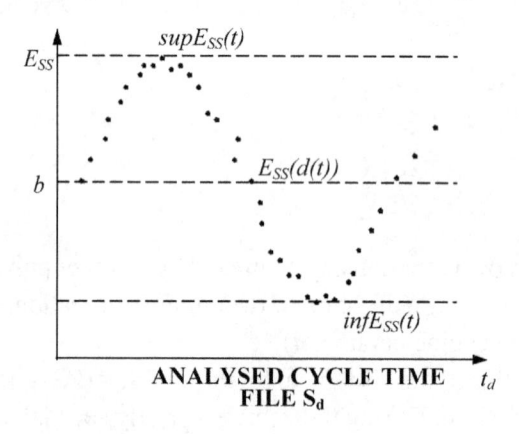

cumulators (e.g. the battery and the flywheel, or the battery and a special ultracapacitor (see Chapter 6). The above–mentioned methods of determination of power distribution could also be effectively used in this procedure. (see Figure 8)

The Equation (12) is the background for the process of energetic optimization of the hybrid drive structure, parameters and the control function. There are two methods:

- Determination of the minimum power of the primary source of power (PS), when its power as a time function is depicted by a continuous line (according to Figure 5);
- Determination of the minimal energy of the PS (Equation 11, Figure 6), when its power is described by the variable time function.

The solution of the basic Equation (12) also presents the opportunity of minimizing the energy stored in the accumulator (the secondary source of power – SS) during the repeated driving cycle.

The determination process of the secondary source (accumulator) energy capacity E_{ss} is illustrated in Figure 8 and by the following equations:

$$E_{ss\,min} = \sup_{d=1,...n} E_{ss}(d) - \inf_{d=1,...n} E_{ss}(d) \tag{16}$$

$$E_{s\sin} = b - \inf_{d=1,...n} E_{ss}(d) = \sup_{d=1,...n} \left(b - E_{ss}(d)\right) \tag{17}$$

Finally, analysis of the power distribution by the use of the depicted computation method based on the energy balance, makes it necessary that the energy level of the secondary source of power (SS) is precisely the same at the beginning and at the end of the driving cycle, which is easy to note.

2. THE MODELING OF THE HYBRID POWER TRAIN ENERGY FLOW

The block diagram of the scheme of the computer analysis-simulation tests is shown in Figure 9. First of all, for the considered hybrid power train architecture its component mathematical models have to be worked out. The control strategy can be defined using the approach depicted below.

The background of the energy flow method is different from the driving cycles statistic, and realistic adopted to the power modeling, as is shown in Figure 9.

Figure 9. The sketch of the scheme of the hybrid drive for the computer simulation (the broken line shows the possibility of different combinations of the drive architecture, including the series and the parallel version)

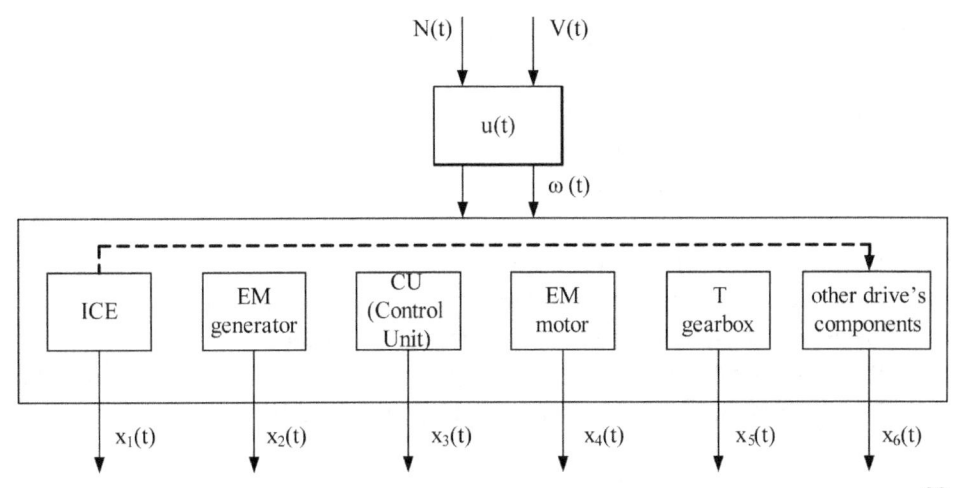

The dynamic functions *N(t)* and *V(t)* referring to traction wheels are generated from the driving cycle directly relying on the vehicle's data.

$N(t)$ and $V(t)$ taken from the driving cycle are the input vector u(t):

$$\mathbf{u}(t) = \begin{bmatrix} N(t) \\ V(t) \end{bmatrix}$$

The model equationss depicting the drive components determine the output vector **x**(t):

$$\mathbf{x}(t) = \begin{bmatrix} x_1(t) \\ x_2(t) \\ \vdots \\ x_n(t) \end{bmatrix}$$

where $x_1, x_2 \ldots x_n$ are the functions of time, such as the flux, current, torque, angular velocity and other variables occurring in the individual model.

What is described above, requires a numerical solution. Gear's method (Simulation Numerical Solution of Differential Algebraic equationss) has been used to solve the differing and nonlinear algebraic differential equationss (Gear, 1971, 1972).

In order to use the discrete variable method, it is necessary to integrate a group of equationss described by a mixture of ordinary differential, nonlinear, and linear equationss. It is especially important during such dynamic simulations to pay attention to the time constants between the mechanical (including the Internal Combustion Engine – ICE) and electrical (including the battery) differences. For this reason, the step of integration during computation has to be properly adjusted – e.g. to 0.01s.

The main target of the simulation study is to compare the performances of the selected hybrid drive architectures for the same driving conditions (cycles). The procedures of dynamic modeling and simulation are very effective tools for the designing of the hybrid power trains.

The designing process of the hybrid drive focuses on the following steps:

1. Selecting and adjusting the best drive architecture considering the following terms: minimum component number, maximum efficiency, minimum fuel and electricity consumption connected with the emissions.
2. Working out the control strategy adjusted to the considered drive architecture.
3. Determining the parameters of the hybrid drive as follows:

The Energy–Power Requirements for HEV Power Train Modeling and Control

    a. The internal combustion engine operating points located on its map regarding the analyzed driving cycle;
    b. The alterations of the ICE torque, angular velocity and power;
    c. The fuel consumption (l/100km) of the internal combustion engine;
    d. The alterations of the motor component's u_q, u_d and i_q, i_q (the vector field oriented control – see Chapters 2 and 3);
    e. The battery state of charge's (SOC) operating range, according to the computed EMF and internal resistance;
    f. The alterations of the battery's current, voltage, power and SOC;
    g. The calculation of the maximal boundary value of the battery, Coulomb capacity (Ah) exchange regarding the analyzed driving cycle;
    h. The final assumption of the battery capacity and the nominal voltage of the battery pack;
    i. The final adjustment of the mechanical transmission ratio.

The first step requires comparative analyses, which can be determined after the simulation results of the obtained different drive architectures. It means that before the comparison of the simulation results for the selected hybrid drives, the optimal value of the parameters mentioned in step 3 for various testing strategies, should be found out.

This book presents some final results from the simulation study. As the global model of the hybrid vehicle is very sensitive to the input data, including the drive architecture and control, the number of simulations is significant. An experienced researcher can cope with this task much more easily: for instance, the hybrid drive is very sensitive to the mechanical ratios (see Chapters 9, 10). Even a small change in this ratio will cause big mistakes, regardless of the proper design target. In general, there are a few mechanical transmissions in the hybrid drive connecting the Internal Combustion Engine (ICE) with the electric motor/generator (see Chapter 10). The ratios of these transmissions have a strong influence on the hybrid power train's efficiency.

The urban bus hybrid drive, meanwhile, should be designed especially for the chosen cities or towns. In this instance, the practical driving cycle in repeated ways, depicts the real driving conditions for a bus.

Passenger cars must meet complex driveability requirements, including acceleration, maximum speed, gradeability, etc., to fulfill these requirements. The hybrid power train design should be adjusted by adopting the proper internal combustion engine and motor size, and/or by adding a gearbox – with two, three or more shifts.

This entire power train design process based on modeling and simulations is described in the following chapters.

Figure 10. The general scheme of the hybrid power train as a system of energy flow and its corresponding control signal

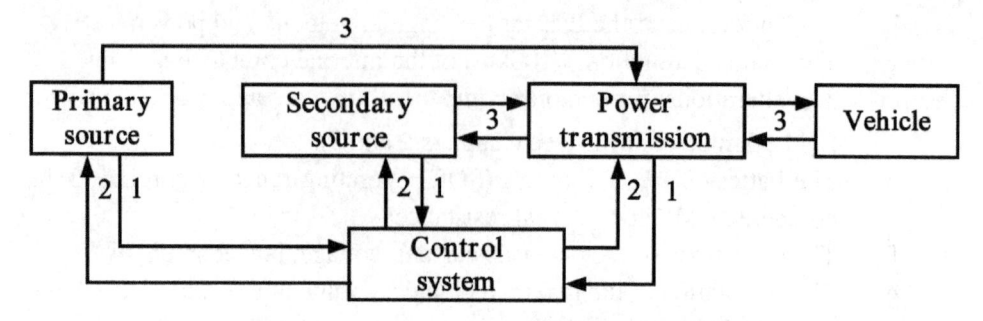

1– energetic information signal, 2– control signal, 3– energy stream

3. AN APPROACH TO THE CONTROL OF HYBRID POWER TRAINS

The sophisticated construction of the hybrid drives requires from designers, both a complex- made technology and determination of the appropriate control strategy. The purpose is just to present an approach to the methods of solving these problems. In order to reach that aim, numerical optimization methods of nonlinear programming are applied (by decomposing dynamical optimization problems into the nonlinear programming problem), as well as application of the methodology of dynamic switching over trajectories.

The general scheme displaying the energetic – informational signal of the hybrid system is shown in Figure 10.

It is possible to achieve the reduction of energy consumption and decrease in the emissions of environmental pollution, only by designing an appropriate dynamic and efficient characteristics of the drive system work.

3.1. Computation of the Control Functions

The control functions which provide the realization of the required vehicle speed distributions by the output shaft torque of the analyzed power train have to minimize the assumed criterion of quality. By this criterion, we should concern ourselves with the dependencies which determine the important quantities for the proper use of the drive system: the fuel consumption and the emission of environmental pollution resulting from it, the energy loss in the system, the energy accumulated in the secondary source of power – the battery SOC minimum alteration, etc. In addition, the computed control function should also factor in the constraints, which follow from the characteristics of the drive system and the receiver system.

The problem of searching for the control functions, which would be able to fulfill the demand presented above, is typically a dynamic optimization problem and the methods of solving it are well known. However, in the process of solving these complex problems, a different approach was assumed and simpler methods were used by the decomposition of the dynamic optimization problem into the Nonlinear Programming Problem (NPP) (Fletcher, 1974; Fletcher & Powell, 1974).

The problem of the decomposition consists of determining the parameters of the control functions (time interval of each phase is divided into a few equidistant or non-equidistant sections, and then cubic or exponential interpolating spline functions are used to approximate the parametric control function over the mesh of grid points), the construction of the cost function and constraints, whose selection determines the functionality of the propulsion system.

For all the phases of the driving cycle, the universal criterion of quality (the cost function) should be taken into account:

$$I = \int_{t_{bf}}^{t_{ff}} (E + M) dt \qquad (18)$$

where:

- t_{bf}, t_{ff}: The initial and terminal time of the considered phase of driving;
- E: The losses of the electric power;
- M: The losses of the mechanical power.

The problem of nonlinear optimization could be solved using Powell's penalty function method.

3.2. The Possibility of Implementation

The most important purpose of proper control of the hybrid drive is to reduce the engine's fuel consumption and the battery's discharging limitation, in effect altering the battery's state of charge (SOC). The real efficiency factors of the drive components strongly depend on torques, shaft speeds, or currents, and voltages, especially in the real driving cycle. For this reason, it is absolutely necessary to include the functions of efficiency of the drive-in simulation studies. The problem is focused on the proper real-time ('on-line') drive control, where the function of the target

The Energy–Power Requirements for HEV Power Train Modeling and Control

is to minimize the fuel and electricity consumption, or minimize the entire power train internal losses. The achievement of this goal is connected with the prediction of the drive load during the vehicle's real driving, as well as for its acceleration, regenerative braking, and steady speed.

The main criteria for the control strategy is as follows:

The fuel consumption of the internal combustion engine should be as low as possible. This could be resolved by the IC engine operating in the area of minimal specific fuel consumption. The balance of the battery's energy should equal zero (only for full hybrid drives), yet the k–factor of the battery state of charge should be the same at the beginning and the end of the driving cycle. The best condition of electricity transfer by the battery is expressed by the Formula (20), and also by the Figure 13. To sum up, the criteria of the control strategy is as follows:

- The electric motor must operate in the area of highest efficiency during the driving cycle;
- The alterations of the battery's internal resistance should be limited (keeping it in the range of minimal value of the analyzed battery internal resistance – see Chapter 5);
- The current of the battery should not exceed the determined limited value.

To fulfill the criteria mentioned above, it is necessary to consider: the vehicle's maximal speed and acceleration, as well as the vehicle's maximal continuous uphill grading.

This is possible to achieve after the proper adjustment of the mechanical ratios of the analyzed power train for different driving conditions. That may be resolved for the static standard driving cycles, such as, for instance, the extended European Driving Cycle (ECE).

We can also consider the vehicle's motion in dynamic conditions, called the 'chaos' driving cycles, and are characterized by probable, undetermined features. Anyway, each element (cycle phase) of the real or static driving cycle can be defined, as is shown in Figure 11, representing the different acceleration/deceleration parameters (the vehicle's steady speed value of acceleration/deceleration is, of course, equal to zero). The number and distribution of these selected elements depends on the application character and the required accuracy of the propulsion system operation.

The vehicle's speed and its momentary acceleration/deceleration determine the total power, which is required to be supplied to the traction wheels, to realize each real driving cycle element in these time points (Figure 11).

The next step is to determine the proper values in relation to the powers produced by the Internal Combustion Engine (ICE) and the motor, regarding the battery state of charge (SOC) changes (the SOC level and its derivative is assumed in Figure 12).

The Energy–Power Requirements for HEV Power Train Modeling and Control

Figure 11. The exemplary vehicle's speed and its momentary acceleration

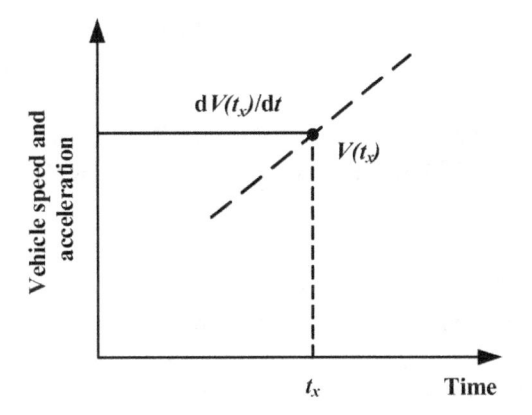

Figure 12. The exemplary battery's SOC level and its derivative – sign = –1; (during the vehicle's regenerative braking this derivative – sign = 1)

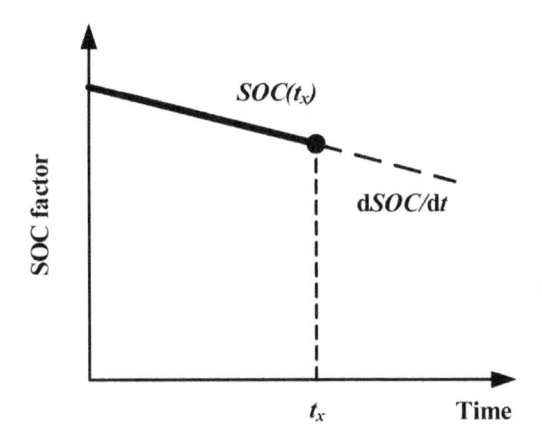

The low level of the SOC and its negative derivative means that the ICE engine has to produce more power to fulfill the requirement of realizing the drive cycle and battery charging. However, a high level of SOC means that it is possible to load the motor more than in ordinary conditions. The IC engine's efficiency should be as high as possible in all cases.

The proper control of ICE and the motor could be applied only when the required powers are determined. It means that the IC engine operation and voltage, and its frequency of the motor must be properly controlled, simultaneously.

The best solution is to shape the control functions in the matrix table form. The data necessary to create this table can be taken only from simulation tests. Multiple tests must be conducted for each operating condition, and the relation between the ICE and the PM motor's power must be determined.

4. THE METHOD OF DETERMINATION OF THE DISCHARGING ACCUMULATOR FACTOR (SOC): MINIMAL INTERNAL LOSSES OF ENERGY

Generally, the quality of the hybrid operation significantly depends on a proper battery or in the entire way, the proper adjustment of the accumulator.

The block of differential equationss in the following form may describe the mathematical model of the systems shown in Figure 9:

$$x' = F(x, u, t) \qquad (19)$$

where:

- **t:** The real time;
- **u(t):** the control vector;

$$u(t) = {}^s(t) = \begin{bmatrix} u_1(t) \\ u_2(t) \end{bmatrix} \begin{array}{l} \text{- traction motor (TM) voltage} \\ \text{- TM current} \end{array}$$

- **γ(t):** The electronic unit (chopper or inverter) control function;
- **x(t):** The state vector;
- **x1(t):** The transmission unit (TU) torque
- **x2(t):** The TU angular velocity
- **x3(t):** The accumulator power
- **x(t) = x4(t):** The accumulator discharging level
- **x5(t):** The primary source (PS) power
- **x6(t):** The primary source (PS) torque
- **xn(t):** Is dependent on the power train architecture
- **Note:** TU includes the traction motor.

Typical for the analyzed driving cycle alteration of k – the SOC battery factor – is shown in Figure 13.

The criterion of energy effectiveness (minimal internal losses) could be described by the following Formula (Szumanowski, 2000) – compare with Figure 8:

The Energy–Power Requirements for HEV Power Train Modeling and Control

Figure 13. The alterations of k battery state of charge factor corresponding to the vehicle's driving cycle

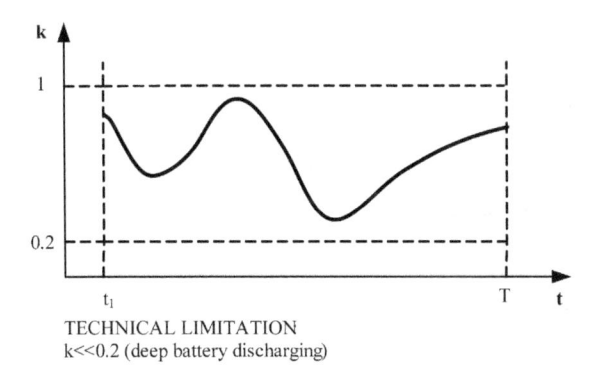

$$\frac{\int_{t_1}^{T} k(t)dt - \inf k(t)(T - t_1)}{(\sup k(t) - \inf k(t))(T - t_1)} \to 1 \qquad (20)$$

In the steady state and for k=1, independently from the driving time, the calculation according to the Equation (20) gives a result, which equals 1. Certainly, this is only a theoretical case. For the strictly repeated cycles (the European ECE or the Federal American FUD, etc.), the initial and final value of the k–factor should be the same. Yet, this does not mean that the Equation (20) reaches the best value from the range < 0, 1 > and it is not adequate to the statement about the high efficiency of energy distribution, either. The conclusion is:

- The dynamic investigation of k (t) function is necessary (see Chapters 5 and 6),
- The value of the calculation according to the Formula (20), for instance, depends very strongly on the battery charging during the vehicle's regenerative braking.

For a comparison of the performances of different hybrid drive architectures, the minimum of fuel consumed by the internal combustion engine should be carefully considered, taking into account, the possible highest effectiveness of the secondary source (the battery or generally the accumulator) during the alteration of its energy.

The criterion of minimal power train internal losses, referring to the battery's SOC alteration amongst the emphasized range of SOC changes, according to Formula 20, in the case of the hybrid urban bus, ensures that driving can only be in such a low range of Coulomb (Ah) battery capacity 1.7– 2.1 (Ah) (see Chapter 7; Figures 14, 16, and 18).

REFERENCES

Beretta, J. (1998). New classification on electric–thermal hybrid vehicle. In *Proceedings of EVS 15*. Brussels: EVS.

Burke, A. F., & Heitner, K. L. (1992). Test procedures for hybrid/electric vehicles using different control strategies. In *Proceedings of EVS 11*. Florence, Italy.

Cackette, T., & Evaoshenk, T. (1995). *A new look at HEV in meeting California's clean air goals*. Paper presented at EPRI North American EV & Infrastructure Conference. Atlanta, GA.

Dietrich, P., Ender, M., & Wittmer, C. (1996). *Hybrid III powertrain update*. Electric & Hybrid Vehicle Technology.

Ehsani, M., Gao, Y., Gay, L., & Emadi, A. (2004). *Modern electronic, hybrid electric and fuel cell vehicles – Fundamentals, theory and design*. Boca Raton, FL: CRC Press. doi:10.1201/9781420037739

Fletcher, R. (1974). Minimization of a quadratic function of many variables subject only to lower and upper hounds. *J. Inst. Maths. Applies*.

Fletcher, R., & Powell, M. J. D. (1974). On the modification of LDLT factorization. *Mathematics of Computation*, 28.

Gear, C. W. (1971). *Simulation numerical solution of differential algebraic equations*. IEEE.

Gear, C. W. (1972). DIFSUB for the solution ordinary differential equationss. *Communications of the ACM*, 14.

Gill, P.E., & Murray, W. (1972). Quasi–Newton methods for unconstrained optimization. *J. Inst. Applies*.

Hayasaki, K., Kiyota, S., & Abe, T. (2009). The potential of parallel hybrid system and nissan's approach. In *Proceedings of Aachen Colloquium*. Aachen Colloquium.

He, X., Parten, M., & Maxwell, T. (2005). Energy management strategies for HEV. In *Proceedings of IEEE Vehicle Power and Propulsion Conference, Illinois Institute of Technology*. Chicago, IL: IEEE.

Hellenbrouch, G., Lefgen, W., Janssen, P., & Rosenburg, V. (2010). New planetary based hybrid automatic transmission with electric torque converter on demand actuation. In *Proceedings of Aachen Colloquium*. Aachen Colloquium.

Hofman, T., & Van Druten, R. (2004). Research overview – Design specification for HV. In *Proceedings of ELE European Drive Transportation Conference*. Estorial, Portugal: ELE.

Jozefowitz, W., & Kohle, S. (1992). Volkswagen golf hybrid – Vehicle result and test result. In *Proceedings of EVS 11*. Florence, Italy: EVS.

Keith, H. (2007). Doe plug in hybrid electric vehicles R&D plan. In *Proceedings of Eoropean ELE – Drive Transportation Conference*. Brussels: ELE.

Martines, J. E., Pires, V. F., & Gomes, L. (2009). Plug-in electric vehicles integration with renewable energy building facilities – Building vehicle interface. In *Proceedings of Powereng IEEE 2nd International Conference on Power Engineering, Energy and Electrical Drives*. Lisbon: IEEE.

Moore, T. (1996). Ultralight HV principles and design. In *Proceedings of EVS 13*. Osaka, Japan: EVS.

Neuman, A. (2004). Hybrid electric power train. In *Proceedings of ELE European Drive Transportation Conference*. Estorial, Portugal: ELE.

Overman, B. (1993). Environmental legislation may initiate the EV and HV industry: Will economics sustain. In *Proceedings of ISATA*. Aachen, Germany: ISATA.

Portmann, D., & Guist, A. (2010). Electric and hybrid drive developed by Mercedes-Benz: Vans and technical challenges to achieve a successful market position. In *Proceedings of Aachen Colloquium*. Aachen Colloquium.

Ruschmayer, R. Shussier, & Biermann, J.W. (2006). Detailed aspects of HV. In *Proceedings of Aachen Colloquium*. Aachen Colloquium.

Sporckman, B. (1992). Comparison of emissions from combustion engines and 'European' EV. In *Proceedings of EVS 11*. Florence, Italy: EVS.

Sporckman, B. (1995). *Electricity supply for electric vehicles in Germany*. Paper presented at the EPRI North American EV & Infrastructure Conference. Atlanta, GA.

Sweet, L.H., & Anhalt, D.A. (1978). Optimal control of flywheel hybrid transmission. *Journal of Dynamic System, Measurement and Control, 100*.

Szumanowski, A. (1993). Regenerative braking for one and two source EV drives. In *Proceedings of ISATA 26*. Aachen, Germany: ISATA.

Szumanowski, A. (1996a). Simulation testing of the travel range of vehicles powered from battery under pulse load conditions. In *Proceedings of EVS13*. Osaka, Japan: EVS.

Szumanowski, A. (1996b). Generalized method of comparative energetic analyses of HEV drives. In *Proceedings of EVS13*. Osaka, Japan: EVS.

Szumanowski, A. (2000). *Fundamentals of hybrid vehicle drives*. Warsaw: ITE Press.

Szumanowski, A. (2006). *Hybrid electric vehicle drives' design*. Warsaw: ITE Press.

Szumanowski, A., & Bramson, E. (1992). Electric vehicle drive control in constant power mode. In *Proceedings of ISATA 25*. Florence, Italy: ISATA.

Szumanowski, A., & Brusaglino, G. (1992). Analysis of the hybrid drive consisting of electrochemical battery and flywheel. In *Proceedings of EVS 11*. Florence, Italy: EVS.

Szumanowski, A., & Hajduga, A. (1998). Energy management in HV drive advanced propulsion systems. In *Proceeding of GPS*. Detroit, MI: GPS.

Szumanowski, A., & Jaworowski, B. (1992). The control of the hybrid drive. In *Proceedings of EVS11*. Florence, Italy: EVS.

Szumanowski, A., & Piórkowski, P. (2004). Ultralight small hybrid vehicles: Why not? In *Proceedings of ELE European Drive Transportation Conference*. Estorial, Portugal: ELE.

Szumanowski, A., Piórkowski, P., Hajduga, A., & Ngueyen, K. (2000). The approach to proper control of hybrid drives. In *Proceedings of EVS 17*. Montreal, Canada: EVS.

Takaoka, T., & Komatsu, M. (2010). Newly-developed toyota plug-in hybrid system and its vehicle performance. In *Proceedings of Aachen Colloquium*. Aachen Colloquium.

Trackenbrodt, A., & Nitz, L. (2006). Two–mode hybrids = adoption power of intelligent system. In *Proceedings of Aachen Colloquium*. Aachen Colloquium.

Vaccaro, A., & Villaci, D. (2004). Prototyping a fussy based energy manager for parallel HEV. In *Proceedings of ELE European Drive Transportation Conference*. Estorial, Portugal: ELE.

Wyczalek, F. A., & Wang, T. C. (1992). Regenerative braking for electric vehicles. In *Proceedings of ISATA 2J*. Florence, Italy: ISATA.

Yamagouchi, K. (1996). *Advancing the hybrid system*. Electric & Hybrid Vehicle Technology.

Yamagouchi, K., Miyaishi, Y., & Kawamoto, M. (1996). Dual system – Newly developed hybrid system. In *Proceedings of EVS 13*. Osaka, Japan: EVS.

Yamamura, H. (1992). Development of power train system for Nissan. In *Proceedings of EVS 11*. Florence, Italy: EVS.

ENDNOTES

[1] N_k is indicated as N_3.

Chapter 3
Electric Machines in Hybrid Power Train Employed Dynamic Modeling Backgrounds

ABSTRACT

The Alternative Current (AC) induction, asynchronous motor, and the Permanent Magnet (PM) synchronous, or the Brushless Direct Current (BLDC) motor, which are types of the Permanent Magnet (PM) synchronous machines, can be applied in hybrid power trains. This chapter presents the fundamental theory as a necessary background to the mentioned motors' generic dynamic nonlinear model determination. The differential equations based on the phase quantities as a complete system of equations describing the transients should include the equations of winding voltages and the equations of motion for the rotating parts of the machine. The phase quantities in terms of the resultant phasors as the background to dynamic modeling are taken into consideration. Introducing a complex (α, β)-plane stationary relative to the stator of a two-pole model equations set is carried out including transformation from the α- and β-axis components of the stator quantities to the d- and q-axis components of rotor quantities. This chapter is a source of the advanced knowledge concerning the principles of electric machine modeling. It might be useful for mechanical engineers engaged in the hybrid vehicle power train design process, but also for electrical engineers, especially those attending master and doctoral courses.

DOI: 10.4018/978-1-4666-4042-9.ch003

Copyright ©2013, IGI Global. Copying or distributing in print or electronic forms without written permission of IGI Global is prohibited.

Electric Machines in Hybrid Power Train Employed Dynamic Modeling Backgrounds

INTRODUCTION

The asynchronous, induction (AC) motors and permanent magnet, synchronous (PM) or permanent magnet, brushless, direct current (BLDC) motors, which are, in fact, a kind of PM synchronous machines, are considered. The fundamental theory necessary for the above-mentioned electric machines' mathematical dynamic modeling in this chapter is presented.

1. AC ASYNCHRONOUS INDUCTION MOTOR MODELING

Transient phenomena may be initiated by a balanced or an imbalanced change in the time phase or peak value of the Alternative Current (AC) and voltages U, fed to the stator winding (e.g. by a change in the peak values of symmetric positive-, negative- phase sequence voltage components, U11 or U21, respectively). A transient phenomenon may be initiated by a sudden change in an external torque, leading to an imbalance of electromagnetic torque, a resultant acceleration, and finally, a new speed value.

1.1. Differential Equations Based on Phase Quantities

The complete system of equations describing transients should include differential equations of winding voltages and equations of motion for the rotating parts of the machine. The 2p-pole, three-phase balanced stator and rotor windings are considered (see Figure 1).

Figure 1. The scheme of an asynchronous machine

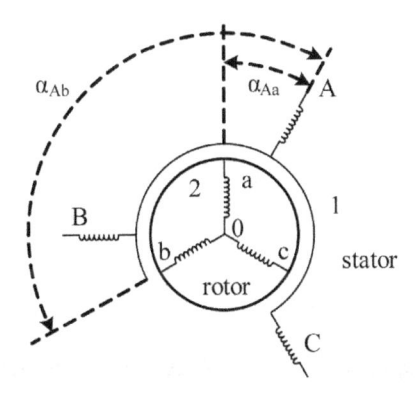

1 – index of stator
2 – index of rotor

Electric Machines in Hybrid Power Train Employed Dynamic Modeling Backgrounds

The fundamental voltage equations are as follows:

stator $\begin{cases} u_{1A} = R_1 i_{1A} + \dot{\psi}_{1A} \\ u_{1B} = R_1 i_{1B} + \dot{\psi}_{1B} \\ u_{1C} = R_1 i_{1C} + \dot{\psi}_{1C} \end{cases}$

rotor $\begin{cases} u_{2a} = R_2 i_{2a} + \dot{\psi}_{2a} \\ u_{2b} = R_2 i_{2b} + \dot{\psi}_{2b} \\ u_{2c} = R_2 i_{2c} + \dot{\psi}_{2c} \end{cases}$ (1)

where: $\dot{\psi} = \dfrac{d\psi}{dt}$

If the magnetic circuit of the machine is unsaturated, the flux linkages with the stator and rotor phases can be written in terms of the phase currents and the corresponding inductances:

stator $\psi_{1A} = L_{AA\Sigma} i_{1A} + L_{AB\Sigma} i_{1B} + L_{AC\Sigma} i_{1C}$
$+ L_{Aa\Sigma} i_{2a} + L_{Ab\Sigma} i_{2b} + L_{Ac\Sigma} i_{2c}$
rotor $\psi_{1a} = L_{aa\Sigma} i_{1a} + L_{ab\Sigma} i_{1b} + L_{ac\Sigma} i_{1c}$
$+ L_{aA\Sigma} i_{2A} + L_{aB\Sigma} i_{2B} + L_{aC\Sigma} i_{2C}$ (2)

where:

$L_{KK\Sigma} = L_{NN\Sigma} = L_{KK} + L_{KKd}$

$L_{KN\Sigma} = L_{NK\Sigma} = L_{KN} + L_{KNd}$

$L_{kk\Sigma} = L_{nn\Sigma} = L_{kk} + L_{kkd}$

$L_{kn\Sigma} = L_{nk\Sigma} = L_{kn} + L_{knd}$

K, N = A, B, C; k, n = a, b, c; d - index of dissipated inductance.

Electric Machines in Hybrid Power Train Employed Dynamic Modeling Backgrounds

For other phases, they can be written in a similar way. The main self-inductances of all phases turn out to be the same (symmetrical machine) and independent of the angular position taken up by the rotor:

$$L_{AA} = L_{BB} = L_{CC} = L_{aa} = L_{bb} = L_{cc} = L_m \tag{3}$$

The main mutual inductances between the stator and rotor phases:

$$L_{AB} = L_{AC} = L_{BC} = L_m \cos\frac{2\pi}{3} = -\frac{L_m}{2} \tag{4}$$

The mutual inductances between the rotor phases are independent of the angular position of the rotor:

$$L_{ab} = L_{bc} = L_{ac} = L_m \cos\frac{2\pi}{3} = -\frac{L_m}{2} \tag{5}$$

The mutual inductance between the rotor and stator phases depends on the angular position of the rotor which is described by the angle between the axes of phases A and a:

$$\begin{aligned} L_{Aa} &= L_{Bb} = L_{Cc} = L_m \cos\alpha_{Aa} \\ L_{Ab} &= L_{Bc} = L_{Ca} = L_m \cos\alpha_{Ab} \\ L_{Ac} &= L_{Ba} = L_{Cb} = L_m \cos\alpha_{Ac} \\ \text{and generally: } L_{kn} &= L_m \cos\alpha_{kn} \end{aligned} \tag{6}$$

where k=A, B, C; n= a, b, c.

In the case when the number of pole pairs p is considered, the electrical angle (see Figure 1) between the stator and rotor phases is defined as:

$$\begin{aligned} \alpha_{Aa} &= \alpha_{Bb} = \alpha_{Cc} = \alpha = p\gamma \\ \alpha_{Ab} &= \alpha_{Bc} = \alpha_{Ca} = \alpha + \frac{2\pi}{3} \\ \alpha_{Ac} &= \alpha_{Ba} = \alpha_{Cb} = \alpha + \frac{4\pi}{3} \end{aligned} \tag{7}$$

When the rotor revolves at a constant ω_0 speed, the angle between A and a axes, including the pairs pole p number, is as follows:

$$\nu = \nu_0 + \omega_0 tb$$

and generally $\quad \nu = \nu_0 + \int_0^t \omega \cdot dt \quad$ (8)

where: t – time; ν_0 – initial value of ν; ω – angular velocity.

The equation of rotor motion could be considered as:

$$M_{em} \pm M_o = J\dot{\omega}$$

and $\omega = \omega_0 + \int_0^t \dot{\omega} \cdot dt \quad$ (9)

where:

- M_{em}: Electromagnetic torque;
- M_o: External load torque;
- J: Moment of inertia reduced on the rotor shaft.

In the other way:

$$M_{em} = \sum_{k=A,B,C} i_{1k\Sigma} \sum_{n=a,b,c} i_{2n\Sigma} \frac{dL_{kn}}{d\nu} \quad (10)$$

The Equations 1-10 give a complete description of the transients in a three-phase induction machine. Their direct solution presents serious difficulties for two reasons:

- They contain too many unknowns (six stator and rotor phases variable).
- They involve periodically varying mutual inductances L_{Aa}, L_{ab}, etc., because some coefficients (Equations 1, 2) are periodic functions of time.

To simplify the solution, the original set of equations is transformed in such a way that the natural variables, currents, flux linkages and phase voltages are expressed in terms of other, more convenient variables.

In the case of a typical asynchronous induction machine (see Figure 1), the set of equations of phase currents, voltages and flux linkages are as follows:

Electric Machines in Hybrid Power Train Employed Dynamic Modeling Backgrounds

$$\left.\begin{array}{l} u_{1A} + u_{1B} + u_{1C} = 0 \\ i_{1A} + i_{1B} + i_{1C} = 0 \\ \psi_{1A} + \psi_{1B} + \psi_{1C} = 0 \end{array}\right\} stator \quad (11)$$

$$\left.\begin{array}{l} i_{2a} + i_{2b} + i_{2c} = 0 \\ u_{2a} + u_{2b} + u_{2c} = 0 \\ \psi_{2a} + \psi_{2b} + \psi_{2c} = 0 \end{array}\right\} rotor \quad (12)$$

When the terms of Equations (1) and (2) are written as sums of the above components, the set of equations divides into two sets of equations, one involving the phase quantities that do not contain zero-sequence terms:

$$\left.\begin{array}{l} \psi_{1A} = L_1 i_{1A} + L_{Aa} i_{2a} + L_{Ab} i_{2b} + L_{Ac} i_{2c} \\ \psi_{1B} = L_1 i_{1B} + L_{Ba} i_{2a} + L_{Bb} i_{2b} + L_{Bc} i_{2c} \\ \psi_{1C} = L_1 i_{1C} + L_{Ca} i_{2a} + L_{Cb} i_{2b} + L_{Cc} i_{2c} \end{array}\right\} stator$$

$$\left.\begin{array}{l} \psi_{2a} = L_2 i_{2a} + L_{aA} i_{1A} + L_{aB} i_{1B} + L_{aC} i_{1C} \\ \psi_{2b} = L_2 i_{2b} + L_{bA} i_{1A} + L_{bB} i_{1B} + L_{bC} i_{1C} \\ \psi_{2c} = L_2 i_{2c} + L_{cA} i_{1A} + L_{cB} i_{1B} + L_{cC} i_{1C} \end{array}\right\} rotor$$

(13)

The Equation (13) represents the flux-linkages which do not contain zero-sequence components.

Also, the flux-linkage equations contain:

$$L_1 = L_{11} + L_{1d} \quad (14)$$

which is the stator-phase inductance for the set of currents i_{1A}, i_{1B}, i_{1C} that does not contain zero-sequence currents (with allowance for the effect of the other phase).

$$L_{11} = L_{AA} - L_{AB} = L_m - \left(-\frac{1}{2} L_m\right) = 3 L_m / 2 \quad (15)$$

which is the stator phase inductance component associated with the fundamental term of mutual field.

$$L_{1d} = L_{AAd} - L_{Abd} \tag{16}$$

which is the stator phase inductance component associated with the leakage fields.

$$L_2 = L_{22} + L_{2d} \tag{17}$$

which is the rotor-phase inductance for the set of currents i_{2A}, i_{2B}, i_{2C} that does not contain zero-sequence currents (which allow the effect of the other phases).

$$L_{22} = L_{aa} - L_{ab} = L_m - \left(-\frac{1}{2}L_m\right) = 3L_m/2 \tag{18}$$

The following fact is taken into account:

$$\begin{aligned} L_{AA} + L_{AB} + L_{AC} &= 0 \\ L_{aa} + L_{ab} + L_{ac} &= 0 \\ L_{Aa} + L_{Ab} + L_{Ac} &= 0 \end{aligned} \tag{19}$$

The above equalities can be verified, using Equations (6-9) recalling that:

$$\begin{aligned} &\cos\alpha_{Aa} + \cos\alpha_{Ab} + \cos\alpha_{Ac} \\ &= \mathrm{Re}\left(\exp(j\alpha_{Aa}) + \exp(j\alpha_{Ab}) + \exp(j\alpha_{Ac})\right) = \\ &= \mathrm{Re}\left(\exp(j\alpha)\left(1 + a + a^2\right)\right) = 0 \end{aligned} \tag{20}$$

where: $a = exp(j2\pi/3)$.

This implies that the system of zero-sequence currents produces only leakage fields and does not set up any mutual fields. That is why $L_{01} \ll L_1$ and $L_{02} \ll L_2$.

For the same reason, the electromagnetic torque receives no contribution from zero-sequence currents, so it can be expressed solely in terms of the components that do not contain zero-sequence currents. This can be verified by writing the currents in Equation (10) as the sums of two components:

$$M_{em} = \sum_{k=A,B,C} i_{1k} \sum_{n=a,b,c} i_{2n} \frac{dL_{kn}}{dv} \tag{21}$$

Electric Machines in Hybrid Power Train Employed Dynamic Modeling Backgrounds

Recalling that:

$$L_{kn} = L_m \cos \alpha_{kn} \tag{22}$$

where:

$$\alpha_{kn} = \alpha + \Delta\alpha_{kn} \rightarrow \alpha = p\nu \quad \text{and} \quad \Delta\alpha_{kn} \neq f(\nu)$$

The electromagnetic torque, in terms of currents and the angles between phases α_{kn} (see Equation (6)), is as follows:

$$M_{em} = -pL_m \sum_{k=A,B,C} i_{1k} \sum_{n=a,b,c} i_{2n} \sin \alpha_{kn} \tag{23}$$

1.2. Phase Quantities in Terms of Resultant Phasors

Introducing a complex (α, β) -plane stationary relative to the stator of two-pole model, its real-α-axis is placed co-linear with the axis of stator phase A. Then the unit phasors a=exp (j2π/3) and a² will line up respectively, with the axes of phases B and C, where the unit phasor exp (jα$_{Aa}$) and α$_{Aa}$=pν, will give the direction of the axis and rotor phase α. The resultant phasor of the stator current can then be written in terms of instantaneous phase currents as:

$$\overline{I}_1 = (2/3)\left(i_{1A} + i_{1B}a + i_{1C}a^2\right) \tag{24}$$

Projection of $\overline{I}_1$ on axis A, B, and C, means the instantaneous values of phase currents (also, in other cases, voltages and flux linkages, in general $\left|\overline{X}\right| = x_{\max} = f(t)_{\max}$ see Figure 2).

Figure 2. The maximal instantaneous value $x(t)=x_{max}\sin\omega t$

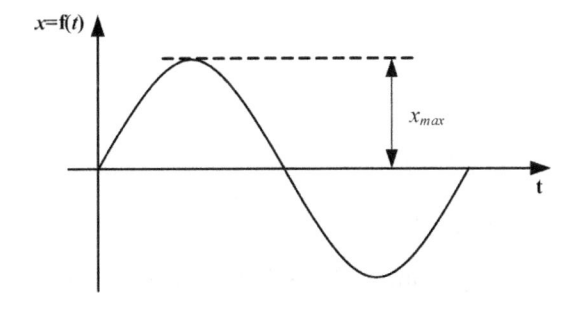

It is necessary to remember the vector sum of projection $\bar{I}_1$ on axis A,B,C is $\frac{3}{2}|\bar{I}_1|$ and its plot is shown in Figure 3.

The phasors of stator voltage and flux-linkage can be expressed by phase quantities in a similar way:

$$\begin{aligned}\bar{U}_1 &= 2\left(u_{1A} + u_{1B}a + u_{1C}a^2\right)/3 \\ \bar{\Psi}_1 &= 2\left(\psi_{1A} + \psi_{1B}a + \psi_{1C}a^2\right)/3\end{aligned} \qquad (25)$$

The resultant phasors of the rotor phase quantities are plotted as a rotating complex (d, q) - plane which is stationary, relative to the rotor of a two-pole model (Figure 1). The aligned real d-axis of this plane with the axis of the rotor a phase, the unit phasors a and a^2 will then line up with the b and c phase axes, respectively, and the unit phasor exp(-jα_{Aa}), where α_{Aa}=pν, will locate the position of the fixed axis of the stator A phase.

The resultant phasor of the rotor current, voltage and flux linkage can be expressed in terms of the respective phase quantities of the rotor as follows:

$$\begin{aligned}\bar{I}_2 &= \left(\tfrac{2}{3}\right)\left(i_{2A} + i_{2B}a + i_{2C}a^2\right) \\ \bar{U}_2 &= \left(\tfrac{2}{3}\right)\left(u_{2A} + u_{2B}a + u_{2C}a^2\right) \\ \bar{\Psi}_2 &= \left(\tfrac{2}{3}\right)\left(\psi_{2A} + \psi_{2B}a + \psi_{2C}a^2\right)\end{aligned} \qquad (26)$$

1.3. The α- and β- Axis Components of Stator Quantities and the d- and q- Axis Components of Rotor Quantities

The resultant phasors of the stator can be written as sums of α- and β-axis components, e.g. - the resultant phasor of the stator current (Equation 25) is:

$$\bar{I}_1 = i_{1\alpha} + ji_{1\beta} \qquad (27)$$

where:

$$\begin{aligned}i_{1\alpha} &= \operatorname{Re}[\bar{I}_1] = \left(\bar{I}_1 + \bar{I}_1^*\right)/2 = i_{1A} \\ i_{1\beta} &= \operatorname{Im}[\bar{I}_1] = \left(\bar{I}_1 - \bar{I}_1^*\right)/2j = \left(i_{1B} - i_{1C}\right)\sqrt{3}\end{aligned} \qquad (28)$$

are the α- and β-axis components of the stator current, respectively.

Electric Machines in Hybrid Power Train Employed Dynamic Modeling Backgrounds

Figure 3. The resultant stator and rotor of asynchronous, alternative current, induction motor current phasors

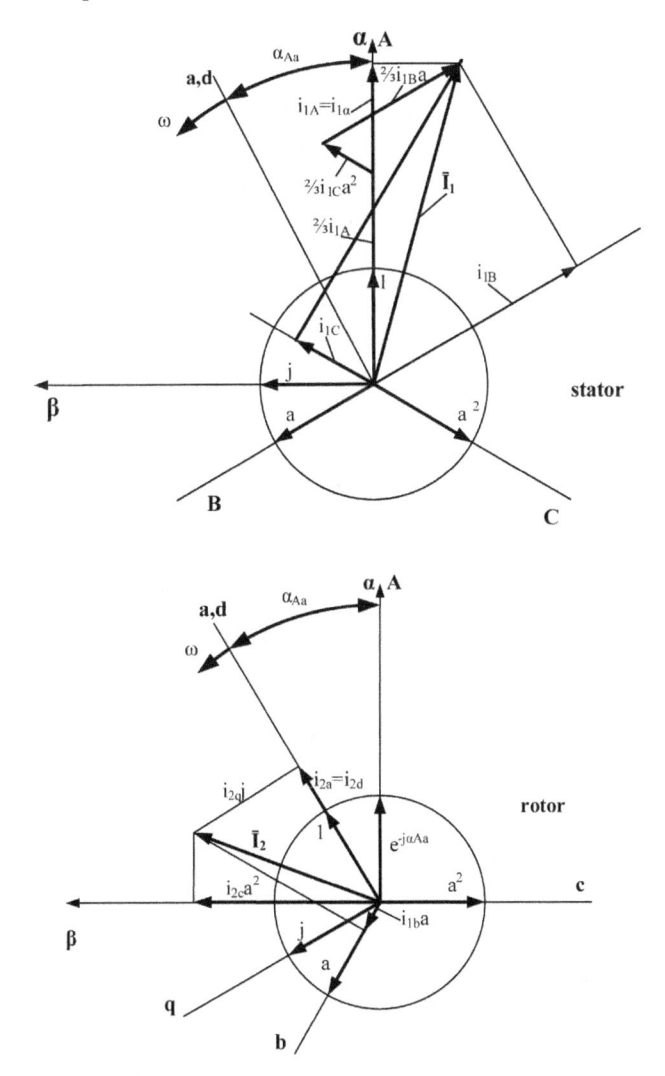

The currents $i_{1\alpha}$ and $i_{1\beta}$ may be visualized as follows in a stationary two-phase winding 1α and 1β phase axes, which are aligned with the α- and β-axes of complex plane, respectively (see Figures 3 and 4). The two phases (1α, 1β) - winding carrying currents $i_{1\alpha}$ and $i_{1\beta}$ (Figure 4) are equivalent to the three-phase (A, B, C) stator winding in Figure 2.3, which carries the currents i_{1A}, i_{1B}, i_{1C}.

The resultant phasors of rotor quantities can be resolved into components directed along the rotating d- and jq- axes.

61

Electric Machines in Hybrid Power Train Employed Dynamic Modeling Backgrounds

Figure 4. The asynchronous, alternative, current induction, motor stator, and rotor current components in stationary α, β co-ordinates

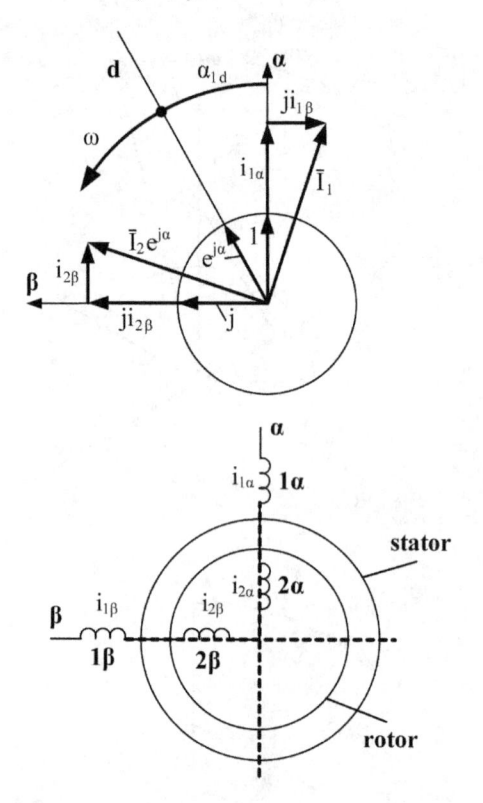

Figure 5. The asynchronous, alternative, current induction, motor stator, and rotor current components in rotating d, q co-ordinates

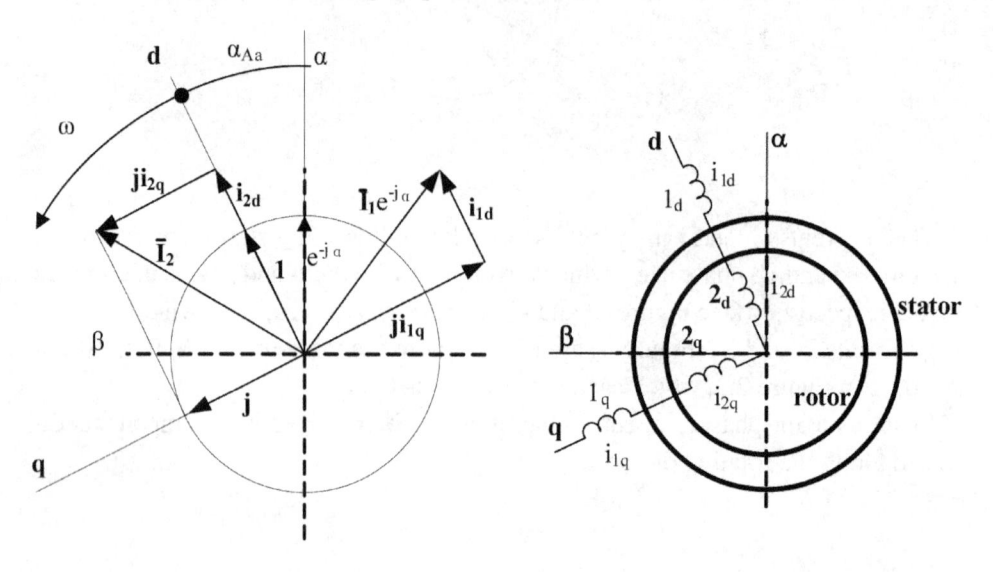

Electric Machines in Hybrid Power Train Employed Dynamic Modeling Backgrounds

As an example, it could be demonstrated how this could be done for the resultant phasor of rotor current.

On the complex (d, jq) plane, the rotor current defined by Equation (25) is represented by a phasor

$$\overline{I}_2 = i_{2d} + ji_{2q} \tag{29}$$

where:

$$i_{2d} = \mathrm{Re}\left[\overline{I}_2\right] = \left(\overline{I}_2 + \overline{I}_2^*\right)/2 = i_{2a}$$
$$i_{2q} = \mathrm{Im}\left[\overline{I}_2\right] = \left(\overline{I}_2 - \overline{I}_2^*\right)/2j = \left(i_{2b} - i_{2c}\right)\sqrt{3}$$

are the d- and q- axis components of rotor current, respectively.

The currents i_{2d} and i_{2q} may be visualized as follows in a rotating two-phase (2d, 2q) winding, the phase axes of which are aligned with the d- and q- axes of a complex plane (see Figures 3 and 5). The rotating two-phase (2d, 2q) winding in Figure 5 carrying the currents i_{2d} and i_{2q} is equivalent to the three-phase (a, b, c) winding in Figure 1, which carries the currents i_{2a}, i_{2b}, i_{2c}. Equations for flux linkages and voltages can be derived in a similar way.

1.4. The d- and q- Axis Components of Stator Quantities and the α- and β- Axis Component of Rotor Quantities

As the (d, q) complex plane is turned through an angle $\alpha = \alpha_{A\alpha}$ from the fixed α, jβ complex plane (see Figure 3), the rotor current will be depicted on the α, jβ plane by a phasor of the form:

$$\begin{aligned}\overline{I}_{2(\alpha,\beta)} &= \overline{I}_2 \exp(j\alpha) = \left(i_{2d} + ji_{2q}\right)\exp(j\alpha) \\ &= i_{2\alpha} + ji_{2\beta}\end{aligned} \tag{30}$$

The α- and β- axis components of the motor current phasor.

$$\begin{aligned}i_{2\alpha} &= \mathrm{Re}\left[\left(i_{2d} + ji_{2q}\right)\exp(j\alpha)\right] \\ &= i_{2d}\cos\alpha - i_{2q}\sin\alpha \\ i_{2\beta} &= \mathrm{Im}\left[\left(i_{2d} + ji_{2q}\right)\exp(j\alpha)\right] \\ &= i_{2d}\sin\alpha + i_{2q}\cos\alpha\end{aligned}$$

Electric Machines in Hybrid Power Train Employed Dynamic Modeling Backgrounds

are the currents in the stationary two-phase (2α, 2β) winding (see Figure 4), that sets up the same magnetic field as the rotating two-phase winding, which carries the currents i_{2d}, and i_{2q}.

A similar transformation is applied to the resultant stator phasors in writing equations in terms of the rotating (d, iq) complex plane. For example, if on the stationary (α, jβ) complex plane, the stator current is depicted in accord with equation (22) by a phasor

$$\bar{I}_1 = i_{1\alpha} + ji_{1\beta}$$

then, on the (d, iq) complex plane turned through angle $-\alpha = -\alpha_{A\alpha}$ from the (α, jβ) plane (see Figure 3), this same current will be represented by the phasor as follows:

$$\bar{I}_{1(d,q)} = \bar{I}_1 \exp(-j\alpha)$$
$$= (i_{1\alpha} + ji_{1\beta})\exp(-j\alpha) = i_{1d} + ji_{1q} \qquad (31)$$

The d- and q- axis components of the stator current phasor

$$i_{1d} = \operatorname{Re}\left[(i_{1\alpha} + ji_{1\beta})\exp(-j\alpha)\right]$$
$$= i_{1\alpha}\cos\alpha + i_{1\beta}\sin\alpha$$
$$i_{1q} = \operatorname{Im}\left[(i_{1\alpha} + ji_{1\beta})\exp(-j\alpha)\right]$$
$$= -i_{1\alpha}\sin\alpha + i_{1\beta}\cos\alpha$$

are the currents in the rotating two-phase (1d, 1q - see Figure 5) winding, which sets up the same magnetic field as the stationary two-phase winding carrying the currents $i_{1\beta}$ and $i_{2\beta}$.

Equations in terms of d- and q- axis components for stator flux linkages and voltages and in terms of α- and β- axis components for rotor flux linkages and voltages can be derived in a similar way.

1.5. The d- and q- Axis Components of Stator Quantities and the α- and β- Axis Component of Rotor Quantities

Because the (d, q) complex plane is turned through an angle $\alpha = \alpha_{A\alpha}$ from the fixed α, jβ complex plane (see Figure 3), the rotor current will be depicted on the α, jβ plane by a phasor of the form:

Electric Machines in Hybrid Power Train Employed Dynamic Modeling Backgrounds

$$\bar{I}_{2(\alpha,\beta)} = \bar{I}_2 \exp(j\alpha)$$
$$= \left(i_{2d} + ji_{2q}\right)\exp(j\alpha) = i_{2\alpha} + ji_{2\beta} \tag{32}$$

The α- and β- axis components of the motor current phasor

$$i_{2\alpha} = \operatorname{Re}\left[\left(i_{2d} + ji_{2q}\right)\exp(j\alpha)\right]$$
$$= i_{2d}\cos\alpha - i_{2q}\sin\alpha$$
$$i_{2\beta} = \operatorname{Im}\left[\left(i_{2d} + ji_{2q}\right)\exp(j\alpha)\right]$$
$$= i_{2d}\sin\alpha + i_{2q}\cos\alpha$$

are the currents in the stationary two-phase (2α, 2β) winding (see Figure 4), which sets up the same magnetic field as the rotating two-phase winding, which carries the currents i_{2d} and i_{2q}.

A similar transformation is applied to the resultant stator phasors in writing equations in terms of the rotating (d, iq) complex plane. For example, if on the stationary (α, jβ) complex plane, the stator current is depicted in accord with the Equation (28) by a phasor

$$\bar{I}_1 = i_{1\alpha} + ji_{1\beta}$$

then, on the (d, iq) complex plane turned through an angle $-\alpha = -\alpha_{A\alpha}$ from the (α, jβ) plane (see Figure 3), this same current will be represented by the phasor as follows:

$$\bar{I}_{1(d,q)} = \bar{I}_1 \exp(-j\alpha)$$
$$= \left(i_{1\alpha} + ji_{1\beta}\right)\exp(-j\alpha) = i_{1d} + ji_{1q} \tag{33}$$

The d- and q- axis components of the stator current phasor

$$i_{1d} = \operatorname{Re}\left[\left(i_{1\alpha} + ji_{1\beta}\right)\exp(-j\alpha)\right]$$
$$= i_{1\alpha}\cos\alpha + i_{1\beta}\sin\alpha$$
$$i_{1q} = \operatorname{Im}\left[\left(i_{1\alpha} + ji_{1\beta}\right)\exp(-j\alpha)\right]$$
$$= -i_{1\alpha}\sin\alpha + i_{1\beta}\cos\alpha$$

Electric Machines in Hybrid Power Train Employed Dynamic Modeling Backgrounds

are the currents in the rotating two-phase (1d, 1q - see Figure 5) winding, which sets up the same magnetic field as the stationary two-phase winding carrying the currents $i_{1\beta}$ and $i_{2\beta}$.

Equations in terms of d- and q- axis components for stator flux linkages and voltages and in terms of α- and β- axis components for rotor flux linkages and voltages can be derived in a similar way.

1.6. The d- and q- Axis Components of Stator Quantities and the α- and β- Axis Component of Rotor Quantities

Because the (d, q) complex plane is turned through an angle $\alpha = \alpha_{A\alpha}$ from the fixed α, jβ complex plane (see Figure 3), the rotor current will be depicted on the α, jβ plane by a phasor of the form:

$$\overline{I}_{2(\alpha,\beta)} = \overline{I}_2 \exp(j\alpha)$$
$$= (i_{2d} + ji_{2q})\exp(j\alpha) = i_{2\alpha} + ji_{2\beta} \tag{34}$$

The α- and β- axis components of the motor current phasor

$$i_{2\alpha} = \operatorname{Re}\left[(i_{2d} + ji_{2q})\exp(j\alpha)\right]$$
$$= i_{2d}\cos\alpha - i_{2q}\sin\alpha$$
$$i_{2\beta} = \operatorname{Im}\left[(i_{2d} + ji_{2q})\exp(j\alpha)\right]$$
$$= i_{2d}\sin\alpha + i_{2q}\cos\alpha$$

are the currents in the stationary two-phase (2α, 2β) winding (see Figure 4), that sets up the same magnetic field as the rotating two-phase winding, which carries the currents i_{2d}, and i_{2q}.

A similar transformation is applied to the resultant stator phasors in writing equations in terms of the rotating (d, iq) complex plane. For example, if on the stationary (α, jβ) complex plane, the stator current is depicted in accord with Equation (28) by a phasor

$$\overline{I}_1 = i_{1\alpha} + ji_{1\beta}$$

then, on the (d, iq) complex plane turned through an angle $-\alpha = -\alpha_{A\alpha}$ from the (α, jβ) plane (see Figure 3), this same current will be represented by the phasor as follows:

Electric Machines in Hybrid Power Train Employed Dynamic Modeling Backgrounds

$$\overline{I}_{1(d,q)} = \overline{I}_1 \exp(-j\alpha)$$
$$= (i_{1\alpha} + ji_{1\beta}) \exp(-j\alpha) = i_{1d} + ji_{1q} \quad (35)$$

The d- and q- axis components of the stator current phasor

$$i_{1d} = \mathrm{Re}\left[(i_{1\alpha} + ji_{1\beta})\exp(-j\alpha)\right]$$
$$= i_{1\alpha}\cos\alpha + i_{1\beta}\sin\alpha$$
$$i_{1q} = \mathrm{Im}\left[(i_{1\alpha} + ji_{1\beta})\exp(-j\alpha)\right]$$
$$= -i_{1\alpha}\sin\alpha + i_{1\beta}\cos\alpha$$

are the currents in the rotating two-phase (1d, 1q - see Figure 5) winding, which sets up the same magnetic field as the stationary two-phase winding carrying the currents $i_{1\beta}$ and $i_{2\beta}$.

Equations in terms of d- and q- axis components for stator flux linkages and voltages and in terms of α- and β- axis components for rotor flux linkages and voltages can be derived in a similar way.

1.7. Flux Linkage Phasors in Terms of Current Phasors

The stator phase flux linkages (ψ_{A1}, ψ_{B1}, ψ_{C1}) appearing in Equation (13) can be expressed on the basis of Equation (6) in terms of phase currents, stator phase inductance L_1 and rotor- stator phase mutual inductance,

$$L_{Aa} = L_m \cos\alpha$$
$$L_{Ab} = L_m \cos\left(\alpha + \frac{2\pi}{3}\right)$$
$$L_{Ac} = L_m \cos\left(\alpha + \frac{4\pi}{3}\right)$$
$$L_{Ba} = L_m \cos\left(\alpha + \frac{4\pi}{3}\right) \quad \text{etc.}$$

which are functions of the angular position of the rotor, that is, the electrical angle between the axes of the stator A phase and rotor a phase, $\alpha = \alpha_{Aa} = p\gamma$.

Let us write the cosines of the angles between phases in exponential form:

$$\cos\alpha = \text{Re}[e] = (e + e^*)/2$$

$$\cos\left(\alpha + \frac{2\pi}{3}\right) = \text{Re}[ea] = (ea + e^*a^2)/2$$

$$\cos\left(\alpha + \frac{4\pi}{3}\right) = \text{Re}[ea^2] = (ea^2 + e^*a)/2$$

where: e=exp(jα), e*=exp(-jα), a=exp(j2π/3) and a*=a²=exp(j4π/3) are unit phasors indicating the direction of the phase axes (Figure 3). Then, to the Equations (25) and (26), for the resultant stator and rotor current phasors, the resultant stator flux linkage phasor can be written in terms of the resultant stator and rotor current phasors as follows:

$$\overline{\Psi}_1 = L_1\overline{I}_1 + L_{12m}\overline{I}_2 \exp(j\alpha) = L_1\overline{I}_1 + L_{12m}\overline{I}_{2(\alpha,\beta)} \qquad (36)$$

where:

- **$L_{12m}=3L_m/2$:** Is the main mutual inductance between a stator phase and rotor phases,
- $\overline{I}_{2(\alpha,\beta)} = \overline{I}_2 \exp(j\alpha)$: Is the resultant rotor current phasor in the stationary (α, β) complex plane of the stator (Figure 3).

Similarly, using Equations (6), (19) and (21) and applying analogous transformations, it can be shown that the rotor flux linkage phasor can be expressed in the form of the stator and rotor current phasors, along with the corresponding inductances:

$$\overline{\Psi}_2 = L_2\overline{I}_2 + L_{12m}\overline{I}_1 \exp(-j\alpha) = L_2\overline{I}_2 + L_{12m}\overline{I}_{1(d,q)} \qquad (37)$$

where:

$$\overline{I}_{1(d,q)} = \overline{I}_1 \exp(-j\alpha)$$

is the resultant stator current phasor on the rotating (d, q) complex plane of the rotor (Figure 3).

Electric Machines in Hybrid Power Train Employed Dynamic Modeling Backgrounds

1.8. Voltage Equation in Terms of the α- and β- Axis Components

The voltage Equation (1), written for the particular stator phase, by an equation written in terms of resultant phasor functions, could be formulated by multiplying the equation for u_{1A} by 2/3; the equation for u_{1B} by 2a/3 and the equation for u_{1C} by $2a^2/3$ and adding together the right- and left-hand sides of these equations, term wisely. Then, in view of Equation (25), the resultant equation will be the voltage equation of the stator in its own (α, β) complex plane.

$$\bar{U}_1 = R_1 \bar{I}_1 + d\bar{\Psi}_1 / dt \tag{38}$$

where: the stator flux-linkage phasor can be defined in terms of currents on the basis of Equation (37).

Applying a similar procedure to the voltage Equations (1), for the individual rotor phase, it is possible to obtain the voltage equation of the rotor in its own (d, q) complex plane

$$\bar{U}_2 = R_2 \bar{I}_2 + d\bar{\Psi}_2 / dt \tag{39}$$

where: the rotor flux linkage phasor can be defined in terms of currents on the basis of Equation (38)

However, the equations for $\bar{U}_1$ and $\bar{U}_2$ are written in different planes and cannot be solved simultaneously. Therefore, one of them, say, the equation for $\bar{U}_2$ must be transformed and written in terms of rotor voltage, current and flux linkage phasors defined in the (α, β) co-ordinates; that is, $\bar{U}_{2(\alpha,\beta)}, \bar{I}_{2(\alpha,\beta)}$ and $\bar{\Psi}_{2(\alpha,\beta)}$. According to Equation (35)

$$\begin{aligned}
\bar{U}_2 &= \bar{U}_{2(\alpha,\beta)} \exp(-j\alpha) \\
\bar{I}_2 &= \bar{I}_{2(\alpha,\beta)} \exp(-j\alpha) \\
\bar{\Psi}_2 &= \bar{\Psi}_{2(\alpha,\beta)} \exp(-j\alpha)
\end{aligned} \tag{40}$$

where: $\alpha = \alpha_{A\alpha} = \alpha(t)$ is the angle between the axes of the stator and rotor phases (Figure 3). After the above transformation, Equation (41) takes the form

$$\bar{U}_{2(\alpha,\beta)} \exp(-j\alpha) = R_2 \bar{I}_{2(\alpha,\beta)} \exp(-j\alpha)$$
$$+ \frac{d}{dt}\left(\bar{\Psi}_{2(\alpha,\beta)} \exp(-j\alpha)\right) \tag{41}$$

In taking the derivative, it is important to remember that the angle between the axes of the stator and rotor phases varies with time, therefore:

$$\bar{U}_{2(\alpha,\beta)} \exp(-j\alpha) = R_2 \bar{I}_{2(\alpha,\beta)} \exp(-j\alpha)$$
$$+ \exp(-j\alpha)\left(\frac{d\bar{\Psi}_{2(\alpha,\beta)}}{dt}\right)$$
$$+ \bar{\Psi}_{2(\alpha,\beta)}\left[-j\exp(-j\alpha)\left(\frac{d\alpha}{dt}\right)\right] \tag{42}$$

Dividing the above equation term wise by exp(-jα) and recalling that dα/dt=ω is the electrical angular velocity of the rotor, the rotor voltage equation may be written as

$$\bar{U}_{2(\alpha,\beta)} = R_2 \bar{I}_{2(\alpha,\beta)} + \frac{d\bar{\Psi}_{2(\alpha,\beta)}}{dt} - j\omega\bar{\Psi}_{2(\alpha,\beta)} \tag{43}$$

where: all quantities reference to the stationary (α, β) complex plane. This applies to the flux linkage as well. It can be found from an equation, which stems from Equations (35), (36), (37) namely:

$$\bar{\Psi}_{2(\alpha,\beta)} = \bar{\Psi}_2 \exp(j\alpha) = L_2 \bar{I}_2 \exp(j\alpha)$$
$$+ L_{12m}\bar{I}_{1(d,q)} \exp(j\alpha) \tag{44}$$
$$= L_2 \bar{I}_{2(\alpha,\beta)} + L_{12m}\bar{I}_1$$

where: the current phasors are written in the (α, β) co-ordinates.

If all quantities reference to the (α, β) complex plane, we can drop the subscripts in the above equation and present the rotor equation as:

$$\bar{U}_2 = R_2 \bar{I}_2 + \frac{d\bar{\Psi}_2}{dt} - j\omega\bar{\Psi}_2 \tag{45}$$

where:

$$\overline{\Psi}_2 = L_2\overline{I}_2 + L_{12m}\overline{I}_1.$$

Using Equations (28), and (35), all phasors could be written in Equations (38), (46), in terms of their α- and β-axis components. Instead of two voltage equations in phasor form, there will be four equations in scalar notation:

$$\begin{aligned}
u_{1\alpha} &= R_1 i_{1\alpha} + \frac{d\psi_{1\alpha}}{dt} \\
u_{1\beta} &= R_1 i_{1\beta} + \frac{d\psi_{1\beta}}{dt} \\
u_{2\alpha} &= R_2 i_{2\alpha} + \frac{d\psi_{2\alpha}}{dt} + \omega\psi_{2\beta} \\
u_{2\beta} &= R_2 i_{2\beta} + \frac{d\psi_{2\beta}}{dt} - \omega\psi_{2\alpha}
\end{aligned} \quad (46)$$

In the above equations, the flux linkage components can be expressed in terms of the corresponding current components

$$\left.\begin{aligned}
\psi_{1\alpha} &= L_1 i_{1\alpha} + L_{12m} i_{2\alpha} \\
\psi_{1\beta} &= L_1 i_{1\beta} + L_{12m} i_{2\beta}
\end{aligned}\right\} \text{stator}$$

$$\left.\begin{aligned}
\psi_{2\alpha} &= L_2 i_{2\alpha} + L_{12m} i_{1\alpha} \\
\psi_{2\beta} &= L_2 i_{2\beta} + L_{12m} i_{1\beta}
\end{aligned}\right\} \text{rotor} \quad (47)$$

1.9. Voltage Equations in Terms of d- and q- Axis Components and in Terms of Components along Axis Rotating at Arbitrary Velocity

Equation (39) should be transformed. The relevant quantities can be expressed in terms d- and q- axis components, using equations of the form of Equation (36)

$$\overline{U}_{1(d,q)} \exp(j\alpha) = R_1 \overline{I}_{1(d,q)} \exp(j\alpha) + \frac{d}{dt}\left[\overline{\Psi}_{1(d,q)} \exp(j\alpha)\right] \quad (48)$$

The angle between the axes of the stator and rotor phases is a time function α=α(t):

$$\bar{U}_{1(d,q)} \exp(j\alpha) = R_1 \bar{I}_{1(d,q)} \exp(j\alpha)$$
$$+ \exp(j\alpha) \frac{d\bar{\Psi}_{1(d,q)}}{dt} \qquad (49)$$
$$+ \bar{\Psi}_{1(d,q)} j \exp(j\alpha) \left(\frac{d\alpha}{dt}\right)$$

to cancel out exp(jα), it is possible to obtain a stator voltage equation:

$$\bar{U}_{1(d,q)} = R_1 \bar{I}_{1(d,q)} + \frac{d\bar{\Psi}_{1(d,q)}}{dt} + j\omega \bar{\Psi}_{1(d,q)} \qquad (50)$$

where: all quantities are referenced to the rotating (d,q) complex plane. Using Equations (35) through (37) the flux linkage can be written in terms of currents on the (d,q) complex plane:

$$\bar{\Psi}_{1(d,q)} = \bar{\Psi}_1 \exp(-j\alpha) = L_1 \bar{I}_1 \exp(-j\alpha)$$
$$+ L_{12m} \bar{I}_{2(\alpha,\beta)} \exp(-j\alpha) \qquad (51)$$
$$= L_1 \bar{I}_{1(d,q)} + L_{12m} \bar{I}_2$$

As all quantities in Equation (51) are referenced to the (d,q) complex plane, the above Equation (51) may be rewritten as:

$$\bar{U}_1 = R_1 \bar{I}_1 + \frac{d\bar{\Psi}_1}{dt} + j\omega \bar{\Psi}_1 \qquad (52)$$
where: $\bar{\Psi}_1 = L_1 \bar{I}_1 + L_{12m} \bar{I}_2$

For Equations (41) and (53) to be recast in scalar form, it is necessary to express all the phasors involved in terms of projections on the d- and q- axes, in accord with Equations (34) and (36), and separate the real and imaginary parts on the right and left-hand sides. This is the way to reach the set of the d- and q- axis stator and rotor voltages equations equivalent to Equations (41) and (53):

Electric Machines in Hybrid Power Train Employed Dynamic Modeling Backgrounds

$$u_{1d} = R_1 i_{1d} + \frac{d\psi_{1d}}{dt} - \omega\psi_{1q}$$

$$u_{1q} = R_1 i_{1q} + \frac{d\psi_{1q}}{dt} + \omega\psi_{1d}$$

$$u_{2d} = R_2 i_{2d} + \frac{d\psi_{2d}}{dt}$$

$$u_{2q} = R_2 i_{2q} + \frac{d\psi_{2q}}{dt}$$

(53)

The flux linkages components in Equation (54) can be written in terms of the corresponding current components as:

$$\psi_{1d} = L_1 i_{1d} + L_{12m} i_{2d}$$

$$\psi_{2d} = L_2 i_{2d} + L_{12m} i_{1d}$$

$$\psi_{1q} = L_1 i_{1q} + L_{12m} i_{2q}$$

$$\psi_{2q} = L_2 i_{2q} + L_{12m} i_{1q}$$

(54)

On differentiating Equation (54) and deeming the rotor to be short-circuited, we can obtain the following:

$$\frac{d}{dt}\psi_{1d} = \frac{d}{dt}\left(L_1 i_{1d} + L_{12m} i_{2d}\right)$$

$$0 = R_2 i_{2d} + \frac{d}{dt}\left(L_2 i_{2d} + L_{12m} i_{1d}\right)$$

which describes the d- axis equivalent circuit in Figure 6a. Applying the same reasoning to the q- axis voltage and flux-linkage equations, there can be obtained the q- axis equivalent circuit in Figure 6b. The two equivalent circuits depict the transformer–inductive coupling between the equivalent stator and rotor windings on the d- and q- axis of the machine, respectively.

In a general case, if the voltage equations are written with reference to some axes rotating at an arbitrary velocity ω_0, it will contain an extra term representing the rotational EMF (electromotive force), which is proportional to the velocity of the axes relative to the winding involved.

Thus, the equation describing the stator voltages referenced to axes rotating at ω_0 will contain an extra EMF proportional to the velocity of those axes relative to stationary stator windings, that is:

Electric Machines in Hybrid Power Train Employed Dynamic Modeling Backgrounds

Figure 6. The equivalent circuits of an asynchronous machine (a) d- axis and (b) q- axis, where L1d and L2d are dissipated (leakage) inductance stator and rotor, respectively

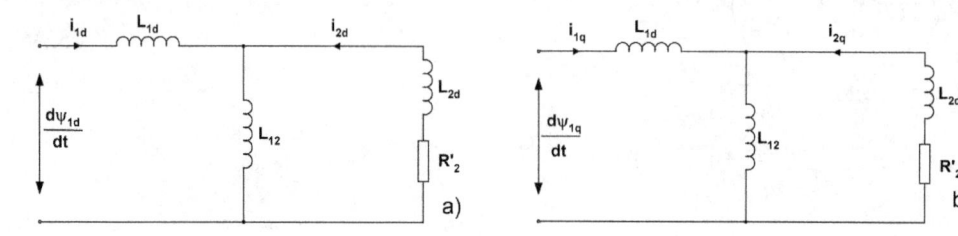

$$\bar{U}_1 = R_1\bar{I}_1 + \frac{d\bar{\Psi}_1}{dt} + j\omega_0\bar{\Psi}_1 \tag{55}$$

Similarly, the equation describing the rotor voltage referenced to axes rotating at ω_0 will contain an extra EMF proportional to $\omega_0-\omega$, which is the velocity of those axes relative to the rotor revolving at velocity ω:

$$\bar{U}_2 = R_2\bar{I}_2 + \frac{d\bar{\Psi}_2}{dt} + j(\omega_0 - \omega)\bar{\Psi}_2 \tag{56}$$

The quantities entering Equations (56), (57) must be referenced to the same complex plane rotating at ω_0. For time, t=0, the real axis of that complex plane is aligned with the axes of stator phase A, and that at time t, it is displaced from stator phase A by an angle α_0, then the currents in Equations (56), (57) will be:

$$\begin{aligned}\bar{I}_1 &= \bar{I}_{1(\alpha,\beta)} \exp(-j\alpha_0) \\ \bar{I}_2 &= \bar{I}_{2(d,q)} \exp[-j(\alpha_0 - \alpha)]\end{aligned} \tag{57}$$

where:
The flux linkages in Equations (56), (57) can be expressed in terms of $\bar{I}_1$ and $\bar{I}_2$ as

$$\begin{aligned}\bar{\Psi}_1 &= L_1\bar{I}_1 + L_{12m}\bar{I}_2 \\ \bar{\Psi}_2 &= L_2\bar{I}_2 + L_{12m}\bar{I}_1\end{aligned} \tag{58}$$

Electric Machines in Hybrid Power Train Employed Dynamic Modeling Backgrounds

1.10. Electromagnetic Torque Expressed in Terms of Resultant Current and Flux Linkage Phasors and their Components

The equation for electromagnetic torque, earlier Equation (24), is too unwieldy for practical purposes. It can be markedly simplified by using the resultant current phasors. Recalling that the resultant current phasors are $\sqrt{2}$ times rms ($D_{rms} = D_{max}/\sqrt{2}$; D- value of current or voltage etc., rms, which means – (root-mean-square) currents used to find the average electromagnetic torque, applied to the rotor, as:

$$M_{em} = (3p/2)\,\mathrm{Im}\left[\bar{\Psi}_2 \bar{I}_2^*\right] = 3p\left(\bar{\Psi}_2 \bar{I}_2^* - \bar{\Psi}_2^* \bar{I}_2\right)/4j \tag{59}$$

The equal electromagnetic torque applied to the stator is as follows:

$$M_{em} = (3p/2)\,\mathrm{Im}\left[\bar{\Psi}_1^* \bar{I}_1\right] = 3p\left(\bar{\Psi}_1^* \bar{I}_1 - \bar{\Psi}_1 \bar{I}_1^*\right)/4j \tag{60}$$

The torque acting on the rotor is assumed as positive, if it is in the direction of rotation (that is acting counter-clockwise). Conversely, the stator torque is taken as positive, if it is against the direction of rotation (that is acting clockwise). Writing the rotor flux linkage $\bar{\Psi}_2$, defined by the Equation (2) as the sum of the self-flux linkage

$$\bar{\Psi}_{22} = L_2 \bar{I}_2$$

and the mutual flux linkage

$$\bar{\Psi}_{21} = L_{12m} \bar{I}_1 \exp(-j\alpha)$$

that is

$$\bar{\Psi}_2 = \bar{\Psi}_{22} + \bar{\Psi}_{21}$$

and noting that in the electromagnetic torque defined by Equation (60) the component due to the interaction of current with self- flux linkage is zero

$$3p\left(\bar{\Psi}_{22}\bar{I}_2^* - \bar{\Psi}_{22}^*\bar{I}_2\right)/4j = 3pL_2\left(\bar{I}_2\bar{I}_2^* - \bar{I}_2^*\bar{I}_2\right)/4j = 0$$

It is possible to write the rotor torque in terms of mutual flux linkage as:

$$M_{em} = (3p/2)\operatorname{Im}\left[\overline{\Psi}_{21}\overline{I}_2^*\right] = 3p\left(\overline{\Psi}_{21}\overline{I}_2^* - \overline{\Psi}_{21}^*\overline{I}_2\right)/4j = \qquad (61)$$
$$= -9pL_m\left[\overline{I}_1\exp(-j\alpha)\overline{I}_2^* - \overline{I}_1^*\exp(-j\alpha)\overline{I}_2\right]/8j$$

Similarly, on writing $\overline{\Psi}_1$ from the Equation (37) as the sum of the self-flux linkage, $\overline{\Psi}_{11} = L_1\overline{I}_1$, and the mutual flux linkage:

$$\overline{\Psi}_{12} = L_{12m}\overline{I}_2\exp(j\alpha)$$

that is $\overline{\Psi}_1 = \overline{\Psi}_{11} + \overline{\Psi}_{12}$ can be written the stator torque from the Equation (61) in terms of the mutual flux linkage:

$$M_{em} = (3p/2)\operatorname{Im}\left[\overline{\Psi}_{12}^*\overline{I}_1\right] = 3p\left(\overline{\Psi}_{12}^*\overline{I}_1 - \overline{\Psi}_{12}\overline{I}_1^*\right)/4j = \qquad (62)$$
$$= 9pL_m\left[\overline{I}_2^*\exp(-j\alpha)\overline{I}_1 - \overline{I}_2\exp(j\alpha)\overline{I}_1^*\right]/8j$$

As it is seen, it does not differ from the electromagnetic torque acting on the rotor.

If the current phasors and their conjugates, in terms of the phase currents from Equations (25) and (27), it is clear that the electromagnetic torque found from Equations (60) and (62), is the same as the torque found from the general Equation (18):

$$M_{em} = (3p/2)\operatorname{Im}\left[\overline{\Psi}_2\overline{I}_2^*\right] = 9pL_m\left(\overline{I}_1 e^{-j\alpha}\overline{I}_2^* - \overline{I}_1^* e^{j\alpha}\overline{I}_2\right)/2j =$$
$$= -pL_m\begin{bmatrix} i_{1A}\left(i_{2a}\sin\alpha_{Aa} + i_{2b}\sin\alpha_{Ab} + i_{2c}\sin\alpha_{Ac}\right) + \\ +i_{1B}\left(i_{2a}\sin\alpha_{Ba} + i_{2b}\sin\alpha_{Bb} + i_{2c}\sin\alpha_{Bc}\right) + \\ +i_{1C}\left(i_{2a}\sin\alpha_{Ca} + i_{2b}\sin\alpha_{Cb} + i_{2c}\sin\alpha_{Cc}\right) \end{bmatrix} = \qquad (63)$$
$$= -pL_m\sum_{k=A,B,C} i_{1k}\sum_{n=a,b,c} i_{2n}\sin\alpha_{kn}$$

Finally, the electromagnetic torque for e.g. Equation (62) can be expressed in terms of the d- and q- axis components of currents and flux linkage as:

$$M_{em} = (3p/2)\operatorname{Im}\left[\overline{\Psi}_2\overline{I}_2^*\right] = (3p/2)\left(\psi_{2q}i_{2d} - \psi_{2d}i_{2q}\right) \qquad (64)$$

Electric Machines in Hybrid Power Train Employed Dynamic Modeling Backgrounds

or in terms of the α- and β- axis components of currents and flux linkages as:

$$M_{em} = (3p/2)\operatorname{Im}\left[\bar{\Psi}_1^* \bar{I}_1\right] = (3p/2)\left(\psi_{1\alpha} i_{1\beta} - \psi_{1\beta} i_{1\alpha}\right) \tag{65}$$

2. PM SYNCHRONOUS MOTOR MODELING

2.1. Operating Principles and Construction Evolution

A conventional synchronous motor has two windings—a three phase winding (the number of phases can be greater) which produces a magnetic field rotating at a constant speed when the winding carries an alternating current, and a field (excitation) winding which carries a DC current to produce a magnetic field of fixed polarity. The operation of a synchronous machine depends on the mutual action of these two fields each upon the other. Under the synchronous-generator performance conditions, the rotating field produced by the field (rotor) winding driven by means of an external torque result in inducing a voltage in the polyphase winding, the voltage frequency depending upon the rotational speed and the number of pole pairs created by the polyphase winding or by dividing the field winding. The generator load results in a current flowing in the polyphase winding to produce a magnetic flux which deflects the magnetic field force lines of the field winding to convert the electric load into a mechanical torque. Under the synchronous-motor performance conditions, the rotating field produced by the polyphase winding drags that produced by the field winding. There is a deflection angle called the load angle and included between the conventional axis of the rotating magnetic field and the axis of field produced by the field winding. This load angle is shown included between the input voltage vector of the input voltage (across the machine terminals) and the vector of the electromotive force induced in the polyphase winding in the synchronous machine vector diagrams. The value of the torque produced by the machine is a function of machine design characteristics (its geometry and materials employed) and sinus of the load angle. Owing to the relation between the synchronous torque and the position of the rotor with respect to the conventional axis of the polyphase winding as well as the rotational speed of the field while the frequency provided by the supply source (power mains) is rigid, the direct-on starting-up of a synchronous motor by applying a voltage at a constant frequency would be difficult and often even impossible. There have been several starting-up methods exercised, say, by means of induction-motor or auxiliary motor, but the problems with the starting-up have been a serious limitation in the use of synchronous motors.

The field winding of a synchronous machine is usually located in the rotor due to the technology and economy reasons. Owing to such an approach it is enough to use only two slip-rings on the machine shaft to connect the current in the winding, and each moving contact carries smaller electric power than that used in the polyphase winding. Thus, the polyphase winding is located in the stator of a conventional synchronous machine. Such a machine employs either salient or non-salient-pole rotor. These structure versions lead to some differences in the dynamic properties of machines as well as enable the machines to rotate at basically different speeds—the salient-pole ones are low speed machines while those with non-salient-pole rotors are high speed machines. The salient-pole rotors are fitted with so called starting squirrel-cages to execute the induction-motor action mentioned above. The cage bars are located in the pole shoes of the field magnet rotor and connected by means of rings at both ends of the rotor. The rings can short-circuit the bars within each pole shoe area or all the bars can be short-circuited together. In situations where any dynamic motor load and/or field current variations occur, such a cage damps the transient disturbances and therefore the term of a damping cage is also used for it in some references. The cage is equivalent to starting cage from the physical point of view. When used it improves the performance of a motor supplied from an adjustable frequency and/or voltage source. The solid material of the rotor where eddy currents are induced plays the role of the cage in the non-salient-pole machines (turbo generators).

By developing the methods of production of better and better permanent magnets, materials employing Samarium and Cobalt have densities of stored energy of about 100-150 kJ/m^3 were obtained in 1974 and the NdBFe (Neodymium-Boron-Iron) of stored energy of an order of 280 kJ/m^3 were available ten years later. Such large densities of energy have resulted in the wider use of these materials in electric machines of higher and higher output power the physical size of the machines being not increased. New construction proposals and patent applications for the field magnet structure have developed rapidly.

The following present various constructions of synchronous machines with permanent magnets.

2.2. Converter Circuits

As a general rule, there are three circuits alternatively employed for continuous adjustment of the machine speed: direct frequency converters for the low-speed drive units supplied from an AC three-phase mains, and the remaining two are basically different being inverters, namely, a voltage-type inverter and a current-type inverter. The inverter circuits require the input voltage V to be controlled. This means that a chopper has to be used in situations where a storage battery or a contact line is used as the power source.

Figure 7. The pictorial drawing of the permanent magnet synchronous motor and its photographs. The stator is a winding, which does not employ a core and is equivalent to a cup rotor (Huang, Cambier, & Geddes, 1992) (Pictures from the author's HEV&EV lab)

The design of a current-type inverter intended to operate with an AC synchronous motor employs the natural feature of synchronous machine to produce the reactive power and the commutation (switching of individual machine phases) is carried out in a natural way - determined by the external commutation (switching of the inverter)

Figure 8. The diagrams of inverters operating with a permanent magnet synchronous machine whose excitation is provided by permanent magnets

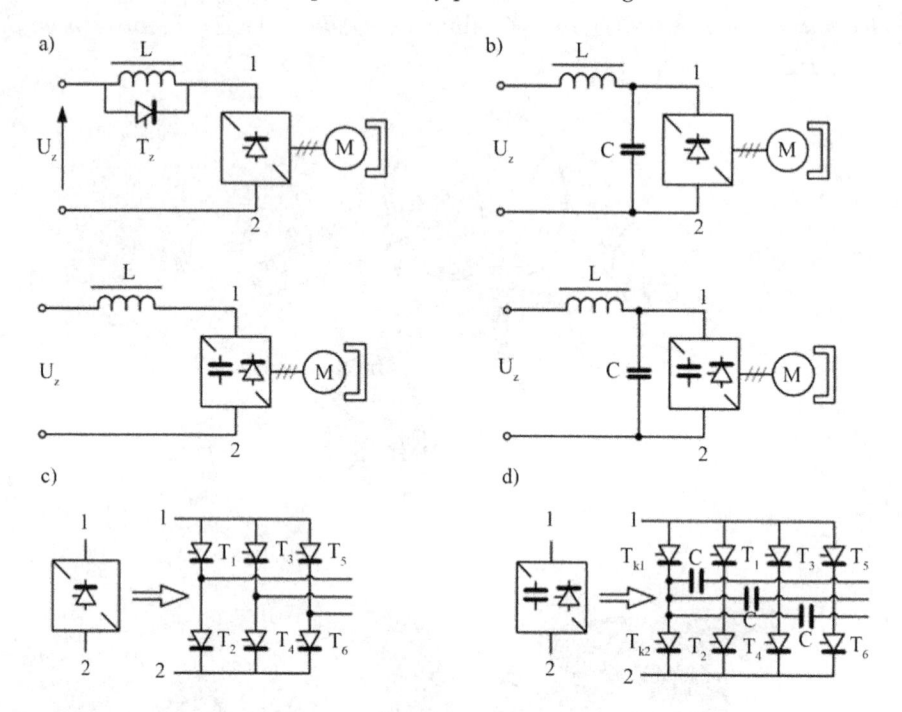

a) Current-type inverters (L - inductance of the intermediate supply circuit);
b) Voltage-type inverters (L - filter inductance of the intermediate supply circuit, and capacitors at the inverter input);
c) Diagram of T1-T6 SCR circuit between the terminals 1 and 2, making an inverter externally switched by the three phase supply to the synchronous motor.
d) Diagram of T1-T6 SCR circuit between the terminals 1 and 2 making an inverter - and capacitors C as well as SCR's Tk1 and Tk2 for internal switching in a three-phase supply circuit

whose solid-state switches are usually silicon controlled rectifiers and turned off by means of the voltage applied to the motor terminals.

The firing circuit for the SCR's T1-T6 is based on any motor position transmitters (both optoelectronic and magnetic ones).

However, this method results in a pulsating torque due to the cycles of three-phase and two-phase performance of the three-phase machine. The amplitude of this hunting is a function of the motor load and can reach more than 20% of the mean torque while its frequency is a multiple of the frequency determined by the motor speed. This effect cannot be removed by engineering methods; there have been proposals of an approach with polyphase motors having a number of phases greater than three to avoid the phenomena of resonance with the drive construction components. An inverter employing the natural switching performance has one more negative feature depending on the need to use a special circuit for controlling the SCR Tz to enable the motor to be accelerated to a speed of about 10-15% of the rated value. As the SCR's are fired by the machine input voltages, the motor has to operate with a lending power factor. In situations where machines with fixed field employing permanent magnets are used, the inverter circuits are simpler because such an inverter operates at a fixed rotor-position firing-angle.

A drive system employing a current-type inverter with the forced (internal) switching is capable of operating at any rotor-position firing-angle. In situations where an excitation control is available, the control can be obedient to any steady-state optimization law and to obtain the desired value either constant or varying with respect to the speed—according to the adjustment range and the initial starting torque value. By maintaining the field nearly constant, the torque adjustment is limited to a certain range.

The drive systems employing voltage-type inverters can be in-group of drive systems where all the controlled motors are to rotate in synchronism. In situations where an external control approach is employed, the voltage input to the inverter is to be controlled. This will result in the motor speed control in the range of constant torque between the speed of zero and the rated synchronous speed. A higher speed can be obtained in the constant power range where the inverter output voltage is limited to its rated value, which is determined by the employed power source.

By the use of power solid-state switches in a voltage-type inverter, their performance in any mode can be under control. To provide for the continuity of current flow in the individual phases, the solid-state switches are bypassed with diodes connected for operation with appropriate polarity. Each inverter leg with a solid-state switch employs a very important device not shown in Figure 8 an over voltage protection device called a snubber diode. These devices protect the main solid-state switches against excessive stresses during switching transients. The conditions of switching orders and timing are based on Pulse Width Modulation (PWM). An advantage of such a circuit consists in the availability of a relatively stable power source such as a storage battery to supply the inverter directly. The pulse width modulation voltage type PWM inverter used in a group drive system enables individual motors to operate in the generator mode and the other motors in their motor-mode. There are a dozen of pulse width modulation motor control method employed not only for traction applications. They have been basically developed owing to the demand for inverters operating with induction motors with an objective to remove some voltage-wave harmonics, to reduce the induction motor losses to a minimum, and to reduce the pulsation of the torque of an asynchronous motor supplied from an inverter. The author has not met any reference presenting the selection of PWM methods for the drives employing synchronous motor yet. It seems interesting to solve the problem of feasible Pulse Width Modulation (PWM) method for a synchronous motor enabling the power consumption to be lowered although it seems to be dependent on the availability of power solid-state devices.

Generally, the inverter role is 3-phases Alternative Current (AC) voltage generation for proper obtains output mechanical speed and torque of a Permanent Machine (PM) machine. The commonly used method is (similar to AC alternative current induction machines) Pulse Width Modulation (PWM). The block scheme of that inverter is shown in Figure 9.

Figure 10 shows the circuit of voltage inverter firing of three-phases PWM modulator.

2.3. Equivalent Circuit of Synchronous Motor

The familiarity with the mathematical model and characteristics of drive systems is a basic problem in the situation where the theoretical analysis of and design work at such system is to be performed. Figure 11 presents electric circuits of a salient-pole machine which is an electromechanical converter with electromagnetic excitation. To make the discussion a general one, a three-phase circuit is taken into consideration. (The present structures employ circuits having a greater number of phases but this does not require any change in the basic mathematical model). Beside the armature windings (those with indices 1 to 3) and the field winding (index 4) there are two additional short-circuited windings in the direct axis d and quadrature axis

Figure 9. Most common block scheme of PWM inverter applied for PM- permanent magnet motor control

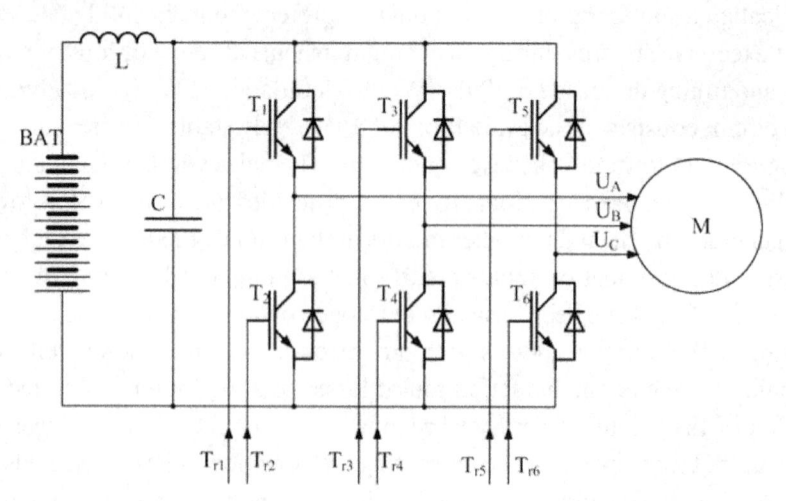

Electric Machines in Hybrid Power Train Employed Dynamic Modeling Backgrounds

Figure 10. The voltage inverter fired by three-phase pulse width PWM modulator

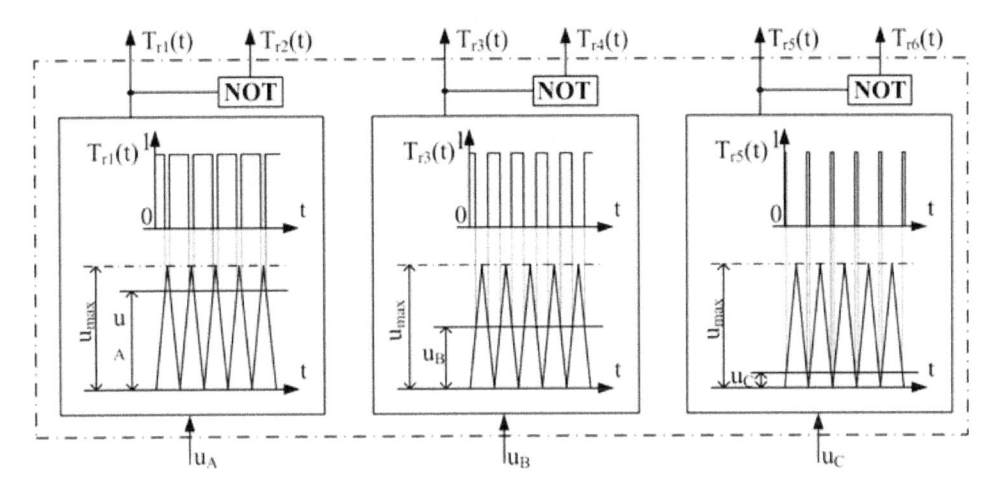

q which represent the damping (starting) cage and eddy currents in the solid and/or laminated parts of the rotor.

The voltage equations can be written for all the circuits with their relations in the form of mutual inductances M_{ij} taken into account. The result can be written in the matrix form shown below:

$$\begin{bmatrix} U_1 \\ U_2 \\ U_3 \\ U_4 \\ 0 \\ 0 \end{bmatrix} = \begin{bmatrix} R_1 & 0 & 0 & 0 & 0 & 0 \\ 0 & R_2 & 0 & 0 & 0 & 0 \\ 0 & 0 & R_3 & 0 & 0 & 0 \\ 0 & 0 & 0 & R_4 & 0 & 0 \\ 0 & 0 & 0 & 0 & R_5 & 0 \\ 0 & 0 & 0 & 0 & 0 & R_6 \end{bmatrix} \cdot \begin{bmatrix} I_1 \\ I_2 \\ I_3 \\ I_4 \\ I_5 \\ I_6 \end{bmatrix}$$

$$+ \frac{d}{dt} \left(\begin{bmatrix} L_{11} & M_{12} & M_{13} & M_{14} & M_{15} & M_{16} \\ M_{21} & L_{22} & M_{23} & M_{24} & M_{25} & M_{26} \\ M_{31} & M_{32} & L_{33} & M_{34} & M_{35} & M_{36} \\ M_{41} & M_{42} & M_{43} & L_{44} & M_{45} & 0 \\ M_{51} & M_{52} & M_{53} & M_{54} & L_{55} & 0 \\ M_{61} & M_{62} & M_{63} & 0 & 0 & L_{66} \end{bmatrix} \begin{bmatrix} I_1 \\ I_2 \\ I_3 \\ I_4 \\ I_5 \\ I_6 \end{bmatrix} \right) \quad (66)$$

The terms of the mutual- and self-inductances of the armature windings are functions of an angle (included between the axis d and the axis of an armature winding - the axis of winding 1 (phase a) is most often selected. The succession of

Figure 11. The model of a salient-pole permanent magnet synchronous machine with short-circuited rotor windings in the axes d and q

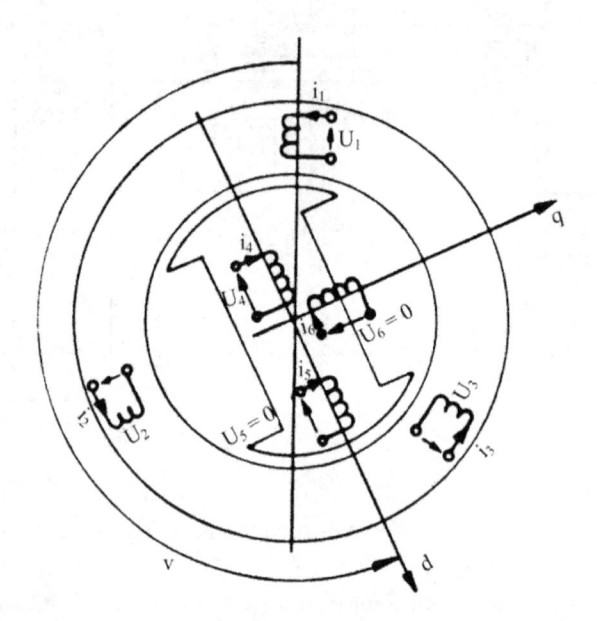

individual axes I d q of the armature windings and the direction in which the angle (rises, i.e. the machine rotates, is significant things. To make the discussion simpler, the following assumptions are made for an ideal model of a synchronous machine:

- The machine under discussion is a symmetrical one (its armature windings are identical and distributed symmetrically: $R_1=R_2=R_3$;
- Only the fundamental harmonic of ampere-turns and magnetic flux density are taken into account and this means that the higher harmonics of the magnetic field distribution in space (due to the discrete form of the windings and the magnetic circuit geometry) occurring in the air-gap are neglected;
- It is assumed that the magnetization curve is linear and therefore the phenomena of saturation and hysteresis are not taken into account; thus, the mutual- and self-inductances do not depend upon the values of currents but on the position of the rotor (co-ordinate ϕ).

By taking the assumptions mentioned above into consideration we can determine the functional relations between individual windings while keeping in mind that the flux of each of the windings consists of two components: the main flux and the leakage flux. The voltage Equations (66) make a set of differential equations with varying factors that depend on the angle ϕ and, in consequence, the time t.

Electric Machines in Hybrid Power Train Employed Dynamic Modeling Backgrounds

To convert this set of equations into a more convenient form of the equations with constant factors, the armature co-ordinates are to be transformed. The most often employed transformation - which is convenient but not the only one - is that reducing to the co-ordinates 0, d, q (called the rotating rectangular coordinates or Park's co-ordinates). The transformation consists in substituting the three-phase armature winding with two windings located in the axes d and q and rotating together with the rotor, and a third winding 0 located outside the machine and not coupled with the remaining machine windings. The equivalence of the windings 1, 2, 3 and 0, d, q is expressed by means of relations between the voltages and currents of the actual and equivalent windings. Let us make an additional assumption that the instantaneous output power and torque are invariant in both d, q co-ordinate systems (transformation matrix was described in this chapter).

By multiplying the left side of the transformation matrix and the matrix (67) and by making the respective substitutions, the voltage and flux equations - called the Park-Goriev equations are obtained in the following form:

$$\begin{bmatrix} U_0 \\ U_d \\ U_q \\ U_4 \\ U_5 \\ U_6 \end{bmatrix} = \begin{bmatrix} R_1 & 0 & 0 & 0 & 0 & 0 \\ 0 & R_2 & 0 & 0 & 0 & 0 \\ 0 & 0 & R_3 & 0 & 0 & 0 \\ 0 & 0 & 0 & R_4 & 0 & 0 \\ 0 & 0 & 0 & 0 & R_5 & 0 \\ 0 & 0 & 0 & 0 & 0 & R_6 \end{bmatrix} \cdot \begin{bmatrix} I_0 \\ I_d \\ I_q \\ I_4 \\ I_5 \\ I_6 \end{bmatrix} + \frac{d}{dt} \begin{bmatrix} 0 \\ \Psi_d \\ \Psi_q \\ \Psi_4 \\ \Psi_5 \\ \Psi_6 \end{bmatrix} + \omega \cdot \begin{bmatrix} 0 \\ -\Psi_q \\ \Psi_d \\ 0 \\ 0 \\ 0 \end{bmatrix} \quad (67)$$

If the flux linkage ψ is the coupled fluxes of respective windings and they are defined by the following equations:

$$\Psi_0 = L_0 \cdot I_0$$
$$\Psi_d = L_d \cdot I_d + M_4^d \cdot I_4 + M_5^d \cdot I_5$$
$$\Psi_q = L_q \cdot I_q + M_6^q \cdot I_6$$
$$\Psi_4 = (L_{s4} + L_4) \cdot I_4 + M_4^d \cdot I_d + M_{45} \cdot I_5$$
$$\Psi_5 = (L_{s5} + L_5) \cdot I_5 + M_5^d \cdot I_d + M_{45} \cdot I_4$$
$$\Psi_6 = (L_{s6} + L_6) \cdot I_6 + M_6^q \cdot I_q$$

where the self-inductances L and the mutual inductances M are to be substituted respectively.

$$L_0 = L_s^z = (z_z \cdot k_{uz})^2 \cdot \Lambda_s^z$$

$$L_d = L_s^z + \frac{3}{2} \cdot (M^z + \Delta M)$$

$$M^z = \frac{1}{2} \cdot (\Lambda_d + \Lambda_q) \cdot (z_z \cdot k_{uz})^2$$

$$\Delta M = \frac{1}{2} \cdot (\Lambda_d - \Lambda_q) \cdot (z_z \cdot k_{uz})^2$$

- z_k: armature winding turns;
- k_{uz}: Armature winding factor;
- Λ_s^z: Armature leakage permeance;
- Λ_d: Maximum permeance value of the main flux, which occurs in the machine owing to the rotation of the rotor (read as a function of the angle of ϕ);
- Λ_q: Minimum permeance value for the main flux, which occurs in the machine owing to the rotation of the rotor.

$$L_q = L_s^z + \frac{3}{2} \cdot (M^z - \Delta M)$$

- M^d_4, M^d_5, M^q_6: Maximum mutual inductances corresponding to the coaxial position of the respective armature and rotor windings;
- L_{s4}, L_{s5}, L_{s6}: Leakage inductances of the respective rotor windings;
- L_4, L_5, L_6: Self-inductances of the respective rotor winding associated with the machine main-flux;
- M_{45}, M_{54}: Mutual inductances of the field and damping windings in the axis d.

The I winding turns, its factor and all the permanencies are involved with the machine geometry. While performing the field analysis of the electromagnetic circuit of the machine by means of, say, final elements, we are capable of determining the Park model parameters of a synchronous machine with great precision. Another problem is how the model parameters are to be identified for existing machines on the basis of laboratory tests.

In the situations where the air-gap is more uniform than that in the salient-pole machine, there is a small difference between the values Λ_d and Λ_q, and this is reflected in the direct-axis and quadrature-axis inductances. The first row in the matrix Equation (67) is neglected independently of the magnetic core construction because - in practice - there are no magnetic couplings between the equivalent winding 0 and the rotary windings I that are in the axes d and q.

Electric Machines in Hybrid Power Train Employed Dynamic Modeling Backgrounds

To provide a full picture of an electromechanical converter such as a synchronous motor, let us write the equation of the electromagnetic torque occurring in the air-gap between the armature and the rotor of the machine. This equation can be derived on the basis of a description of the magnetic field energy occurring in the air-gap by differentiating the energy with respect to the angle ϕ for a defined speed - or on the basis of Euler-Lagrange equation for a mechanical co-ordinate. Thus, the result can be written in the form of the following equation:

$$M_{em} = -\frac{1}{2} \cdot 3 \cdot \Delta M \cdot \begin{bmatrix} I_0 & I_d & I_q \end{bmatrix} \cdot \begin{bmatrix} 0 & 0 & 0 \\ 0 & 0 & 1 \\ 0 & 1 & 0 \end{bmatrix} \cdot \begin{bmatrix} i_0 \\ i_d \\ i_q \end{bmatrix}$$

$$-\begin{bmatrix} I_0 & I_d & I_q \end{bmatrix} \cdot \frac{3}{2} \cdot \begin{bmatrix} 0 & 0 & 0 \\ 0 & 0 & -M_6^q \\ M_4^d & M_5^d & 0 \end{bmatrix} \cdot \begin{bmatrix} I_4 \\ I_5 \\ I_6 \end{bmatrix} \quad (68)$$

By carrying out the multiplication and substituting $3 \cdot \Delta M = L_d - L_q$ we obtain:

$$M_{em} = -(L_d - L_q) \cdot I_d \cdot I_q - \left(\frac{3}{2} \cdot M_4^d \cdot I_4 + \frac{3}{2} \cdot M_5^d \cdot I_5 \right) \cdot I_q$$

$$+ \frac{3}{2} \cdot M_6^q \cdot I_q \cdot I_d \quad (69)$$

By re-arranging the terms in respect to the currents in both axes, the following equation can be written:

$$M_{em} = \frac{3}{2} \left(\Psi_q \cdot I_d - \Psi_d \cdot I_q \right) \quad (70)$$

where Ψ_d and Ψ_q are the coupled fluxes discussed in the description of the torque Equations (68) and (69).

The sets of Equations (67) and (71) are non-linear as before because the rotational speed occurs in the voltage equation and, in addition, the products of time functions present in the Equation (70) are a source of non-linearity.

General principles of the motor design and its equivalent circuit reduced to the axes d and q are presented in Figure 12. New equation involving the permanent magnet and its magnetizing force has occurred when compared with the model discussed above. The Park-Goriev Equation (59) for the motor presented in Figure 12 can be written as follows:

Figure 12. The equivalent circuit diagram reduced to axes q and d of permanent magnet synchronous machine (refer to the winding designations in the text)

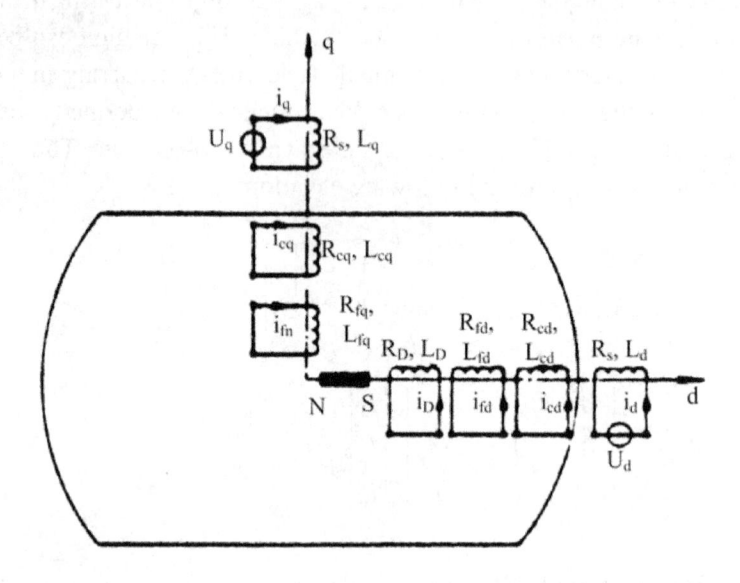

$$\begin{bmatrix} U_d \\ U_q \\ 0 \\ 0 \\ 0 \\ 0 \\ 0 \end{bmatrix} = \begin{bmatrix} R_s & 0 & 0 & 0 & 0 & 0 & 0 \\ 0 & R_s & 0 & 0 & 0 & 0 & 0 \\ 0 & 0 & R_{cd} & 0 & 0 & 0 & 0 \\ 0 & 0 & 0 & R_{cq} & 0 & 0 & 0 \\ 0 & 0 & 0 & 0 & R_{fd} & 0 & 0 \\ 0 & 0 & 0 & 0 & 0 & R_{fq} & 0 \\ 0 & 0 & 0 & 0 & 0 & 0 & R_D \end{bmatrix} \cdot \begin{bmatrix} I_d \\ I_q \\ I_{cd} \\ I_{cq} \\ I_{fd} \\ I_{fq} \\ I_D \end{bmatrix} + \frac{d}{dt} \begin{bmatrix} \Psi_d \\ \Psi_q \\ \Psi_{cd} \\ \Psi_{cq} \\ \Psi_{fd} \\ \Psi_{fq} \\ \Psi_D \end{bmatrix} + \omega \cdot \begin{bmatrix} -\Psi_q \\ \Psi_d \\ 0 \\ 0 \\ 0 \\ 0 \\ 0 \end{bmatrix}$$

(71)

where:

- R_s: Armature winding resistance; R_{cd} and R_{cq} - Damping (starting) cage resistance reduced to the axes d and q, respectively: R_{fd} and R_{fq} - Damping circuit resistances (eddy currents) created in the rotor iron and reduced to the axes d and q, respectively,
- R_D: Field winding resistance;
- Ψ_D: Coupled fluxes determined by the following formulae:

Electric Machines in Hybrid Power Train Employed Dynamic Modeling Backgrounds

$$\Psi_d = L_d I_d + M_{dc} I_{cd} + M_{df} I_{fd} + M_{dD} I_D + M_{da} \theta_s$$
$$\Psi_q = L_q I_q + M_{qc} I_{cq} + M_{qf} I_{fq}$$
$$\Psi_{cd} = L_{cd} I_{cd} + M_{cd} I_d + M_{cfd} I_{fd} + M_{cD} I_D + M_{cs} \theta_s$$
$$\Psi_{cq} = L_{cq} I_{cq} + M_{cq} I_q + M_{cfq} I_{fq} \quad (72)$$
$$\Psi_{fd} = L_{fd} I_{fd} + M_{fd} I_d + M_{fdc} I_{cd} + M_{fD} I_D + M_{fa} \theta_s$$
$$\Psi_{rq} = L_{fq} I_{fq} + M_{fq} I_q + M_{fqc} I_{cq}$$
$$\Psi_D = L_D I_D + M_{Dd} I_d + M_{Dc} I_{cd} + M_{Df} I_{fd} + M_{Ds} \theta_s$$

- θ_s: Magnetic potential of the permanent magnet;
- M_{is}: Mutual inductance of the ith winding in the axis d and the equivalent electric circuit of the magnet structure

The description does not take into account the internal symmetry of the machine, which provides the values of mutual inductances equal each to the other $M_{ij} = M_{ji}$.

The mutual- and self-inductances are determined in a similar way to that presented in the discussion to the Equation (67). The model developed in this way enables an analysis of the effects of individual machine parts during the selected operational duties to be taken into account. To reduce the number of equations that are necessary for the solution of the problem in the analysis of dynamic states, the rotor circuits can be reduced to a form of two windings in the axes d and q representing the damping of cage currents, eddy currents and the currents in the short-circuited field windings. By rearranging the equations and differentiating the coupled fluxes, the set of Equation (72) can be written in the following form:

$$U_d = R_s I_d + L_d \cdot \frac{dI_d}{dt} + M_{dD} \cdot \frac{dI_D}{dt}$$
$$+ M_{ds} \cdot \frac{d\theta_s}{dt} - \omega \cdot L_q \cdot I_q - \omega \cdot M_{qQ} \cdot I_Q$$
$$U_q = R_s I_q + L_q \cdot \frac{dI_q}{dt} + M_{qQ} \cdot \frac{dI_Q}{dt} \quad (73)$$
$$+ \omega \cdot L_d \cdot I_D + \omega \cdot M_{dD} \cdot I_d + \omega \cdot M_{ds} \cdot \theta_s$$
$$0 = R_D I_D + L_D \cdot \frac{dI_D}{dt} + M_{Dd} \cdot \frac{dI_d}{dt} + M_{Ds} \cdot \frac{d\theta_s}{dt}$$
$$0 = R_Q I_Q + L_Q \cdot \frac{dI_Q}{dt} + M_{Qq} \cdot \frac{dI_q}{dt}$$

where the indices written in capital letters are related to the damping parts in the rotor. These equations do not take any possible symmetry of the mutual inductances in individual circuits into account as before.

To achieve a complete dynamic model of an electromechanical converter, it is necessary to include the motion Equation (72) and the position Equation (74) into the Equations (73):

$$J\frac{d\omega}{dt} = \frac{3}{2}p\left[\begin{array}{c}(L_d I_d + M_{dD} I_D + M_{Ds}\theta_s)I_q \\ -(L_q I_q + M_{qQ} I_Q)I_d\end{array}\right] - M_{obc} \quad (74)$$

where:

- ω: Angular speed of the machine shaft;
- p: Number of pole pairs in the machine;
- M_{obc}: Machine load torque;
- v: Angle of rotor position.

A rotor position equitation:

$$\frac{dv}{dt} = \omega \quad (75)$$

The Equations (72), (73), and (74) can be reduced to a form of a set of state variables where the state of the machine is by the derivatives of the following variables I_d, I_q, I_D, I_Q, ω and v with respect to time. The resulting equations enable an analysis of converter dynamic including the effects of individual (reluctance, asynchronous and synchronous) torque upon the transient performance of the motor to be executed. It is also possible to develop a control circuit for a synchronous motor operation, which is supplied from a PWM inverter, on the basis of the state equations as carried out in Abdulaziz and Jufer (1974).

The references Abdulaziz and Jufer (1974), Anderson and Cambier (1990), and Baucher (2007) present an experimental car with an electric drive system employing a synchronous permanent-magnet motor. The vehicle weight is 403 kg and its capacity is 200 kg. It employs a motor having the full-load power of 3 kW and the rated speed of 4800 rpm. Its rotor design is similar to that of a commutator motor with special air-gaps that separate individual made of those of DC motors. Therefore among other reasons, an equivalent model circuit reduced to a single axis d might be used for the initial analysis of the power of a drive system employing a NdFeB magnet. These air-gaps affect the differences between the reactances profitably in

Electric Machines in Hybrid Power Train Employed Dynamic Modeling Backgrounds

the direct axis d and the quadrature axis q. The magnetic flux density in the air-gap of the motor amounts to 0.8T. The motor is supplied from a lead-acid battery via a Pulse Width Modulation (PWM) inverter. The motor input-frequency is adjusted in 0.02 Hz steps in a range of 2-2000 Hz. The authors propose the pulse ON time P; and the OFF time N; for the input voltage and the rotation angle υ to be involved in the following relation:

$$P_{ON} = \frac{1}{\omega} \cdot \left(\frac{\upsilon_{j+1} - \upsilon_j}{2} + \frac{U}{U_b} \cdot \frac{\cos\upsilon_j - \cos\upsilon_{j+1}}{2} \right)$$
$$N_{OFF} = \frac{1}{\upsilon} \cdot \left(\frac{\upsilon_{j+1} - \upsilon_j}{2} + \frac{U}{U_b} \cdot \frac{\cos\upsilon_j - \cos\upsilon_{j+1}}{2} \right) \tag{76}$$

where:

- P_{ON}: Pulse ON time in the jth interval;
- P_{OFF}: Pulse OFF time in the jth interval;
- $\upsilon_j, \upsilon_{j+1}$: Range of the rotor angle variation in the jth time interval;
- ω: Required angular speed of the rotor;
- U: Required peak amplitude of the output voltage;
- U_b: DC voltage provided by the battery.

As the philosophy of the drive control in this car is directed to a maximum efficiency of the energy conversion, it has been suggested to use the machine characteristics for the determination of the load loss ΔP_{Cu} and the iron loss ΔP_{Fe} per machine phase for the controlling the solid-state switches in PWM control system.

$$\Delta P_{Cu} = \left[\left(I_{2d} - \frac{P_p \cdot X_q}{E_0 \cdot R_c} \right)^2 + \left(\frac{E_0 + I_{2d} \cdot X_d}{R_c} + \frac{P_p}{E_0} \right)^2 \right] \cdot R_1$$
$$\Delta P_{Fe} = \left[\left(E_0 + I_{2d} \cdot X_d \right)^2 + \frac{P_p^2 \cdot X_q^2}{E_0^2} \right] \cdot \frac{1}{R_c} \tag{77}$$

where:

- I_{2d}: Root-mean square value of the current provided by the inverter;
- P_p: Output power per phase;
- X_q: Quadrature-axis reactance of synchronous motor;

- E_0: Electromotive force induced in the armature winding;
- R_c: Damping circuit resistance reduced to a phase;
- X_d: Direct-axis reactance of synchronous motor;
- R_1: Phase resistance of the armature windings

By adding both terms, the motor loss per phase is determined for a defined load. The speed is covered under the term E_0 and for a fixed field flux (ampere-turns) is proportional to the rotor speed.

A significantly simplified model of a synchronous machine, in which the electromotive forces induced in the individual phases of the machine as well as the self-inductances of the motor windings are identified is presented in the references . The authors investigated the configurations of the switching of PWM inverter operation with a motor operating at a constant speed under steady-state conditions. The results were presented in the form of the motor current and torque vs. the pulse duty factor with the battery ratio of the battery voltage to the voltage induced in a machine phase taken as a parameter. The presented waveforms obtained by means of digital simulation suggest that their hunting of both current and torque have occurred and this was proved in experiments in the case of the motor phase current.

The drive systems employing induction and synchronous motors operated via PWM inverters discussed in the references Abdulaziz and Jufer (1974), Anderson and Cambier (1990), Baucher (2007), Ortmeyer (2005), Unique Mobility Inc. (n.d.), and Veinger (1985) direct the attention of the reader to the possibility of occurrence of an effect similar to the decreasing of the field current in a DC motor, i.e. the speed control in the second zone by field weakening although in our case synchronous permanent-magnet machines make the case. The author's proposal consists in the use of specially shaped waveform of the voltage applied to the motor. His proposal results from the vector analysis of the equivalent circuit corresponding to a system in which a synchronous machine is operated from a PWM inverter. The engineering approach is left open. The waveform for supplying a synchronous motor taken as a standard for determination of the timing of solid-state switch operation based on a PWM method employing a sawtooth wave as proposed by the author increases the value of fundamental harmonic of the supply voltage waveform produced in this way.

REFERENCES

Abdulaziz, M., & Jufer, M. (1974). Magnetic and electric model of synchronous permanent – Magnet machines. *Bull. SEV, 74*(23), 1339–1340.

Anderson, W. M., & Cambier, C. (1990). An advanced electric drivetrain for EVs. In *Proceedings of EVS 10*. Hong Kong: EVS.

Ashihaga, T., Mizuno, T., Shimizu, H., Natori, K., Fujiwara, N., & Kaya, Y. (1992). Development of motors and controllers for electric vehicle. In *Proceedings of EVS 11*. Florence, Italy: EVS.

Baucher, J. P. (2007). Online efficiency diagnostic method of three phases asynchronous motor. In *Proceedings of Powereng IEEE*. Setubal, Portugal: IEEE.

Braga, G., Farini, A., Fuga, F., & Manigrasso, R. (1991). Synchronous drive for motorized wheels without gearbox for light rail systems and electric cars. In *Proceedings of EPE'91 European Conference on Power Electronics*. EPE.

Brusaglino, G., & Tenconi, A. (1992). System engineering with new technology for electrically propelled vehicles. In *Proceedings of EVS 11*. Florence, Italy: EVS.

Burke, A. F., & McDowell, R. D. (1992). The performance of advanced electric vans – Test and simulation. In *Proceedings of EVS 11*. Florence, Itlay: EVS.

Chan, C. C. (1994). The development of an advanced electric vehicle. In *Proceedings of EVS 12*. Los Angeles, CA: EVS.

Chan, C. C., Jiang, G. H., Chen, X. Y., & Wong, K. T. (1992). A novel high power density PM motor drive for electric vehicle. In *Proceedings of EVS 11*. Florence, Italy: EVS.

Chan, C. C., & Lueng, W. S. (1990). A new permanent magnet motor drive for mini electric vehicles. In *Proceedings of EVS 10*. Hong Kong: EVS.

Chris, C., & Luo, L. (2005). Analitical design of PM traction motors. In *Proceedings Vehicle Power and Propulsion Conference VPPC*. Chicago, IL: IEEE.

Datla, M., & High, A. (2007). Performance decoupling control of induction motor with efficient flux estimator. In *Proceedings of Powereng IEEE*. Setubal, Portugal: IEEE.

Ferraris, P., Tenconi, A., Brusaglino, G., & Ravello, V. (1996). Development of a new high performance induction motor drive train. In *Proceedings of EVS 13*. Osaka, Japan: EVS.

Gosden, D. F. (1992). Wide speed range operation of an AC PM EV drive. In *Proceedings of EVS 11*. Florence, Italy: EVS.

Henneberger, G., & Lutter, T. (1991). Brushless DC – Motor with digital state controller. In *Proceedings of the EPE'91 European Conference on Power Electronics an Application*. EPE.

Huang, H., Cambier, C., & Geddes, R. (1992). High constant power density wide speed range PM motor for EV application. In *Proceedings of EVS 11*. Florence, Italy: EVS.

Jezernik, K. R. (1994). Induction motor control for electric vehicle. In *Proceedings of EVS 12*. Los Angeles, CA: EVS.

Kenjo, T., & Nagamori, S. (1985). *Permanent magnet and brushless DC motors*. Oxford, UK: Claderon Press.

King, R. D., & Konrad, C. E. (1992). Advanced on–board EV AC drive – Concept to reality. In *Proceedings of EVS 11*. Florence, Italy: EVS.

Ledowskij, A. N. (1985). *Electrical machines with high coercive force permanent magnets*. Moscow: Energoatimizdat.

Matsusa, K., & Katsuta, S. (1996). Fast rotor flux control of vector contolled induction motor operating at maximum efficiency for EV. In *Proceedings of EVS 13*. EVS.

Mc Cann, R., & Domagatla, S. (2005). Analyzes of MEMS based rotor flux sensi g in a hybrid reluctance motor. In *Proceedings of Vehicle Power and Propulsion Conference VPPC*. Chicago, IL: IEEE.

Ortmeyer, T. (2005). Variable voltage variable frequency options for series HV. In *Proceedings of Vehicle Power and Propulsion Conference VPPC*. Chicago, IL: IEEE.

Schofield, M., Mellor, P. H., & Howe, D. (1002). Field weakening of brushless PM motors for application in a hybrid electric vehicle. In *Proceedings of EVS 11*. Florence, Italy: EVS.

Szumanowski, A., & Bramson, E. (1992). Electric vehicles drive control in constant power mode. In *Proceedings of ISATA*. ISATA.

Szumanowski, A., & Brusaglino, G. (1992). Analyses of the hybrid drive consisting of electrochemical battery and flywheel. In *Proceedings of EVS 11*. EVS.

Veinger, A. H. (1985). *Adjustable synchronous drive system*. Moscow: Energoatimizdat.

Yamura, H. (1992). Development of powertrain system for Nissan FEV. In *Proceedings of EVS11*. EVS.

Chapter 4
Generic Models of Electric Machine Applications in Hybrid Electric Vehicles Power Train Simulations

ABSTRACT

Chapter 4 presents an approach to obtain the power simulation model of electric machines that would be practically useful in hybrid power train simulation studies. The induction motor (AC) and the permanent magnet motor's (PM) mathematical dynamic models are based on the necessary and fundamental knowledge conveyed in the previous chapter. These generic models are here adapted to the hybrid power train requirements, while the mechanical characteristics of the vehicle's driving system are relegated to the background. The vector field oriented control of induction and permanent magnet motors is applied in the conducted mathematical modeling. The influence of the controlled voltage frequency is discussed as well. In the case of permanent magnet motors, the adjusted method of magnetic field weakening is very important during pulse modulation (PWM) control. The chapter presents the model of synchronous permanent motor magnetic field weakening. The basic simulation studies' results dedicated especially to the above-mentioned electric motors are included. One of the targets of these simulations is the determination of these electric machines' static characteristics (motor's map) as the function: output mechanical torque versus the motors' shaft rotational speed. This feature is indicated as the map of electric machines connected with its efficiency in a four quarterly operation (4Q), which means the operation of the motor/generator mode in two directions of the shaft rotational speed, which appears very useful in practice.

DOI: 10.4018/978-1-4666-4042-9.ch004

1. APPROACH TO A POWER SIMULATION MODEL OF A DRIVING SYSTEM WITH AN AC INDUCTION MOTOR

1.1. Inverter/Motor Control Strategy

Before presenting the relationships between motor performance and motor design parameters, a description of the inverter/ motor control strategy is presented. This description is the fundamental basis for most thoughts involved in the selection and design of the motor for the particular application of the motor in the traction drive system.

The inverter/motor control strategy for maximum motor output is shown in Figure 1.

The motor torque-speed curve has the expected profile for a wide-speed range traction drive. Three regions of operation characterize this profile: constant torque, constant power and slip limited.

The constant torque region lies from a standstill point to the corner point speed (base speed). In this region, the inverter operates in the PWM mode to supply the adjustable voltage and frequency of the motor.

The alternative current AC voltage is adjusted as speed (frequency) changes to maintain constant flux density in the motor, as shown in Figure 1. The alternative current voltage, therefore, basically increases proportionally with speed (frequency). The frequency of the voltage inducted in the rotor (sleep frequency) is held constant and the AC current is nearly constant. This produces a nearly constant torque. Of

Figure 1. The operational strategy of the traction motor controlled by an inverter, according to the pulse-width-modulation method

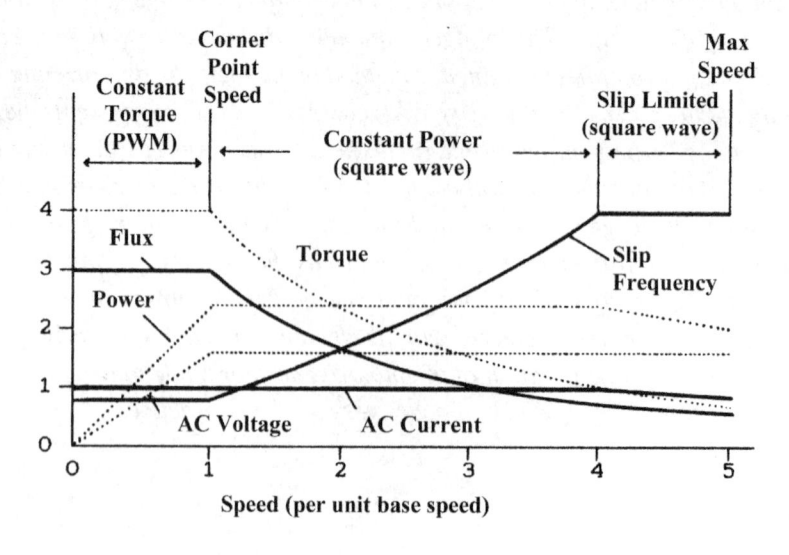

Generic Models of Electric Machine Applications in Hybrid Electric Vehicles

course, the output power of the motor is proportional to speed, so power increases linearly, with speed up to the corner point speed. This method of control is possible until the AC voltage reaches the maximum available from the inverter. This juncture is the corner point and is the end of the constant torque region and the beginning of the constant power region. In this case, the behavior of the alternative current AC motor is the same as a separately excited, direct current DC motor.

The constant power region lies from the corner point speed to maximum speed (speed at the end of the constant power region).

In this region, the inverter supplies adjustable frequency to the motor. The alternative current AC voltage is no longer adjustable, as the inverter is producing the maximum voltage, so AC voltage is constant. The results in the flux density decrease as the reciprocal of speed (frequency). The slip frequency is increased and alternative current voltage (AC) is nearly constant. This produces a torque, which decreases as the reciprocal of speed, while power out of the motor is nearly constant. This method of control is possible until the motor reaches the breakdown at its slip limit. The important juncture is the maximum speed or the end of the constant power region and the beginning of the slip limited region, which lies behind the constant power region. Any further increases in speed must be carried out at a constant slip frequency, with alternative (AC) current decreases. The voltage and the flux density behave as in the constant power region. This results in the torque, which decreases as the reciprocal of speed squared, and a resulting decrease in power.

1.2. Effect of Transmission

The torque can be transmitted to the wheels via the gearbox with two or more speed transmissions. In time, the gears shift from a low gear (high transmission ratio) to a high gear (low transmission ratio), and the motor will operate more easily in the constant power range. The drive in the city traffic cycle up to a maximum speed, for example 70 km/h (or more), is only possible using one basic gear step ratio. The best solution is permanent use of CVT (Continuously Variable Transmission) between the motor and traction wheels. This way is costly, but it offers the possibility of minimizing the motor copper losses, and better adjusts the motor parameters to vehicle driving requirements.

1.3. The AC Alternative Current Induction Motor for Hybrid Power Train Application Modeling

For practical reasons, it is better to present the previously depicted AC Induction Motor modeling in a more useful form. It means that for a hybrid vehicle drive application, where an AC Induction Motor is controlled, it is necessary to transform instantaneous phase currents, voltages or flux linkages to the complex plane (α_x, β_x), rotating with arbitrary angular velocity ω_{ex}. In this case, angles between the axis α_x and the axis of three-phase windings A, B, C is presented as follows:

$$\alpha_{\alpha_x A} = \alpha_{\alpha_x A0} + \int_0^t (\omega_x - \omega)dt$$

$$\alpha_{\alpha_x B} = \alpha_{\alpha_x A0} + \frac{4\pi}{3} + \int_0^t (\omega_x - \omega)dt$$

$$\alpha_{\alpha_x C} = \alpha_{\alpha_x A0} + \frac{2\pi}{3} + \int_0^t (\omega_x - \omega)dt \qquad (1)$$

where:

- $\alpha_{\alpha x A0}$: Angle between axis α_x and A for $t=0$;
- ω_x: Angular velocity of the complex plane (α_x, β_x) – most often, this is the synchronous angular velocity referring to the rotating magnetic field of stator, depending on supplying voltage frequency;
- ω: Angular velocity of three-phase windings A, B, C (for stator $\omega=0$, for rotor ω means the rotor's electrical velocity $\omega=p\omega_m$ (ω_m –mechanical rotor's velocity).

Then, instantaneous phase values are projected on the axis α_x (multiplied by cosines of the correctly-mentioned above angles) and on the axis β_x (multiplied by the –sinus of the proper angle).

Finally α_x and β_x components (for currents) are as follows:

$$\begin{bmatrix} I_{\alpha_x} \\ I_{\beta_x} \end{bmatrix} = \frac{2}{3} \cdot \begin{bmatrix} \cos\alpha_{\alpha_x A} & \cos\alpha_{\alpha_x B} & \cos\alpha_{\alpha_x C} \\ -\sin\alpha_{\alpha_x A} & -\sin\alpha_{\alpha_x B} & -\sin\alpha_{\alpha_x C} \end{bmatrix} \cdot \begin{bmatrix} i_A(t) \\ i_B(t) \\ i_C(t) \end{bmatrix} \qquad (2)$$

and reverse transformation for $i_a+i_b+i_c=0$:

$$\begin{bmatrix} i_A(t) \\ i_B(t) \\ i_C(t) \end{bmatrix} = \begin{bmatrix} \cos\alpha_{\alpha_x A} & -\sin\alpha_{\alpha_x A} \\ \cos\alpha_{\alpha_x B} & -\sin\alpha_{\alpha_x B} \\ \cos\alpha_{\alpha_x C} & -\sin\alpha_{\alpha_x C} \end{bmatrix} \cdot \begin{bmatrix} I_{\alpha_x} \\ I_{\beta_x} \end{bmatrix} \qquad (3)$$

Generic Models of Electric Machine Applications in Hybrid Electric Vehicles

In the same way, the transformation can be done for voltages and flux linkages. The transformations may also be used in the case of the Permanent Magnet (PM) Synchronous Motor.

The phasors Equations (45), (46), (48) from Chapter 3 can be practically used for numerous simulations, but at first, they should be written in scalar form:

$$\begin{aligned}
u_{1d} &= R_1 i_{1d} + \frac{d\psi_{1d}}{dt} - \omega_0 \psi_{1q} \\
u_{1q} &= R_1 i_{1q} + \frac{d\psi_{1q}}{dt} + \omega_0 \psi_{1d} \\
u_{2d} &= R_2 i_{2d} + \frac{d\psi_{2d}}{dt} - (\omega_0 - \omega) \psi_{2q} \\
u_{2q} &= R_2 i_{2q} + \frac{d\psi_{2q}}{dt} + (\omega_0 - \omega) \psi_{2d}
\end{aligned} \quad (4)$$

For the squirrel cage rotor, there exists the following condition; $u_{2d}=u_{2q}=0$.

$$\begin{aligned}
\psi_{1d} &= L_1 i_{1d} + L_{12m} i_{2d} \\
\psi_{1q} &= L_1 i_{1q} + L_{12m} i_{2q} \\
\psi_{2d} &= L_2 i_{2d} + L_{12m} i_{1d} \\
\psi_{2q} &= L_2 i_{2q} + L_{12m} i_{1q}
\end{aligned}$$

$$M_{em} = (3p/2)(\psi_{1d} i_{1q} - \psi_{1q} i_{1d})$$

$$J \frac{d\omega}{dt} = M_{em} - M_o \text{ - for motoring mode} \quad (5)$$

$$J \frac{d\omega}{dt} = M_0 + M_{em} \text{ - for generating mode}$$

$$P = (3/2)(u_{1d} i_{1d} + u_{1q} i_{1q})$$

where

99

$$\overline{U} = u_{1d} + ju_{1q} \text{ etc.}$$

$$L_1 = \frac{3}{2}L_m$$

$$L_2 = \frac{3}{2}L_m$$

$$L_{12m} = \frac{3}{2} \cdot \frac{z_2}{z_1}(L_m - L_{1d})$$

where:

- R_1: Resistance of single winding of stator;
- R_2: Resistance of single winding of rotor;
- L_1: Inductance of single winding of stator;
- L_2: Inductance of single winding of rotor;
- L_m: Mutual inductance;
- M_{em}: Electromagnetic torque;
- M_o: External load torque;
- J: Moment of inertia reduced on the rotor shaft;
- Z_1: Number of stator windings;
- Z_2: Number of rotor windings.

This is the complete set of differential equations, which allow identifying the state of the dynamic balance, and should be solved by proper integral procedure, for example Runge-Kutt's method. All variables in the mentioned equations are in the form of the amplitudes of appropriate values, and may be shown in rms-form also, which can be achieved by dividing the amplitude values by $\sqrt{2}$. The equations are solved in a (d, q) rotating complex plane and do not show the real, sinusoidal functions. The maximum or rms-values are obtained as the computing results, as the function of time. Using rms-values for calculation gives us the possibility to conduct energetic calculations, with a special emphasis on determination of the motor torque, power and watt efficiency, which is very important for proper adjustment of hybrid drive structures consisting of many components. This is very useful, because the equations are easier to solve, and the obtained functions give clear information about the dynamic processes during the vehicle drive time, in which case, the stator and rotor currents are as follows:

$$\begin{aligned} i_1 &= \sqrt{i_{1d}^2 + i_{1q}^2} = I_1\sqrt{2} \\ i_2 &= \sqrt{i_{2d}^2 + i_{2q}^2} = I_2\sqrt{2} \end{aligned} \quad (6)$$

Generic Models of Electric Machine Applications in Hybrid Electric Vehicles

where: I_1, I_2 – rms-value of stator and rotor currents.

If necessary, all variables can be presented in real, sinusoidal form. For example, the current on phase A of the stator is:

$$i_A(t) = \sqrt{i_{1d}^2 + i_{1q}^2} \cos(\omega_0 t - \varphi) \tag{7}$$

On phase B:

$$i_B(t) = \sqrt{i_{1d}^2 + i_{1q}^2} \cos\left(\omega_0 t + \frac{2\pi}{3} - \varphi\right) \tag{8}$$

and on phase C:

$$i_C(t) = \sqrt{i_{1d}^2 + i_{1q}^2} \cos\left(\omega_0 t + \frac{4\pi}{3} - \varphi\right) \tag{9}$$

If the induction AC motor is operating in generating mode, e.g. during vehicle regenerative braking, in the above-mentioned equations, it is important to notice that angle ϕ is more than $\pi/2$ and $\cos\phi$ changes its sign. It means the phasor of the current is in a different quarter in a coordinate system. Voltages and fluxes can be determined in this same way.

The exemplary, asynchronous, alternative, current induction motor voltage, stator and rotor currents, synchronous rotary flux's velocity and rotor's angular velocity alterations, according to Equation (5) for motoring and generator braking modes, are shown in Figure 2.

Taking into consideration the Equations (4), (5), and (6) the exemplary asynchronous current induction motor test is shown in Figure 3.

1.4. Vector-Field Oriented Control of AC: The Alternative Current Induction Motor

The phasor of the stator current is described by Equation (24) in Chapter 3 or in Euler form by equation (3.27). Figure 4 shows the stator phasor i_1 rotated in the referred to stator and immovable α, β axis with angular speed ω_0 respectively, to feed voltage frequency. Phasor i_1 is projected on an immovable axis in a Cartesian coordinates system α, β connected with motor stator (e.g. α with winding of phase A) and also projected on rotated axis d, q stiffened, connected with the rotor windings (e.g. d with a rotor phase).

Figure 2. Exemplary test of traction AC motor for acceleration steady speed and braking of the vehicle for data: motors' shaft moment of inertia J – 8 kgm²; motor's shaft output torque M_o – 25 Nm (Equation (5))

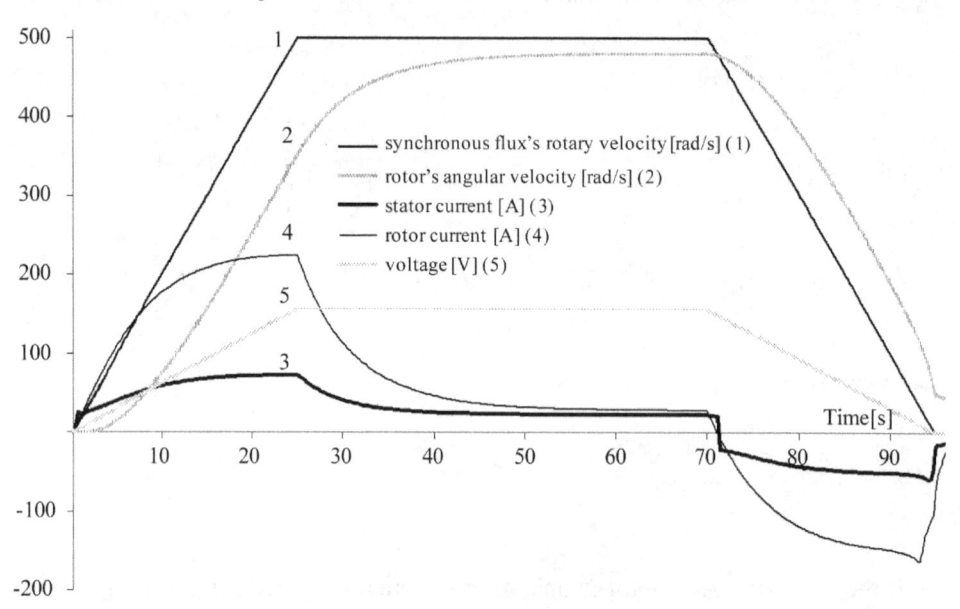

Figure 3. The exemplary simulation test of the AC alternative current induction motor for Equations (5) – (9)

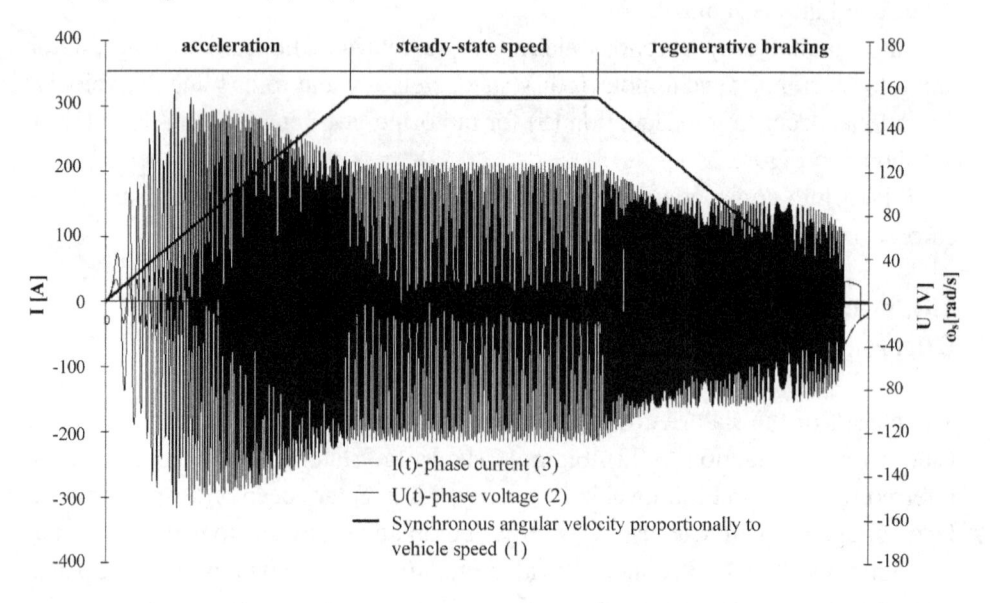

Figure 4. The illustration of vector-field oriented transformation

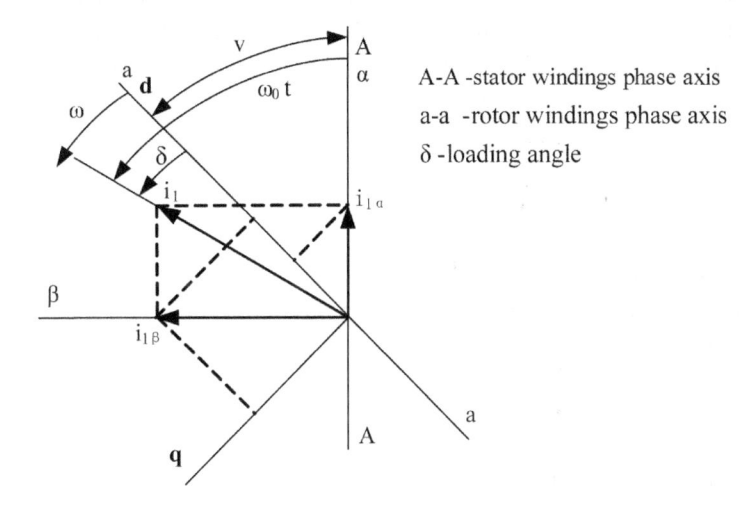

A-A - stator windings phase axis
a-a - rotor windings phase axis
δ - loading angle

$$i_{1\alpha} = i_A; \quad i_{1\beta} = \frac{2}{\sqrt{3}}\left(\frac{1}{2}i_A + i_B\right) \qquad (10)$$

$$\bar{i}_1 = i_1 e^{-j\omega_s t} \qquad (11)$$

where: $i_A(t)$, $i_B(t)$ - respectively instantaneous phases currents.
For immovable α, β axis

$$\bar{i}_1 = i_{1\alpha} + ji_{1\beta} \qquad (12)$$

According to Figure 4:

$$\left.\begin{array}{l} i_{1\alpha} = i_1 \cos\omega_0 t \rightarrow i_{1\alpha} = |i_1|\cos\omega_0 t \\ i_{1\beta} = i_1 \sin\omega_0 t \rightarrow i_{1\beta} = |i_1|\sin\omega_0 t \end{array}\right\} \qquad (13)$$

Generally,

$$\left.\begin{array}{l} i_{1d} = |i_1|\cos\delta \\ i_{1q} = |i_1|\sin\delta \end{array}\right. \qquad (14)$$

where: δ - is the loading angle.

The component of current i_{1d} refers to the flux, and in this case, is the magnetizing current (field current); i_{1q} component refers to load current (torque current). The relationship between the mentioned current components could be described by the general electromagnetic torque equation:

$$M_{em} = c i_d i_q \tag{15}$$

where: c - constant of the analyzing motor.

For further transformation, it is necessary to write the Equation (11) in the following form:

$$\begin{aligned} i_{1d} &= i_{1\alpha} \cos \nu + i_{1\beta} \sin \nu \\ i_{1q} &= -i_{1\alpha} \sin \nu + i_{1\beta} \cos \nu \end{aligned} \tag{16}$$

Or, in another way:

$$\begin{Bmatrix} i_{1d} \\ i_{1q} \end{Bmatrix} = \begin{bmatrix} \cos \nu & \sin \nu \\ -\sin \nu & \cos \nu \end{bmatrix} \begin{bmatrix} i_{1\alpha} \\ i_{1\beta} \end{bmatrix} \tag{17}$$

The reverse transformation can be obtained as:

$$\begin{aligned} i_{1\alpha} &= i_{1d} \cos \nu - i_{1q} \sin \nu \\ i_{1\beta} &= i_{1d} \sin \nu + i_{1q} \cos \nu \end{aligned} \tag{18}$$

Considering Equations (11) and (12) in Chapter 3, it is possible to obtain

$$\begin{aligned} i_{1d} &= i_1 (\cos \omega_0 t \cos \nu + \sin \omega_0 t \sin \nu) \\ i_{1q} &= -i_1 (\cos \omega_0 t \sin \nu - \sin \omega_0 t \cos \nu) \end{aligned} \tag{19}$$

and

$$\begin{aligned} i_{1d} &= i_1 \cos(\omega_0 t - \nu) \\ i_{1q} &= i_1 \sin(\omega_0 t - \nu) \end{aligned} \tag{20}$$

The coordinate systems for determination of current components are called: for i_α, i_β – the constant-current, for i_d, i_q – the variable-current.

Generic Models of Electric Machine Applications in Hybrid Electric Vehicles

Figure 5. The block diagram of the AC alternative current induction motor

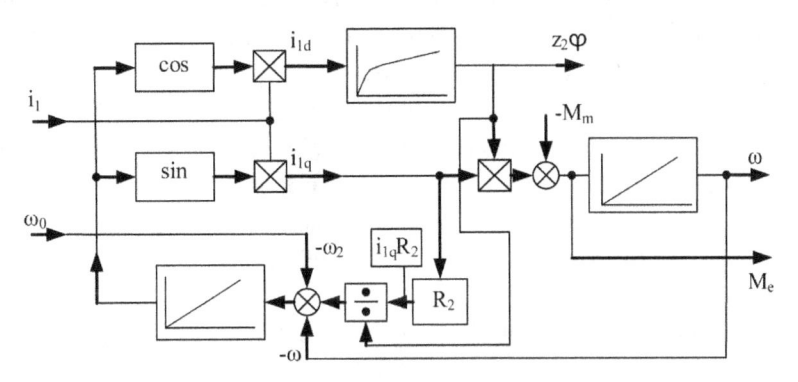

- z_2: Number of rotor windings;
- ϕ: The main flux;
- ω_2: Related to slip frequency (rotor);
- ω_0: Rotational speed of stator flux (stator frequency);
- ω: Angular speed of rotor shaft;
- M_m: Mechanical load torque;
- M_e: Electromagnetic torque;
- R_2: Rotor windings resistance;
- $i_{1q} = -I_2$: Rotor current

The passing from a two-phase to a three-phase system is as follows:

$$i_{1\alpha} = i_A \quad ; \quad i_{1\beta} = \frac{2}{\sqrt{3}}\left(\frac{1}{2}i_A + i_B\right)$$

In the opposite direction:

$$i_B = -\frac{1}{2}i_{1\alpha} + \frac{\sqrt{3}}{2}i_{1\beta}; i_C = -\frac{1}{2}i_{1\alpha} - \frac{\sqrt{3}}{2}i_{1\beta}.$$

The electromagnetic torque could be expressed in another form:

$$M = c|z_2\varphi||i_1|\sin\delta \tag{21}$$

The best operation of the AC motor is to slip, δ load angle and cos ϕ (ϕ – angle between stator voltage and current) – constants.

Generic Models of Electric Machine Applications in Hybrid Electric Vehicles

For instance, the IGBT-based inverter circuit is shown in Figure 6. The high-speed IGBT transistors and vector-field oriented, specially adopted, control system is the reason of the very efficient drive. The Automatic Speed Regulator (ASR), using the encoder sensor Automatic Current Regulator (ACR) and Pulse Width Modulation (PWM) method, provides the possibility to obtain high efficiency of the drive.

The vector control block diagram is shown in Figure 7. The concept of proper control of the alternative current AC induction motor is to minimize the motor current by calculating the torque current i_{1q} and field current i_{1d}, which are the vector components of motor current i_1.

The vector of current is controlled, so as to set the angle of torque current and field current perpendicular, as illustrated in Figure 8.

Figure 6. The inverter circuit diagram adapted to the pulse-width-modulation motor control method

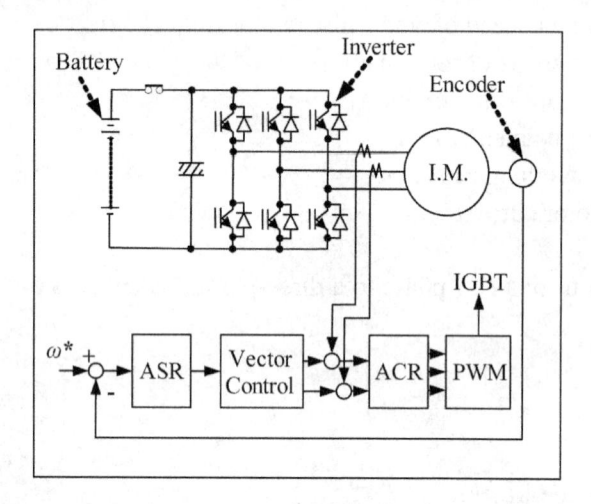

Figure 7. The vector control block diagram

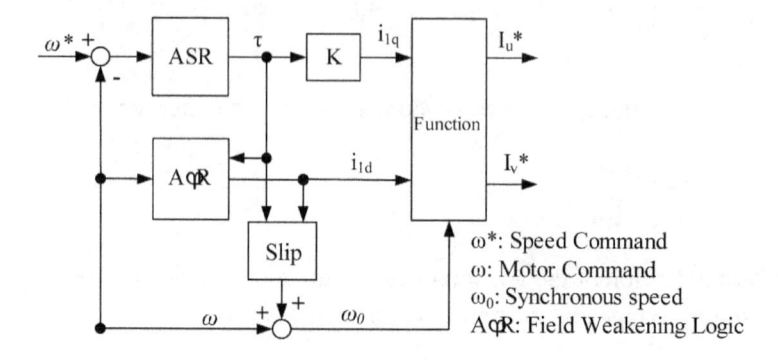

ω^*: Speed Command
ω: Motor Command
ω_0: Synchronous speed
AφR: Field Weakening Logic

Generic Models of Electric Machine Applications in Hybrid Electric Vehicles

Figure 8. The vector decomposition of motor stator current

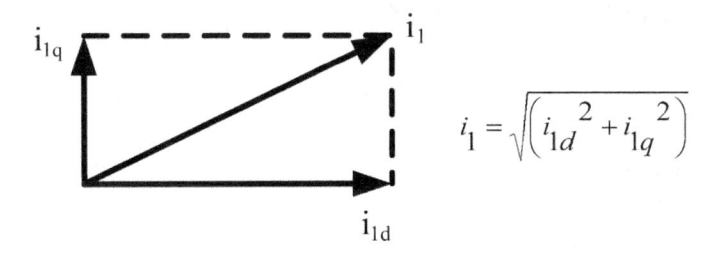

$$i_1 = \sqrt{\left(i_{1d}^2 + i_{1q}^2\right)}$$

The current commands such as I_u^* and I_v^* (Figure 7) are a sinusoidal wave, at whose base is a calculation of functions of i_{1d}, i_{1q} and primary frequency. The primary frequency is calculated by adding slip frequency, with motor revolution frequency.

The field current i_{1d} is obtained by calculating the field weakening logic of torque and speed. The torque current i_{1q} is also calculated by torque, through the automatic speed regulator, ASR. The motor current i_1 wave form in the switching frequency 10 kHz at 386 Hz operation is shown in Figure 9.

The motor current is very close to the sinusoidal current command: that is, within target and this reduces harmonic losses which increases efficiency. The braking current is controlled, so as to regenerate the vehicle kinetic energy to the battery.

Figure 9. The speed change step command and battery current, motor speed and current response

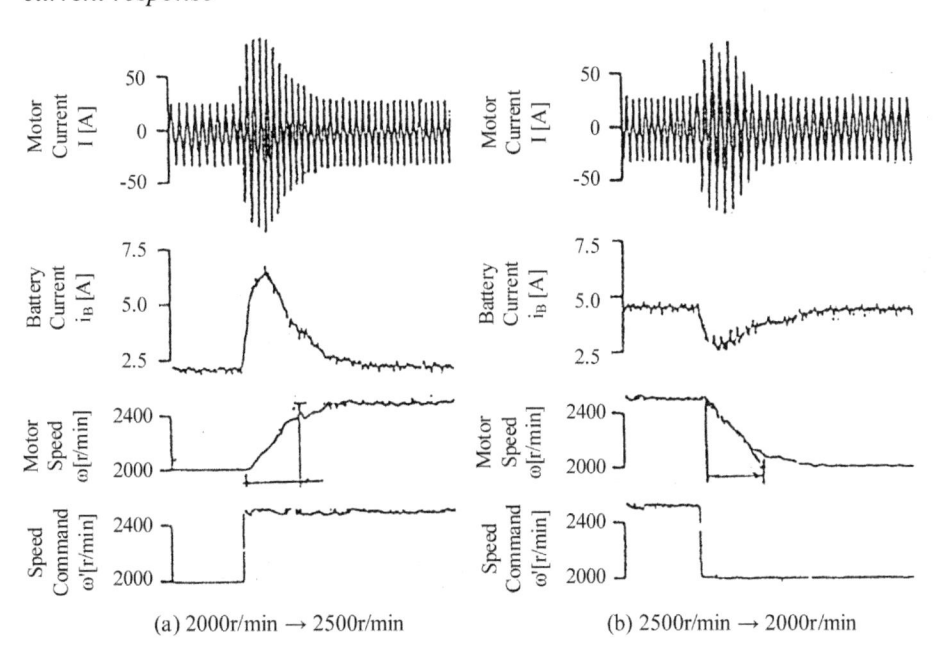

(a) 2000r/min → 2500r/min

(b) 2500r/min → 2000r/min

1.5. Influence of Frequency on the Asynchronous Current Induction Motor Field Weakening

Mechanical characteristics of the vehicle's drive system must be similar to this, which is shown in Figure 10.

The characteristic has two zones. In the first zone, the value of the torque is constant and the value of the motion's power linearly reinforces with the rotary speed. In the second zone, the motion's power is constant and the torque hyperbolically decreases with the reinforcing rotary speed. In the first zone, the rotary speed is small (from 0 to ω_b): and at this stage, vehicles usually speed up – and the driving system has to overcome the resistance of inertia. In the second zone (between ω_b to ω_{max}), the motion is more uniform - there are no major accelerations - the torque can be smaller and adequate only to driving resistances. Such a mechanical characteristic of the driving system is achieved in various ways.

In the case of the driving system with the asynchronous motor characteristic, it is achieved by proper control of voltage and frequency.

It is known that the torque of the asynchronous motor is:

$$M = \frac{3}{2} \cdot p \cdot \left(\overline{\Psi}_1 \times \overline{I}_1 \right) \tag{22}$$

where:

- ***M*:** Mechanical torque of motor;

Figure 10. The traction motor torque controlled by the pulse-width-modulation method (PWM)

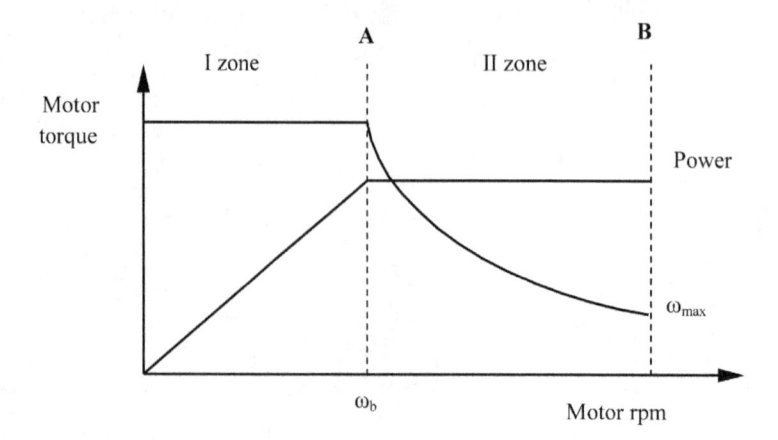

Generic Models of Electric Machine Applications in Hybrid Electric Vehicles

- p: Number of pole's pair;
- Ψ_1: Magnetic flux of stator;
- I_1: Current of stator.

The magnetic flux of stator is:

$$\Psi_1 = L_1 I_1 + L_{12m} I_2 \tag{23}$$

where:

- L_1: Self-inductance of stator;
- L_{12m}: Mutual inductance between stator and rotor

The voltage at the ends of the windings is:

$$U_1 = R_1 I_1 + L_1 I_1 \omega_s + L_1 \frac{dI_1}{dt} \tag{24}$$

where:

- U_1: Voltage between ends of the windings;
- R_1: Resistance of the winding;
- ω_s: Synchronous rotary speed.

For steady-state:

$$L_1 \frac{dI_1}{dt} = 0 \tag{25}$$

Equation (23) has the form:

$$U_1 = R_1 I_1 + L_1 I_1 \omega_s \tag{26}$$

In the other form:

$$U_1 = (R_1 + \omega_s L_1) \cdot I_1 \tag{27}$$

In such conditions, the Equation (25) in Chapter 3, may be written like this:

$$M = \frac{3}{2} \cdot p \cdot \left[\left(L_1 \bar{I}_1 + L_{12m} \bar{I}_2 \right) \times \bar{I}_1 \right] \qquad (28)$$

or:

$$M = \frac{3}{2} \cdot p \cdot \left[\left(L_1 \bar{I}_1 + L_{12m} \bar{I}_2 \right) \times \frac{\bar{U}_1}{\left(R_1 + \omega_s L_1 \right)} \right] \qquad (29)$$

Equation (28) is difficult to analyze, so a few simplifications are made:

- Vectors are analyzed like scalars,
- The vector's product is analyzed like a scalar's product,
- The element $L_{12m} I_2$ is neglected,
- Resistance R_1 is neglected because $\omega L_1 \gg R_1$.

In this way, the Equation (28) has a form as follows:

$$M \cong \frac{3}{2} \cdot p \cdot L_1 \cdot \frac{U_1^2}{\omega_s^2 L_1^2} = \frac{3}{2} \cdot p \cdot \frac{1}{L_1} \cdot \frac{1}{(2\Pi)^2} \cdot \frac{U_1^2}{f^2} \qquad (30)$$

and finally:

$$M = c \cdot \left(\frac{U_1}{f} \right)^2 \qquad (31)$$

In the first zone, see Figure 10 of control, there is the condition that M=*const*, which is possible to obtain for flux Ψ_1 and I_1 equal constants, so the Equation (30) takes the form:

$$M = const \Rightarrow c \cdot \left(\frac{U_1}{f} \right)^2 = const \Rightarrow$$
$$\Rightarrow \frac{U_1}{f} = const \qquad (32)$$

Generic Models of Electric Machine Applications in Hybrid Electric Vehicles

In the second zone of control, there is the condition that mechanical power N=*const* Equation (31) takes the form:

$$N = const \Rightarrow M \cdot \omega = const \Rightarrow M \cdot \omega_s \cdot (1-s)$$
$$= const \Rightarrow M \cdot 2\Pi f \cdot (1-s) = const \Rightarrow$$
$$\Rightarrow c \cdot \left(\frac{U_1}{f}\right)^2 \cdot 2\Pi f \cdot (1-s) \quad (33)$$
$$= const \Rightarrow c_1 \cdot (1-s) \cdot \frac{U_1^2}{f} = const$$

because in steady-state (1-s)=*const*, then:

$$\frac{U_1^2}{f} = const \quad (34)$$

Finally, to achieve a proper mechanical characteristic with two zones (constant torque and constant power) there is needed a proper form of control, as shown in Figure 11.

Sometimes, it may be difficult or even impossible to increase the voltage value in the second zone, because the top voltage is limited.

It is easy to prove that:

Figure 11. The strategy of frequency and voltage control in the case of an asynchronous, alternative, current, induction motor

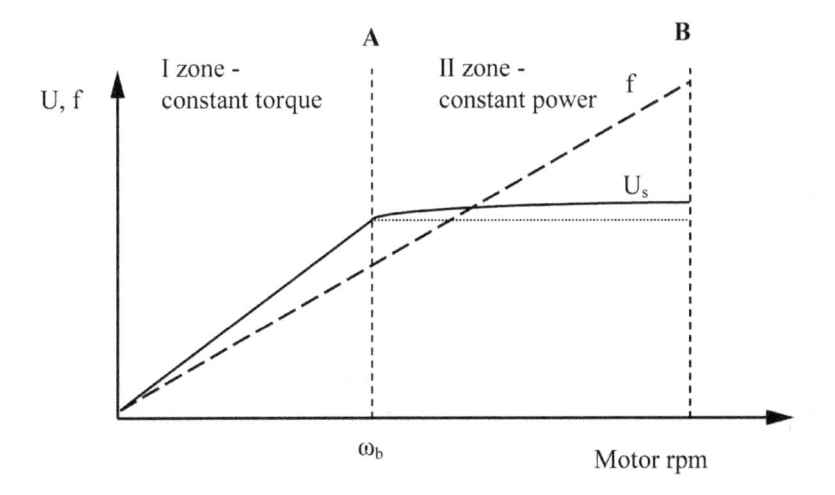

$$\frac{dU_1}{df} = \frac{1}{2} \cdot (f \cdot const)^{-\frac{1}{2}} \tag{35}$$

and

$$\lim_{f \to \infty} \frac{dU_1}{df} = 0 \tag{36}$$

Equations (35) and (36) show that the voltage's function in the second zone can be simplified to linear form:

$$U = const = U_{max} \tag{37}$$

where: U_{max} - is the top voltage's value that can be obtained from the frequency inverter.

This simplification (35) ensures that power in the second zone is not exactly constant - it slowly decreases for the highest rotary speed.

Figure 12 shows the stable parts of some characteristics for various feeding conditions of an asynchronous motor with the following parameters:

L_s = 0.0204 H

L_r = 0.0025 H

L_μ = 0.0069 H

R_s = 0.015 Ω

R_r = 0.015 Ω

$p = 2$

1.6. Exemplary Simulation Study of an AC Motor

The suggested block diagram of AC, the asynchronous current induction motor drive for computer simulation purposes is shown in Figure 13.where:

- M_k, ω_k, P_k: Torque, rotary speed and power on wheels;
- M_m, ω_m, P_m: Torque, rotary speed and power on motor shaft;
- $M_m = M_k/k_g$, $\omega_s = \omega_k k_g$, k_g: Gearbox ratio;

Generic Models of Electric Machine Applications in Hybrid Electric Vehicles

Figure 12. The mechanical characteristics of an asynchronous, alternative, current, induction motor for two control strategies: motor constant voltage - current relation U/f (constant torque) and motor constant voltage U (constant power)

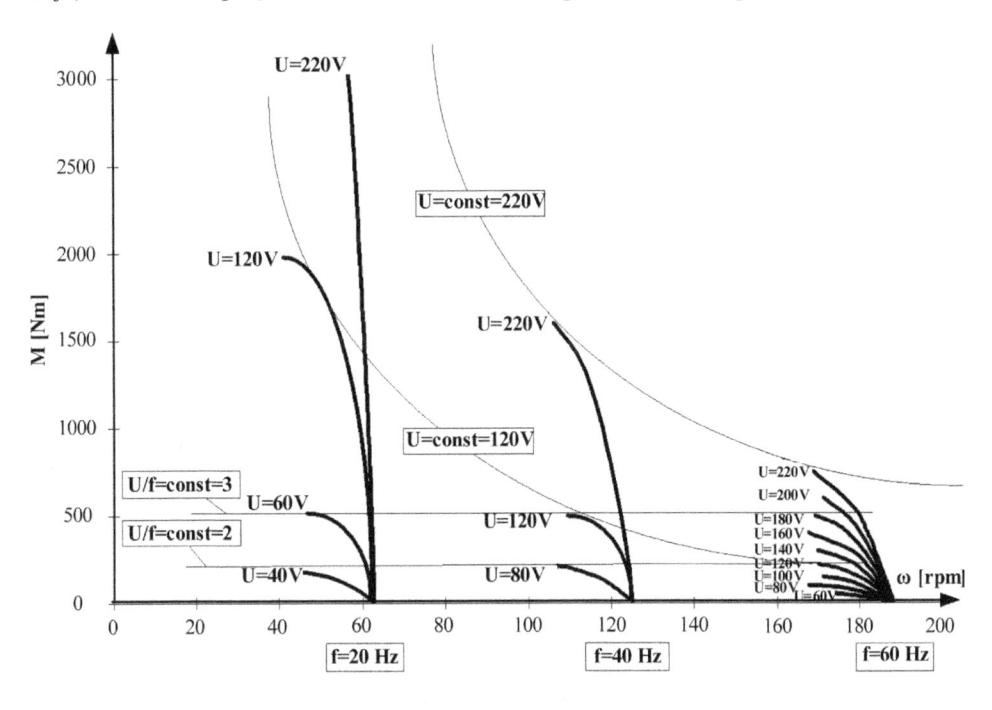

Figure 13. The block diagram of an AC asynchronous, alternative, current, induction motor drive for battery-powered electric vehicles (two-axis motor drive)

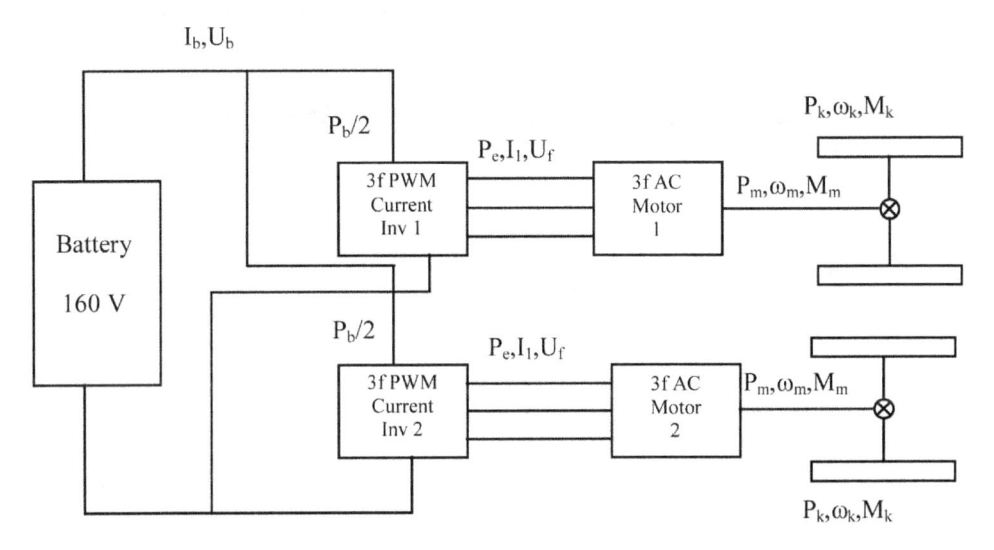

- **P_e=P_i**: Electric power of AC motor.

$$P_e = \mp P_m + "P_{Cu1} + "P_{Fe} + "P_{Cu2} + "P_d \mp "P_m$$

- $\Delta P_{Cu1} = 3R_s I_1$: Stator copper losses;
- ΔP_{Fe}: Iron losses;
- $\Delta P_{Cu2} = 3R_r I_2$: Rotor copper losses;
- $\Delta P_d = 0{,}005 P_s$: Stray load losses;
- ΔP_m: Mechanical losses $\left(M_{m0} + \lambda |\omega_m(t)|\right)\omega_m(t)$;
- $U_f = \dfrac{2P_e}{3I_1 \cos Æ}$: phase voltage of AC motor;
- I_1, I_2: Currents determined from Equation (6) in Chapter 3;
- M_{m0}: Fixed losses torque;
- λ: Coefficient connected with the losses, depending on motor speed.

$$P_b = \begin{cases} P_e \cdot \eta_i - \text{for braking} \\ \dfrac{P_e}{\eta_i} - \text{for motoring} \end{cases}$$

- η_i: Efficiency coefficient of current inverter

The power and SOC of battery are used as input data for calculating battery current I_b and battery voltage U_b, according to the battery model described in Chapter 5 (see Figures 14-21).

Simulation Study: Assumed Values of the Drive System Parameters

Battery
- **Nominal Capacity:** 7.616 kWh
- **Nominal (1 Hour) Current:** 47.6 A
- **Nominal Voltage:** 160 V
- **Approximate Mass:** 274 kg
- **Initial Coefficient of Battery Charge:** SOC1

Figure 14. The alternative, current, asynchronous, induction motor stator frequency versus the exemplary vehicle driving time

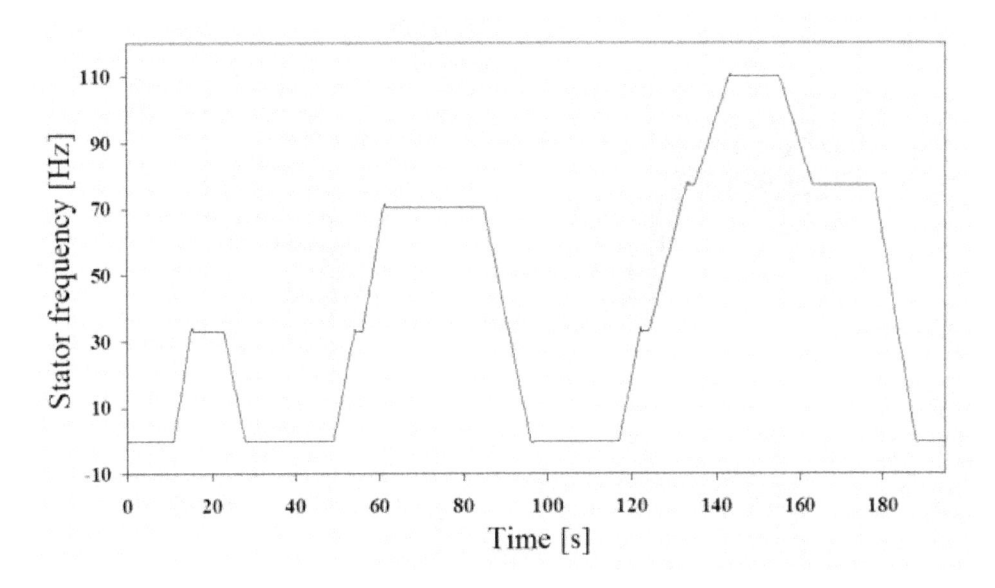

Figure 15. The alternative, current, asynchronous, induction motor output shaft velocity versus the exemplary vehicle driving time

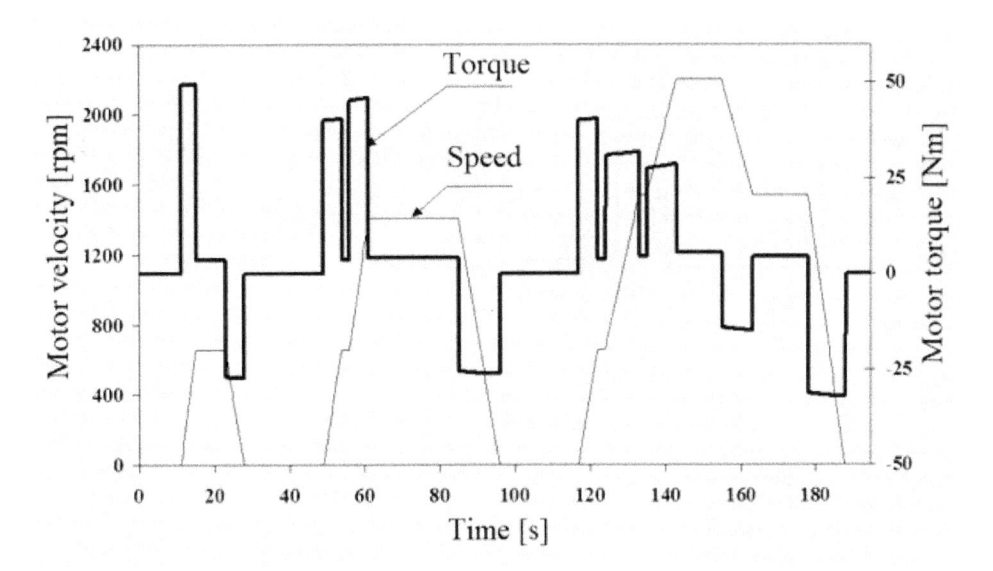

Generic Models of Electric Machine Applications in Hybrid Electric Vehicles

Figure 16. The analyzed power train; battery, and AC motor powers versus the exemplary vehicle driving time

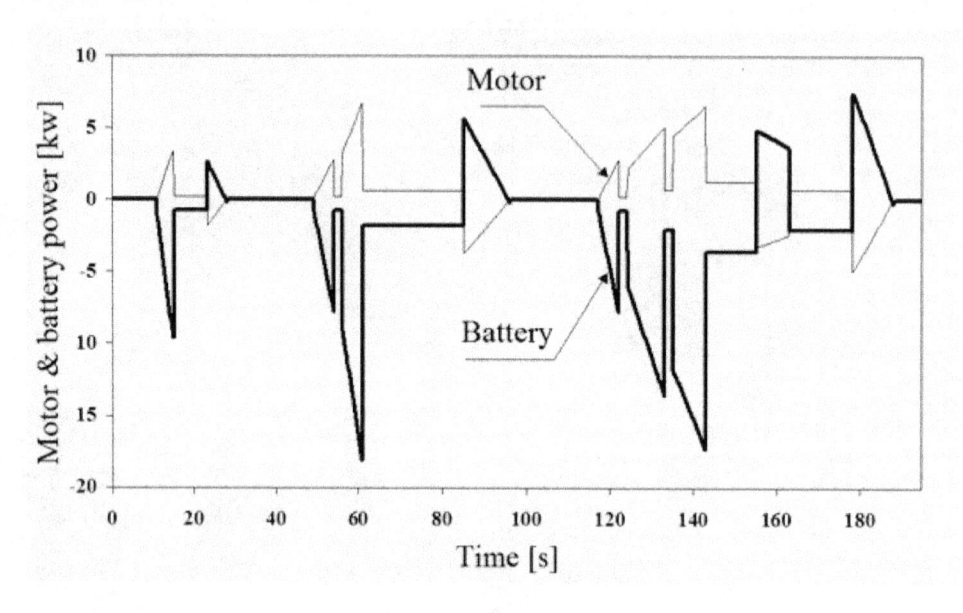

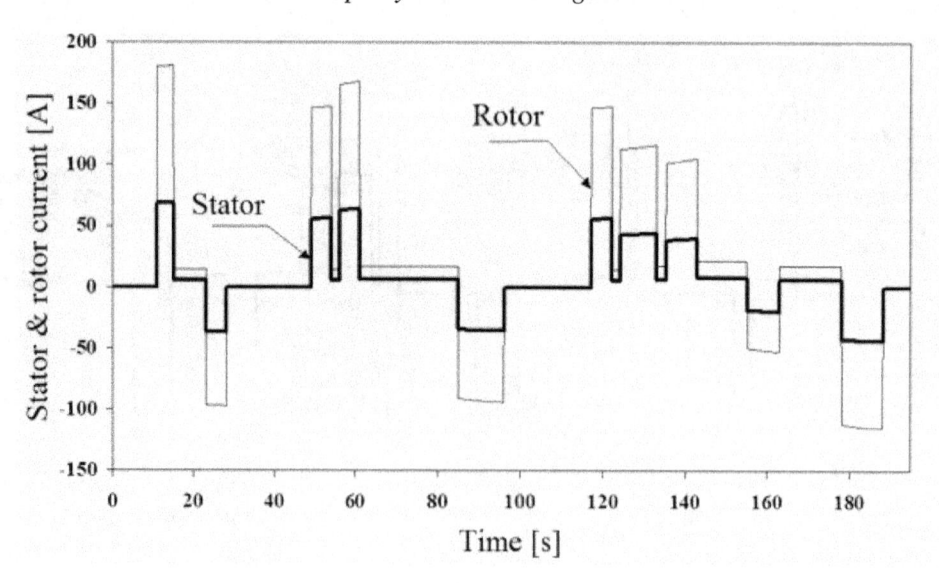

Figure 17. The alternative, current, asynchronous, induction motor's stator and rotor current versus the exemplary vehicle driving time

Generic Models of Electric Machine Applications in Hybrid Electric Vehicles

Figure 18. The alternative, current, asynchronous, induction motor current and battery current versus the exemplary vehicle driving time

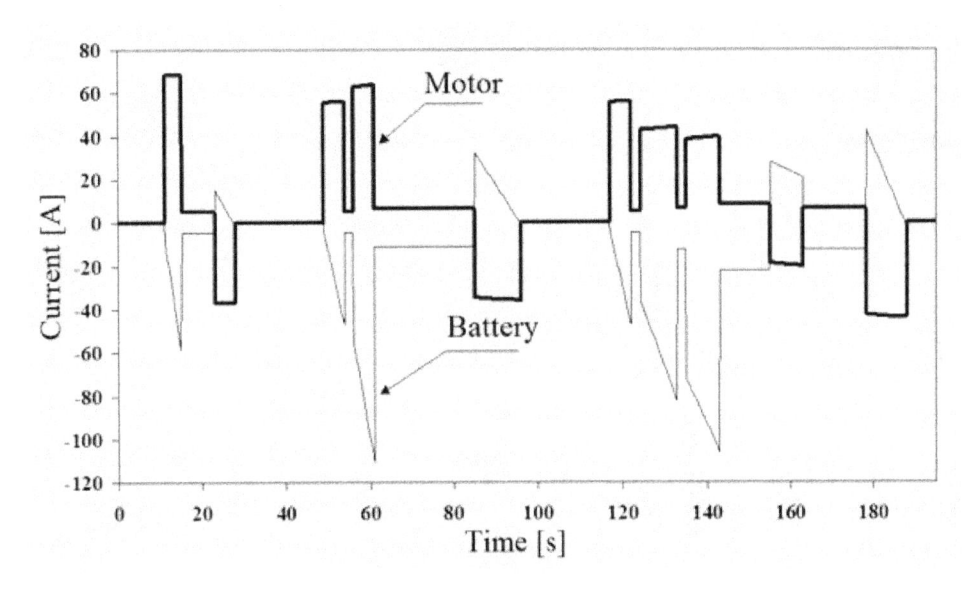

Figure 19. The battery's current and voltage versus the exemplary vehicle driving time

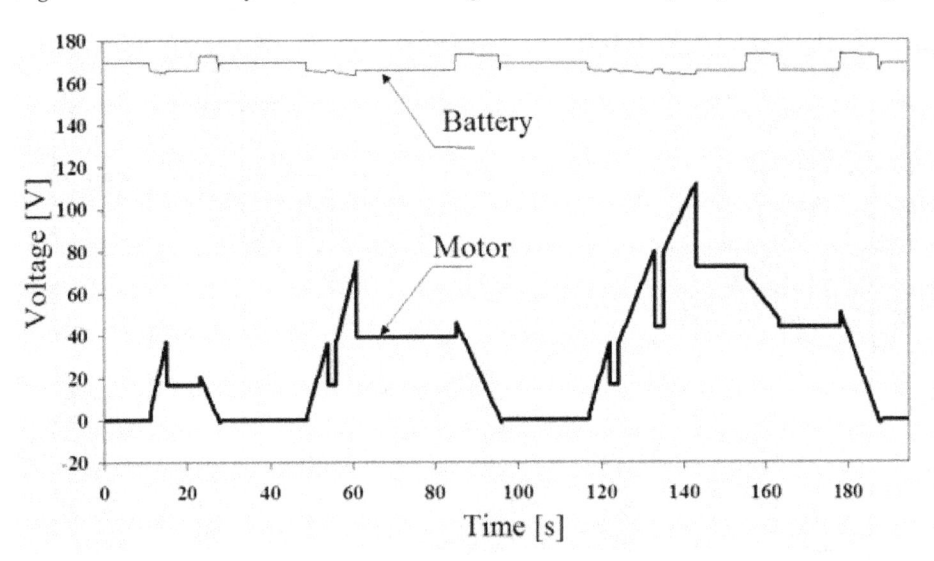

Figure 20. The entire power train's power efficiency versus the exemplary vehicle driving time

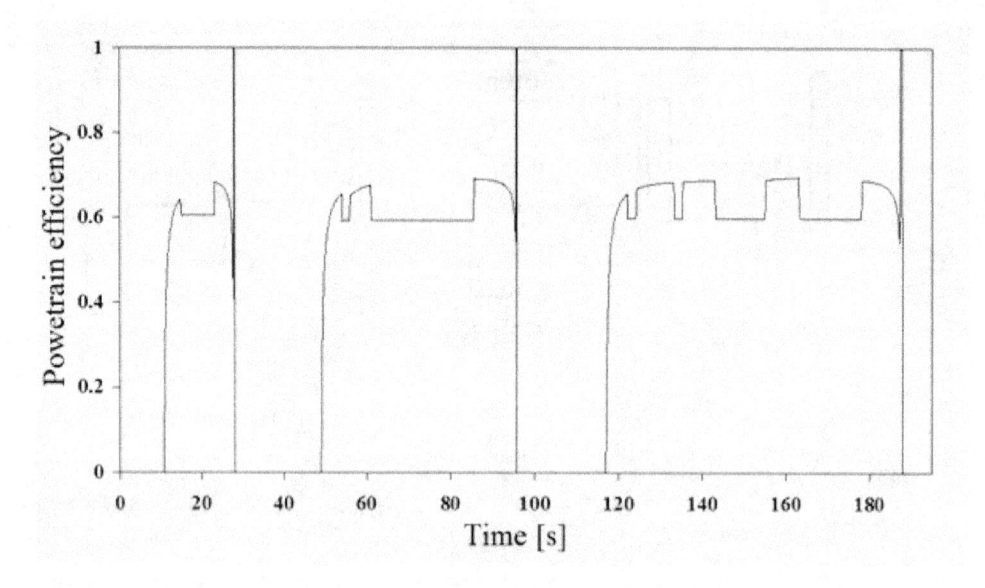

Figure 21. The exemplary map of an asynchronous, alternative, current, induction motor drive obtained from its mathematical model resolution, and identified during lab-bench tests. Isoclines depict its power efficiency.

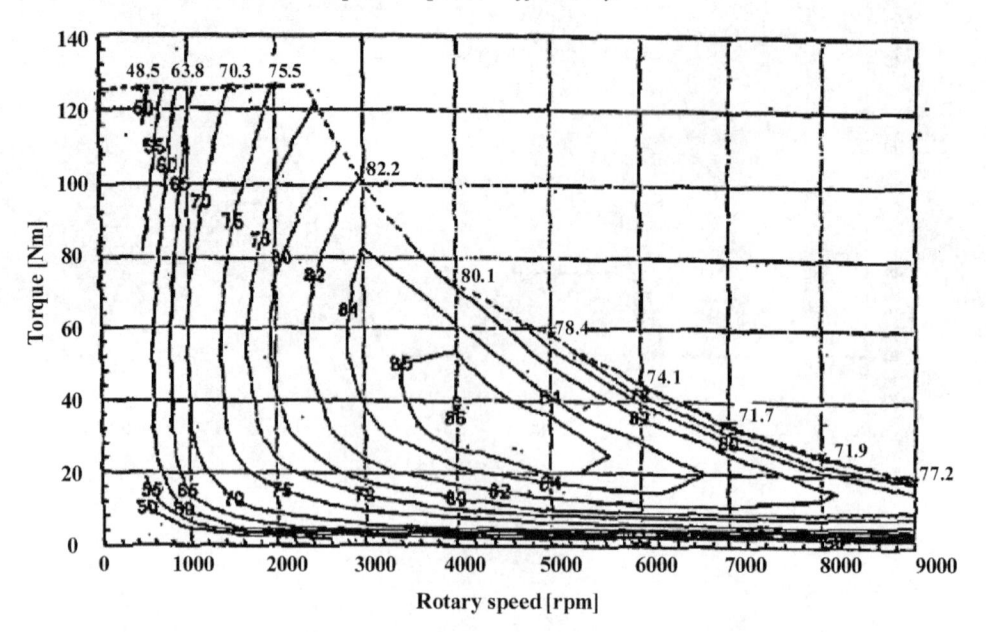

Generic Models of Electric Machine Applications in Hybrid Electric Vehicles

AC Induction Motor
- **Rated Power (Continuous):** 13 kW
- **Maximum Power (5 Min.):** 21 kW
- **Torque Nominal Rating:** 55 Nm
- **Maximum Torque (5 Min.):** 90 Nm
- **Rated Speed:** 2200 rpm
- **Maximum Speed:** 8000 rpm
- **Nominal Phase Voltage:** 115 V
- **Nominal Phase Current** [1]**:** 49.26 A
- **Nominal Stator Frequency:** 110 Hz
- **Stator Phase Resistance:** 10.5 mΩ
- **Rotor Phase Resistance:** 0.07 mΩ
- **cosΦ**[2]**:** 0.83
- **winding ratio** [2]**:** 2.883
- x_r **inductance** [2]**:** 2.237 [p.u.]
- x_s **inductance** [2]**:** 2.194 [p.u.]
- x_m **inductance** [2]**:** 2.133 [p.u.]

Mechanical Transmission Ratio
- Ratio between AC motor shaft and vehicle road wheels [3]: 4.76

Note

[1] The nominal current value has been calculated from the following formula:

$$I_n = \frac{P_n}{3 U_n \cos\varphi \eta_m}$$

where:
- P_n: Rated power;
- U_n: Nominal phase voltage;
- η: Motor mechanical efficient was an assumed 0.9.

[2] The data taken was, for example, similar to the Brown Bovery AC motor and calculated for the assumed nominal slip equal to 0,01.

[3] The mechanical ratio was assumed, so that the vehicle achieves 50 km/h velocity for 2200 rpm motor speed.

The flux $\{\phi_r\}$ value was assumed for each simulation case individually, in order to achieve the instantaneous motor voltage equal to the nominal value 110 V.

In particular, for an ECE cycle of drive – $\phi_r = 0{,}961$ [p.u. – per unit].

ECE Cycle of Drive
- **The Total Cycle Time:** 195 s
- **The Covered Distance:** 1.016 km
- **Maximum Power at the Acceleration Time:** 11.614 kW
- **Maximum Power at the Braking Time:** 10.652 kW
- **Total Acceleration Energy:** 208.813 kJ
- **Total Braking Energy:** 162.684 kJ
- **Total Constant Speed Energy:** 82.725 kJ

Note: The above parameters relate to the axles of the road wheels.

Results of Energetic Analysis
- **Acceleration Phases:**
  - Energy Absorbed from Battery: 340.11 kJ
- **Braking Phases:**
  - Energy Absorbed from Battery: 0.55 kJ
  - Energy Supplied to Battery: 106.83 kJ
- **Constant Speed Motion Phases:**
  - Energy Absorbed from Battery: 140.12 kJ
- **Net Expended Energy from the Battery:** 373.95 kJ
- **The Final Coefficient of the Battery Charge:** 0.987
- **Maximum Power on the AC Motor Shaft:** 6.11 kW
- **Maximum AC Motor Speed:** 2200 rpm

2. PM PERMANENT MAGNET MOTORS MODELING

Permanent Magnet motors are rapidly gaining in popularity. Two types of these motors are especially common: the permanent magnet synchronous (PMS), and permanent magnet, brushless, direct current (BLDC). The magnetic field generated by the stator, and the magnetic field generated by the rotor, rotate at the same frequency. This means both are a type of synchronous motor. For automotive application, 3-phase motors are used. A theoretical difference between PMS and BLDC motors is the shape of the Back EMF (Electromotive Force). In the case of BLDC Back EMF, it is trapezoidal; for PMS – sinusoidal, as shown in Figure 22. In practice, it is used in three basic PM motor constructions shown in Figure 23, and its classification features are included in Table 1.

In the case of Insert Magnets and Buried Magnet PM motor rotors, their angular velocity can be increased more than with Mounted Magnets. This is caused by stronger mechanical magnets fixed in the rotor body which permit the acquisition of higher velocity, accompanying its significant value of central fugal force. This

Generic Models of Electric Machine Applications in Hybrid Electric Vehicles

Figure 22. The back electromotive force EMF's typical shapes: a) trapezoidal – brushless, direct, current, permanent, magnet motor (BLDC); b) sinusoidal – permanent magnet, synchronous motor (PMS)

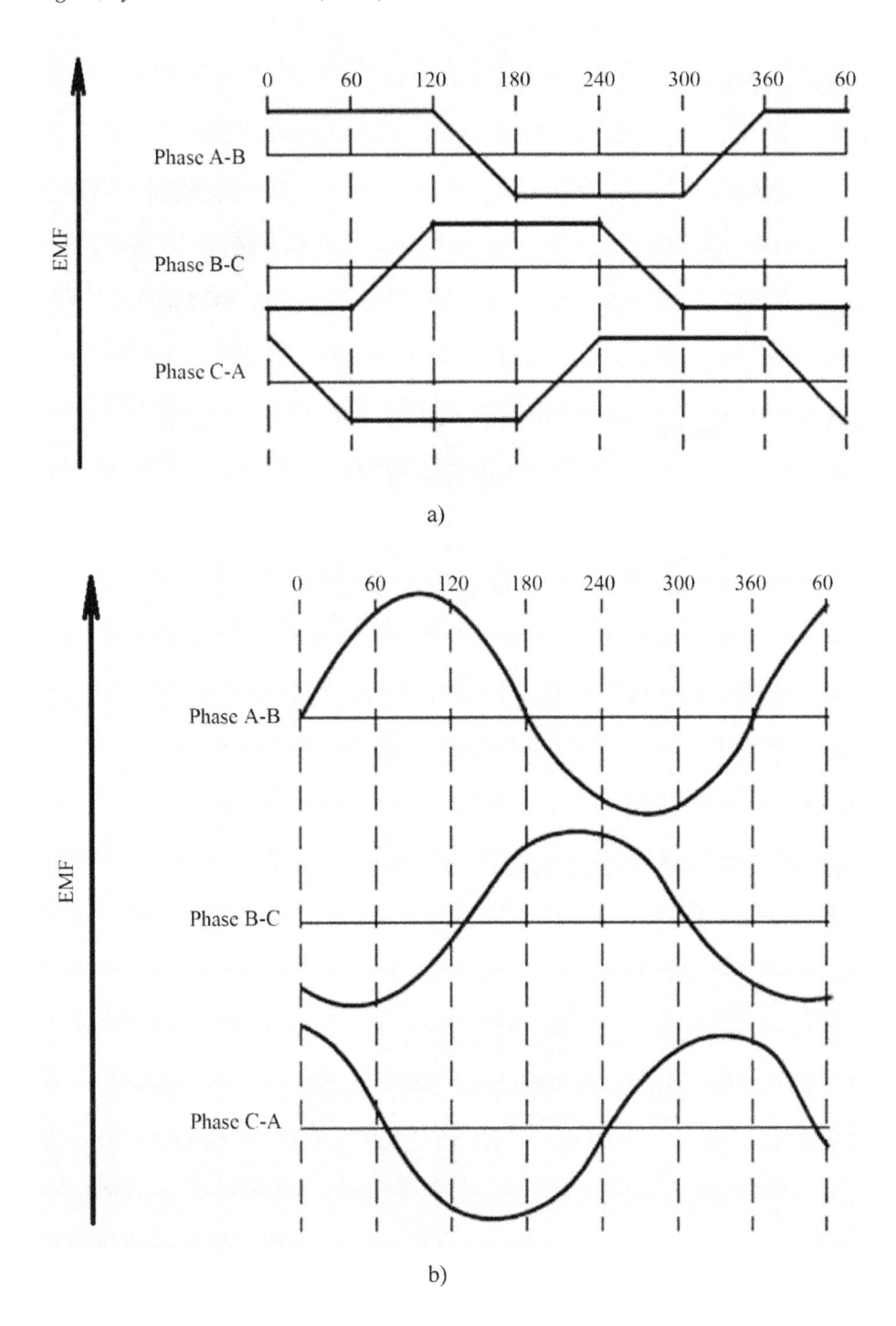

Figure 23. The rotor configuration of a permanent magnet motor (PM) synchronous and brushless: a) surface mounted magnets, only for brushless permanent magnet motor construction (BLDC); b) insert magnets, as well as for brushless or synchronous permanent magnet motor construction (BLDC / PMSM); c) buried magnets, only for synchronous, permanent, magnet motor construction (PMSM) (Padmaraja, 2003; Wu, n.d.)

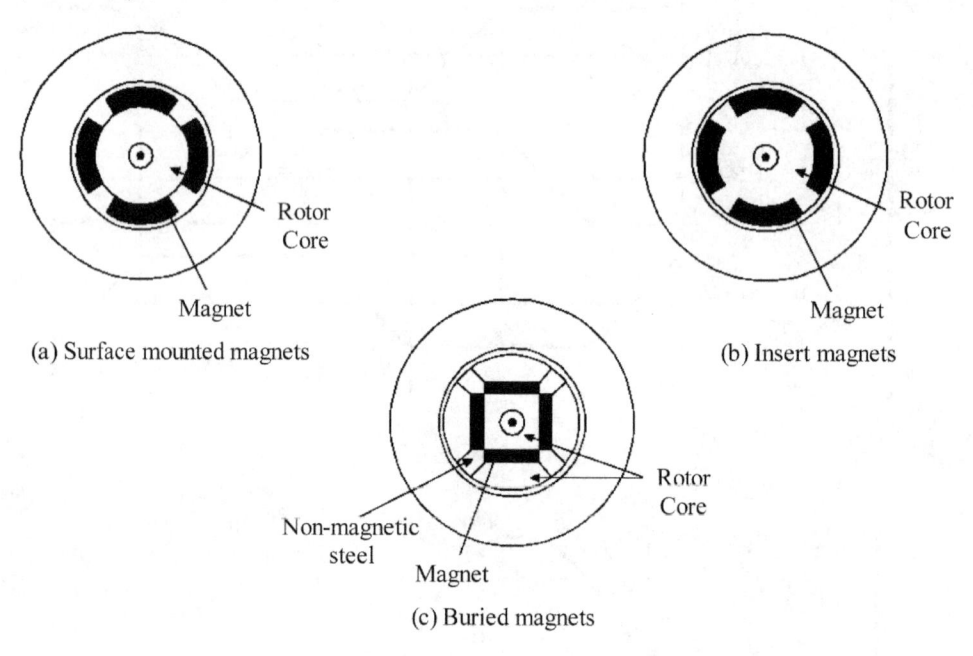

Table 1. PM rotor classification comparison (Padmaraja, 2003; Wu, n.d.)

	Surface Mounted Magnet	Insert Magnet	Buried Magnet
Suitability	BLDC	BLDC/PMSM	PMSM
Rotor Complexity	Low	Medium	High
Flux Distribution	Square Wave	Square Wave or Sinusoidal	Sinusoidal
Speed Limit	1.2 × Rated speed	1.5 × Rated speed	(2~3) × Rated speed

Generic Models of Electric Machine Applications in Hybrid Electric Vehicles

mechanical construction is directly connected with the PM's motor features. This can include mixed features of the BLDC/PMS, in the case of the Insert Magnet – which means this Back EMF shape is 'semi-sinusoidal' or this is more sinusoidal for the Buried Magnet rotor (PMSM – features only). The special winding, energizing, frequency sequence focused at this moment on two (from three) stator coils connected in the series, gives the Back EMF of BLDC/PMS motors a shape which could be called 'semi-sinusoidal' – in-between trapezoidal and sinusoidal. This will be further depicted.

For this reason, the general modeling theory of PM motors can be based on the analysis of the Synchronous Motor – corresponding to its traditional, electromagnetic, rotor, magnetic field generation.

3. APPROACH TO A POWER SIMULATION MODEL OF A DRIVE SYSTEM WITH A PM SYNCHRONOUS MOTOR

Imposed operation cycles are carried, into effect, by means of power simulation of a vehicle employing a multiple-source drive system with electromechanical and electrochemical power converters. The desired velocity and acceleration waveforms vs. time is defined for each cycle. Our goal is to determine the electric energy demand which meets the cycle conditions, when the vehicle motion resistance characteristics and the electrochemical and mechanical source specifications are known. It has been decided to adjust the procedure describing the model of a permanent magnet motor to the input and output variables, because there have already been available computing procedures for the models of vehicles fitted with a direct current DC or synchronous motor drive system, operating with various supply sources. The current and voltage are applied to the electric machine terminals, via a solid-state converter (chopper or inverter) from an electrochemical battery or a generator operating with a flywheel, and are determined in successive time steps. Also, the power of the electric machine occurring in the braking mode at its terminals, can be delivered to the supply sources with a defined lag.

The following basic assumptions for a synchronous motor were accepted on the basis of the above conditions:

1. The motor is controlled by means of applied voltage and the switching frequency of the inverter solid-state switches;
2. Mean resulting torque of the motor is equal to the synchronous torque produced by the machine;
3. Current derivatives and the permanent magnet 'ampere-turns' derivatives are equal to zero.

Generic Models of Electric Machine Applications in Hybrid Electric Vehicles

By taking these assumptions into account, the following set of equations is obtained:

$$\left.\begin{array}{l}\dfrac{d\psi_d}{dt} - \omega\psi_q + Ri_{1d} = u_{1d} \\ \dfrac{d\psi_q}{dt} + \omega\psi_d + Ri_{1q} = u_{1q}\end{array}\right\} \tag{38}$$

where:

$$\psi_d = L_d i_{1d} + \psi_{fd};$$

$$\psi_q = L_q i_{1q};$$

$$\psi_{fd} = M_{df}\theta_f$$

constant flux of permanent magnet connected with d - axis; M_{df} - appropriate mutual inductance; θ_f - magnetic potential of the permanent magnet; L_d, L_q - d - and q- axis inductance ; i_d, i_q - d - and q- axis currents; R - stator resistance, ω - rotor angular velocity.

The stator voltage resultant phasor $\overline{U}$ in Euler's shape is as follows:

$$\overline{U} = u_1 e^{-j\omega_0 t}; \quad u_1 = |\overline{U}| \tag{39}$$

For stationary α, β- axis:

$$\overline{U} = u_{1\alpha} + j u_{1\beta} \tag{40}$$

According to Figure 24:

$$\begin{cases} u_{1\alpha} = u_1 \cos\omega_0 t \\ u_{1\beta} = u_1 \sin\omega_0 t \end{cases} \tag{41}$$

For transformation, it is necessary to write the above equations in the following form:

Generic Models of Electric Machine Applications in Hybrid Electric Vehicles

$$\begin{cases} u_{1d} = u_{1\alpha}\cos\nu + u_{1\beta}\sin\nu \\ u_{1q} = -u_{1\alpha}\sin\nu + u_{1\beta}\cos\nu \end{cases} \tag{42}$$

or in another way:

$$\begin{Bmatrix} u_{1d} \\ u_{1q} \end{Bmatrix} = \begin{bmatrix} \cos\nu & \sin\nu \\ -\sin\nu & \cos\nu \end{bmatrix} \begin{bmatrix} u_{1\alpha} \\ u_{1\beta} \end{bmatrix} \tag{43}$$

Considering the Equations (40), (41) it is possible to obtain:

$$\begin{aligned} u_{1d} &= u_1\left(\sin\omega_0 t \sin\nu + \cos\omega_0 t \cos\nu\right) \\ u_{1q} &= u_1\left(\sin\omega_0 t \cos\nu - \cos\omega_0 t \sin\nu\right) \end{aligned} \tag{44}$$

and finally:

$$\begin{aligned} u_{1d} &= u_1 \cos\left(\omega_0 t - \nu\right) \\ u_{1q} &= u_1 \sin\left(\omega_0 t - \nu\right) \end{aligned} \tag{45}$$

The angle ν of rotor rotation, expressed by rotor angular velocity, ω is as:

Figure 24. The illustration of voltage transformation from α, β axes to rotating d, q axes

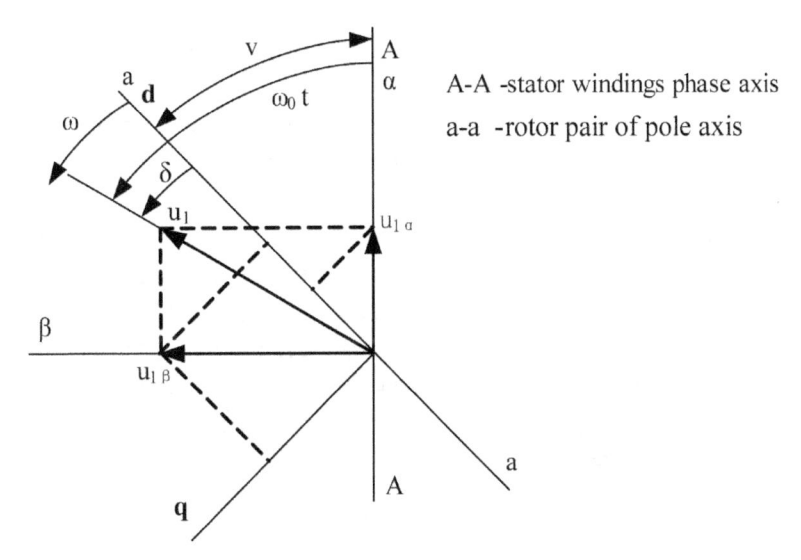

$$\nu = \int_0^t \omega dt + \nu_o \rightarrow (\nu_o - \text{initial value})$$

and the motor loading angle δ expressed by the stator synchronous velocity is defined as:

$$\omega_0 t - \nu = \delta$$

Torque (M) and electrical power (P) equations are as follows:

$$J\frac{d\omega}{dt} = M_e - M_l \qquad (46)$$

$$M_e = \frac{3}{2}p\left(\left(L_d i_{1d} + M_{df}\Theta_f\right)i_{1q} - L_q i_{1d} i_{1q}\right)$$

where:

- **p:** Pole pair number,
- **M_e:** Motor electromagnetic torque,
- **M_l:** External loading torque and

$$P = \frac{3}{2}\left(u_{1d} i_{1d} + u_{1q} i_{1q}\right) \qquad (47)$$

For computer analysis, the following assumptions were made:

- The component current i_q is determined from the torque equation for the average value of electromagnetic torque - it corresponds to mechanical torque – whilst it is also connected with effective current value.
- The value of current component i_d is calculated from the voltage component equations for condition cosϕ=const according to an assumed value, where: "ϕ"- angle between two vectors: u_1 - line-to-line voltage (3 phases, its components are u_{1q} and u_{1d}) and i_1 - phase current (star connection, its components are i_{1q} and i_{1d}). By using the PWM method during control converter operation, it is possible to keep a constant value of cosϕ. The optimal value –the most effective- is cosϕ=1.

3.1. Vector-Field Oriented Control of a PM Synchronous Motor

For PM motors with permanent sinusoidal shape flux, the equations expressed in stator instantaneous current, take the simplified form:

$$i_{1\alpha} = -i_{1q} \sin \nu$$
$$i_{1\beta} = i_{1q} \cos \nu \tag{48}$$

and finally,

$$i_{1q} = -i_{1\alpha} \sin \nu + i_{1\beta} \cos \nu \tag{49}$$

However, it is possible to express the vector control of the PM motor in the following way.

Figure 25 shows the vector diagram for the PMS motor in motoring mode. Figure 26 shows simplified diagrams (winding resistance is neglected) for motoring and generating respectively, to different relations between Ψ_s – magnetic flux connected with phasor I_s (stator current) and Ψ_f connected with the d axis, which means a constant flux of permanent magnets.

Figure 27 shows the PMSM operation for maximum torque of a motor in terms of I_s = const. This means the value of quantity M_e/Ψ_s is maximal. As was mentioned before, this takes place on condition of $\cos\varphi = 1$. Meanwhile, Figure 28 shows an illustration of PMSM vector torque control in terms of I_q = const. where: ω_x – motor stator frequency, Ls – motor stator inductance, Is – motor stator current, E – back EMF, Ψ_s – stator flux, Ψ_f - permanent magnet flux, connected with d axis, R_s – motor stator winding resistance, U – motor stator voltage.

Note:

$$M_o = \frac{3}{2} p \left(\bar{\Psi}_s \times \bar{I}_s \right); \quad p - \text{number of poles}$$

3.2. PM Brushless DC Motor (BLDC)

A typical BLDC motor construction is shown in Figure 29.

The equivalent circuit of that machine is shown in Figure 30.

The stator of a Brushless Direct Current (BLDC) motor consists of stacked steel laminations with windings placed in the slots that are axially out along the inner periphery. The stator resembles that of an induction motor, however, the windings

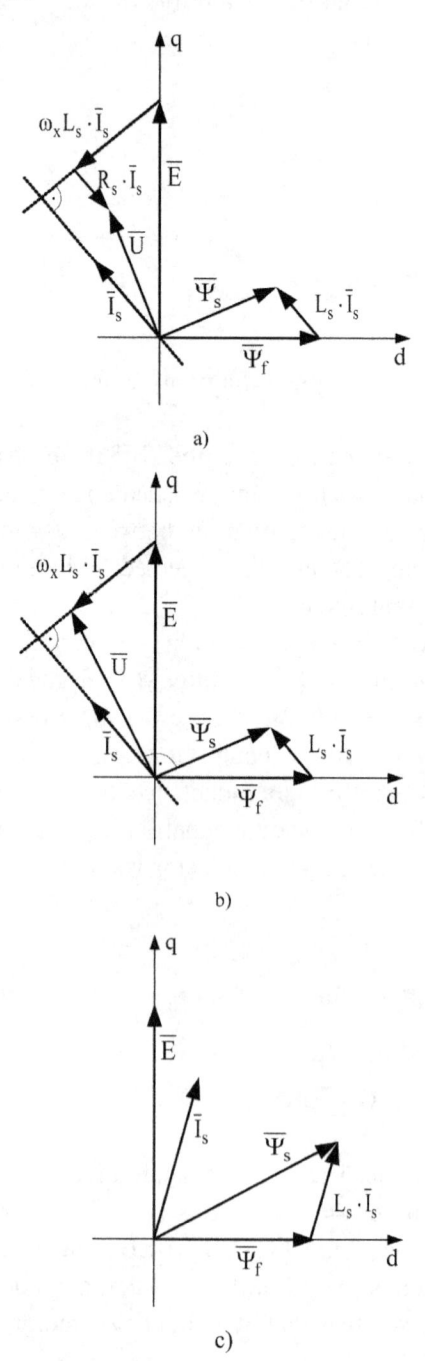

Figure 25. The vector diagrams for the permanent magnet synchronous motor PMSM operating in motoring mode: a) completed vector diagram for $\Psi_s<\Psi_f$ and motoring; b) simplified vector diagram for $\Psi_s<\Psi_f$ and motoring; c) simplified vector diagram for $\Psi_s>\Psi_f$ and motoring (Yamura, 1992)

a)

b)

c)

Generic Models of Electric Machine Applications in Hybrid Electric Vehicles

Figure 26. The vector diagram for a permanent magnet, synchronous, motor PMSM operation in generating mode: a) $\Psi_s < \Psi_f$; b) $\Psi_s > \Psi_f$

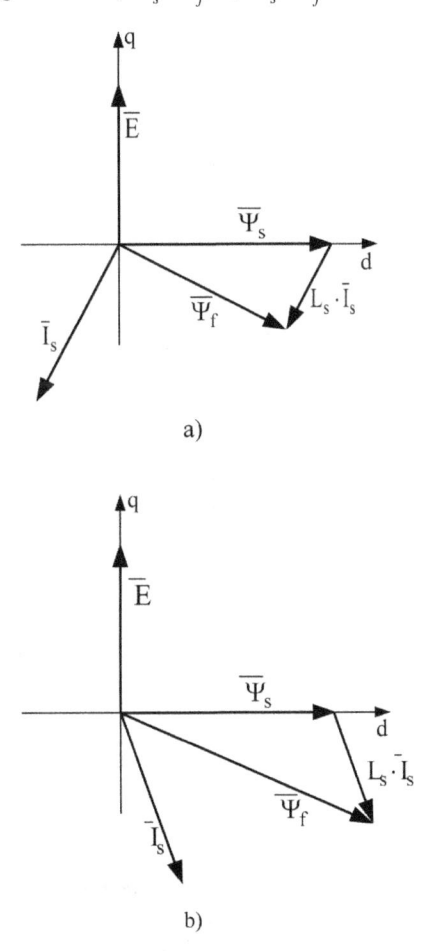

are distributed in a different manner. The brushless, direct current, motor has three stator windings connected in star fashion. One or more coils are placed in the slots, making a winding. Each of these windings is distributed over the stator periphery, to form an even numbers of poles. There are two types of stator windings: trapezoidal and sinusoidal motors. The differentiation is made on the basis of the interconnection of coils to obtain, in an alternative way, the back electromotive force, the EMF – which is more similar to the trapezoidal or sinusoidal forms. The 'sinusoidal' motor torque output is smoother than the 'trapezoidal' one. Sinusoidal motors take extra winding interconnections, because of the coils distributed on the stator periphery, thereby increasing copper intake by stator windings, causing increasing motor cost, also. The rotor is made of permanent magnet and can vary from two to eight pole pairs.

Figure 27. The vector diagram of a permanent magnet, synchronous motor (PMSM) operating in the motoring mode for maximum value of M_e torque – for I_s = const and $\cos\phi = 1$

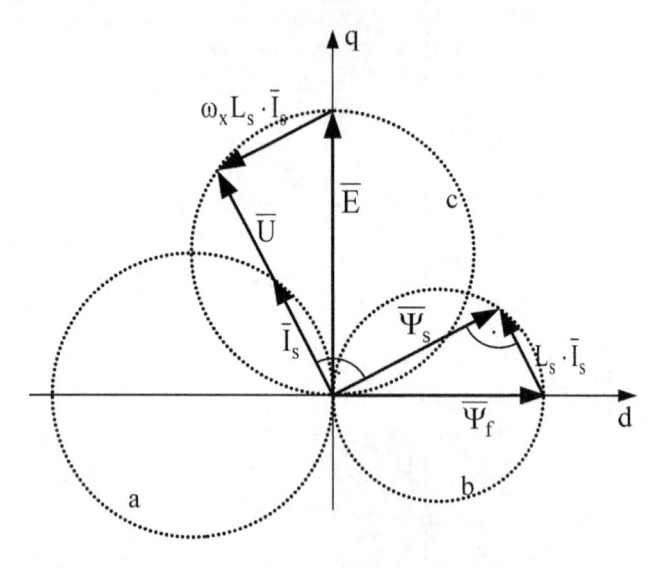

Figure 28. The illustration of a permanent magnet, synchronous motor (PMSM) shaft output torque vector M_e controlled for different values of I_s motor stator current in terms of the same current of the motor stator I_s vector projection on the q axis

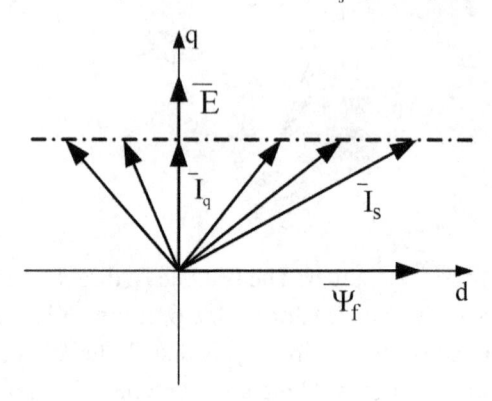

The brushless, direct current, BLDC motors are controlled electronically. To rotate the motor, the stator windings should be energized in a sequence. For this reason, to know the rotor's geometrical position is very important. Most brushless, direct current, BLDC motor stators are equipped with Hall sensors. The rotor magnetic

Generic Models of Electric Machine Applications in Hybrid Electric Vehicles

Figure 29. The exemplary, brushless, direct current BLDC motor construction (Szumanowski & Bramson, 1992)

Field windings:

Rotating magnetic field created by the sequential excitation of the pole pairs by a DC pulse.

Brushless DC Motor

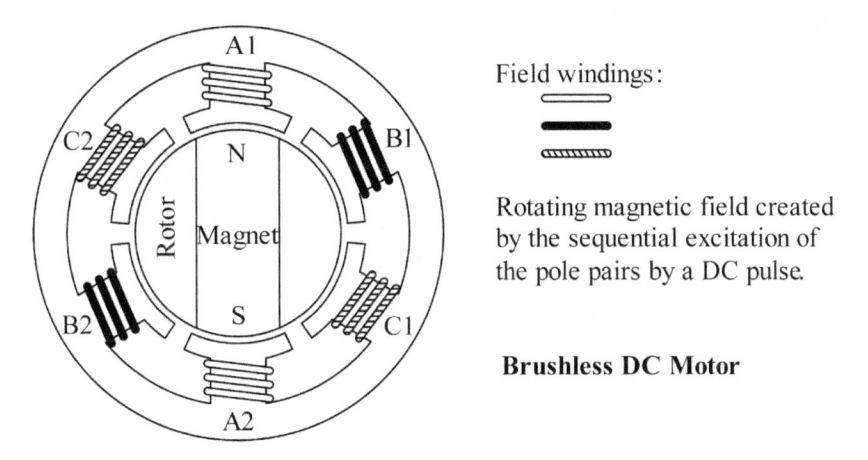

Figure 30. The brushless, direct current, motor BLDC circuit and its mathematical depiction (Kenjo & Nagamori, 1985; Padmaraja, 2003; Unique Mobility Inc., n.d.; Wu, n.d.)

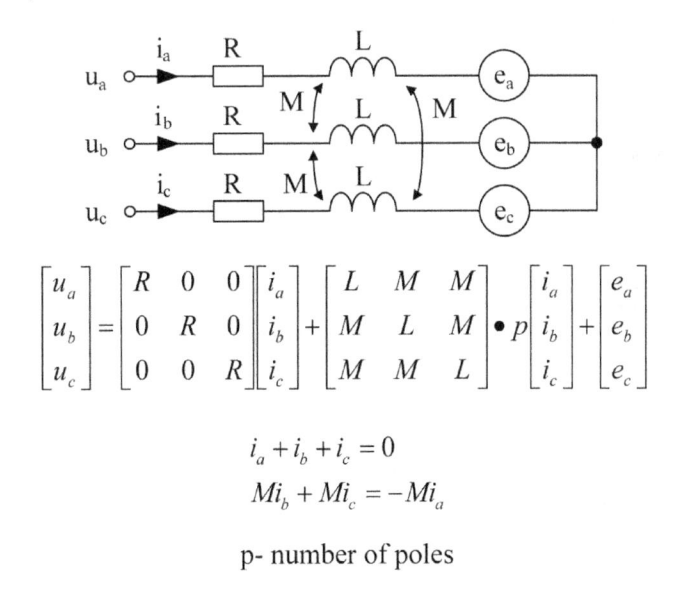

$$\begin{bmatrix} u_a \\ u_b \\ u_c \end{bmatrix} = \begin{bmatrix} R & 0 & 0 \\ 0 & R & 0 \\ 0 & 0 & R \end{bmatrix} \begin{bmatrix} i_a \\ i_b \\ i_c \end{bmatrix} + \begin{bmatrix} L & M & M \\ M & L & M \\ M & M & L \end{bmatrix} \bullet p \begin{bmatrix} i_a \\ i_b \\ i_c \end{bmatrix} + \begin{bmatrix} e_a \\ e_b \\ e_c \end{bmatrix}$$

$$i_a + i_b + i_c = 0$$
$$Mi_b + Mi_c = -Mi_a$$

p- number of poles

poles pass near the Hall sensor giving a signal, indicating the N or S pole is just passing. Based on the combination of these Hall sensor signals, the exact sequence of commutation can be determined (shown in Figure 31). Figure 32 shows the coil switching sequence, in respect to the Hall sensor signal.

Figure 31. The commutation of brushless, direct current, BLDC motor stator wiring, according to the hall sensor signals (Kenjo & Nagamori, 1985; Padmaraja, 2003; Unique Mobility Inc., n.d.; Wu, n.d.)

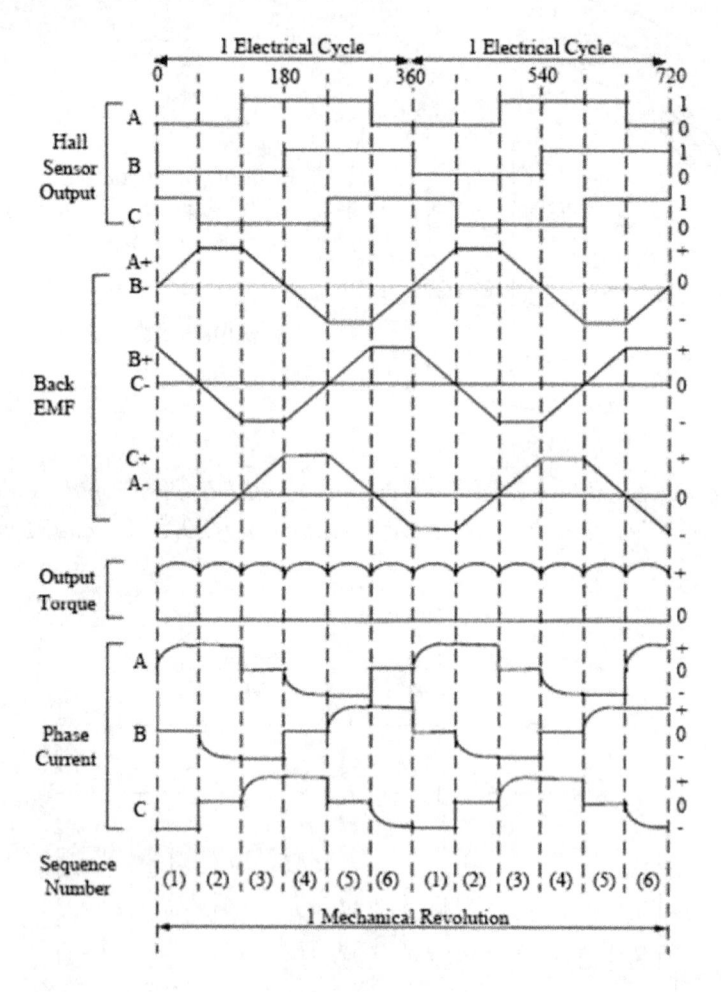

3.3. Brushless, Permanent Magnet, Motor (PM) Field Weakening Modeling

The torque versus speed characteristics of a typical road vehicle are depicted by the motor controlled by the pulse-width-modulation method (PWM), a typical characteristic in the shape of its torque and power versus the motor output, shaft rotational speed corresponding to vehicle road speed, as shown in Figure 33.

Generic Models of Electric Machine Applications in Hybrid Electric Vehicles

Figure 32. Switching sequence in respect to the hall sensor signals (Kenjo & Nagamori, 1985; Padmaraja, 2003; Unique Mobility Inc., n.d.; Wu, n.d.)

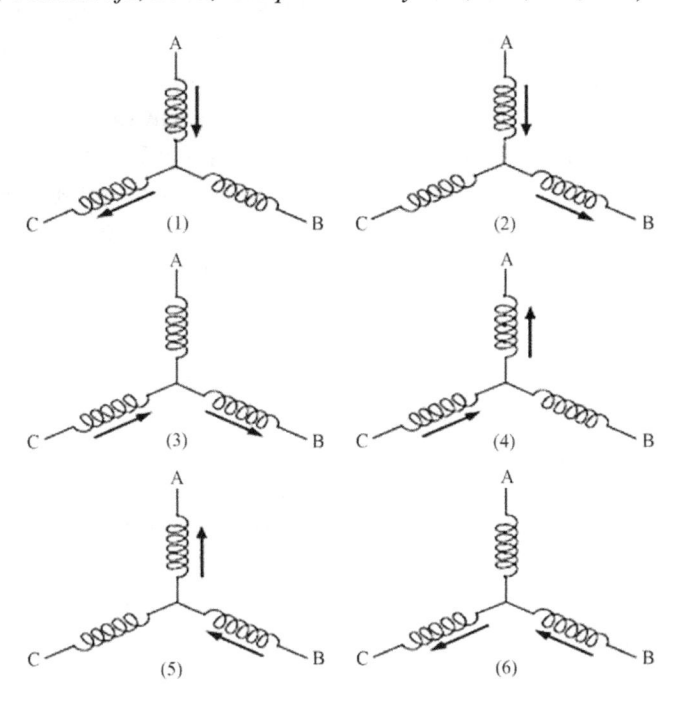

Figure 33. The three hundred volts, brushless, direct current, permanent magnet (BLDC) motor time characteristics (Kenjo & Nagamori, 1985; Padmaraja, 2003; Unique Mobility Inc., n.d.; Wu, n.d.)

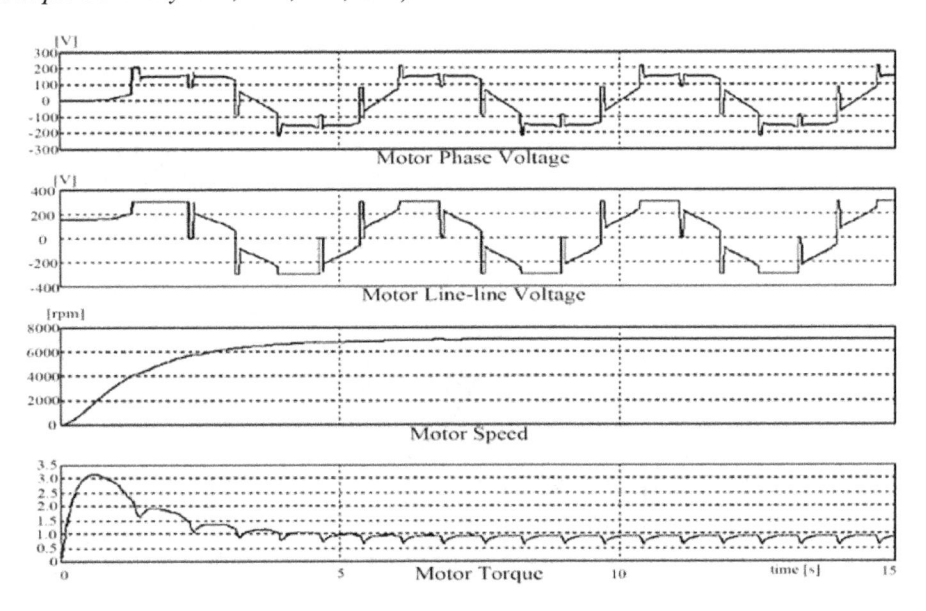

133

Generic Models of Electric Machine Applications in Hybrid Electric Vehicles

The electric traction motor torque required to achieve the desired vehicle wheel torque depends upon the gear reduction ratio between this motor and the wheels. The speed at point A (see Figure 34) could be called 'base speed'. However, the torque versus speed curve of the brushless, permanent motor is generally flat, as shown in Figure 34 (Bin, n.d.). The capacity requirement for the power amplifier at point A, is much greater than the requirement at point B, assuming the battery-converter (controller) unit source voltage at point A, is the same as that at point B. In the design of an optimal, electric vehicle drive system, high power density over a large speed range above the base speed, is preferred, as shown in Figure 34.

There are three basic methods of permanent magnet PM motor field weakening:

- The motor stator vector control
- The motor rotor's additional winding feeds, by controlling a current creating the magnetic field oriented in the opposite direction to the main magnetic field, from permanent magnet units located on the rotor.
- The combination of both the above-mentioned methods.

The per unit system is used for defining parameters in the presentation. For d- and q- axis voltage, the basic equations are as follows:

$$u_d = R \cdot i_d - \omega \psi_q + \frac{d\psi_d}{dt}$$
$$u_q = R \cdot i_q + \omega \psi_d + \frac{d\psi_q}{dt} \tag{50}$$

whereas

Figure 34. The traction motor's typical characteristic in the shape of its torque and power versus the motor output, shaft rotational speed corresponding to vehicle road speed; the method of torque – speed control PWM

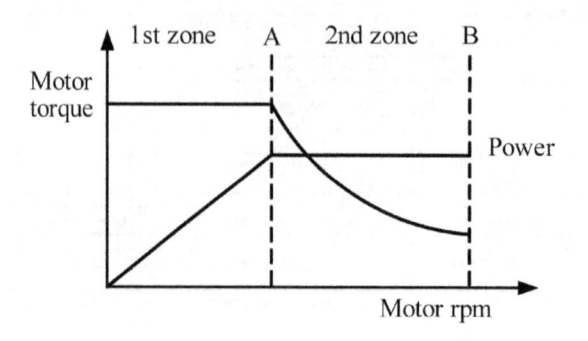

Generic Models of Electric Machine Applications in Hybrid Electric Vehicles

$$u_d = R \cdot i_d - L_q i_q \omega + L_d \frac{di_d}{dt}$$
$$u_q = R \cdot i_q + \omega(L_d i_d + \psi_f) + L_q \frac{di_q}{dt} \tag{51}$$

where:

- ψ_f: The flux linkage per phase of the stator;
- L_d, L_q: The d- and q- axis inductances respectively

If there are additional values involved:

$$R_1 = R_b + r_p + r_m$$

$$X_1 = X_p + X_m$$

- R_b: DC source battery resistance;
- r_p: Power amplifier resistance;
- r_m: Motor stator winding resistance;
- X_1: Power amplifier X_p and motor X_m stator inductive leakage reactance;
- E: Induced EMF (electromagnetic force) in stator motor windings;
- R_m: Stator motor iron resistance;
- i_q^m, i_d^m: Energy conversion components of currents (d-, q- axes) i_d and i_q

It is possible to obtain a system of the following equations:

$$u_d = \left(R_1 - \frac{X_{ad} X_1}{R_m}\right) i_d^m$$
$$- \left(X_q + \frac{X_{aq} R_1}{R_m}\right) i_q^m - \frac{X_1 E}{R_m}$$
$$u_q = \left(X_d + \frac{X_{ad} R_1}{R_m}\right) i_d^m \tag{52}$$
$$+ \left(R_1 - \frac{X_{aq} X_1}{R_m}\right) i_q^m + \frac{(R_1 + R_m) E}{R_m}$$

Neglecting the iron and leakage losses and for $i_d = i_d^m$, $i_q = i_q^m$:

$$u_d = R_1 i_d - X_q i_q$$
$$u_q = X_d i_d + R_1 i_q + E \quad (53)$$

It is possible to obtain the same form of voltage equations, by transformation of the Equation (51) for $(di_q/dt)=0$ and $(di_d/dt)=0$ in the current's steady state

$$u_d = R i_d - L_q i_q \omega$$
$$u_q = R i_q + L_d i_d \omega + \psi_f \omega$$

and for angular rotor velocity $\omega = const$ (armature frequency respectively) and motor torque $M = const$.

$$u_d = R i_d - X_q i_q$$
$$u_q = R i_q + X_d i_d + E \quad (54)$$

Also neglecting copper resistance losses in Equations (53) and (54), it is possible to express:

$$u_d = -X_q i_q$$
$$u_q = X_d i_d + E \quad (55)$$

Yet, for the per unit system:

$$u_d^2 = X_q^2 i_q^2$$
$$u_q^2 = X_d^2 i_d^2 + 2 X_d i_d E + E^2 \quad (56)$$

Adding on both equations' sides:

$$1 = X_q^2 i_q^2 + X_d^2 i_d^2 + 2 X_d i_d E + E^2 \quad (57)$$

and

Generic Models of Electric Machine Applications in Hybrid Electric Vehicles

$$\frac{1}{X_d^2} = \frac{X_q^2}{X_d^2} i_q^2 + \left(i_d^2 + \frac{2 i_d E}{X_d} + \frac{E^2}{X_d^2} \right)$$

$$\frac{1}{X_d^2} = \frac{X_q^2}{X_d^2} i_q^2 + \left(i_d + \frac{E}{X_d} \right)^2$$

and finally:

$$1 = \frac{\left(i_d + \dfrac{E}{X_d} \right)^2}{\dfrac{1}{X_d^2}} + \frac{i_q^2}{\dfrac{1}{X_q^2}} \tag{58}$$

The Equation (58) is the background for field weakening analyses.

Including once again, the power amplifier and additional parameters (Equation (51)), the power relationship is:

$$P = \frac{3}{2}\left(u_d i_d + u_q i_q \right) = \frac{3}{2}\left[\left(X_d - X_q \right) i_d^m + E \right] i_q^m$$
$$+ R_1 \left(i_d^2 + i_q^2 \right) + \frac{1}{R_m} \left[\left(E + X_{ad} i_d^m \right)^2 + \left(X_{aq} i_q^m \right)^2 \right] \tag{59}$$

After neglecting the copper and iron motor losses, we can obtain:

$$P = \frac{3}{2}\left(\left(-X_q i_q^m \right) i_d^m + \left(X_d i_d^m + E \right) i_q^m \right)$$
$$= \frac{3}{2}\left(X_d i_d^m i_q^m - X_q i_q^m i_d^m + E \cdot i_q^m \right) \tag{60}$$
$$= \frac{3}{2}\left[\left(X_d - X_q \right) i_d^m + E \right] i_q^m$$

The mechanical torque is as follows:

$$M = \frac{P}{\omega}$$

where ω - rotational rotor speed, and

$$M = \frac{1}{\omega}\left(\frac{3}{2}\left[\left(L_d\omega - L_q\omega\right)i_d^m + \psi_f \omega\right]i_q^m\right)$$

$$M = \frac{3}{2}\left[\left(L_d - L_q\right)i_d^m + \psi_f\right]i_q^m$$

(61)

The above equations are the background determination of the control range of field weakening methods: stator vector control, rotor field weakening by an additional (coil) magnetic field and a combination of both methods mentioned.

The graphic interpretation of Equations (58) and (61) is shown in Figure 35.

At a speed above the base speed, there exists a corresponding ellipse on which the operating points (i_d^m, i_q^m) are located at the speed characteristics. The dotted lines are the torque curves (Equation 60). The intersection point between a torque curve with the particular value on an ellipse at a certain speed, determines the required currents i_d^m and i_q^m. If the Equation 62 is presented in a per unit system, it has another form:

$$1 = \frac{\left(i_d + \frac{\omega \psi_f}{\omega L_d}\right)^2}{\left(\frac{1}{\omega L_d}\right)^2} + \frac{\left(i_q\right)^2}{\left(\frac{1}{\omega L_q}\right)^2}$$

(62)

when we assume $\psi_f \cong L_d$, and neglect the displacement along the i_d axis, the new shape of the Equation 60 is as follows:

$$1 \cong \frac{\left(i_d\right)^2}{\left(\frac{1}{\omega L_d}\right)^2} + \frac{\left(i_q\right)^2}{\left(\frac{1}{\omega L_q}\right)^2}$$

(63)

According to the above equation, the motor's maximum rpm speed n_m can be expanded substantially, without suffering significant current or copper losses' increases. The theoretic maximum is:

Generic Models of Electric Machine Applications in Hybrid Electric Vehicles

Figure 35. The graphic interpretation of Equations (58), (61), and (63) for stator field weakening; note: Equation 64 responds to Equation 63 and Equation 61 responds to Equation 58

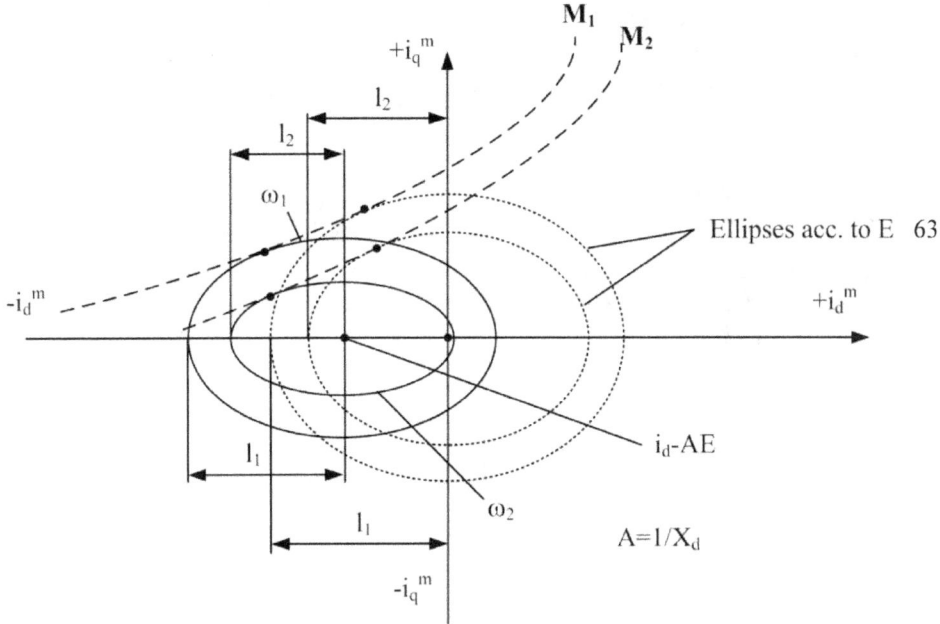

For: $\omega_2 > \omega_1$; $\omega_1 M_1 = const$; $\omega_2 M_2 = const \rightarrow M_2 < M_1$

$$\begin{cases} n_m = \dfrac{\sqrt{\sqrt{(R_1^2 + 2R_m)^2 - 4cR_1^2 R_m} - (R_1^2 + 2R_1 R_m)}}{L_1} \\ c = 1 - R_1 \end{cases} \quad (64)$$

(according to UM technical papers Huang, Cambier, & Geddes, 1992)

Rotor field weakening could be achieved by installation of a coil turned in the rotor, which creates a magnetic flux in the opposite direction to the main flux of the permanent magnets on the rotor (see Figure 36).

For the motor rotor field weakening, the centers of the ellipses migrate to the point (0,0), when the air gap field is weakened (see Figure 37).

The torque value, according to Figure 37, is determined by the intersection point between the ellipse and i_q^m axis, provided that the commutation phase advance (done by control - unit - converter) can guarantee $i_d^m = 0$. In another case (see Figure 36), when the magnet flux is generated by a positive i_d^m, depending on the rotor position and motor speed, it will be reduced similarly, to the method depicted by Figure 37.

Figure 36. The motor rotor field weakening by using an additional rotor coil for a one pair-pole theoretical machine; d-axis position: a) +d-d; b) -d+d

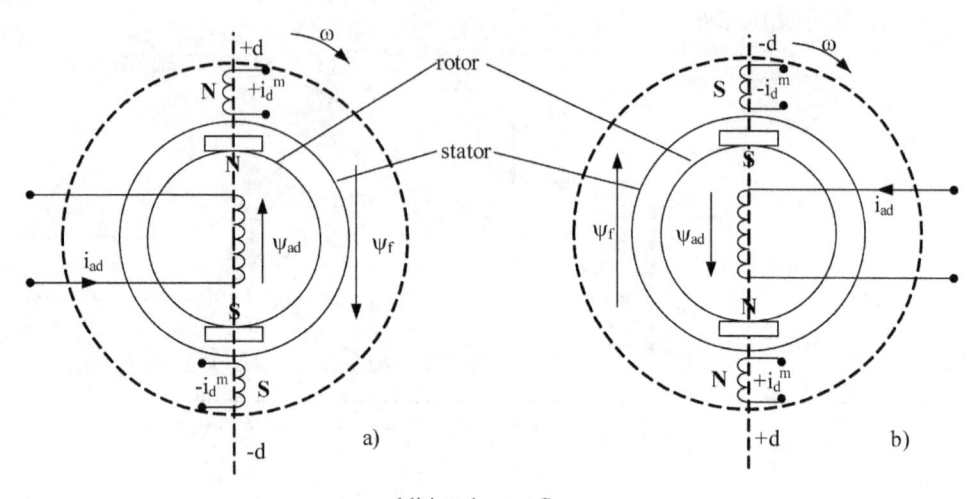

ψ_{ad} - additional rotor flux
i_{ad} - current of additional rotor coil

Figure 37. The illustration of permanent magnet synchronous motor field weakening a) ellipses for motor rotor field weakening, b) illustration of a field weakening mechanism, using additional motor stator coils

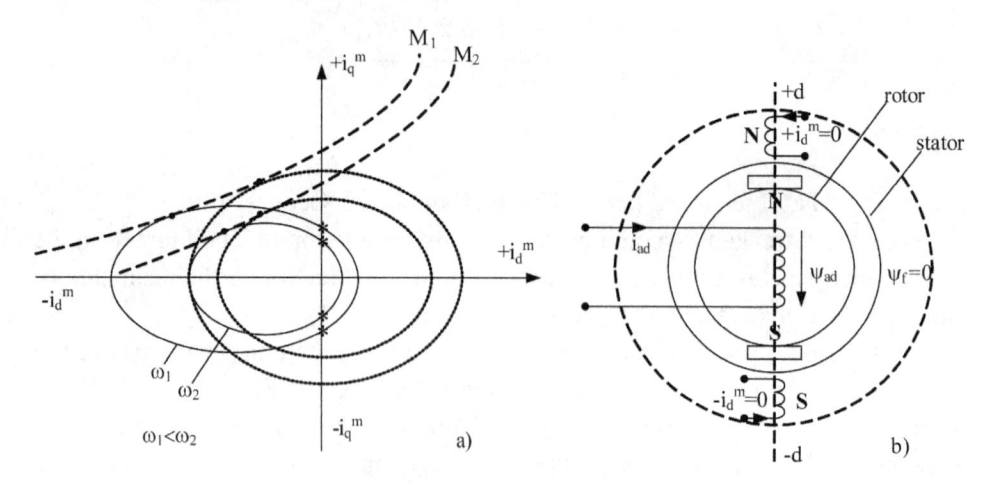

Figure 38. The torque and efficiency alterations of a permanent magnet, synchronous motor operating in the field-weakening mode for the pulse-width-modulation PWM motor control method (Bin, n.d.; Huang et al., 1992)

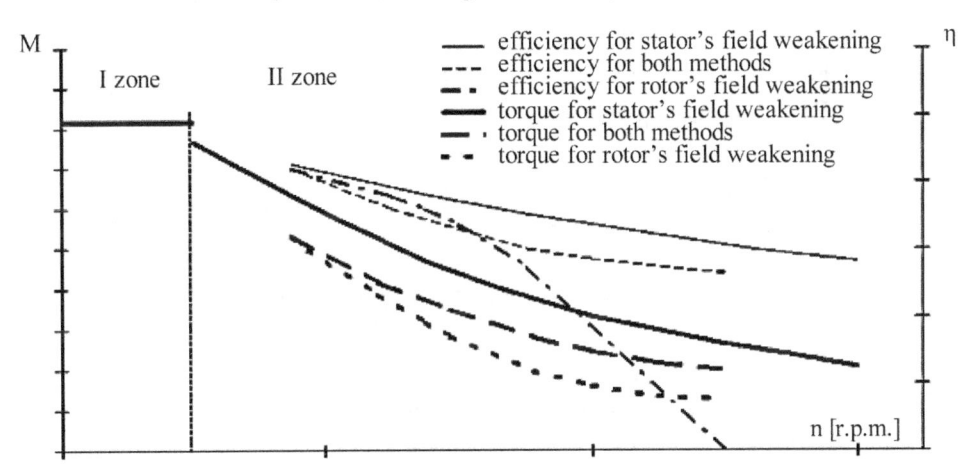

The American company, Unique Mobility, obtained some interesting control results, which are shown in Figure 38.

3.4. Exemplary Simulation Study of a PM Synchronous Motor

For computer analysis, the transformation of the Equation (39) was made. The following assumptions are considered:

- The component current I_q is determined from the torque equation for the average value of an electromagnetic torque - it corresponds to a mechanical torque and, meanwhile, is connected with an effective current value.
- The value of the current component I_d is calculated from the voltage component equations for the condition $\cos\phi = $ const., according to an assumed value, where: "ϕ" - angle between two vectors: U_s - line-to-line voltage (3 phases, its components are U_q and U_d) and I_s - phase current (star connection, its components are I_q and I_d).

By using a PWM methodology during an inverter operation, it is possible to keep a constant value of $\cos\phi$. This optimal value - the most effective - is $\cos\phi = 1$.

- The loading angle d - between the inducted in armature EMF (EMF = $\omega*k$) in axis q, and the vector U_s is equal to $\pi/2$. The fulfillment of this condition offers the best possibilities for a synchronous PM motor operation.

Generic Models of Electric Machine Applications in Hybrid Electric Vehicles

For preliminary analysis, a Unique Mobility PM motor characterized by the following data was taken:

- PM Motor type SR180/2.8L:
  - **Rated Power:** 25 kW @ 3750 rpm
  - **Rated Speed:** 4000 rpm
  - **EMF Constant "K":** 0.0561 v/rpm
  - **Winding Resistance "R_s":** 0.054 Ohm
  - **Winding Inductance "L":** 0.072 mili H

For calculations, there are some additional assumptions:

- Number of points per pair p = 1,
- The inductance of the direct axis equals the inductance of the quadrature axis i.e. Ld=Lq=L=0.072 mg,
- Mechanical losses (refers to the torque) $M = \alpha_2 \omega + \alpha_1$,
- Iron losses are dependent on angular velocity, according to:

$$\Delta P_{Fe} = \alpha_3 \omega$$

where:

- The value of α_3 was assumed, corresponding to a power loss for rated power = 0.03 P_N,
- Inverter average efficiency = 0.9,
- Mechanical ratio = 7.29.
- The following calculating formulas are used:
- The torque equation for first approximation of determination of current I_q (in the first stage of calculation, copper losses are neglected)

$$J \frac{d\omega}{dt} \pm (M_o + \Delta M + \frac{\Delta P_{Fe}}{\omega}) = \frac{3}{2} pk I_q$$

where:

Generic Models of Electric Machine Applications in Hybrid Electric Vehicles

M_o – torque of vehicle resistance

$$\begin{cases} +1 \text{ for motor mode} \\ -1 \text{ for generator mode} \\ \text{(regenerative braking)} \end{cases}$$

- Voltage Equations

$$U_d = R_s I_d - \omega p L I_q$$
$$U_q = R_s I_q + \omega p L I_d + \omega p k$$

- Motor Current

$$I_s = \sqrt{I_d^2 + I_q^2}$$

- Motor Voltage

$$U_s = \sqrt{U_d^2 + U_q^2}$$

- $\cos\phi$

$$\cos\varphi = \frac{U_d I_d + U_q I_q}{U_s I_s}$$

- Power Equation

$$P = \frac{3}{2}(U_d I_d + U_q I_q)$$

for $\cos f = 1$ $\quad P = \dfrac{3}{2} U_s I_s$

- Main Parameters of a Drive System
  ○ Battery

Generic Models of Electric Machine Applications in Hybrid Electric Vehicles

- **Nominal Battery Voltage:** 260 V
- **Nominal Battery Capacity:** 26.36 Ah
- **Nominal One-Hour Current:** 26.36 A
- **Number of Electric Motors:** 1
- **Motor Rated Power:** 25 kW
- **Motor Rated Speed:** 3750 rpm
- **Motor Nominal Current:** 106 A
- **Machine Constant:** 0.536 V/s
- **Mechanical Ratio Motor/Vehicle Wheels:** 7.294

- Simulation Results
  - Battery Discharge Energy:
    - **During Acceleration Phases:** -267.337 kJ
    - **During Steady Speed Phases:** -130.207 kJ
    - **During Braking Phases:** -0.084 kJ
  - Battery Charge Energy:
    - **During Braking Phases:** 133.899 kJ
    - **Total Energy Battery Discharge:** -397.628 kJ
    - **Total Energy Battery Charge:** 133.899 kJ
    - **The Energy Battery Discharges:** -263.729 kJ
    - **Maximum Power on the Motor Shaft:** 12.229 kW

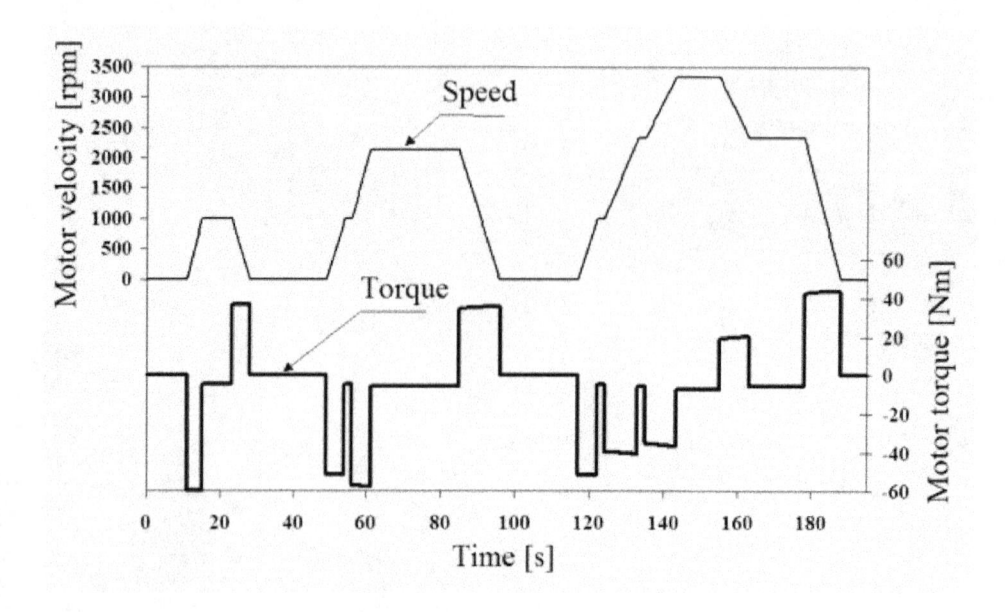

Figure 39. The permanent magnet synchronous motor output shaft rotational velocity versus vehicle driving time

Generic Models of Electric Machine Applications in Hybrid Electric Vehicles

Figure 40. The permanent magnet synchronous motor output shaft power versus vehicle driving time

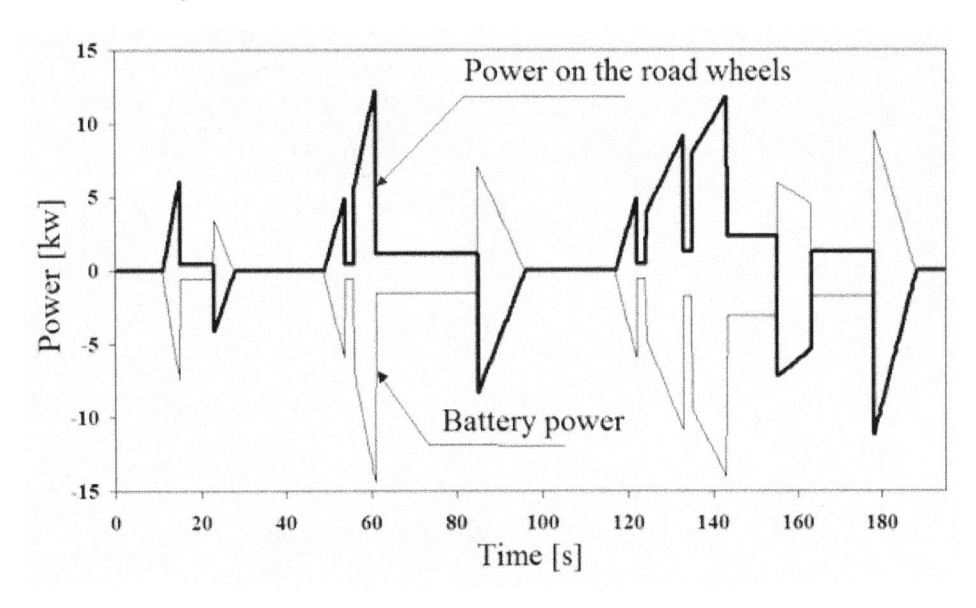

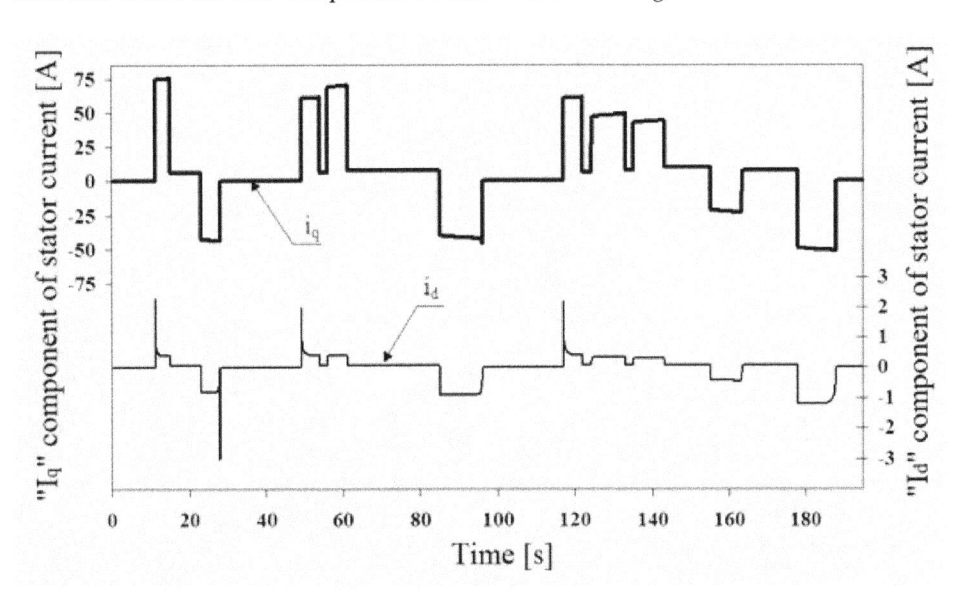

Figure 41. The permanent magnet synchronous motor stator q axis current component and d axis current component versus vehicle driving time

Figure 42. The permanent magnet synchronous motor stator q axis voltage component and d axis voltage component versus vehicle time driving

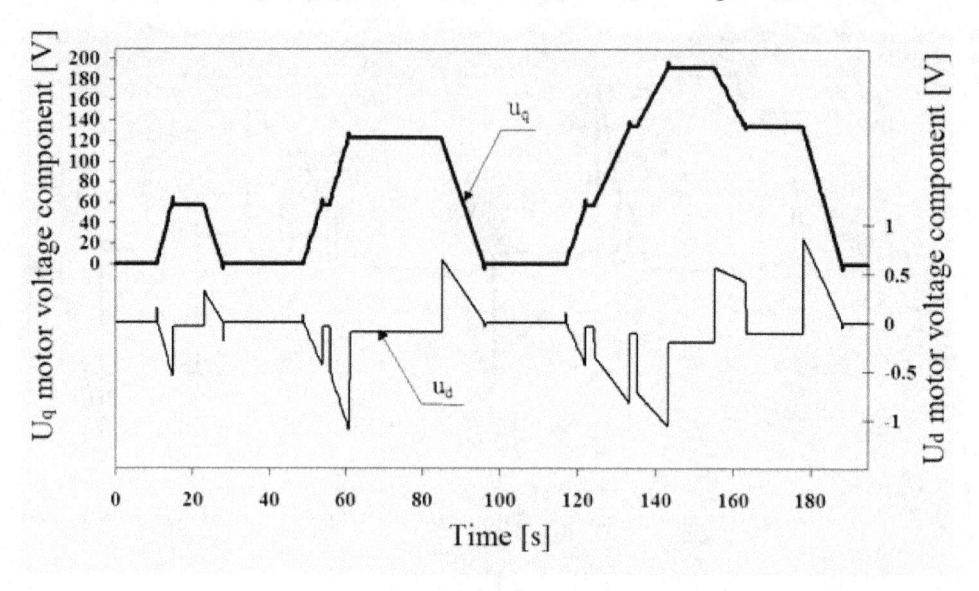

Figure 43. The permanent magnet synchronous motor stator voltage and battery voltage versus vehicle driving time

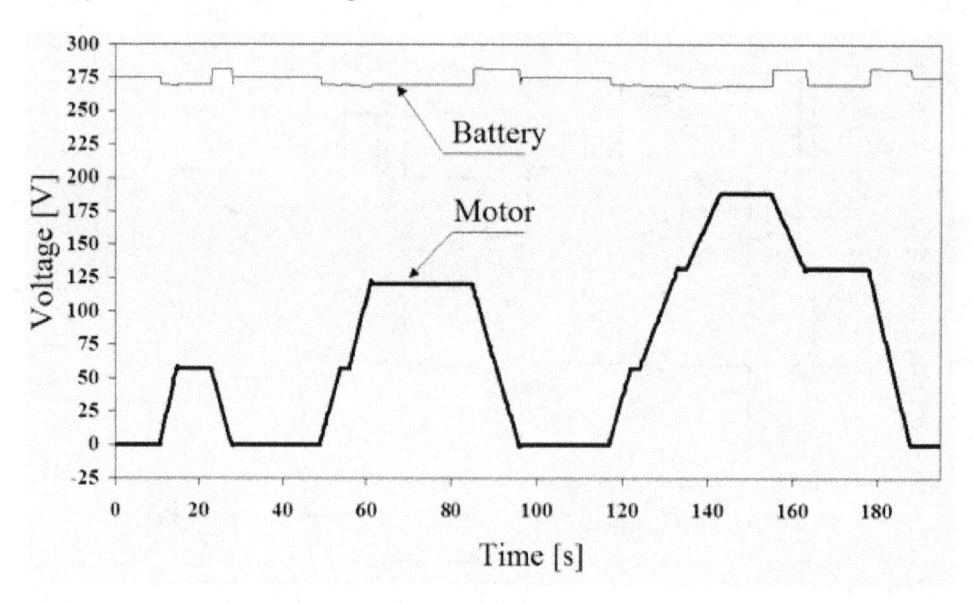

Figure 44. The battery current and voltage versus vehicle driving time

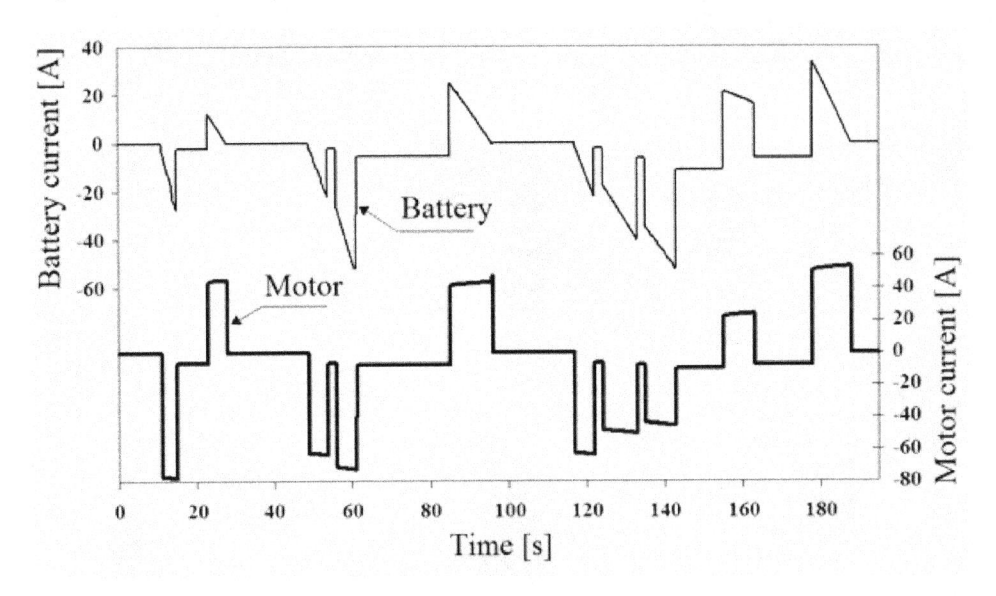

Figure 45. The entire power train's power efficiency versus vehicle driving time

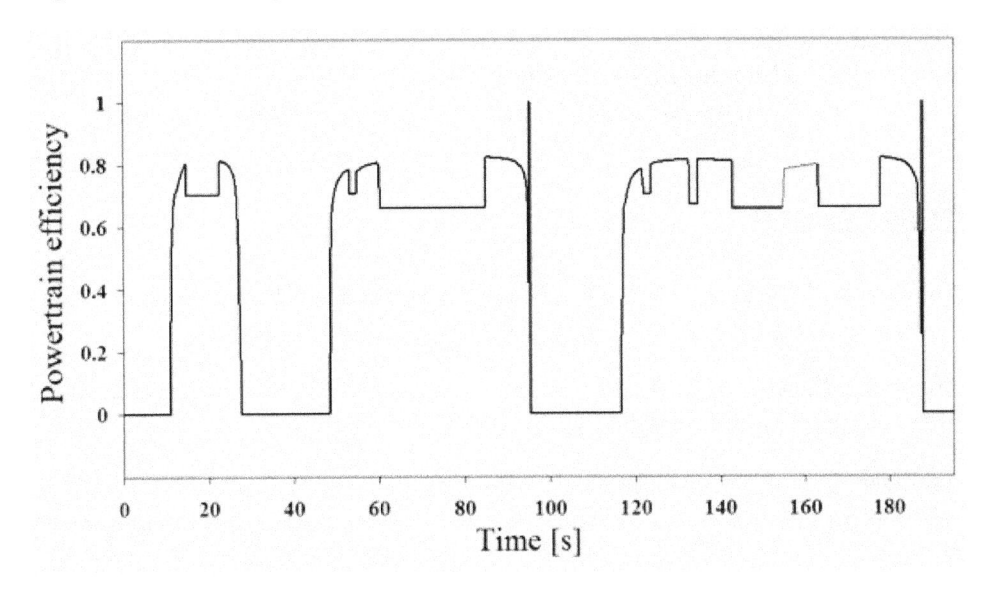

Figure 46. The exemplary map of a permanent magnet, synchronous motor (PM) drive (4Q-motor and generator modes operation) obtained from its mathematical model resolution and identified during lab-bench tests (declines depict power efficiency)

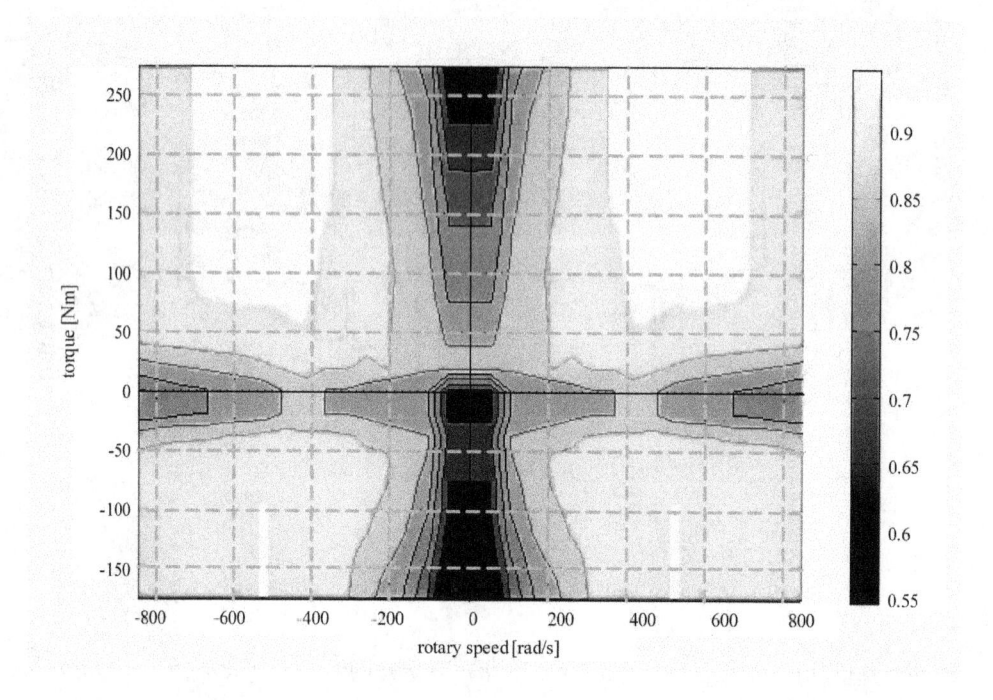

- **Maximum Speed of the Motor:** 3368.864 rpm
- **The Final Value of Battery Capacity Factors:** 0.990116

REFERENCES

Abdulaziz, M., & Jufer, M. (1974). Magnetic and electric model of synchronous permanent – magnet machines. *Bull. SEV, 74*(23), 1339–1343.

Anderson, W. M., & Cambier, C. (1990). An advanced electric drivetrain for EVs. In *Proceedings of EVS 10*. Hong Kong: EVS.

Ashihaga, T., Mizuno, T., Shimizu, H., Natori, K., Fujiwara, N., & Kaya, Y. (1992). Development of motors and controllers for electric vehicles. In *Proceedings of EVS 11*. EVS.

Baucher, J. P. (2007). Online efficiency diagnostic method of three phases asynchronous motor. In *Proceedings of Powereng IEEE*. IEEE.

Bin, W. (n.d.). Brushless DC motor speed control. *Ryerson Polytechnic University.*

Braga, G., Farini, A., Fuga, F., & Manigrasso, R. (1991). Synchronous drive for motorized wheels without a gearbox for light rail systems and electric cars. In *Proceedings of EPE'91 European Conference on Power Electronics.* EPE.

Brusaglino, G., & Tenconi, A. (1992). System engineering with new technology for electrically propelled vehicles. In *Proceedings of EVS 11.* EVS.

Burke, A. F., & McDowell, R. D. (1992). The performance of advanced electric vans – Test and simulation. In *Proceedings of EVS 11.* EVS.

Chan, C. C. (1994). The development of an advanced electric vehicle. In *Proceedings of EVS 12.* EVS.

Chan, C. C., Jiang, G. H., Chen, X. Y., & Wong, K. T. (1992). A novel high power density PM motor drive for electric vehicle. In *Proceedings of EVS 11.* EVS.

Chan, C. C., & Lueng, W. S. (1990). A new permanent magnet motor drive for mini electric vehicles. In *Proceedings of EVS 10.* EVS.

Chris, C., & Luo, L. (2005). Analytical design of PM traction motors. In *Proceedings of Vehicle Power and Propulsion Conference VPPC.* IEEE.

Datla, M., & High, A. (2007). Performance decoupling control of an induction motor with efficient flux estimator. In *Proceedings of Powereng.* IEEE.

Ferraris, P., Tenconi, A., Brusaglino, G., & Ravello, V. (1996). Development of a new high- performance induction motor drive train. In *Proceedings of EVS 13.* EVS.

Gosden, D. F. (1992). Wide speed range operation of an AC PM EV drive. In *Proceedings of EVS 11.* EVS.

Henneberger, G., & Lutter, T. (1991). Brushless DC – Motor with digital state controller. In *Proceedings of EPE '91 European Conference on Power Electronics and Application.* EPE.

Hirschmann, D., Tissen, D., Schroder, S., & De Donecker, R. W. (2005). Inverter design for HEV considering mission profile. In *Proceedings of Vehicle Power and Propulsion Conference VPPC.* IEEE.

Huang, H., Cambier, C., & Geddes, R. (1992). High constant power density wide speed range PM motor for EV application. In *Proceedings of EVS 11.* EVS.

Jezernik, K. R. (1994). Induction motor control for electric vehicle. In *Proceedings of EVS 12.* EVS.

Kenjo, T., & Nagamori, S. (1985). *Permanent magnet and brushless DC motors*. Oxford, UK: Claderon Press.

King, R. D., & Konrad, C. E. (1992). Advanced on–board EV AC drive – Concept to reality. In *Proceedings of EVS 11*. EVS.

Ledowskij, A. N. (1985). *Electrical machines with high coercive force permanent magnets*. Moscow: Energoatimizdat.

Matsusa, K., & Katsuta, S. (1996). Fast rotor flux control of vector contolled induction motor operating at maximum efficiency for EV. In *Proceedings of EVS 13*. EVS.

Mc Cann, R., & Domagatla, S. (2005). Analyses of MEMS-based rotor flux sensing in a hybrid reluctance motor. In *Proceedings of Vehicle Power and Propulsion Conference VPPC*. IEEE.

Ortmeyer, T. (2005). Variable voltage variable frequency options for series HV. In *Proceedings of Vehicle Power and Propulsion Conference VPPC*. IEEE.

Padmaraja, Y. (2003). *(BLDC) motor fundamentals*. Brushless, DC: Microchip Technology Inc.

Schofield, M., Mellor, P. H., & Howe, D. (1002). Field weakening of brushless PM motors for application in a hybrid electric vehicle. In *Proceedings of EVS 11*. EVS.

Schofield, N. (2006). Hybrid PM generators for EV application. In *Proceedings of Vehicle Power and Propulsion Conference VPPC*. IEEE.

Szumanowski, A. (2006). *Hybrid electric vehicle drives design*. Warsaw: ITE Press.

Szumanowski, A., & Bramson, E. (1992). Electric vehicles drive control in constant power mode. In *Proceedings of ISATA*. ISATA.

Szumanowski, A., & Brusaglino, G. (1992). Analyses of the hybrid drive consisting of electrochemical battery and flywheel. In *Proceedings of EVS 11*. EVS.

Veinger, A. H. (1985). *Adjustable synchronous drive system*. Moscow: Energoatimizdat.

Yamura, H. (1992). Development of a powertrain system for Nissan FEV. In *Proceedings of EVS11*. EVS.

Chapter 5
Nonlinear Dynamic Traction Battery Modeling

ABSTRACT

The role of the battery as the source of power in Hybrid Electric Vehicles (HEVs) is basic and significant. The process of battery adjustment and its management is crucial during the hybrid and electric drive design. The approach to battery modeling based on the linear assumption (as the Thevenin model) and then adopted for the data obtained in experimental tests, is here ignored, because the dynamic nonlinear modeling and simulations are the only tools for the optimal adjustment of the battery's parameters according to the analyzed vehicle driving cycles. The battery's capacity, voltage, and mass should be minimized, considering its overload currents. This is the way to obtain the minimal cost of the battery. Chapter 5 presents the method of determining the Electromotive Force (EMF) and the battery internal resistance as time functions, which are depicted as the functions of the State Of Charge (SOC). The model is based on the battery's discharge and charge characteristics under different constant currents that are tested in a laboratory experiment. The algorithm of battery's State-Of-Charge (SOC) indication is depicted in detail. The algorithm of battery State-Of-Charge (SOC) "online" indication considering the influence of temperature can be easily used in practice. The Nickel Metal Hydride (NiMH) and Lithium ion (Li-ion) batteries are taken into consideration and thoroughly analyzed. In fact, the method can also be used for different types of contemporary batteries, if the required test data is available.

DOI: 10.4018/978-1-4666-4042-9.ch005

Copyright ©2013, IGI Global. Copying or distributing in print or electronic forms without written permission of IGI Global is prohibited.

INTRODUCTION

Certainly, the battery modeling is of major concern for HEV and EV power trains. The basic data obtained from laboratory bench tests as a voltage for different values constant currents versus delivered to, or taken from, the battery charge (Ah), are commonly available. These static battery's characteristics are designated by producers and are the fundamental real data strictly depicted considered battery type, during its charging and discharging. For this reason, this data, after its proper transformation, can be used as the background to battery modeling. The dynamic nonlinear battery modeling related to a vehicle driving cycle is the target. The most important result of this process, based on the measurements of the real time battery voltage, current, and temperature, is the indication online of its state-of-charge (SOC) momentary value.

Battery modeling is of major concern for HEV and EV modeling. Accurate battery modeling is very important and necessary because of three reasons:

1. Simulation research for dynamic behaviour, considering power train architecture, is impossible without an accurate battery model;
2. An accurate battery model is the base of methodology to design battery monitoring and the management system (BMS), especially for the state of charge (SOC) calculation.
3. An accurate battery SOC indication is the background to a generation of proper feedback signals transmitted to the power train's master controller, using the most efficient online operation.

For the Ni-MH and Li-ion batteries, especially for HEV application, researchers in the world are working hard in order to find good and accurate models for these batteries. The major issue lies in the highly nonlinear characteristics of the battery in HEV applications. The challenge is associated, in particular, with the difficulty that the characteristic parameters of the battery, namely, the accurate data of the electromotive force (EMF) and the internal resistance are hardly obtainable in practical conditions. For this reason, most authors negate determination of EMF and the internal resistance of the battery, and their real and only direct influence on SOC in dynamic conditions. Some authors use a simplified model, but it's not useful for dynamic simulation of HEV - especially for optimization of drive operation. Some models are of sufficient enough complexity and accuracy, but they are difficult to use in practice, as the basis of methodology for BMS design. The HEV environment requires that the BMS has no direct control over current and voltage

Nonlinear Dynamic Traction Battery Modeling

experienced by the battery pack, because it's in the domain of the vehicle controller and inverter. This requirement implies that the study must depend on such measurements as instantaneous battery terminal voltage, current and external temperature.

This study adopts the method of modeling EMF and the internal resistance of Ni-MH and Li-ion battery as a function of SOC. The author's evaluated original model is based on the discharge/charge characteristics under a different discharge/charge rate tested by experiments in the laboratory. Accuracy of the model in the battery's operating range in a hybrid vehicle is higher than 98% (the missing 2% is the difference between real and computed characteristic values). In this study, SOC is defined as the ratio of quantity of output or input Coulomb capacity (in Ah) to the battery's nominal capacity.

In modern, full hybrid power trains, the high power (HP) Ni-MH battery in the majority of cases is used. For pure electric (EV) and plug-in hybrid (PHEV) power trains, Li-ion batteries are applied. The reason is that the Li-ion battery has more energy density in comparison with a Ni-MH. Anyhow, the cost of the Li-ion battery is higher than in the case of Ni-MH. The necessity of decreasing costs caused the appearance of the new Li-ion battery constructions based, not only on classical materials as Mn or Co, but also on 'three metals' technology, such as LiFePh and

Figure 1. The comparison of energy and power density for different batteries. Note: HP – high-power battery; HE – high-energy battery.

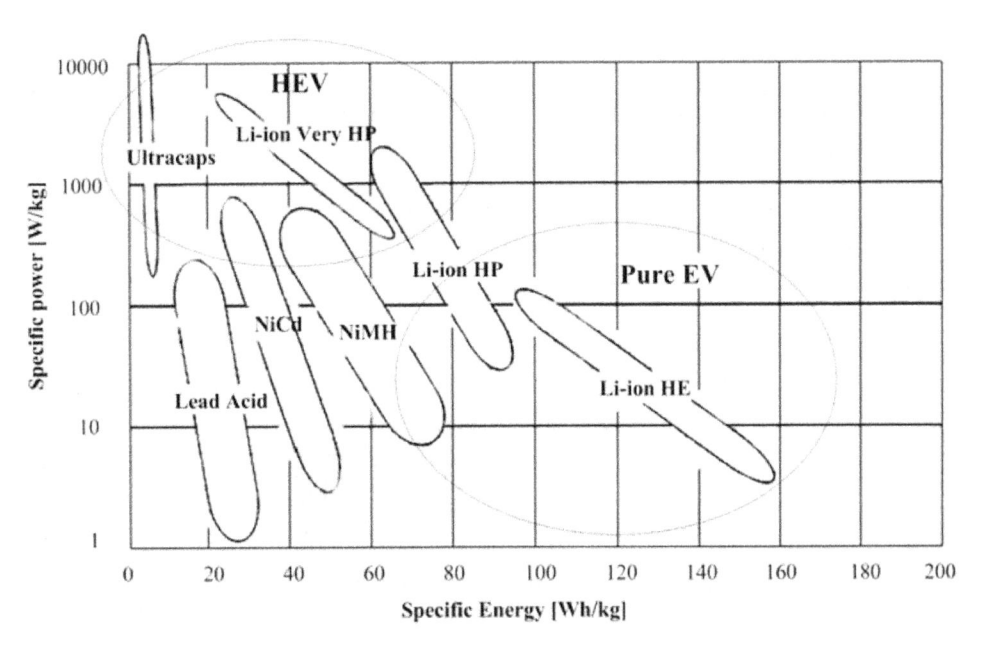

Table 1. The power density of different battery types

Battery Type	Energy density (Wh/kg)	Power density (W/kg)	Cycle life
Lead-acid	25~35	75~130	200~400
Advanced Lead- acid	35~42	240~421	500~600
Ni-MH	50~80	150~250	600~1500
NiCad	35~57	50~200	1000~2000
Li-ion	100~150	300	400~1200
Li-ion Polymer	100~155	100~315	400~600

Table 2. The data of different (existing and newly-designed) lithium ions Li-ion battery constructions

Cathode material	Average potential (V)	Gravimetric capacity (mAh/g)	Gravimetric energy (kWh/kg)
$LiCoO_2$	3.7	140	0.518
$LiMn_2O_4$	4.0	100	0.400
$LiNiO_2$	3.5	180	0.630
$LiFePO_4$	3.3	150	0.495
Li_2FePO_4F	3.6	115	0.414
$LiCo_{1/3}Ni_{1/3}Mn_{1/3}O_2$	3.6	160	0.576
$Li(Li_aNi_xMn_yCo_z)O_2$	4.2	220	0.920
Anode material	**Average potential (V)**	**Gravimetric capacity (mAh/g)**	**Gravimetricenergy (kWh/kg)**
Graphite (LiC_6)	0.1-0.2	372	0.0372-0.0744
Titanate ($Li_4Ti_5O_{12}$)	about 1	160	0.16-0.32
Si ($Li_{4.4}Si$)	0.5-1	4212	2.106-4.212
Ge ($Li_{4.4}Ge$)	0.7-1.2	1624	1.137-1.949

LiTiPh technology. In the last two cases, the basic voltage is lower than in classic Li-ion batteries, which means the energy density is about 15% lower, but the cost of manufacturing is significantly lower. Additionally, the thermal operating conditions are much safer than in the case of classic Li-ion batteries, which need special temperature management and complicated, also costly, voltage equalization (see the next chapter).

Figure 2. The data comparison of contemporary advanced lithium batteries. Note: The new lithium titanate nLiT is a super high-powered solution, applied successfully for fast charging.

LiMn
- specific energy 122-140 Wh/kg
- energy density 222 Wh/L
- Cell weight 2,9kg
- Nominal cell voltage 3,8V

nLiT
- specific energy 86 Wh/kg
- energy density 166 Wh/L
- Cell weight 1,6kg
- Nominal cell voltage 2,3V

LiFePO$_4$
- specific energy 90–110 Wh/kg
- energy density 220 Wh/L (790 kJ/L)
- specific power >300 W/kg
- Time durability >10 years
- Cycle durability 2,000 cycles
- Nominal cell voltage 3,3V

1. MAIN FEATURES OF MOST COMMON BATTERIES APPLIED IN HEV AND EV POWER TRAINS

1.1. The Nickel Metal Hydride Battery

The nickel metal hydride NiMH battery has been on the market since 1992. This kind of battery characteristic is similar to that of the metal hydride (MH) battery technology. It has the nominal voltage of 1.2 V and attains the specific energy of 65 Wh/kg, energy density of 150 Wh/l and specific power of 200 W/kg. Its active materials are hydrogen in the form of a metal hydride for the negative electrode, and nickel oxy-hydroxide for the positive electrode. The metal hydride is capable of undergoing a reversible hydrogen desorbing-absorbing reaction as the battery is discharged and recharged. An aqueous solution of potassium hydroxide is the major component of the electrolyte. The overall electro-chemical reactions are:

$$MH + NiOOH \leftrightarrow M + Ni(OH)_2$$

When the battery is discharged, metal hydride in the negative electrode is oxidized to form metal alloy, while nickel oxy-hydroxide in the positive electrode is reduced to nickel hydroxide. During charging, the reverse reactions occur.

A key component of the nickel metal hydride Ni-MH battery is the hydrogen storage metal alloy which should be formulated to obtain a material that is stable over a large number of cycles. There are two major types of these metal alloys being used for the nickel metal hydride battery. These are the rare-earth alloys based around lanthanum nickel, known as the AB_5, and alloys consisting of titanium and zirconium, known as the AB_2. The AB_2 alloys, typically, have a higher capacity than the AB_5 alloys. However, the trend is to use the AB_5 alloys, because of a better charge retention and stability characteristics.

Since the nickel metal hydride battery is still under continual development, its advantages based on present technology are summarized as follows: the highest specific energy and energy density of nickel-based batteries (65 Wh/kg and 150 Wh/l), environmental friendliness (cadmium free), flat discharge profile (similar to Ni-Cd), and rapid recharge capability (similar to Ni-Cd). However, it still suffers from the problem of high initial costs. Also, it may have a memory effect and be exothermic on charge.

1.2. The Lithium Ion Battery

Since the first announcement of the lithium ion (Li-Ion) battery in 1991, the lithium ion battery technology has seen an unprecedented rise to what is now considered to be the most promising rechargeable battery of the future. Although still in the stage of development, the Li-Ion battery has already gained acceptance for electric vehicles (EVs) and hybrid vehicle (HEVs) applications.

The lithium ion (Li-Ion) battery uses a lithiated carbon intercalation material (Li_xC) for the negative electrode instead of metallic lithium, a lithiated transition metal intercalation oxide ($Li_{1-x}M_yO_z$) for the positive electrode and a liquid organic solution or a solid polymer for the electrolyte. Lithium ions are swinging through the electrolyte between the positive and negative electrodes, during discharge and charge. The general electro-chemical reactions are described as:

$$Li_xC + Li_{1-x}M_yO_z \leftrightarrow C + LiM_yO_z$$

On discharge, lithium ions are released from the negative electrode, migrate via the electrolyte, and are taken up by the positive electrode. On charge, the process is reversed. Possible positive electrode materials include $Li_{1-x}CoO_2$, $Li_{1-x}NiO_2$, $Li_{1-x}Mn_2O_4$ or others (see Table 2) which have the advantages of stability in the air, high voltage and reversibility for the lithium intercalation reaction.

Nonlinear Dynamic Traction Battery Modeling

The $Li_xC/Li_{1-x}NiO_2$ type, loosely written as $C/LiNiO_2$ or simply called the nickel-based lithium ion (Li-Ion) battery, has the nominal cell voltage of 4 V, specific energy of 120 Wh/kg, energy density of 200 Wh/l, and specific power of 260 W/kg. The cobalt-based type has a higher specific energy and energy density, but with a higher cost and a significant increase of the self-discharge rate. The manganese-based type has the lowest cost and its specific energy and energy density lie between those of the cobalt-based and nickel-based types. It is anticipated that the development of the Li-Ion battery will ultimately move to the manganese-based type because of the low cost, abundance, and environmental friendliness of the manganese-based material. The general advantages of the Li-Ion battery are highest cell voltage (as high as 4 V), high specific energy and energy density (90-130 Wh/kg and 140-200 Wh/l), the safest design of lithium batteries (absence of metallic lithium), and long life-cycle (about 1,000 cycles). However, it still suffers from the drawback of a relatively high self-discharge rate (as high as 10% per month).

2. FUNDAMENTAL THEORY OF BATTERY MODELING

2.1. Electric Work and Gibe's Free Energy

Energy takes many forms: mechanical work (potential and kinetic energy), heat, radiation (photons), chemical energy, nuclear energy (mass), and electric energy. A summary is given regarding the evaluation of electric energy, as this is related to electro-chemistry.

Gibe's free energy ΔG is the negative value of maximum electric work,

$$\Delta G = -L_{max} \tag{1}$$

and, it is energy needed to transport electrical charges to a higher E volt-level of electrical potential, so:

$$L_{max} = nFE \tag{2}$$

where: n – quantity of grams (equivalent in electro-chemical process), F – Faraday constant, E – electromotive force.

2.2. The General Nernst Equation

The general Nernst equation correlates Gibe's Free Energy ΔG and E of a chemical system known as the galvanic cell. For the reaction

$$aA+bB = cC+dD \tag{3}$$

and:

$$Q = \frac{[C]^c[D]^d}{[A]^a[B]^b} \tag{4}$$

It has been shown that:

$$\Delta G = \Delta G^0 + RT \ln Q \tag{5}$$

and:

$$\Delta G = -nF\Delta E \tag{6}$$

where: ΔG^0 – battery thermodynamic potential change of the cell for nominal conditions.

Therefore:

$$nF\Delta E = -nF\Delta E^0 + RT \ln Q \tag{7}$$

where: ΔE^0 is the electromotive force for nominal conditions; R, T, Q and F are gas constant (8.314 J mol^{-1} K^{-1}), temperature (in K), reaction quotient, and Faraday constant (96485 C), respectively.

Thus, the following equation can be obtained:

$$\Delta E = \Delta E^0 - \frac{RT}{nF} \ln \frac{[C]^c[D]^d}{[A]^a[B]^b} \tag{8}$$

This is well-known as the Nernst equation. The equation provides the opportunity to calculate the cell potential of any galvanic cell for any concentrations. As we can see, the electromotive force depends on internal resistance, the capacity of a cell and on temperature alternation.

2.3. The Third Law of Thermodynamic

The third law of thermodynamics tells us about the relationship of the free energy change, the enthalpy change, and the entropy change:

$$\Delta G = \Delta H - T \cdot \Delta S \qquad (9)$$

The original energy change that comes out of the reaction in the system is ΔG. Since it is the only event that occurs in the universe, and energy flow divided by T is entropy, $\Delta G / T$ is equal to the total entropy change, and due to this reaction:

$$\Delta S = \Delta G / T \qquad (10)$$

where:

- ΔG : Battery thermodynamic potential change of the cell;
- ΔH : Enthalpy change;
- T : Temperature;
- ΔS : Entropy change.

2.4. Peukert's Formula

Peukert presented the relationship describing changes of the battery's capacity as a function of load at the beginning of the 20th century. The Peukert formula is also known as the battery's capacity decay equation,

$$K_w = i_a^{n(\tau)} t \qquad (11)$$

where:

- K_w: Discharge capacity of the battery;
- t: Duration of battery loading;
- n: Peukert's constant, varying in different types of batteries;
- τ: Transitory temperature

3. THE BASIC BATTERY DYNAMIC MODELING

One of the dynamic battery parameters has an identification approach which is based on the Thevenin generic model shown in Figure 3. The assumption is that battery's equivalent circuit is linear. In this case, the Laplace transformation can be used.

It is possible to obtain the battery's terminal voltage by determination of the operator's impendence Z(s) as follows:

$$Z(s) = \frac{Ua(s)}{Ia(s)} = R_1 + \frac{R_2}{R_2 C + 1}$$
$$U_a = U_{oc} - U_1 - U_2 \qquad (12)$$
$$U_a = U_{oc} - I_a Z(s)$$

where:

- U_{oc}: Open circuit voltage;
- R_1: Referring electrodes, separators, electrolyte resistances;
- R_2: Polarization resistance
- C: Polarization capacitance
- s: Operator;
- I_a: Battery's load current.

$$U_a = U_{oc} - R_1 i_a(t) - R_2(1 - e^{-t/T_a}) i_a(t) \qquad (13)$$

The Equation 13 is based on the following necessary assumptions:

Figure 3. The battery's substitution scheme, according to the Thevenin method for terminal voltage determination, respectively, to load step

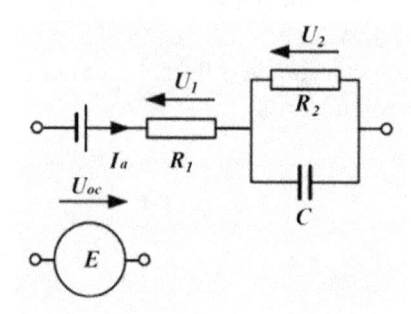

Nonlinear Dynamic Traction Battery Modeling

$$i_a(t) = I_{a1}(t)$$

$$T_a = R_2 C$$

$$\frac{dE}{dt} = 0$$

$$\frac{dU_2}{dt} \neq 0$$

where:

- I_{a1} : The battery's maximum current under pulse load conditions;
- E : Defined in steady-state conditions electromotive battery force (EMF);
- t : Time.

Note: Time constant T_a is determined by experimental test.

The application of linear equations is possible, only when the battery's accumulated energy is assumed not changed. For a very short time, when the battery's current peak jumps (pulse condition), the battery's terminal voltage identified by Equation (13) can be accepted, e.g. when the controller's parameters are adjusted.

However, in practice, the battery's average load current (see Figures 5, 7) is alternated in about a second. During this period of time, the physical assumption that the battery's electromotive force (battery's energy) is not changeable is a false approach.

Anyway, the electro-chemical accumulator is so significantly nonlinear, that the battery's energetic analysis in EV and HEV driving conditions, is better carried out it using another method based on common battery laboratory tests. For this reason, the original nonlinear dynamic method is proposed.

The basis upon which we can formulate the energy model of the electro-chemical battery, is the battery substitute circuit, as shown in Figure 4.

3.1. State of Charge

The available capacity of a battery does not only depend on temperature, but also is influenced by the load current and the duration of the current's flow. The residual capacity Q_u is a function of temperature, current and time, for the case when $i_a = f(t)$ can be described by a monotonic, non-decreasing function which can be derived from the equation:

Figure 4. The substitute circuit of an electrochemical battery expressed by battery capacity, corresponding to battery electromotive force E and nonlinear battery electrodes R_e and its electrolyte R_{el} nonlinear resistance ; u (i_a, t, k) is the battery's terminal momentary voltage, t is time, k – battery's state-of-charge factor, and i_a is the battery's momentary current

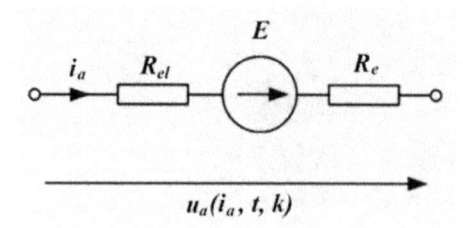

$$Q_u(i_a, t, \tau) = Q_\tau(\tau) - K_w(i_a(t), t) \tag{14}$$

or

$$Q_u(i_a, t, \tau) = Q_\tau(\tau, i_a) - \int_0^t i_a(t)\, dt \tag{15}$$

where:

- $K_w(i_a(t), t)$: A nonlinear function which determines the discharged capacity
- $\int_0^t i_a(t)\, dt$: Discharged capacity, which has been drawn from the battery from $t=0$ till the time t
- $Q_\tau(\tau)$: Battery capacity depends only on temperature for $i_a = 0$
- $Q_\tau(\tau, i_a)$: Battery's available capacity as a function of temperature and load current.

In practice, the residual capacity Q_u is usually determined for any currents and for any assumed, but constant temperature τ. As a result, characteristic families $Q_u = f(i_a, t)$ for a fixed range of temperature variations are obtained.

The Peukert formula is well-known as the battery's capacity decay equation:

$$K_w = i_a^{n(\tau)} t \tag{16}$$

where:

- K_w: Discharged capacity;
- t: Duration of battery loading;
- n: Peukert's constant, varying with types of battery.

It can be easily noticed that in the Equation (16), for $n=1$ the relationship between the change of the battery's capacity and the load current is linear. The higher the n value is, the higher capacity decay occurs during the flow of the same current, i_a. Usually, the Peukert equation is used in a different form, i.e. for $\tau = \tau_n = $ const

$$Q(i_a) = Q_{\tau n}\left(\frac{i_a}{I_n}\right)^{-\beta} \tag{17}$$

where:

- I_n: Battery-rated current;
- β: A constant, determined according to Peukert's equation;
- $Q_{\tau n}$: Battery's nominal capacity.

When the i_a current for determined periods $0 \leq t \leq tm$ is constant, the discharged capacity can be derived from Equations (14) and (16), the following equations can be used for numerical calculations, with the function $i_a = f(t)$ approximated by a staircase function, while:

$$\int_0^{t_m} i_a(t)\,dt = \sum_{t_i=0}^{t_m} i_{ai} t_i \tag{18}$$

For $\tau = $ const

$$Q_u(i_a, t) = Q_\tau - \sum_{t_i=0}^{t_m} i_{ai}^n t_i \tag{19}$$

Assuming temperature influence:

$$Q_u(i, t, \tau) = c_\tau(\tau) Q_{\tau n}\left(\frac{i_a(t)}{I_n}\right)^{-\beta} - \int_0^t i_a(t)\,dt \tag{20}$$

where $c_\tau(\tau)$ coefficient can be defined as the temperature index of nominal capacity,

$$c_\tau(\tau) = \frac{Q_\tau}{Q_{\tau n}} = \frac{1}{1+\alpha\left|(\tau_n - \tau)\right|} \qquad (21)$$

where: α- is the temperature capacity index (we can assume $\alpha \approx 0.01$ deg-1).

Both the numerator and the denominator of the equation's (17) left side shall be multiplied by the value of average terminal voltage of the battery under load current i_a.

$$\frac{Q(i_a)\overline{U}}{Q_{\tau n}\overline{U}} = \left(\frac{i_a(t)}{I_n}\right)^{-\beta(\tau)} \qquad (22)$$

The left side of the equation is the quotient of electric power drawn from the battery during the flow of $i_a \neq I_n$ current, and the electric power drawn from the battery, during discharging at the rated current. The quotient mentioned above defines the usability index of the accumulated power:

$$\eta_A(i_a, \tau) = \left(\frac{i_a(t)}{I_n}\right)^{-\beta(\tau)} \qquad (23)$$

When $i_a < I_n$, the value of the index can exceed 1.

During further solution of the Equation (20), it can be transformed by means of the Equation (23):

$$Q_u(i, t, \tau) = c_\tau(\tau)\eta_A(i_a, \tau)Q_{\tau n} - \int_0^t i_a(t)\mathrm{d}t \qquad (24)$$

Therefore, the real battery's state of charge can be expressed in the following way:

$$k = \frac{Q_u}{Q_{\tau n}} = \frac{c_\tau(\tau)\eta_A(i_a, \tau)Q_{\tau n} - \int_0^t i_a(t)\mathrm{d}t}{Q_{\tau n}} \qquad (25)$$

where $k=1$ for a nominally-charged battery, $0 \leq k \leq 1$, thus

$$k = c_\tau(\tau)\eta_A(i_a, \tau) - \frac{1}{Q_{\tau n}}\int_0^t i_a(t)\mathrm{d}t \qquad (26)$$

Nonlinear Dynamic Traction Battery Modeling

We can solve Equations (24) and (26) for any values of τ, i_a and t. Nominal capacity is the equation's parameter and it is constant for a given type of battery. The state of charge (SOC) k of battery can also be expressed in percentage terms.

3.2. The Battery's Internal Resistance

The internal resistance can be expressed in an analytical way, where:

$$R_w(i_a, \tau, Q) = R_{el}(\tau, Q) + R_e(Q) + bE(i_a, \tau, Q) I_a^{-1} \qquad (27)$$

$bE(i_a, \tau, Q) I_a^{-1}$

is the resistance of polarization.

b is the coefficient which expresses the relative change of the polarization's electromotive force on the cell's terminals during the flow of I_a current in relation to the electromotive force E for nominal capacity. Electrolyte resistance R_{el} and electrode resistance R_e are inversely proportional to temporary capacity of the battery. The Equation (27) can be transformed into a function of the SOC of the battery.

$$R_w(i_a, \tau, Q) = \frac{l_1}{Q_u(i_a, t, \tau)} + \frac{l_2}{Q_u(i_a, t, \tau)} + \frac{bE(i_a, \tau, Q)}{i_a(t)} \qquad (28)$$

$$R_w(i_a, t, \tau) = lk^{-1} + b\frac{E(k, \tau)}{i_a(t)} \qquad (29)$$

where: $l = (l_1 + l_2) Q_{rn}^{-1}$, is piecewise constant, assuming that the temporary change of the battery's capacity is significantly smaller than its nominal capacity. The coefficient l is determined experimentally under static conditions. $E(k)$ is the temporary value of the electromotive force and depends on SOC of battery.

According to the Equation (29), the b coefficient can be defined as:

$$b(k) = \frac{E(k) - E_{min}}{E_{max}} \qquad (30)$$

where:

- E_{max}: Is the local (see Figure 8) maximum EMF value
- E_{min}: Is the local (see Figure 8) minimum EMF value.

3.3. Terminal Voltage of Battery

According to the assumed equivalent circuit diagram (see Figure 4), the external battery's voltage characteristics measured at the battery's terminals, are determined by the equation:

$$u(i_a, t, k) = E(i_a, \tau, k) - i_a(t) R_w(i_a, t, \tau, k) \tag{31}$$

where:

- $E(i_a, \tau, k)$: Temporary electromotive force;
- $R_w(i_a, t, \tau, k)$: Temporary internal resistance.

The battery's state of charge factor k, which in the Equation (31) is the independent variable, has been previously defined for conditions when the battery was discharged with a continuous current, described by a monotonic, non-decreasing function of time. In reality, in applications of vehicle drives, the electro-chemical battery load current - as a function of time - is rather a piecewise monotonic function.

3.4. Characteristic of Battery Load Current in Electric Vehicles

In electric vehicle drive, discharging conditions are mostly influenced by the method of control of the electric traction motor, which acts as the battery load. When an AC traction motor is employed, and the frequency of powering the traction machine is varied, in principle, the battery load current is continuous due to the structure of the inverter and smoothing networks. Typical current waveforms for an AC asynchronous motor and a battery for a step-load under both motor and generator operation conditions are presented in Figure 5 (Piller, Perrin, & Jossen, 2001). Thus, the battery current is a continuous function in a time interval of 100 ms and this is quite sufficient for power calculations.

Similar statements are true for both synchronous and brushless direct current permanent magnet motors (Moseley & Cooper, 1998).

However, a drive employing a direct current (DC) permanent magnet (PM) motor creates a different situation. There are two control circuit concepts – with, or without, an input filter. The battery load waveforms are different from these approaches: the latter provides a pulse load waveform, while the former provides a continuous load waveform corresponding to a mean value over the pulse repetition period T (see Figure 5, 6, 7). This battery's current waveform depends directly on the control function, which is defined by the ratio of the conduction time t_{BATT} of

Nonlinear Dynamic Traction Battery Modeling

the semi-conductor switch to the pulse repetition period T. A pulse load causes, in the battery's circuit, temporary current decays, correlated with the pulsation period. However, from the energy point of view, we can disregard this phenomenon, as we know that the pulsation frequency is very high. Thus, it is possible to assume that the $i_a = f(t)$ is a monotonic function.

For the purposes of power calculations, the battery's current throughout the pulsation period is substituted by its average value (see Figure 7), which can be influenced by the flux of energy flowing through the battery. The amount of energy is known and can be derived from the vehicle's driving cycle, i.e. from the distribution of changes of vehicle propelling power, in the function of time. The function, similar to the current, is a piecewise monotonic function.

3.5. The State of Charge (SOC) of Battery with the Load Current Described by a Piecewise Monotonic Function

For the definitions of the battery's discharged capacity, we have assumed that it should be possible to determine its value at any moment during discharging. We assume $i_a(t)$ current is continuous and its value does not decrease. Thus, we have

Figure 5. The exemplary alternative current (AC) motor and battery currents versus time for a step load

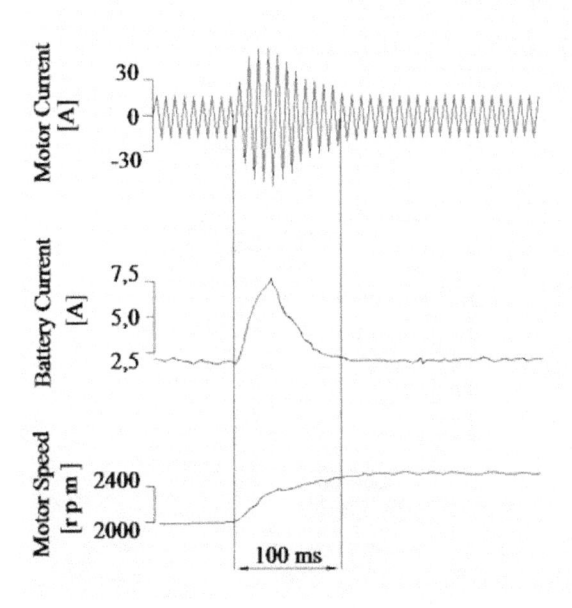

Figure 6. The exemplary brush direct current (DC) permanent magnet (PM) motor and battery currents versus time under real pulse load conditions

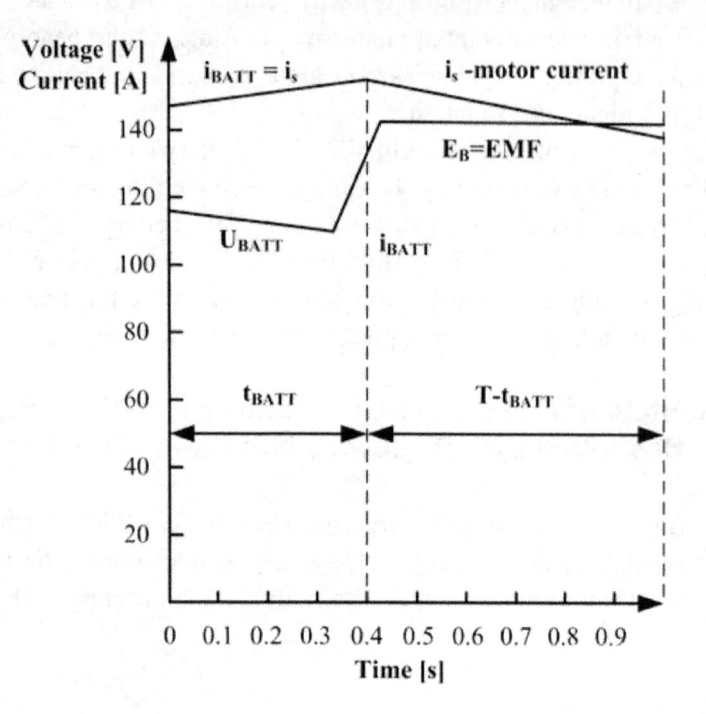

Figure 7. The battery current and voltage waveform under pulse load conditions

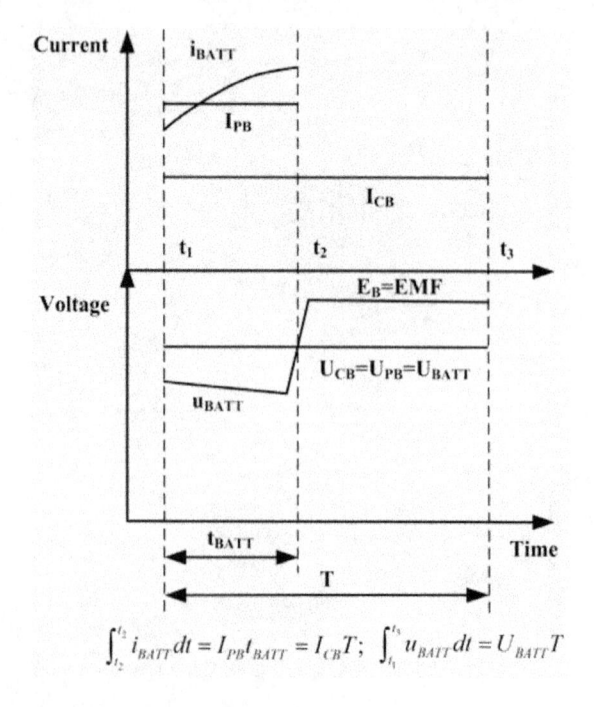

$$\int_{t_2}^{t_2} i_{BATT} dt = I_{PB} t_{BATT} = I_{CB} T; \quad \int_{t_1}^{t_3} u_{BATT} dt = U_{BATT} T$$

Nonlinear Dynamic Traction Battery Modeling

to assume that at any moment $t_i = 0$ the initial battery capacity is described by the Equation (26).

According to the Equation (28), the battery's state of charge equals k; after a period of time $t_i \neq 0$ the battery discharges, $i_a = f(t)$ is a piecewise monotonic function. In this case, the capacity drawn from the battery is equal to:

$$\Delta Q = \int_{t_i}^{t_{i+1}} i_a(t) \, dt \tag{32}$$

Under such conditions, the residual capacity of the battery can be derived from the equation

$$Q'_u(i_a, t, \tau) = c_\tau(\tau) \eta_A(i_a, t, \tau) Q_{T\tau n} - \int_0^t i_a(t) dt - \int_{t_i}^{t_{i+m}} \eta_A(i_a, t, \tau) i_a(t) dt \tag{33}$$

where m=1, 2, 3...

When the load current is described by a piecewise monotonic function, the state of charge can be described by the equation:

$$k' = \frac{Q'_u(i_a, t, \tau)}{Q_{\tau n}} = k - Q_{\tau n}^{-1} \int_{t_i}^{t_{i+m}} \eta_A(i_a, t, \tau) i_a(t) dt \tag{34}$$

We can neglect the increase of the battery's capacity during the decay of the current under a pulse load, because it has no influence on the electro-chemical characteristic of the battery. In real conditions, the residual capacity is slightly higher than the calculated result, because during braking, the battery takes in some regenerative energy. If we neglect the regeneration, the computed result is slightly less favorable than the real case, which from a technical point of view, substantiates the assumption of such simplifications. Therefore, in practice, the battery's state of charge can be derived from the equation:

$$k' = \frac{Q'_u(i_a, t, \tau)}{Q_{\tau n}} = k - Q_{\tau n}^{-1} \int_{t_i}^{t_{i+m}} \eta_A(i_a, t, \tau) i_a(t) dt \tag{35}$$

The internal resistance of the battery is described by the equation:

$$R_w'(i_a,t,\tau,k') = b\frac{E(k')}{i_a(t)} + l\left[k - Q_{TN}^{-1}\int_{t_i}^{t_{i+m}}\eta_A(i_a,t,\tau)^{-1}i_a(t)\,\mathrm{d}t\right]^{-1} \qquad (36)$$

The voltage at the battery's terminals during discharging, with the current described by a piecewise monotonic function, can be expressed by the equation:

$$u(i_a,t,k') = E(i_a,t,\tau,k') - i_a(t)R_w'(i_a,t,\tau,k') \qquad (37)$$

3.6. The Battery Charging

The accumulation of the electric energy in the electro-chemical traction battery takes place during the generating operation of the traction motor (i.e. regenerating the kinetic energy during the vehicle's braking) or during charging of the battery with a current from a primary source (an Internal Combustion Engine [ICE] – alternator set). In general, the charging current of the battery is a piecewise monotonic time function, non-increasing or non-decreasing. Change of traction motor flux value in the stage of its generating operation, counteracts any decrease of the angular velocity of the motor during vehicle braking, and is usually forced by pulse control.

The state of charge of the battery, which has been charged with a current being a piecewise monotonic time function, can be determined from the equation:

$$k'' = k' + Q_{TN}^{-1}\int_{t_i}^{t_{i+m}} i_a(t)\,\mathrm{d}t \qquad (38)$$

where k' is the battery's state of charge at the instant when the charging starts.

The internal resistance of the battery, at the time of starting the charging process, is described by the equation:

$$R_w''(i_a,t,\tau,k'') = b\frac{E(k'')}{i_a(t)} + l\left[k' + Q_{TN}^{-1}\int_{t_i}^{t_{i+m}} i_a(t)\,\mathrm{d}t\right]^{-1} \qquad (39)$$

In Equations (38) and (39), the charging current's direction is in reverse to the discharging directional current, which means the following condition is assumed $i_a < 0$.

Nonlinear Dynamic Traction Battery Modeling

Battery charging is limited at the moment when the threshold voltage at the battery's terminals is reached. At such voltage, the phenomenon of electrolyte gassing takes place for Lead-acid and Ni-MH batteries. For the Li-ion battery, the overcharge will result in permanent damage.

The voltage characteristics of battery charging can be described by the equation:

$$u(i_a,t,k'') = E(i_a,t,\tau,k'') + i_a(t)R_w''(i_a,t,\tau,k'') \tag{40}$$

3.7. Electro-Chemical Battery Power Efficiency

The power efficiency of an electrochemical battery is defined in the following way:

$$\eta_E = \frac{\int_0^{t_d} u_d(i_a,t,k')i_{ad}(t)\,dt}{\int_0^{t_c} u_c(i_a,t,k'')i_{ac}(t)\,dt} \tag{41}$$

where d subscript is used to denote time, current and the voltage of discharge, and c subscript is used to mark time, current and voltage of charging.

For most general conditions of battery loads, the Equation (41) can be transformed into the form:

$$\eta_E = \frac{\int_0^{t_d} \begin{pmatrix} E(i_a,t,\tau,k') \\ -i_{ad}(t)R_w'(i_a,t,\tau,k') \end{pmatrix} i_{ad}(t)\,dt}{\int_0^{t_c} \begin{pmatrix} E(i_a,t,\tau,k'') \\ +i_{ac}(t)R_w''(i_a,t,\tau,k'') \end{pmatrix} i_{ac}(t)\,dt} \tag{42}$$

Electromotive force E is an indicator of the level of energy, which has been stored in the battery. During charging, the energy which is supplied to the battery, initiates electro-chemical reactions, which result in the increase of the potential difference between negative and positive electrode plates. Obviously, certain losses take place during the process. From the energy point of view, we can see the losses by measuring the voltage drop at the internal battery's resistance. The energy supplied to the battery can be expressed by the equation:

$$E_c = \sum_{i=1}^{m} u_i\left(i_a, t, \tau, k''\right) i_{ai}\left(t\right) t_i = \sum_{i=1}^{m} \left(E\left(i_a, t, \tau, k''\right) + i_{ac}\left(t\right) R_w''\left(i_a, t, \tau, k''\right)\right) i_{ac}\left(t\right) t_i \tag{43}$$

$i = 1, 2, 3, ..., m$

The energy, which is stored in the battery, is directly proportional to the electromotive force. From the Equation (40) we can determine the quantitative dependency between the energy supplied to the battery and the real energy stored in the effect of charging. The charging efficiency factor η_{AC} can be defined as follows:

$$\eta_{Ac}\left(i_a, t, \tau, k''\right) = \sum_{i=1}^{m} \frac{E\left(i_a, t, \tau, k''\right)}{E\left(i_a, t, \tau, k''\right) + i_{ai}\left(t\right) R_w''\left(i_a, t, \tau, k''\right)} \tag{44}$$

For numerical calculations of temporary values of the efficiency factor η_{AC} it is convenient to use the Equation (44).

During discharging of the battery, the electro-chemical reaction is inverted; however, during the quantitative valuation of the energy drawn, we shall consider the battery's decay of capacity according to Peukert's theory. In this case, the discharge efficiency η_{Ad} of the battery takes the form of:

$$\eta_{Ad}\left(i_a, t, \tau, k'\right) = \left(\sum_{i=1}^{m} \frac{E\left(i_a, t, \tau, k'\right)}{E\left(i_a, t, \tau, k'\right) - i_{ai}\left(t\right) R_w'\left(i_a, t, \tau, k'\right)}\right)^{-1} \tag{45}$$

The purpose of the Equation (45) is quite similar to that of the Equation (42). The discharge efficiency of the battery can also be determined by means of the stored energy usability factor, according to the Equation (21). In that case, the exponent β shall be determined for a temporary load current and temperature from the following equation:

$$\beta(\tau) = \frac{m \ln Q_{\tau n} - \sum_{1}^{m} \ln Q_\tau\left(i_a\right)}{\sum_{1}^{m} \ln\left(\frac{i_a}{I_n}\right)} \tag{46}$$

3.8. The Method of Determination of Electromotive Force by Experimental Data

The above model depends on time, so it has a dynamic character enabling the mean time to analyze a great deal of physical quantities, relatively, to either traction battery or the overall propulsion structure, of which the battery is one of the elements.

Nevertheless, the vehicle's driving power change is closely connected with the instantaneous load current change, the voltage, and the internal resistance, and most of all, with the state of charge of battery.

For the given sort of battery, the above-mentioned model is basically used to determine the following functions of the respective state of charge of battery:

Internal resistance:

$$R_w(t,k) = b(k)\frac{E(k)}{i_a(t)} + \frac{l(k)}{k(t)} \quad k(t) \in (0,1\rangle \tag{47}$$

Electromotive force:

$$E(t) = E(k) \tag{48}$$

For the complete solution of the previously mentioned equations set, it is necessary to determine the following functions:

$$\begin{aligned} b(t) &= b(k) \\ l(t) &= l(k) \end{aligned} \tag{49}$$

It is impossible to evaluate all the above-considered quantities, directly or experimentally. It is impossible to record their real value (monitoring by measure) in a real vehicle. It can only be evaluated by resolving the described battery model, at the input function in the shape of the battery current and voltage. Determining the dynamic state of charge (k) is certainly necessary information for hybrid vehicle designing and maintenance.

3.9. The Determination of Internal Resistance and Electromotive Force

According to laboratory tests from nickel metal hydride (Ni-MH) battery GAIA and SAFT lithium ion (Li-Ion) batteries, which were given the battery terminal voltage as the functions of the state of charge (SOC) or its Coulomb capacity (Ah) for different constant load currents, the obtained experimental data was taken under a proper modeling process. To determine internal resistance R_w, it is necessary to assume R_w is not only the function of load current. It results in additional inexactitude in terms of numerological calculation, but to an extent, it is irrelevant. The value of R_w is obtained as an arithmetical average of R_w from the following equation:

$$R_w = (U_n - U_{n+1})/(I_{n+1} - I_n) \tag{50}$$

where:

$$U_n > U_{n+1}, \ I_{n+1} > I_n$$

After this step, we can determine the EMF, according to the following equation.

$$\begin{aligned} U_a &= E - i_a R_w \quad \text{discharge} \\ U_a &= E + i_a R_w \quad \text{charge} \end{aligned} \tag{51}$$

and then, it is possible to determine the coefficient b(k), using the iteration method described below.

Determination of the coefficients b(k) and l(k) is as follows:

Real electromotive force E's approximated curve is a straight line (see Figure 8), according to the equation:

$$\begin{aligned} E(k) &= E_{min} + \Delta U(k) \\ \Delta U &= E_{max} - E_{min} \end{aligned} \tag{52}$$

and

$$b(k) = \frac{E(k) - E_{min}}{E_{max}} \tag{53}$$

is determined for the whole range of

$$E(k) \in \langle E_{min}, E_{max} \rangle,$$

Nonlinear Dynamic Traction Battery Modeling

$k \in \langle 0, 1 \rangle$

In practice, the Equation 53 takes the form expressed by the Equation 55, whilst graphical interpretation is shown in Figure 8.

The coefficient l(k) in regard to linear relation E(k) approximation, takes values in ranges as a constant or as a little variable. Hence, we can make the assumption that l(k) = 1 = const for the whole alteration range,

$E(k) \in \rangle E_{min}, E_{max} \langle$

and $k \in \langle 0, 1 \rangle$ which is validated by the verifications, e.g. electric vehicle mileage and other laboratory researches.

4. NONLINEAR DYNAMICS TRACTION BATTERY MODELING

This chapter presents a method of determining Electromotive Force (EMF) and battery internal resistance as time functions, which are depicted as functions of the state of charge (SOC). The model is based on battery discharge and charge characteristics under different constant currents, tested by laboratory experiments. Another method of determining the battery's SOC, according to the battery's modeling result is also considered. The influence of temperature on battery performance is analyzed, according to laboratory-tested data, and the theoretical background for calculating the SOC is obtained. The algorithm of the battery's SOC indication is depicted in detail. The algorithm of the battery's SOC 'On-line' indication considering the influence of temperature, can be easily used in practice by microprocessors. Ni-MH and Li-ion battery are taken under analysis. In fact, the method also can be used for different types of contemporary batteries, if the required test data is available.

Hybrid Electric Vehicles (HEVs) are remarkable solutions for the worldwide environmental and energy problem caused by automobiles. The research and development of various technologies in HEVs is being actively conducted. The role of battery as power source in HEVs is significant. Dynamic nonlinear modeling and simulations are the only tools for the optimal adjustment of battery parameters according to analyzed driving cycles. The battery's capacity, voltage and mass should be minimized, considering its over-load currents. This is the way to obtain the minimum cost of the battery according to the demands of its performance, robustness, and operating time.

The process of battery adjustment and its management is crucial during the hybrid's drive design. The generic model of the electro-chemical accumulator, which can be used in every type of battery, is carried out. This model is based on the physical and mathematical modeling of the fundamental electrical impacts during energy conservation by the battery. The model is oriented to the calculation of the parameters - EMF and internal resistance. It is easy to find direct relations between SOC and these two parameters. If the battery's electromotive force EMF is defined, and the function versus the SOC ($k \in <0,1>$) is known, it is simple to depict the discharge/charge state of a battery.

The model is really nonlinear because the correlative parameters of the equations are functions of time (or functions of the SOC because $SOC = f(t)$) during battery operation. The modeling method presented in this chapter must use the laboratory data (for instance, voltage for different constant currents or internal resistance versus the battery SOC) that are expressed in a static form. This data has to be obtained by discharging and charging tests. The considered generic model is easily adapted to different types of battery data, and is expressed in a dynamic way, using approximation and iteration methods.

An HEV operation puts unique demands on the battery when it operates as the auxiliary power source. To optimize its operating life, the battery must spend minimal time in overcharge and/or over-discharge. The battery must be capable of furnishing or absorbing large currents, almost instantaneously, while operating from a partial-state-of-charge baseline of roughly 50% (Ovshinski, Dhar, Venkatesan, Fetchenko, Gifford, & Corrigan, 1992; Pang, Farrell, Du, & Barth, 2001; Piller et al., 2001; Plett, 2003a; 2003b; Rodrigues, Munichandraiah, & Shukla, 2000; Salkind, Atwater, Singh, Nelatury, Damodar, Fennie, & Reisner, 2001; Sato, & Kawamura, 2002; Shen, Chan, Lo, & Chau; 2002). For this reason, especially significant is knowledge about the battery's internal loss (efficiency), because its strong influence on the battery's state of charge (SOC) takes place.

There are many studies dedicated to determining the battery's SOC (Piller et al., 2001; Plett, 2003a; Szumanowski, 2010; Tojura, & Sekimori); however, these solutions have some limitations with regard to practical application (Szumanowski, 2006). Some solutions for practical application are based on a loaded terminal voltage (Sonnenschein, Varta, Ovonic, Horizon, & Trajan, n.d.; Stempel, Ovshinsky, Gifford, & Corrigan, 1998; Szumanowski & Brusaglino, 1999; Szumanowski, 2000) or a simple calculation of the flow of charge to/from a battery (Szumanowski, Dębicki, Hajduga, Piórkowski, & Chang, 2003; Szumanowski, Chang, Piórkowski, Jankowska, & Kopczyk, 2005), which is the integral factor that is based on current and time. Both solutions do not consider the strong nonlinear behavior of a battery. It is possible to determine transitory value of the SOC 'on-line' in real driving conditions with proper accuracy, considering the nonlinear characteristic of a battery by resolving the mathematical model that is presented in this paper.

Nonlinear Dynamic Traction Battery Modeling

This is the background for optimal battery parameters, as well as the proper battery management system (BMS) design - particularly in the case of SOC indication (Szumanowski, 2010). The high power (HP) Ni-MH and Li-ion batteries so commonly used in HEV, were considered.

Finally, the plots of battery voltage, current and SOC as alterations, in time for a real experimental hybrid drive equipped with BMS, especially designed, according to the presented original battery modeling method, are attached.

4.1. The Mathematical Battery Modeling Based on Equations Outlined in Paragraph 3

The basis behind the formulation of the energy model of an electro-chemical battery is the battery's physical model, as shown in Figure 4.

The electromotive force (EMF) as a function of k is deduced from the well-known battery voltage equation, including the momentary value of voltage and internal resistance, because the values of the battery's internal resistance R_w and its electromotive force EMF are unknown. The solution can be obtained by a linearization and iterative method, which is explained as follows:

For each value of k it is possible to obtain:

$$\begin{cases} b(k) = \dfrac{E(k) - E^*_{min}}{E^*_{max}} \\ R_w(k_n) = \dfrac{E(k_n) - E^*_{min}}{E^*_{max}} \dfrac{E(k_n)}{I_{a(n)}} + \dfrac{l(k_n)}{k_n} \\ R_w(k_{n-1}) = \dfrac{E(k_{n-1}) - E^{**}_{min}}{E^{**}_{max}} \dfrac{E(k_{n-1})}{I_{a(n-1)}} + \dfrac{l(k_{n-1})}{k_{n-1}} \\ l(k_n) \cong l(k_{n-1}) \end{cases} \quad (54)$$

where: $I_{a(n)}$ – discharge or charge currents, for which the experimental characteristics are determined (see Figures 8, 9, 10).

In this figure, the curve of the electromotive force E is presented discreetly, by use of straight broken lines, determining the inclination angle, which is dependent on the relative changes of the internal resistance R_w measured values. For each relative line segment, an electromotive force of local maximal and minimal values can be correspondingly determined as E^*_{max} and E^*_{min}. Where the indices * denote the analyzed range $\Delta k = k_n - k_{n-1}$, $n = 1,2,3...$ where n is the range number of the set $k \in (0,1\rangle$.

Figure 8. The battery electromotive force EMF versus its state of charge factor SOC depicted by straight broken lines

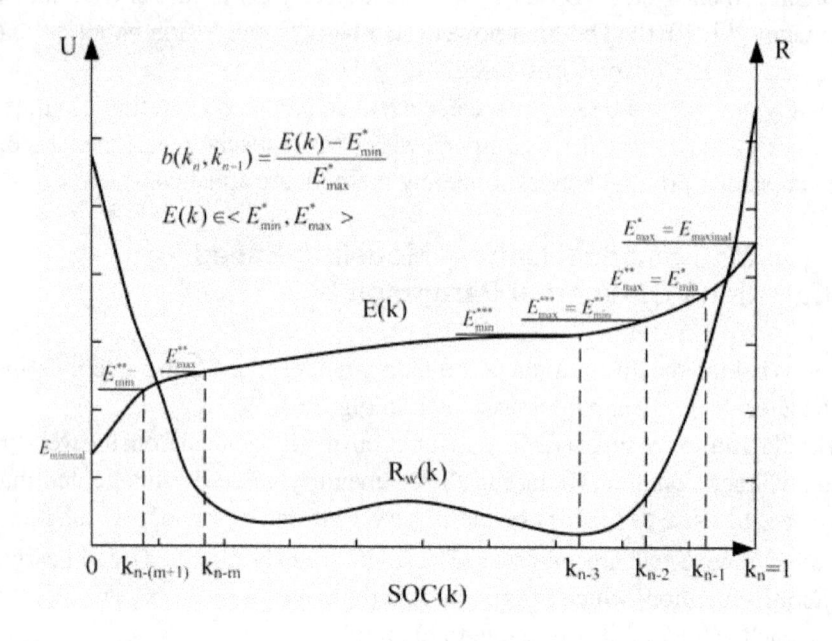

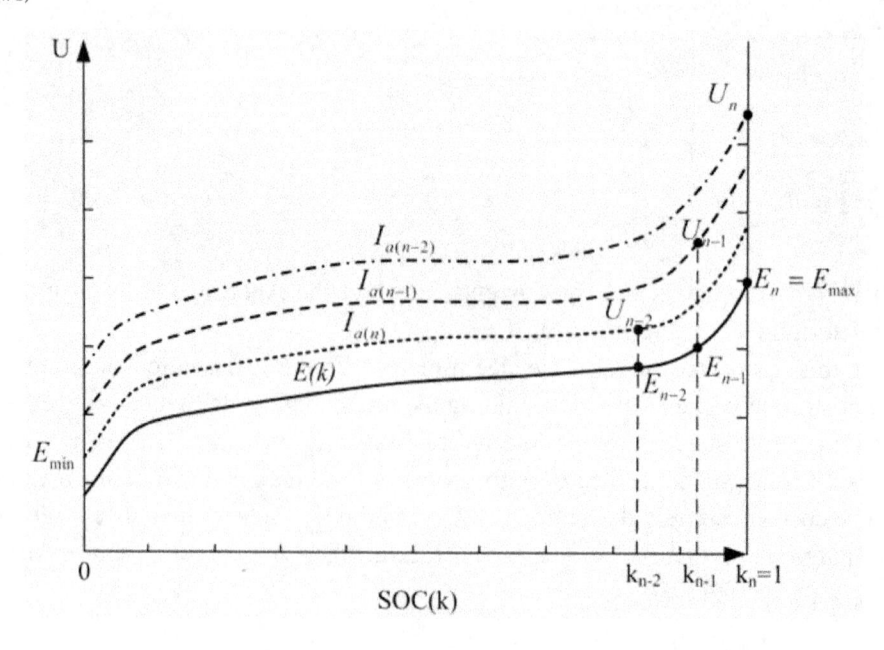

Figure 9. The exemplary experimental battery voltage versus the battery's state-of-charge factor, SOC, data for different discharging constant current values ($I_{a(n)}, I_{a(n-1)}, I_{a(n-2)}$)

Nonlinear Dynamic Traction Battery Modeling

Figure 10. Exemplary experimental voltage versus battery state of charge SOC data for different discharging constant current values ($I_{a(n)}$, $I_{a(n-1)}$, $I_{a(n-2)}$) as a background for battery electromotive force EMF E(k) calculation

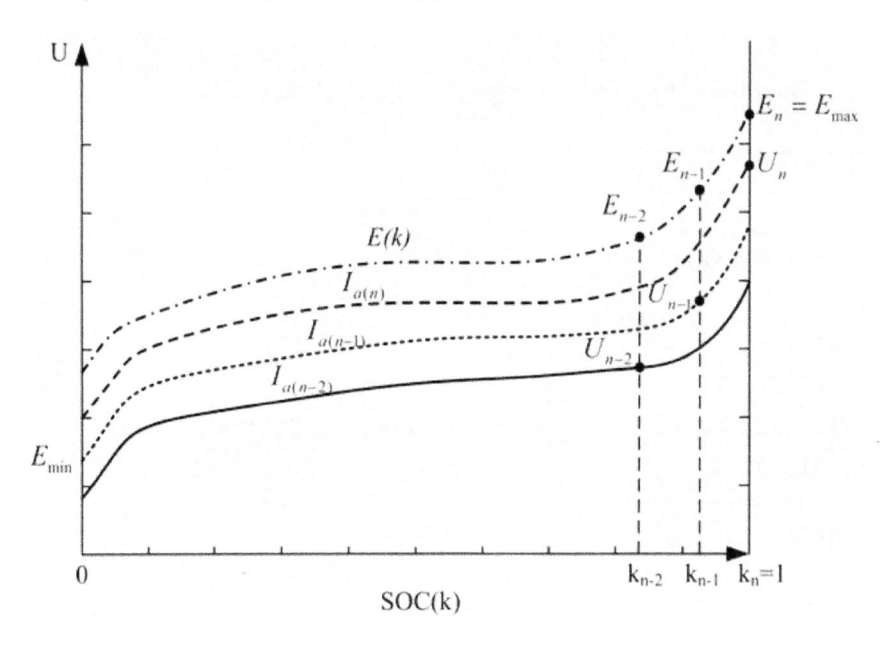

If increasing (or decreasing) the value of k is sufficiently small, the following equations can be written:

$$E^*_{min} \cong E^{**}_{max}$$
$$l(k_n) = l(k_{n-1}) = const$$
$$b(k_n) = \frac{E(k_n) - E^*_{min}}{E^*_{max}};$$

for the upper range limit $k \in \langle k_{n-1}, k_n \rangle$ (55)

$$E(k_n) = E^*_{max} \quad \text{hence} \quad b(k_n) = \frac{E^*_{max} - E^*_{min}}{E^*_{max}}$$

for the lower range limit $k \in \langle k_{n-1}, k_n \rangle$

$$E(k_{n-1}) = E^*_{min} \quad \text{hence} \quad b(k_{n-1}) = \frac{E^*_{min} - E^*_{min}}{E^*_{max}} = 0$$

where

$$E(k_n) = E^*_{max}$$

is determined from the previous range

$$k \in \,]k_n, k_{n+1}[\,$$

and within this range

$$E(k_n) = E^*_{min}$$

hereafter, the equation system for each range of k is an equation system with two unknowns:

$$E^*_{min}, l(k_n) = l(k_{n-1}) = const.$$

For

$$\Delta k \to 0, E^*_{max} \to E^*_{min} \to E(k),$$

the value of $b(k)$, $l(k)$ is calculated using the above presented method.

As the determined-for-the-range- values are the output values, determined discreetly, within the next range

$$k \in \langle k_{n-2}, k_{n-1}\rangle, \left(E(k_{n-1}) = E^*_{max}\right)$$

then, the described method above, can be expressed as an iteration method.

The obtained values by using this method of $E(k_n)$, $b(k_n)$, $l(k_n)$ for ranges

$$k \in \sum_{n=1}^{m}\langle k_{n-1}, k_n\rangle \Rightarrow k \in (0,1),$$

where $k_n = k_{n-1} + k$, are sets of points. These sets should be approximated for a more comfortable form to use in the mathematical model. So, the described mathematical model formulating method can be termed an iteration-approximation method.

Nonlinear Dynamic Traction Battery Modeling

In a series of simulation analyses for various values of k it has been determined that a satisfactorily accurate imaging of the experimental internal resistance data could be gained at $k = 0.01$.

Similarly as in the case of Figure 8, the following equations can be generated (Figure 9 and 10):

$$\begin{cases} u(k_n) = E(k_n) \pm I_a R_w(k_n) \\ u(k_{n-1}) = E(k_{n-1}) \pm I_a R_w(k_{n-1}) \end{cases} \quad (56)$$

$u(k_n)$ and $u(k_{n-1})$ are known from the family of voltage characteristics that are obtained by laboratory tests. $I_{a(n)}$ is also known because $u(k_n)$ is determined for $I_{a(n)} = const$, $I_{a(n-1)} = const$.

$+$ is for discharge

$-$ is for charge

$k \in <0,1>$

Using the above-presented approach, based on experimental data (shown in Figures 9 and 10), it is possible to construct a proper equation set as in the shape of Equations (55), (56), and resolve it in an iterative way.

If

$k_n = k_1, E_n = E_1 = E_{max}, I_{a(n)} = I_1, U_n = U_1$
$k_{n-1} = k_2, E_{n-1} = E_2, I_{a(n-1)} = I_2, U_{n-1} = U_2$
$k_{n-2} = k_3, E_{n-2} = E_3, I_{a(n-2)} = I_3, U_{n-2} = U_3$

hence,
First step:

$$\begin{cases} U_1 = E_1 \pm I_1 R_{w1} \\ U_2 = E_2 \pm I_2 R_{w2} \\ R_{w1} = \dfrac{E_1 - E_2}{E_1} \dfrac{E_1}{I_1} + \dfrac{l_1}{k_1} \\ R_{w2} = \dfrac{E_1 - E_2}{E_1} \dfrac{E_2}{I_2} + \dfrac{l_1}{k_2} \end{cases} \quad (57)$$

Second step:

$$\begin{cases} U_2 = E_2 \pm I_2 R_{w2} \\ U_3 = E_3 \pm I_3 R_{w3} \\ R_{w2} = \dfrac{E_2 - E_3}{E_2} \dfrac{E_2}{I_2} + \dfrac{l_2}{k_2} \\ R_{w3} = \dfrac{E_2 - E_3}{E_2} \dfrac{E_3}{I_3} + \dfrac{l_2}{k_3} \end{cases}$$

……more steps.

Obviously, the battery's electromotive force versus battery state-of-charge factor function $E(k)$ is the function that we need. To obtain it, it is necessary to use the known functions $u_a(k)$, which are obtained by laboratory tests (Figure 11), exemplary and based on nickel metal battery Ni-MH data (further will be continued for lithium ion. Li-ion battery data). This computation was done by the author and his former PhD student, Dr. Y. Chang.

Last, the equations of R_w and EMF take the shape of the following polynomial:

$$\begin{aligned} R_w(k) &= A_r k^6 + B_r k^5 + C_r k^4 \\ &+ D_r k^3 + E_r k^2 + F_r k + G_r \\ E(k) &= A_e k^6 + B_e k^5 + C_e k^4 \\ &+ D_e k^3 + E_e k^2 + F_e k + G_e \\ b(k) &= A_b k^6 + B_b k^5 + C_b k^4 \\ &+ D_b k^3 + E_b k^2 + F_b k + G_b \\ l(k) &= A_l k^7 + B_l k^6 + C_l k^5 \\ &+ D_l k^4 + E_l k^3 + F_l k^2 + G_l k + H_l \end{aligned} \quad (58)$$

Nonlinear Dynamic Traction Battery Modeling

Figure 11. The discharging data of a 14-Ah nickel metal hydride Ni-MH battery versus battery state-of-charge factor. Note: The constant values of discharging currents corresponds to a nominal battery's Coulomb capacity (Ah). For nominal capacity, a nominal current is defined. For instance: a1h nominal current is equal to 1C, which is typical for HP (high power) batteries. A 0.5h nominal current is equal to 0.5C, which is typical for HE (high energy) batteries.

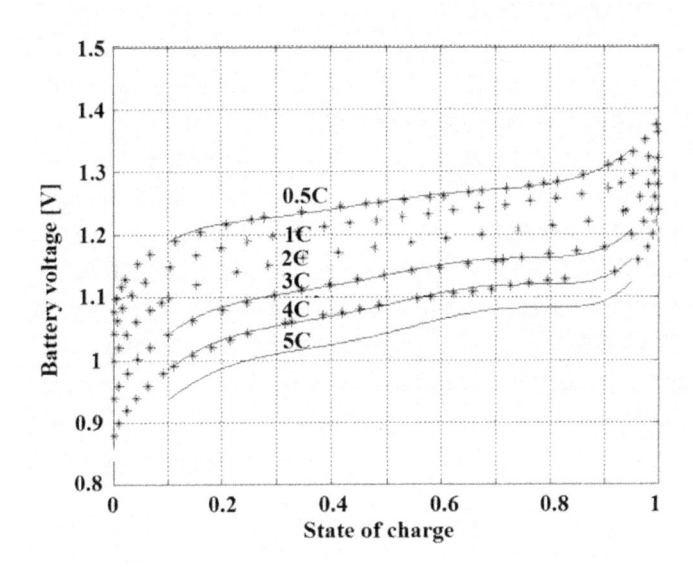

Figure 12. The charging data of a 14-Ah nickel metal hydride Ni-MH battery versus battery state-of-charge factor. Note: see note in Figure 11

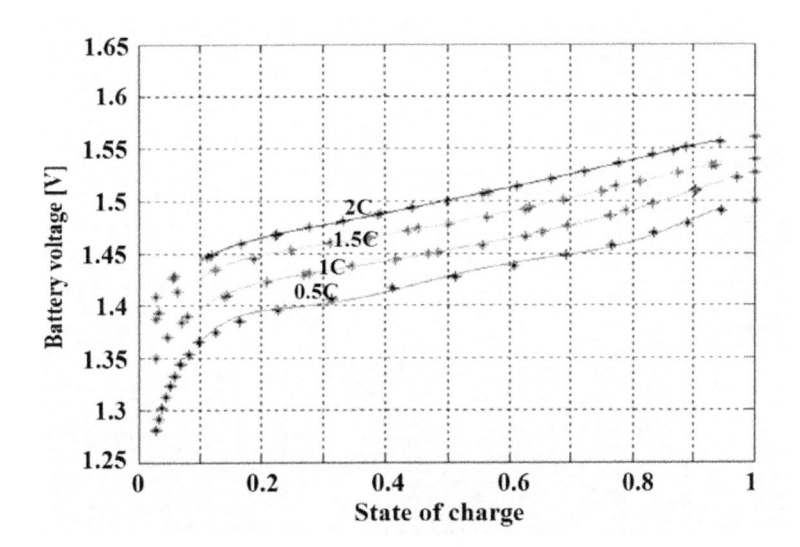

4.2. Battery Modeling Results

The basic elements that are used to formulate the mathematical model of a Ni-MH battery are the described iteration-approximation method and the approximations based on the battery discharging and charging characteristics that are obtained by an experiment. Experimental data is approximated to enable determination of the internal resistance in a small-enough range k=0.001. The modeling results (Figures 13-15) in the battery SOC operating range of 0.1-0.95 show a small deviation (less than 1%) from the experimental data (Figures 16 and 17). The Ni-MH battery that is used in the experiment and the modeling is an HP battery for HEV application. The nominal voltage of the battery is 1.2V, and the rated capacity 14Ah.

After approximation, according to the computed results, the approximated equations of (58) for the tested 14-Ah nickel metal hydride (Ni-MH) battery can be obtained. These factors in the Equations (58) are revealed in Table 3.

The basic elements used to formulate the mathematical model of Li-ion battery modules from the SAFT Company, are the previously described iteration-approximation method and the approximations based on the battery discharging characteristics obtained by experiment. The experimental data is approximated to enable determination of the internal resistance in a small enough range k = 0.001. The

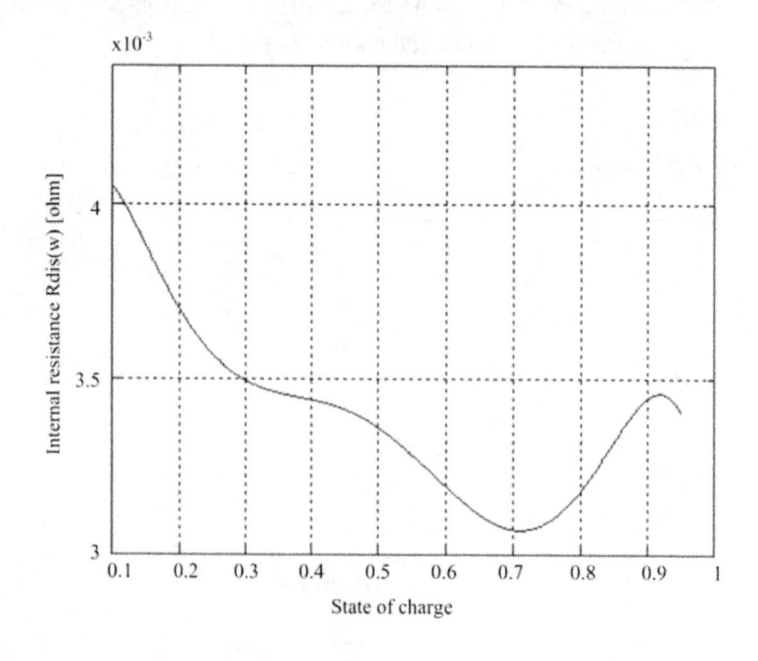

Figure 13. The computed internal resistance characteristics of a 14-Ah nickel metal hydrideNi-MH battery for discharging versus battery state-of-charge factor

Nonlinear Dynamic Traction Battery Modeling

Figure 14. The computed internal resistance characteristics of a 14-Ah nickel metal hydride Ni-MH battery for charging versus battery state-of-charge factor

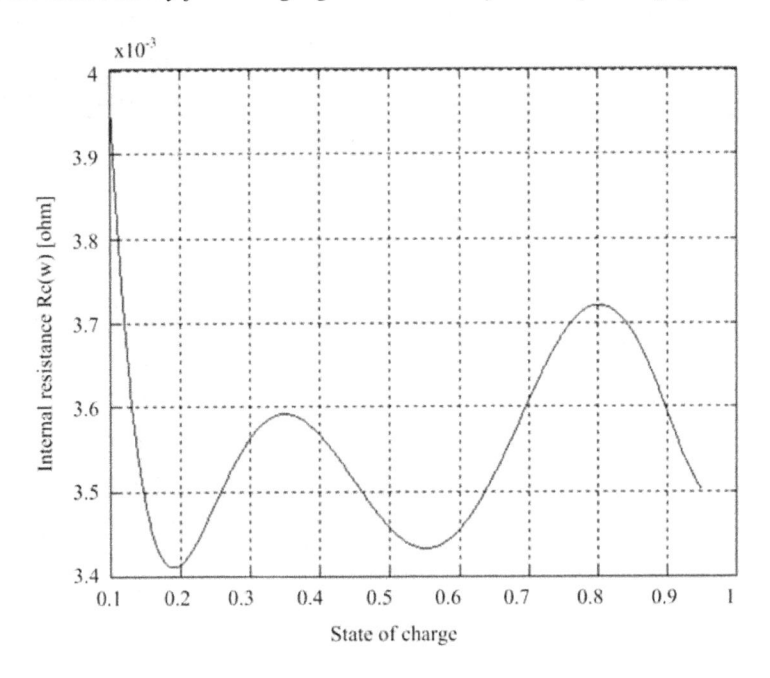

Figure 15. The exemplarycomputed electromotive force EMF of the 14-Ah nickel metal hydride Ni-MH batteryversus battery state-of-charge factor

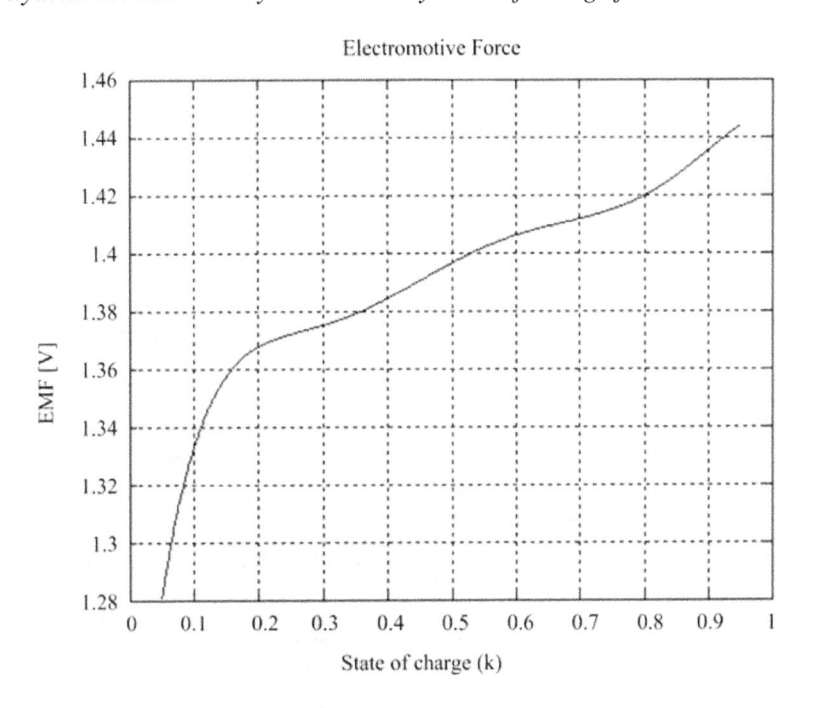

Table 3. The factors of Equation (58) for a 14-Ah Ni-MH battery

Factors of Equation (58)	Internal resistance R_w during discharging	Internal resistance R_w during charging	Electromotive Force	Coefficient b Discharging Charging	Coefficient l Discharging Charging
A	0.65917	0.42073	13.504	-0.015363	0.65917
				0.015341	0.42073
B	-2.0397	-1.4434	-36.406	0.10447	-2.0528
				-0.10661	-1.4376
C	2.4684	1.9362	36.881	-0.18433	2.4978
				0.22702	1.9195
D	-1.4711	-1.2841	-17.198	0.13578	-1.495
				-0.21788	-1.2661
E	0.44578	0.43809	3.5264	-0.045129	0.45416
				0.10346	0.42896
F	-0.065274	-0.071757	-0.10793	0.0059814	-0.066422
				-0.023367	-0.06961
G	0.0099109	0.0078518	1.234	-9.416e-005	0.0099289
				0.0020389	0.0076585
H					-1.2154e-015
					1.9984e-008

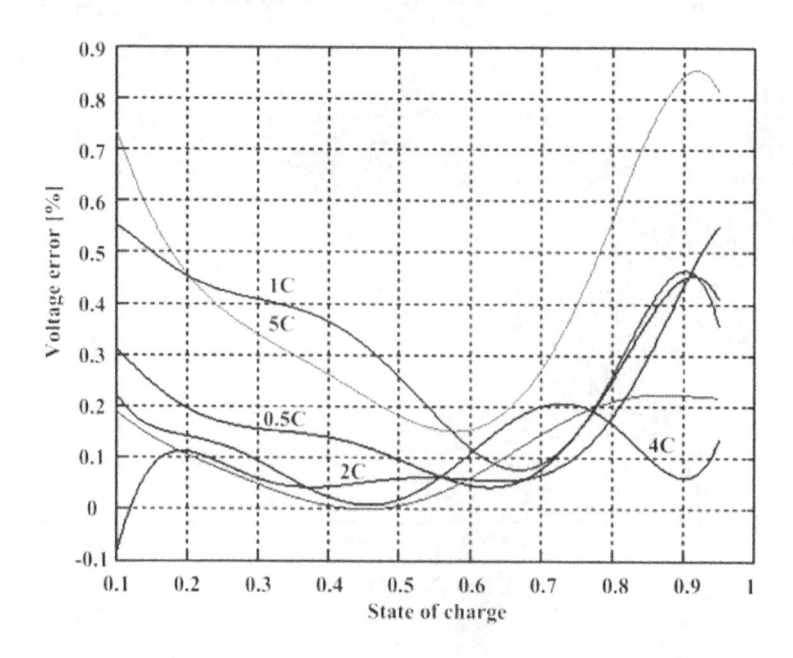

Figure 16. The error of experimental data and the computed voltage at different battery discharge currents versus battery state-of-charge factors

Nonlinear Dynamic Traction Battery Modeling

Figure 17. The error of experiment data and computed voltage at different charge battery currents versus battery state-of-charge factors

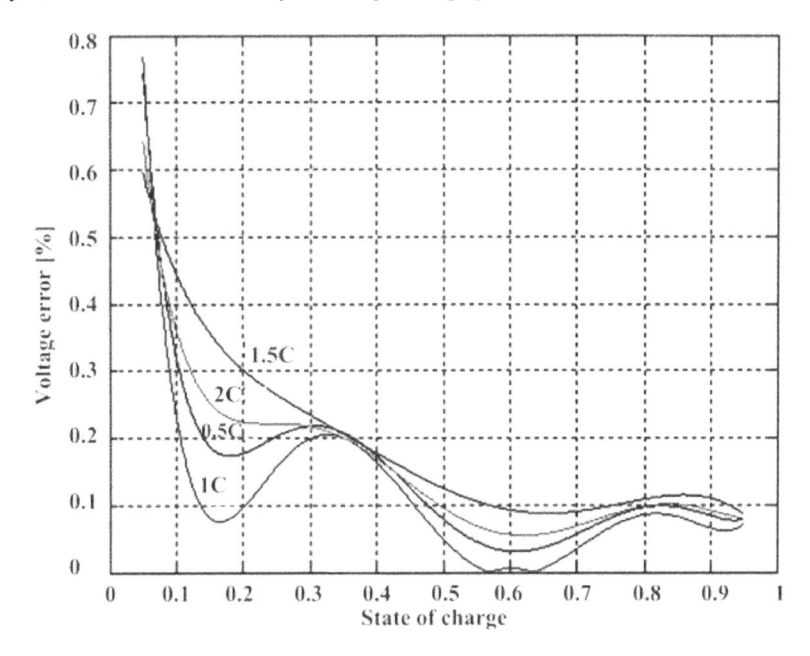

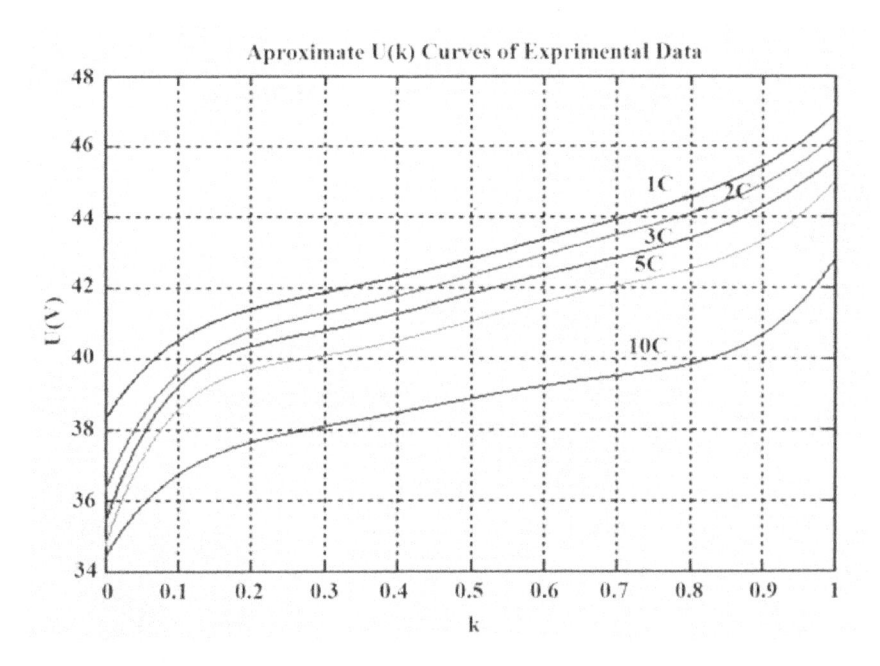

Figure 18. The discharging voltage characteristics of a SAFT 30Ah Li-ion module versus battery state of charge factor

Figure 19. The computed internal resistance of SAFT 30Ah Li-ion module versus battery state-of-charge factors

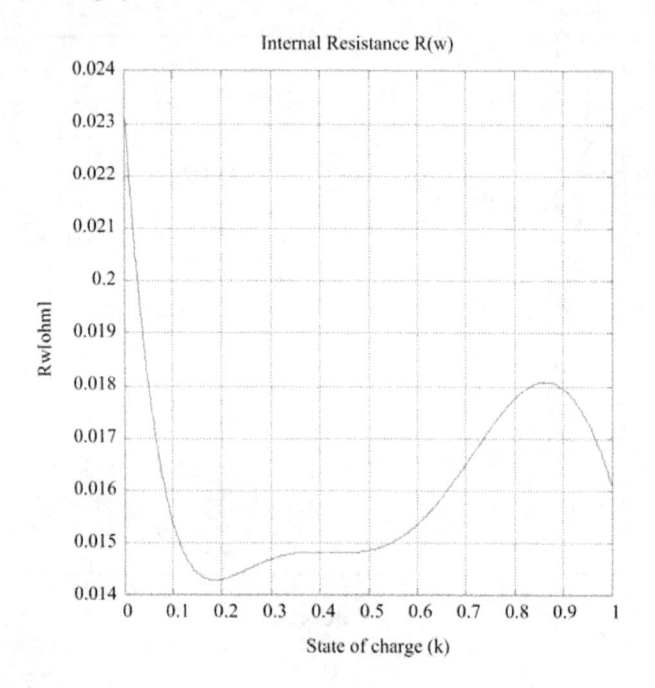

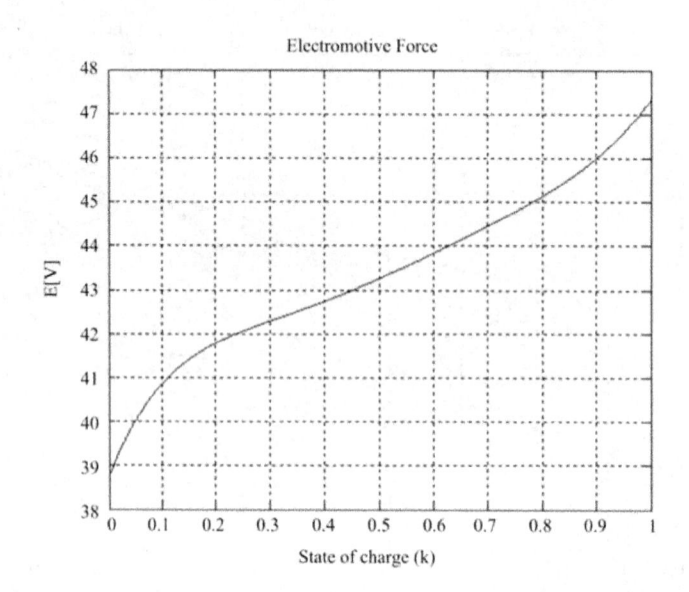

Figure 20. The computed electromotive force EMF of a SAFT 30Ah Li-ion module versus battery state-of-charge factors

Nonlinear Dynamic Traction Battery Modeling

Figure 21. The computed coefficient b of a SAFT 30Ah Li-ion module versus battery state-of-charge factors

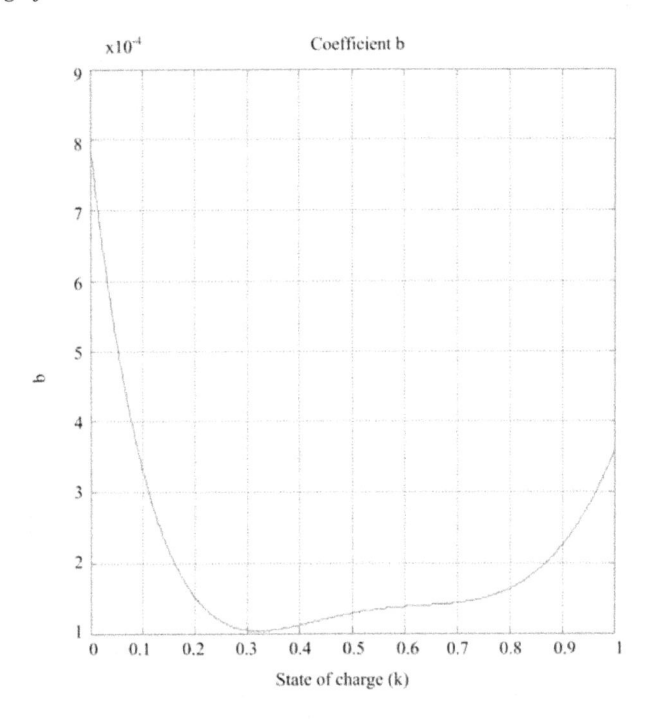

Figure 22. The computed coefficient l of a SAFT 30Ah Li-ion module versus battery state-of-charge factors

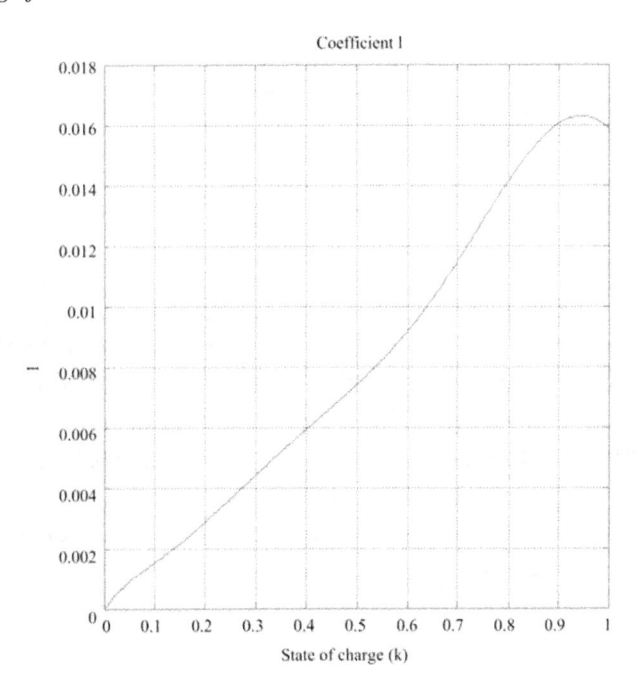

Figure 23. The errors between testing data and the computed result of a SAFT 30Ah Li-ion module versus battery state-of-charge factors

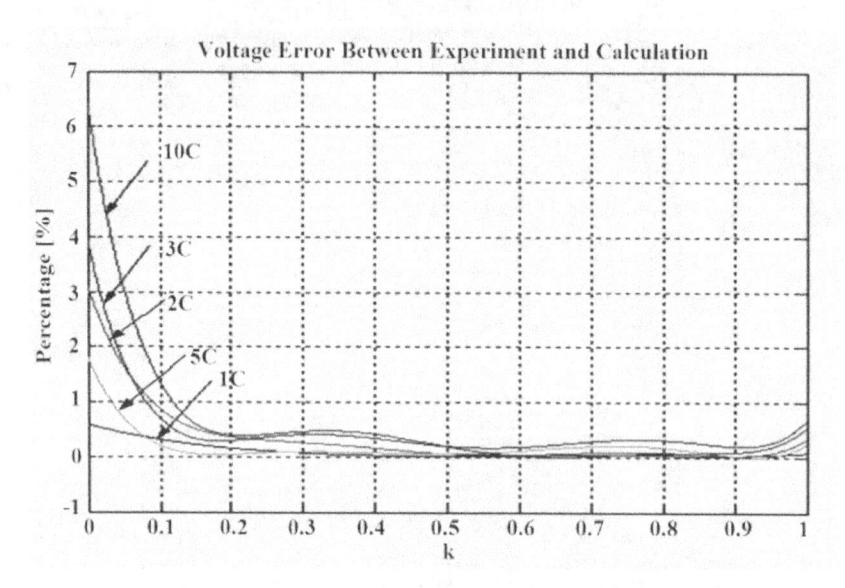

Table 4. The Equation (58) factors for a 30-Ah Li-ion module

Factors of equations (4.58)	Internal resistance R_w	Electromotive force E	Coefficient b	Coefficient l
A	0.71806	-28.091	0.0032193	0.71806
B	-2.6569	157.05	-0.016116	-2.6545
C	3.7472	-296.92	0.036184	3.736
D	-2.5575	2634	-0.040738	-2.5406
E	0.8889	-119.29	0.023539	0.87755
F	-0.14693	30.476	-0.0065159	-0.14352
G	0.023413	38.757	0.00078501	0.022978
H				-1.7916e-015

analyses, in the operating range battery state-of-charge SOC between 0.01~0.95, gives us a small deviation (less than 2%) by using the iteration-approximation method of the experimental data. The VL30P-12S module has a 30Ah rated capacity, and it is specially designed for hybrid electric vehicle HEV application.

After approximation, according to the computed results, approximated equations of (58) for a 30-Ah Li-ion module can be obtained. These factors of the Equation (58) can be seen in Table 4.

REFERENCES

Butler, K. L., Ehsani, M., & Kamath, P. (1999, Nov.). A matlab-based modeling and simulation package for electric and hybrid electric vehicle design. *IEEE Transactions on Vehicular Technology*, *48*(6), 1770–1778. doi:10.1109/25.806769

Caumont, O., Moigne, P. L., Rombaut, C., Muneret, X., & Lenain, P. (2000). Energy gauge for lead acid batteries in electric vehicles. *IEEE Transactions Energy Conservation*, *15*(3), 354–360. doi:10.1109/60.875503

Ceraol, M., & Pede, G. (2001). Techniques for estimating the residual range of an electric vehicle. *IEEE Transactions on Vehicular Technology*, *50*(1), 109–111. doi:10.1109/25.917893

Chan, C. C. (2002). The state-of-the-art of electric and hybrid vehicles. *Proceedings of the IEEE*, *90*(2), 247–270. doi:10.1109/5.989873

Chan, C. C., & Chau, K. T. (2001). *Modern electric vehicle technology*. Oxford, UK: Oxford University Press.

Do, D., Forgez, C., Benkara, K., & Friedrich, G. (2009). Imedance observer for a li-ion battery using kalman filter. *IEEE Transactions on Vehicular Technology*, *58*(8).

Fujioka, N., Ikona, M., Kiruna, T., & Konomaro, K. (1998). Nickel metal-hydride batteries for hybrid vehicle. In *Proceedings of EVS 15*. EVS.

Giglioli, R., Salutori, R., & Zini, G. (1992). Experience on a battery state of charge observer. In *Proceedings of EVS 11*. EVS.

Gu, W., & Wang, C. (2000). Thermal-electrochemical modeling of battery systems. *Journal of the Electrochemical Society*, *147*(8), 2910–2922. doi:10.1149/1.1393625

Johnson, V. H., & Pesaran, A. A. (2000). Temperature-dependent battery models for high-power lithium-ion batteries. In *Proceedings of International Electric Vehicle Symposium*. EVS.

Jyunichi, L., & Hiroya, T. (1996). Battery state-of-charge indicator for electric vehicle. In *Proceedings of International Electric Vehicle Symposium*. EVS.

Karden, E., Buller, S., & De Doncker, R. W. (2002). A frequency-domain approach to dynamical modeling of electrochemical power sources. *Electrochimica Acta*, *47*(13–14), 2347–2356. doi:10.1016/S0013-4686(02)00091-9

Kuhn, B., Pitel, G., & Krein, P. (2005). Electrical properties and equalization of li-ion cells in automotive application. In *Proceedings of VPPC*. IEEE.

Malkhandi, S., Sinha, S. K., & Muthukumar, K. (2001). Estimation of state-of-charge of lead-acid battery using radial basis function. In *Proceedings of Industrial Electronics Conference*, (vol. 1, pp. 131–136). IEC.

Marcos, J., Lago, A., Penalver, C. M., Doval, J., Nogueira, A., Castro, C., & Chamadoira, J. (2001). An approach to real behavior modeling for traction lead-acid batteries. In *Proceedings of Power Electronics Specialists Conference*, (vol. 2, pp. 620–624). PES.

Morio, K., Kazuhiro, H., & Anil, P. (1997). Battery SOC and distance to empty meter of the Honda EV plus. In *Proceedings of International Electric Vehicle Symposium*, (pp. 1–10). EVS.

Moseley, P., & Cooper, A. (1998). Lead acid electric vehicle batteries – Improved performance of the affordable option. In *Proceedings of EVS 15*. EVS.

Nelson, R. F. (2000). Power requirements for battery in HEVs. *Journal of Power Sources*, *91*, 2–26. doi:10.1016/S0378-7753(00)00483-3

Ovshinski, S. R., Dhar, S. K., Venkatesan, S., Fetchenko, M. A., Gifford, P. R., & Corrigan, D. A. (1992). Performance advances in ovonic NiMH batteries for electric vehicles. In *Proceedings of EVS 11*. EVS.

Pang, S., Farrell, J., Du, J., & Barth, M. (2001). Battery state-of-charge estimation. In *Proceedings of American Control Conference*, (vol. 2, pp. 1644–1649). ACC.

Piller, S., Perrin, M., & Jossen, A. (2001). Methods for state–of–charge determination and their applications. *Journal of Power Sources*, *96*, 113–120. doi:10.1016/S0378-7753(01)00560-2

Plett, G. (2003a). LiPB dynamic cell models for kalman-filter SOC estimation. In *Proceedings of EVS-20*. EVS.

Plett, G. (2003b). LiPB dynamic cell models for Kalman-filter SOC estimation. In *Proceedings of International Electric Vehicle Symposium*. EVS.

Rodrigues, S., Munichandraiah, N., & Shukla, A. (2000). A review of state-of-charge indication of batteries by means of A.C. impedance measurements. *Journal of Power Sources*, *87*(1-2), 12–20. doi:10.1016/S0378-7753(99)00351-1

Salkind, A., Atwater, T., Singh, P., Nelatury, S., Damodar, S., Fennie, C., & Reisner, D. (2001). Dynamic characterization of small lead-acid cells. *Journal of Power Sources*, *96*(1), 151–159. doi:10.1016/S0378-7753(01)00561-4

Sato, S., & Kawamura, A. (2002). A new estimation method of state of charge using terminal voltage and internal resistance for lead acid battery. *Processing Power*, *2*, 565–570.

Shen, W. X., Chan, C. C., Lo, E. W. C., & Chau, K. T. (2002). Estimation of battery available capacity under variable discharge currents. *Journal of Power Sources*, *103*(2), 180–187. doi:10.1016/S0378-7753(01)00840-0

Shen, W. X., Chau, K. T., Chan, C. C., & Lo, E. W. C. (2005). Neural network-based residual capacity indicator for nickel-metal hydride batteries in electric vehicles. *IEEE Transactions on Vehicular Technology*, *54*(5), 1705–1712. doi:10.1109/TVT.2005.853448

Stempel, R. C., Ovshinsky, S. R., Gifford, P. R., & Corrigan, D. A. (1998). Nickel-metal hydride: Ready to serve. *IEEE Spectrum*. doi:10.1109/6.730517

Szumanowski, A. (2000). *Fundamentals of hybrid vehicle drives*. Warsaw: ITE Press.

Szumanowski, A. (2006). *Hybrid electric vehicle drives design-ed. based on urban buses*. Warsaw: ITE Press.

Szumanowski, A. (2010). *Nonlinear dynamics traction battery modeling* (pp. 199–220). INTECH.

Szumanowski, A., & Brusaglino, G. (1999). Approach for proper battery adjustment for HEV application. In *Proceedings of EVS 16*. EVS.

Szumanowski, A., & Chang, Y. (2008). Battery management system based on battery nonlinear dynamics modeling. *IEEE Transactions on Vehicular Technology*, *57*(3), 1425–1432. doi:10.1109/TVT.2007.912176

Szumanowski, A., Chang, Y., Piórkowski, P., Jankowska, E., & Kopczyk, M. (2005). Performance of city bus hybrid drive equipped with li-ion battery. In *Proceedings of EVS 21*. EVS.

Szumanowski, A., Dębicki, J., Hajduga, A., Piórkowski, P., & Chang, Y. (2003). Li-ion battery modeling and monitoring approach for hybrid electric vehicle applications. In *Proceedings of EVS-20*. EVS.

Tojura, K., & Sekimori, T. (1998). Development of battery system for hybrid vehicle. In *Proceedings of EVS 15*. EVS.

Chapter 6
Basic Design Requirements of an Energy Storage Unit Equipped with Battery

ABSTRACT

The storage unit is understood as the battery. Practically, it is true in the majority of cases. However, another type of electro-chemical energy storage unit can be considered, which is the capacitor. The most important is of course the battery and the emphasis is put on the battery's thermal behavior, its State-Of-Charge (SOC) indication and monitoring as the background of the Battery Management System's (BMS) design. The chapter discusses the original algorithm base of the nonlinear dynamic traction battery's modeling, which includes the battery temperature impact factor. The battery State Of Charge (SOC) coefficient presented in this chapter has to be determined in terms of its maximal accuracy. This is very important for the control of the entire hybrid power train. The battery state of charge signal is the basic feedback in power train online control in every operation mode: pure electric, pure engine, or in the majority, the hybrid drive operation. Electro-chemical capacitors applied in hybrid power trains are commonly called super or ultra capacitors. The application of ultra capacitors in hybrid electric vehicle power trains does not seem to be a strong alternative to the batteries. The exemplary complex solution of the parallel connection of the battery and the capacitor as a means of increasing the cell's lifetime and decreasing its load currents is also discussed in this chapter. Voltage equalization for both energy storage devices is depicted. For the ultra capacitor this is necessary.

DOI: 10.4018/978-1-4666-4042-9.ch006

Basic Design Requirements of an Energy Storage Unit Equipped with Battery

INTRODUCTION

In the majority of cases, the energy storage unit in a HEV consists of a battery only. In full hybrid (HEV) power trains, this is a HP high-power battery. In a PHEV, the battery type is closer to a HE high-energy, similar to what is in pure electric vehicles. The main difference is the Coulomb capacity (Ah), which is connected with different discharging and charging maximal currents values. In the case of the HP battery, the capacity is smaller, but both currents' values are higher than in the case of the HE. Anyway, in both battery applications, the same approach to the battery management system design is required. The temperature influence on energy accumulated in the battery and its state-of-charge (SOC) indication and monitoring, are the main targets of this process design. The inclusion of these impacts in battery modeling is necessary, as well as in the battery management system design.

The ultra capacitors application in HEV power trains seems not to be a strong alternative to batteries. Nevertheless, the exemplary complex solution of parallel connected battery and capacitor is considered. An additional problem which appears in energy storage system design is cell voltage equalization. The newest Li-ion battery construction characterizes high quality, which means every single cell has the same parameters as others, and this, in consequence, permits the avoidance of costly and complex electronic cell voltage balance devices. This system is discussed, however.

1. BATTERY MANAGEMENT SYSTEM DESIGN REQUIREMENTS

1.1. Temperature Influences Analysis of Battery Performance

The determination of the battery EMF and internal resistance gives unlimited possibilities for calculating the battery's voltage versus SOC (k) relationship for a different value of discharge-charge current. For real driving conditions, the battery discharge or charge, depends on the drive architecture influencing the respective power distribution. In the majority of cases, battery charging takes place during vehicle regenerative braking, which means that this situation lasts for a relatively short time, with a significant peak-current value. In this short period of time, a discharging or charging current that is too high, results in a rapid increase of the temperature.

The main role of this study is to find a theoretical background for calculating the temperature influence on the battery SOC. The presented method is more accurate and complicated compared with other methods, which does not mean that it is more difficult to apply. First of all, it is necessary to make the following assumptions:

Basic Design Requirements of an Energy Storage Unit Equipped with Battery

The considered battery is fully charged in nominal conditions: nominal current, nominal temperature and nominal capacity (i_b =1C, τ_b =20°C, the capacity is designed for nominal parameters, respectively).

The EMF for the considered battery is defined as its nominal condition in the nominal SOC alteration range $k \in\, <1,0>$. The assumption is taken that the EMF value of k=0.15 is the minimum EMF. For k=0, the EMF is defined as the 'minimum-minimorum', in a practice which should not be obtained. The same assumption is recommended for a value that is different from the nominal temperature for the k_τ (SOC) definition. As shown in Figure 2, the starting point value of the EMF, for a different value from the nominal temperature, can be higher or lower, which means that the extension alteration of the SOC could be longer or shorter. For instance (see Figure 1), in the case of the Ni-MH battery for a value that is higher than the nominal temperature, the discharge capacity is smaller than the nominal, which means that for a certain temperature, the battery capacity corresponding to this temperature is also changed on file $k_\tau \in\, <1,0>$. However, the full battery capacity indicated by the proper value of battery state-of-charge factor k_τ does not mean the same discharge capacity, as in the case of nominal temperature, but does mean the maximum discharge capacity at this temperature. For this reason, in fact, k_τ for

Figure 1. The temperature dependence of the discharge capacity of the nickel metal hydride Ni-MH battery

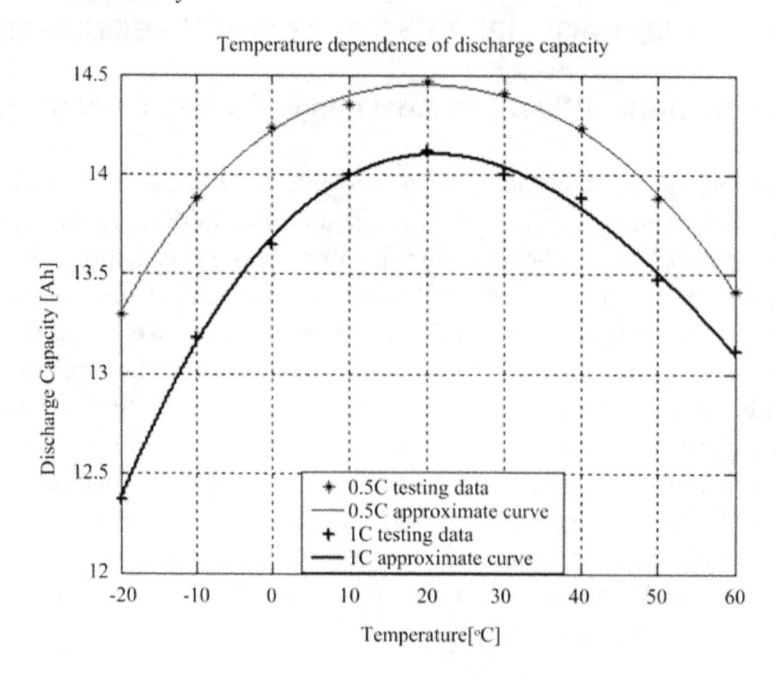

Basic Design Requirements of an Energy Storage Unit Equipped with Battery

Figure 2. The temperature dependence of the battery's usable discharge capacity and the EMF's starting point

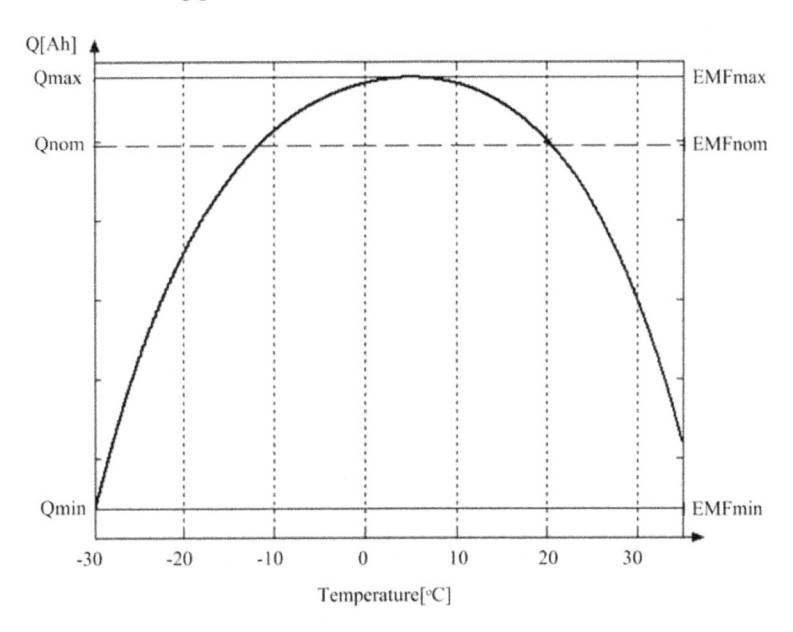

this temperature is only $k(t) > k_\tau(t)$, (in some cases, $k(t) < k_\tau(t)$, where $k(t)$ is connected only with nominal conditions).

From Figure 2, it is easy to note that the EMF (in the case of this battery type) value in the nominal conditions, is smaller than the EMF value for a temperature that is lower than 20°C (the nominal temperature), which means that for a maximum EMF value, the available battery capacity is higher than in the case of the nominal terms. For nominal conditions, the SOC can be defined by a k factor ($k \in\, <1, 0>$). If the EMF for the non-nominal conditions reaches its highest value, the available charge (in ampere-hours) will be also greater. It is easy to note the relation $Q_\tau = Q_{max}$ and Q_{nom} is defined as follows:

$$\frac{Q_\tau = Q_{max}}{Q_{nom}} > 1$$

If

$$Q_\tau < Q_{nom} \rightarrow \frac{Q_\tau}{Q_{nom}} < 1,$$

correspondingly,

$$EMF_\tau < EMF_{nom} \to \frac{EMF_\tau}{EMF_{max}} < 1$$

This corresponds to:

$$\frac{EMF_\tau = EMF_{max}}{EMF_{nom}} > 1$$

On the other hand, for Q_{nom}, $k \in <1,0>$, but relating it to $Q_\tau > Q_{nom}$ in τ condition, the file $<1,0>$ means file $<0, Q_{max}>$. Transforming k in nominal terms to k_τ is necessary to use the general relation $\frac{Q_\tau}{Q_{nom}}$. Theoretically, the product $k_{nom} \frac{Q_\tau}{Q_{nom}}$ transfers the SOC factor into something other than nominal temperature conditions. The same transformation can be obtained for

$$k_{nom} \frac{EMF_\tau}{EMF_{nom}}$$

where

$$k_{nom} \in <1,0>$$

Using the transformation factor $k_{nom} \frac{EMF_\tau}{EMF_{nom}}$ or

$$k * s_\tau \left(k_{nom} = k, s_\tau = \frac{EMF_\tau}{EMF_{nom}} \right)$$

it is possible to relate the SOC of the battery that is determined for the nominal temperature to other different temperatures.

1.2. Algorithm of Battery SOC Indication

The algorithm is given as follows:

Basic Design Requirements of an Energy Storage Unit Equipped with Battery

Figure 3. The relation of $\dfrac{EMF_\tau}{EMF_{nom}}$ versus battery temperature

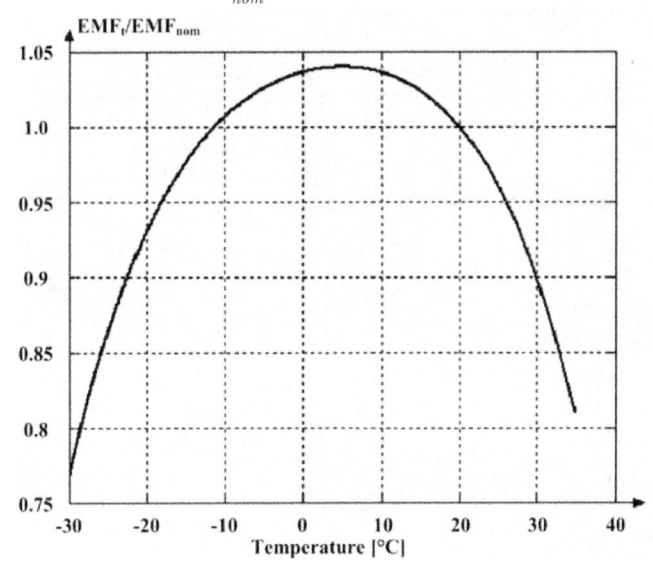

1. By simulation, the family of $u_b(k)$ for different constant currents $i_b \in\ <0.5C, 6C>$ and nominal temperature (e.g. 20°C) can be obtained according to battery modeling results (EMF and internal resistance as functions of the SOC).
2. From Figure 3,

$$s_\tau = \frac{EMF_\tau}{EMF_{nom}}$$

is defined for $\tau \in <$-30°C, +35°C$>$

From Figure 4, for k=0.9, ...0.2, the following look-up table can be obtained:

$$k = 0.9 \Rightarrow \begin{bmatrix} u_{11}, i_{11}, E_1 \\ u_{12}, i_{12}, E_1 \\ ... \\ u_{1n}, i_{1n}, E_1 \end{bmatrix} \cdots\cdots k = 0.2 \Rightarrow \begin{bmatrix} u_{81}, i_{81}, E_8 \\ u_{82}, i_{82}, E_8 \\ ... \\ u_{8n}, i_{8n}, E_8 \end{bmatrix}$$

Figure 4. The battery cell electromotive force EMF and calculated, discharging battery voltage characteristics of different, discharging battery currents and at its nominal temperature versus battery state-of-charge factors

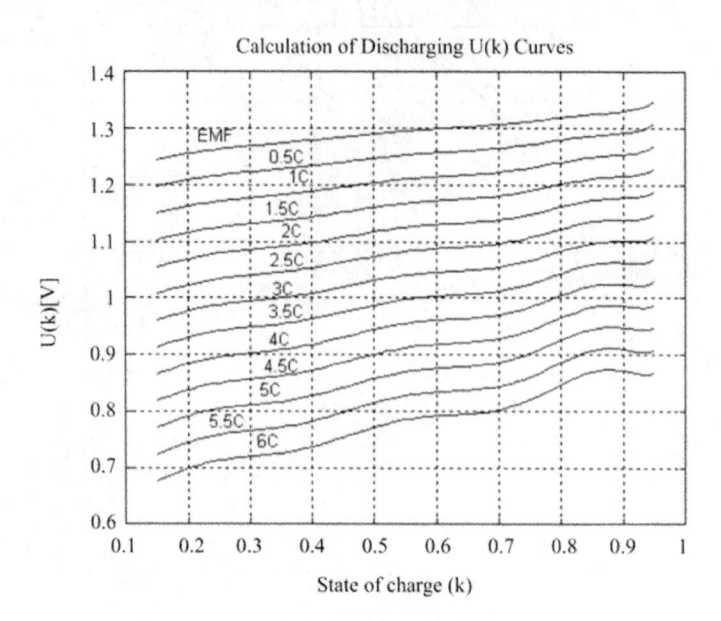

Due to the practical limitation of the SOC alteration of the battery that is applied in hybrid drives, the range of k changes can be expressed as $<0.9, 0.2>$ for the nominal temperature.

Considering the real temperatures, the SOC of the battery in relation to the nominal temperature can be defined as $s_\tau * k = k_\tau$. For instance, at a temperature of +5°C,

$$\frac{EMF_\tau}{EMF_{nom}} = 1.06;$$

hence, $k_{+5°C} = 1.06\,k$, which means that, at this moment, and this temperature, the available capacity is 1.06 times that of the nominal temperature. At a temperature +30°C,

$$\frac{EMF_\tau}{EMF_{nom}} = 0.89;$$

hence, $k_{+30°C} = 0.89\,k$, which means that, at this moment, and this temperature, the available capacity is 0.89 times that of the nominal temperature.

Basic Design Requirements of an Energy Storage Unit Equipped with Battery

Figure 5. The battery electromotive force EMF and calculated battery charging voltage characteristics at different battery charging currents, and at its nominal temperature, versus battery state-of-charge factors

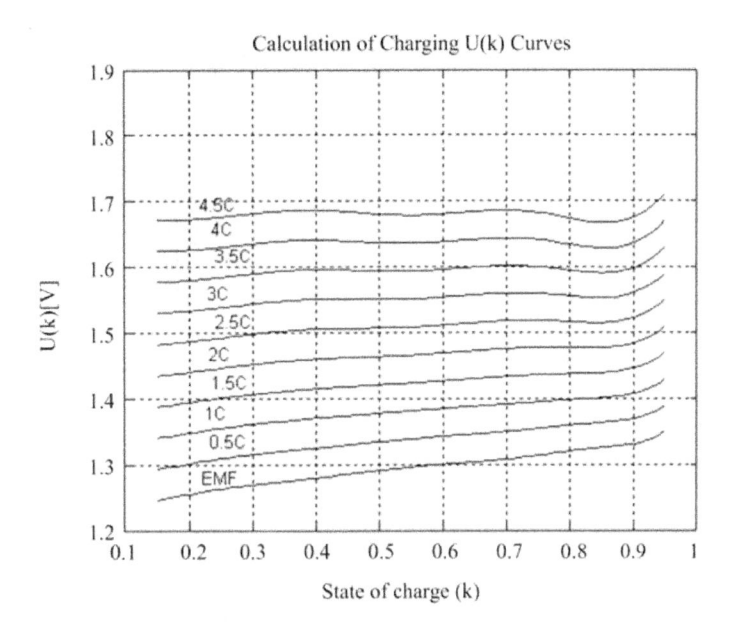

A similar method and process can be used in the battery charging process (see Figure 5).

The above-depicted method can be used in the Battery Management System (BMS) for the SOC determination, especially in hybrid (HEV) and electric (EV) vehicle drives. Based on the aforementioned steps (1 - 4), the SOC indication algorithm can be depicted, as is shown in Figure 6.

In a hybrid electric vehicle, HEV, the battery's state-of-charge value, SOC, changes faster (because a HP high-power battery is used) but not so deeply, as in pure electric vehicles, equipped with a high-energy HE battery. It means that the indication of the battery's state-of- charge displayed process may not be realized as frequently. It is not necessary to display the value of the battery's state-of-charge factor every second. Certainly, the previous value of the battery's SOC factor has to be remembered by a microprocessor.

High accuracy of determination of the battery's state-of-charge factor, SOC, is first of all necessary for the entire drive system's control. In opposition to 'indicate – display', the feedback signals from the battery must be available online.

The presented original method of battery electromotive force EMF (as a function of k - battery state-of-charge factors) calculation is the background for constructing the battery management system, BMS. This procedure is easily adopted for control

Basic Design Requirements of an Energy Storage Unit Equipped with Battery

Figure 6. The battery state-of-charge SOC indication algorithm for its discharging and charging

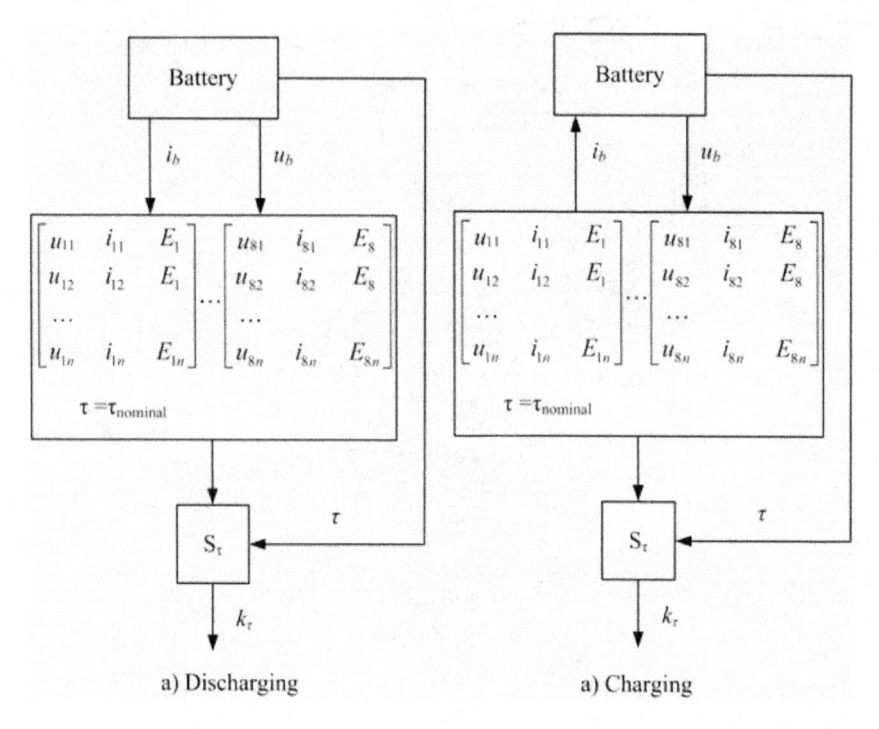

a) Discharging   a) Charging

application in the hybrid electric vehicle, HEV, and the battery-powered vehicle, EV. Its high accuracy is very important for control of the drive systems (master controller), based on feedback signals from the battery management system, BMS.

The following equation is the background to determine the accurate value of SOC (k) for dynamic conditions:

$$u(t) = E(k) \pm R(k)i(t)$$
$$k = k_{nom} s_\tau \qquad (1)$$

+ is for discharge; - is for charge; where E(k) and R(k) are taken from the equation (5.58) for the real battery module.

Based on the Equation (1), the battery state-of-charge factor calculation can be obtained in a direct way via 'on-line' dynamic battery voltage and current alteration. The solving of Equation (58) from chapter 5 as a high-power factor polynomial is really possible 'on-line' by using two procedures: look-up table (dividing the polynomial function in shaped-line ranges) or 'bi-section' numerical iterative computation. In some cases, when the accuracy of the battery's state-of-charge factor (can also

Basic Design Requirements of an Energy Storage Unit Equipped with Battery

be expressed by percentage figures), the indication can be lower (about 5%), which is accepted in hybrid electric vehicle power trains and electric vehicle drives, the power factor of a polynomial can be decreased by additional approximation E(k) and R(k). The accuracy of real time calculation is about 100 μs.

The second method is 'bi-section' iterative calculation.

The exemplary plots of battery voltage, current and battery state-of-charge, SOC, is shown in the following Figures 7, 8, 9. As the SOC of the battery is much slower and changeable than its voltage and current, the SOC indication is computed and indicated by using the 'moving average' procedure.

The assumed method and effective model are very accurate, according to error checking results of the nickel metal hydride Ni-MH batteries. The modeling method is valid for different types of batteries. The model can be conveniently used for vehicle simulation, because the battery model is accurately approximated by mathematical equations. The model provides the methodology for designing a battery management system and calculating the battery's state-of- charge factors. The influence of temperature on the battery performance is analyzed, according to laboratory-tested data, and the theoretical background for the SOC calculation is obtained. The algorithm of the battery's state-of-charge factor indication 'on-line' considering the influence of temperature, can be easily used in practice by the microprocessor and its proper programming.

Figure 7. The exemplary test of battery load in hybrid drive; battery current and voltage versus real time

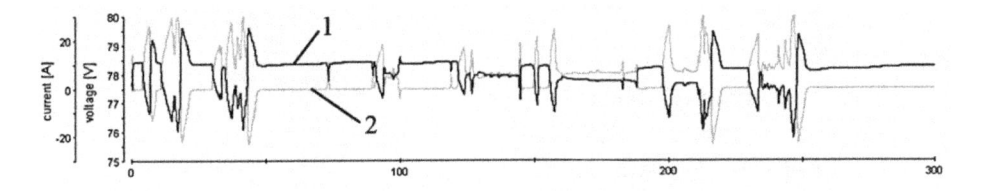

1 – battery current; 2 – battery voltage

Figure 8. The exemplary test of the battery's state-of-charge factor k indication in real driving conditions, corresponding to the battery current load shown in Figure 7

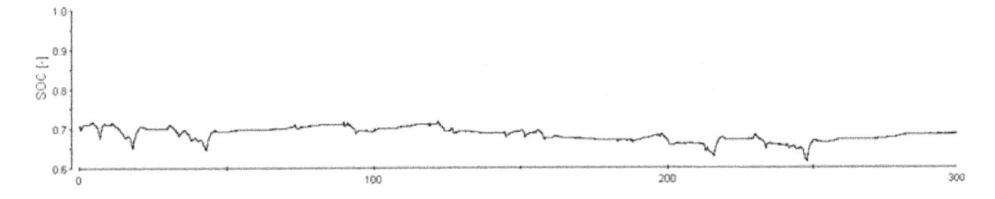

Basic Design Requirements of an Energy Storage Unit Equipped with Battery

Figure 9. The screen of the control-monitoring system device which operates, is based on the d'Space programming in the tested battery's state-of-charge factor, SOC indication

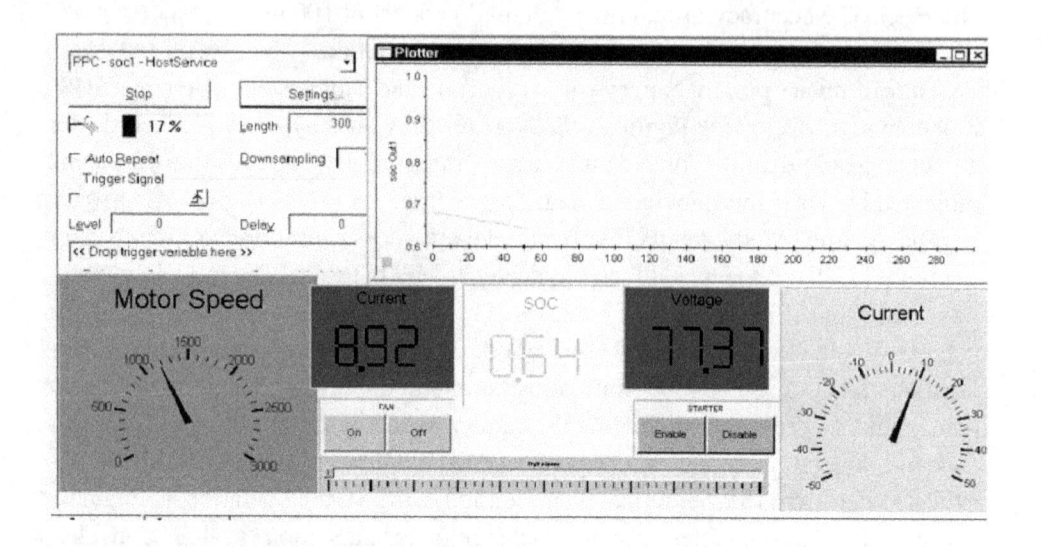

1.3. Monitoring of State-of-Charge of the Exemplary Nickel Metal Hydride Battery

The nickel metal hydride Ni-MH battery as used under laboratory tests, is depicted by the following parameters:

- **Number of Cells:** 60
- **Nominal Capacity:** 27Ah
- **Nominal Voltage:** DC 72V
- **Range of Working Voltage:** 60.0~88.8V
- **Max Current for Discharging:** 135A
- **Range of Temperature for Charging:** -10~35°C
- **Range of Temperature for Discharging:** -20~55°C

The entire SOC Monitoring system includes voltage and current transducers, a signal communicating circuit, d'Space and personal computer PC (Matlab/Simulink modeling).

The electromotive force and internal resistance of the battery are described by 6th degree polynomials (see Chapter 5).

Basic Design Requirements of an Energy Storage Unit Equipped with Battery

$$E(k) = -370.83x^6 + 1480.32x^5 - 2150.04x^4$$
$$+1476.20x^3 - 511.48x^2 + 92.33x + 68.21$$
$$R(k) = -4.5107x^6 + 13.6086x^5 - 15.7488x^4 \quad (2)$$
$$+8.7774x^3 - 2.4155x^2 + 0.2974x + 0.05036$$

However, the polynomials, in some cases, can be simplified to the 2nd degree:

$$E(k) = -8.0736x^2 + 21.5994x + 71.0442 \quad (3)$$
$$R(k) = 0.0038334x^2 - 0.00041826x + 0.06138$$

For instance, the discharging relationship of E, U, I and R is as below.

$$E(k) = U - IR(k) \quad (4)$$

Voltage U and current I are measured online. Thus, the state-of-charge (k) could be solved from this equation.

According to the experimental data, the EMF and internal resistance for charging and discharging are different. As the influence of internal resistance is relatively small, the range of the battery's state-of-charge, SOC, is due to proper design. It can be assumed that the internal resistance for charging and discharging, approximately, follows the same function (see figures included in Chapter 5; max. error is about 19%, which is from the technical point of view, seen as acceptable). In real conditions, the factor k value of the battery's state of charge operates between 0.2 and 0.9.

Based on Equations (3-4), the eclectic curve is chosen to describe the battery electromotive force, *EMF* (see Figure 10).

To increase the efficiency of online calculation, the bi-section method was used to find the solution of the Equation (3).

The method is applicable when we wish to solve the equation f(x) = 0 for the scalar variable x, where f is a continuous function.

The bi-section method requires two initial points, a and b, such that *f(a)* and *f(b)* have opposite signs. This is called a bracket of a root, for by the intermediate value theorem, the continuous function f must have at least one root in the interval (a, b). The method now divides the interval in two by computing the midpoint *c=(a+b)/2* of the interval. Unless c is itself a root which is very unlikely, but possible - there are now two possibilities: either *f(a)* and *f(c)* have opposite signs, or *f(c)* and *f(b)* have opposite signs and bracket a root. We select the subinterval that is a bracket, and apply the same bi-section step to it. In this way, the interval that might contain

Figure 10. The electromotive force of tested nickel metal hydride battery versus its state-of-charge factor

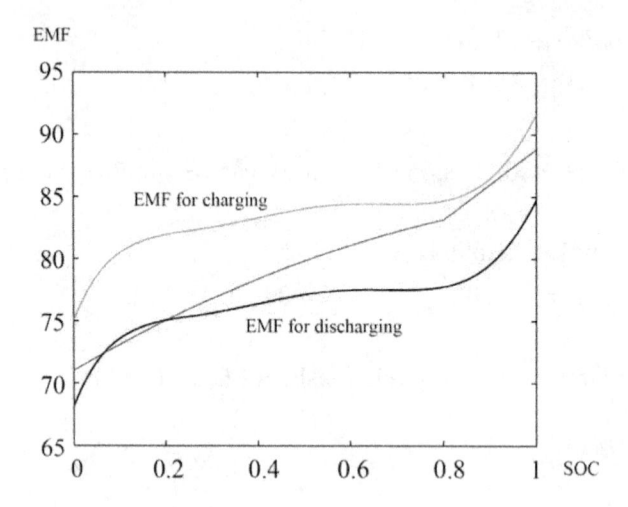

a zero of f is reduced in width by 50% at each step. We continue until we have a bracket sufficiently small for our purposes.

Explicitly, if $f(a) f(c)<0$, then the method sets b equal to c, and if $f(b) f(c)<0$, then the method sets a equal to c. In both cases, the new $f(a)$ and $f(b)$ have opposite signs, so the method is applicable to this smaller interval. A practical implementation of this method must guard against the uncommon occurrence that the midpoint is indeed a solution.

The results of the battery's state-of-charge, SOC, expressed in its factor value, battery voltage and current alteration during the monitoring test, are shown in Figures 11 and 12. The three curves shown in the above-mentioned figures express the relationship among current, voltage and state-of-charge factor values of the battery. Compared with theoretic values, the results display the battery's state-of-charge, SOC, in an absolutely proper way with very correct accuracy.

2. BATTERY AND ULTRA CAPACITOR SET IN A HYBRID POWER TRAIN

2.1. Electrochemical Capacitors

Electro-chemical capacitors applied in hybrid drives are commonly called 'super'- or 'ultra' capacitors. Their role is the same as with HP batteries. The peaks of power during vehicle acceleration or climbing are taken from the capacitors.

Basic Design Requirements of an Energy Storage Unit Equipped with Battery

Figure 11. The result of the battery's state-of-charge monitoring: test number 1: a) current and terminal voltage of battery versus real time; b) SOC of battery versus real time

a)

b)

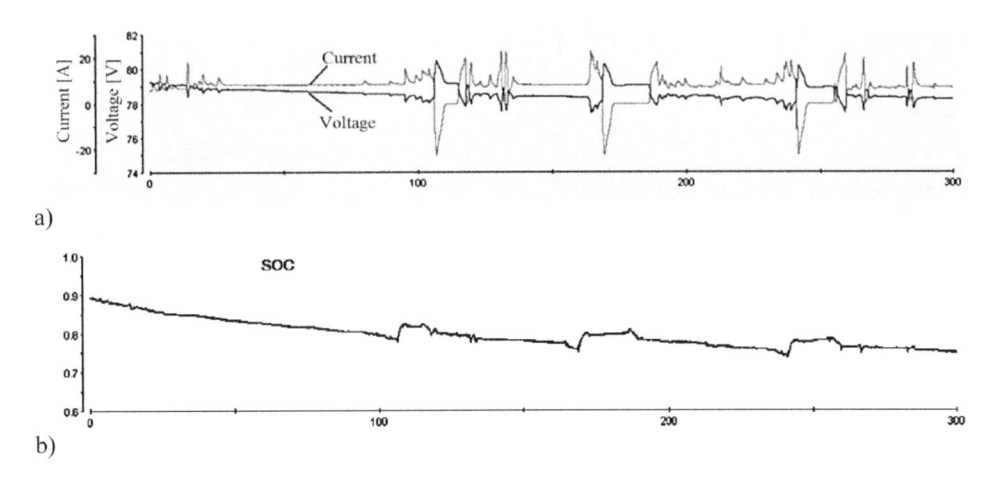

Figure 12. The result of the battery's state-of-charge monitoring: test number 2: a) current and terminal voltage of battery versus real time ; b) SOC of battery versus real time

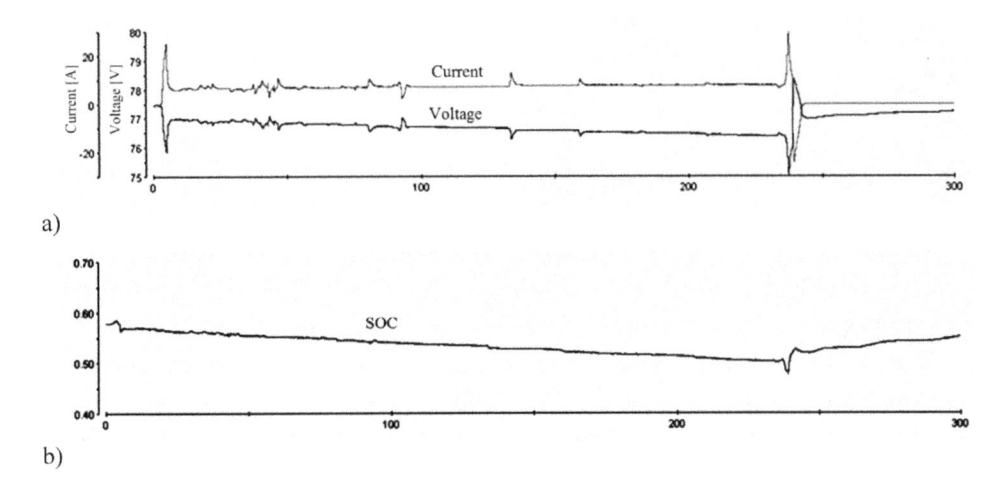

a)

b)

The vehicle's kinetic energy during its recovery braking is easily transferred in the capacitors. This process is further discussed.

The double layer capacitor technology is illustrated in Figure 13. Chargeable and dischargeable double layers build up on the contact surface between the solid electrode (electron conductor) and the liquid electrolyte (ion conductor).

207

Figure 13. The double-layered capacitor's basic construction

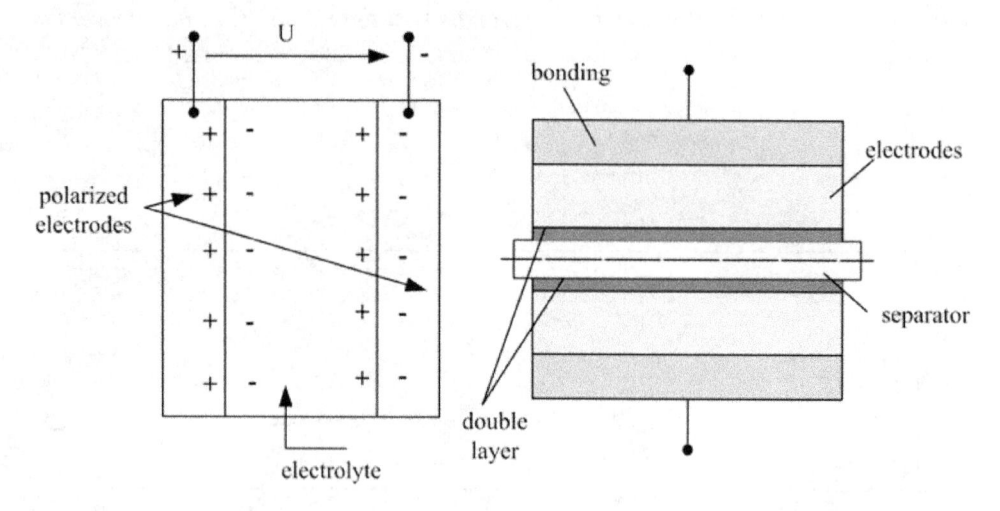

The capacitor's capacity is defined, as is well-known, by the following formula:

$$C = \frac{\varepsilon A}{d};$$

$$E = \frac{1}{2}CU^2 \qquad (5)$$

where: A is the electrode surface area, d is separation distance, ε is an effective dielectric constant.

At present, the electrodes are made with carbon-metal fiber composites, doped conducting polymer films, carbon cloth or mixed metal oxide coatings on metal foil. The aqueous organic or solid polymer as an electrolyte is used.

The high-dielectric materials permit the design of short separation distances and a large electrode surface area. It is the reason for the great increase of its capacitance. The basic feature of ultra capacitors is both high-power and low-energy densities. Due to construction features, the shortest separation distance between electrodes, causes a decrease of the mass, but the self-discharge of capacitors is a significant problem. For this reason, the application of ultra capacitors are only possible for very fast charging and discharging cycles, which take place during city driving. During this driving of a hybrid vehicle, energy stored in the capacitor is changed in the following ranges:

Basic Design Requirements of an Energy Storage Unit Equipped with Battery

$$\begin{cases} \Delta E_{dicharge} = \dfrac{1}{2}C(V_{max}^2 - V_{min}^2) \\ \Delta E_{charge} = \dfrac{1}{2}C\left|V_{min}^2 - V_{max}^2\right| \end{cases} \qquad (6)$$

A typical capacitor is characterized by the following exemplary data:

- **Voltage Rating (V):** 2.5
- **Capacity Density (F/CC):** 20.0
- **Capacity (KF/l):** 10
- **Specific ESR (ΩF):** 20.0
- **ESR (mΩl):** 1.25
- **Energy Density (Wh/l):** 14
- **Energy Density (Wh/kg):** 11
- **Power Density (W/l):** 304
- **Power Density (W/kg):** 243
- **Peak Current (A/l):** 2000

Table 1 shows the basic data of capacitors produced by selected developers.

The interesting solution seems to be the Maxwell Lab. 24V ultra capacitor, which contains bipolar electrode stacks with an active area of approximately 20 cm^2, made with mixtures of organic liquids, containing a conductive additive.

The goal of 24V eight-cell design was to achieve 5Wh/kg and high-power 1kW/kg. During testing, the performances were lower, falling slightly below the goals at 4.5Wh/kg and 500W/kg. The resistance of the package 24V devices, using an organic electrolyte, is in the range 0.4~0.6, with specific ESR (ΩF) at room temperature

Table 1. Review of ultra capacitor technology

Developer	Electrode/Electrolyte Material	Wh/kg	Wh/l	W/kg	Voltage
Panasonic	Carbon/Organic[2]	2.2	2.9	400	3V
Pinnacle Research Institute	Mixed Metal Oxides/Aqueous[1]	0.8	3	500	28V
Maxwell Lab.	Carbon/Organic[1]	6	9	2,500	24V
Maxwell Lab.	Carbon/Organic[1]	7	9	2,000	3V
Maxwell Lab.	Carbon/Organic[2]	4.5	5	1,000	24V
LLNL	Aerogel Carbon/Aqueous[2]	<2	1.5	2,000	5V
LANL	Polymer/Aqueous[1]	<2	-	>500	0.75V

[1] Weight based on cell, including current collector, active material, separator, and electrolyte
[2] Packaged weight

25°C, and did not change significantly at +60°C. For temperature -20°C, resistance increased substantially to 1.5~2.0, with specific ESR (ΩF). The range of resistance changes is connected with constant power charge/discharge. Specific ESR in this case can be calculated as follows: for one capacitor a basic element of 24V ultra capacitor ESR is equal to 0.7 mΩ and capacity 2,600F. For 24V devices, energy density is 5Wh/l. Considering the series connection of elementary capacitors, the resultant capacity of a 24V device is 325F. Further, available energy for nominal values is 26Wh, which means the total volume is 5.2l. Hence, capacity is 62.5F/l. Finally, a specific ESR is obtained as 0.43 ΩF at 25°C temperature.

Figure 14 shows a significant loss of available energy when the temperature falls below -20°C.

The next figure shows a self-discharge test result. Certainly, at minus temperatures, the self-discharge process is slower because of ultra capacitor internal resistance increases.

Table 2 shows a comparison of batteries and ultra capacitors.

The advantage of ultra capacitors is its lower mass, which means higher power density. The energy creating this power density strongly depends on the ultra capacitors' voltage. The storage unit consisting of a series of connected ultra capacitors, applied in a hybrid bus, achieves 400~600V. It will result in the increase of internal resistance. The shortage of ultra capacitors applied in a hybrid bus determines the impossibility of the pure electric drive.

Figure 14. The constant power ultra capacitor discharge at different temperatures

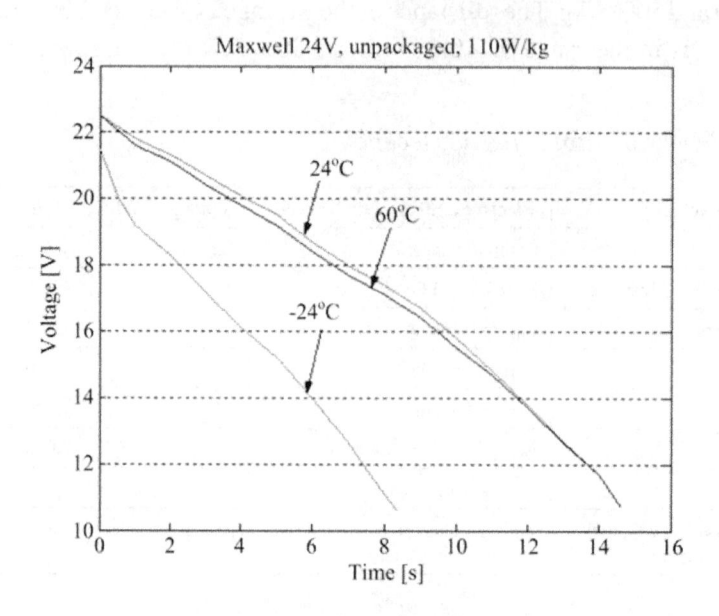

Basic Design Requirements of an Energy Storage Unit Equipped with Battery

Table 2. Comparison of the performance characteristics of various ultra capacitors and HP batteries

Ultra capacitor Devices	Voltage (V)	Ah	Weight (kg)	Resistance (mΩ)	Wh/kg	W/kg 95% Effic.	W/kg Match. Imped.
SKT 47F	3	0.038	0.005	5.2	10 (unpack.)	9735	>80K
Ness 2600F	3	2.2	0.65	0.25	5.1	1558	13850
Panasonic1200F 800F	3	1.0	0.34	1.0	4.2	744	6618
	3	0.67	0.32	2.0	3.1	392	3505
Maxwell 2700F	3	2.25	0.70	0.5	4.8	723	6428
Montena 1800F	3	1.5	0.40	1.0	5.6	632	5625
Battery Devices							
Panasonic Ni-MH (Spirwd)	7.2	5	1.1	18	42	124	655
Ni-MH(Prismt.)	7.2	5	0.92	10	50	218	1152
Ovonic Ni-MH	12	20	5.2	11	46	120	628
Hawker Pb-acid	12	13	4.9	15	29	93	490
Optima Pb-acid	6	15	3.2	4.4	28	121	635
Bolder Tech. Pb-acid	2.1	1.05	0.083	5.7	25	442	2330
Shin Kobe Li-ion	4	4.4	0.3	3.2	55	792	4166

Figure 15. The ultra capacitor self-discharge tests

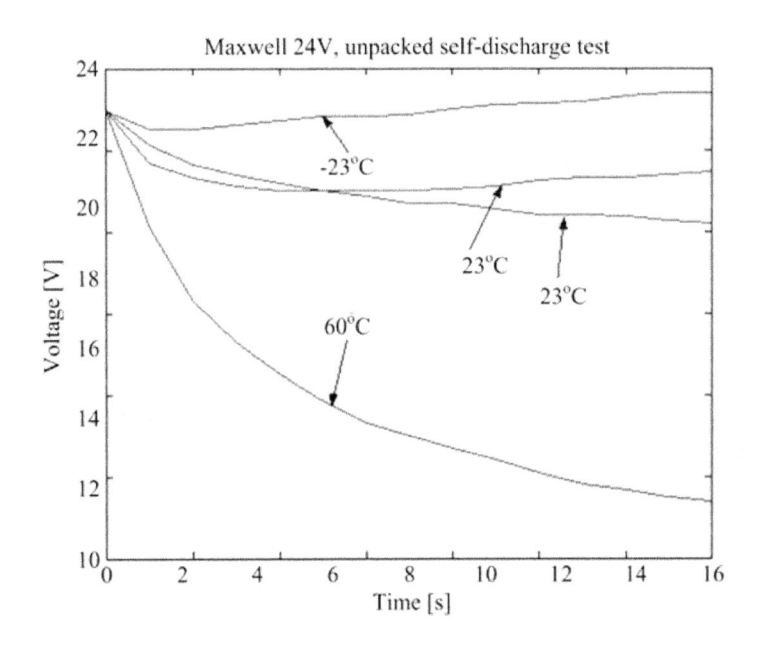

Basic Design Requirements of an Energy Storage Unit Equipped with Battery

Figure 16. Maxwell's ultra capacitors. a) overview; b) data of a package; c) data of an individual component (European 6F project HYHEELS with the author's collaboration)

a)

Nominal Voltage	360V*
Peak Voltage	403V**
Rated Current	400A
Capacitance	18.05F
Total Energy Stored nominal/max.	0.325kWh/0.407kWh
Leakage Current	5mA nominal
Operating Temperature	-35 to 65°C
Weight	220lbs
Dimensions w x l x h	24"x40"x12"
Standard Pack	144 Capacitors
Optional Fire System	Heat Activated Halotron System

* at 2.5 Volts nominal per capacitor
** at 2.8 Volts nominal per capacitor

b)

Model	BCAP0010
Capacitance	2600F
Series Resistance	0.7mΩ at DC
Voltage-Continuous	2.5V
Voltage Peak	2.8V
Rated/Max Current	400A/600A
Long Cycle Life	>5x10^5 cycles

c)

Figure 16 shows the Thunderpack of Maxwell ultra capacitors and its system specifications.

Basic Design Requirements of an Energy Storage Unit Equipped with Battery

2.2. Ultra Capacitor and Battery Set Modeling

It is common that batteries are used as the secondary energy source in a hybrid propulsion system, but technological progress has produced new opportunities. The novel technology of materials, and other new technologies, enable us to develop new, high capacitance capacitors called 'ultra' capacitors or 'super' capacitors. The capacitance depends on the area of the electrodes, and the distance between electrodes. The porosity of electrodes and polarization of an electrolyte allows us to increase the capacitance of a cell to thousands of Farads. The ultra capacitor has the chance to be the alternative of electro-chemical batteries in the hybrid propulsion system. The best results can be obtained by connecting ultra capacitors and batteries together.

There are some advantages of ultra capacitors in comparison with electrochemical batteries. Lower internal resistance and the electro-static energy storage method of the ultra capacitors causes higher energy efficiency during charging and discharging, because the hysteresis loop is very thin. The power density can exceed 3 kW/kg (Bartley, 2005; Burke, 2002; Miller, McCleer, & Everett, 2005), which means the acceptable current of one cell can be in hundreds of amperes. A special electrolyte (AcetoNitrile) makes the operation of ultra capacitors less sensitive to the temperature changes. Ultra capacitors can operate with almost the same efficiency in a wide temperature range (-40°C/ +70°C) (Miller et al., 2005).

The change of ultra capacitor parameters in relation to temperature is very small in comparison with batteries. There are also some disadvantages with ultra capacitors. The main issues are lower energy density and higher price. The specific energy of the ultra capacitor is about 10 times smaller than the battery. If ultra capacitors and batteries are connected together in a hybrid propulsion system, they can be complementary. This solution decreases battery load currents, making its operational live time longer, and besides, it is technically complicated and costly. The battery /ultra capacitor set is recommended only for very special applications.

Contemporarily, the lithium titanate super HP battery can replace the capacitors.

2.2.1. Ultra Capacitor Model

There are many different ultra capacitor models; some are very simple, some very complicated (Miller, McCleer, Everett, & Strangas, 2005; Piórkowski, 2004; Schupbach, Balda, Zolot, & Kramer, 2003; Szumanowski, Piorkowski, & Chang, 2007). Most useful models are based on the substitute circuit. Depending on complexity, models differ in the number of the substitute circuit's elements. Some additional elements are used for considering the nonlinear characteristics of an ultra capacitor. In general, the capacitance and internal resistance depend on temperature, voltage,

age, frequency. It is possible to build a very sophisticated model (see Figure 17), consisting of many elements, and then determine all their parameters, characteristics, and dependent factors, but such researches are costly and time-consuming.

In some cases, it is possible to neglect some factors that have a small influence on the analyzed processes. In case of power split and energy flow analysis in hybrid drives, where time constants are measured in seconds, it is possible to eliminate almost all elements, except the main capacitance *C* and internal resistance – called *ESR* (Equivalent Series Resistance). Parameters *C* and *ESR* are depicted as the function of temperature, voltage and age.

After detailed discussion of both separated models, for simulation of the storage unit in HEV application, it is possible to simplify the ultra capacitor model, when it is parallel and connected with a battery unit:

$$u = u_C - i_C \cdot ESR - L \frac{di_C}{dt} \tag{7}$$

where:

- u_C: Ultra capacitor voltage,
- i_C: Ultra capacitor current,
- **L**: Inductance;
- **ESR**: Equivalent series resistance.

In the above-mentioned model, *C* is depicted as the function of temperature, voltage and age, and the equivalent series resistance *ESR* is the function of temperature and age. The exemplary laboratory test is shown in Figure 18 and 19. It is possible to determine *C* and the equivalent series resistance *ESR* (as a function of voltage and temperature) based on the test result.

The ultra capacitor model based on the scheme in Figure 17, is accurate enough for the power train simulation application. The average possible error in relation to experimental data is less than 2%.

2.2.2. Battery Ultra Capacitor Set Model

A pure electric start is a typical operating mode in the HEV. In this case, a single ultra capacitor is not useful. The solution is (if we do not want to use a huge ultra capacitor) to use the combination of ultra capacitor and battery. If the battery and ultra capacitor are connected in parallel (Figure 20), both energy storage components

Basic Design Requirements of an Energy Storage Unit Equipped with Battery

Figure 17. The complex substitute circuit of ultra capacitor

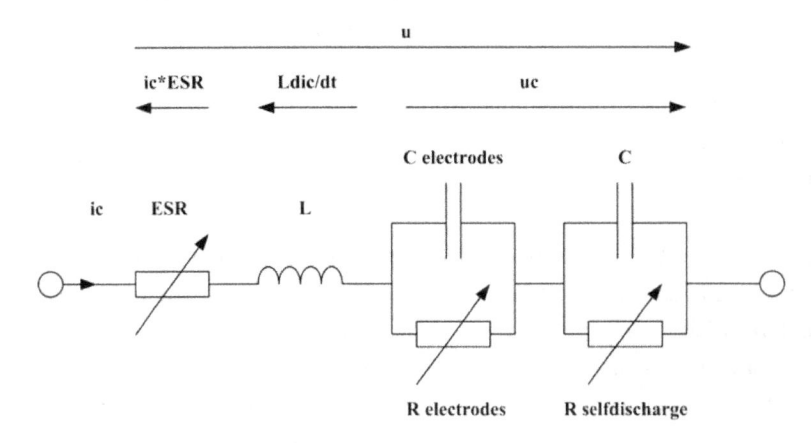

Figure 18. The exemplary laboratory test results of an ultra capacitor produced at 20°C

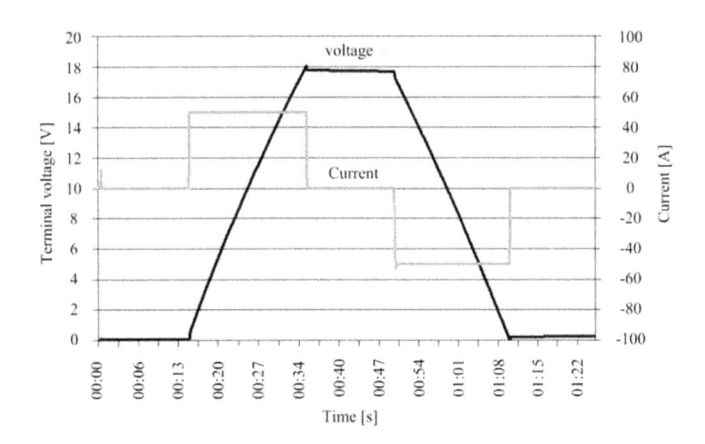

Figure 19. The comparison of UC ultra capacitor voltage alterations, obtained in laboratory and in simulation tests

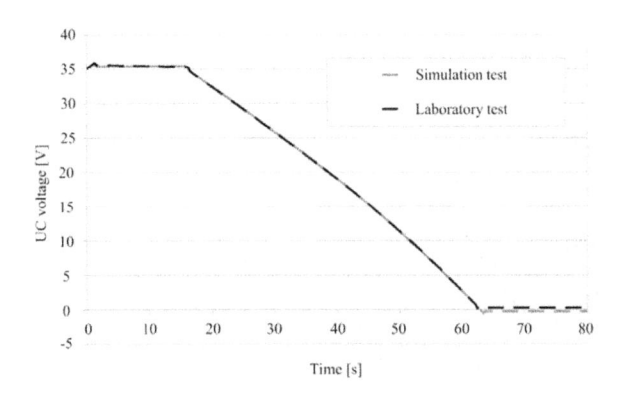

can co-operate very efficiently. Therefore, the following features of this energy storage set can be obtained in a hybrid drive system:

- High current, power, and energy.
- High efficiency.
- Low voltage variation and temperature-sensitive.

The method of battery modeling (determining the battery's electromotive force *EMF* and its internal resistance as functions of the state-of-battery-charge, *SOC*) is depicted in Chapter 5. This model is based on the charge and discharge battery characteristics under different battery current rates, tested in the laboratory.

The above-mentioned features are consequences of a battery's ultra capacitor system operation, which can be described by the following equations:

$$i = i_a + i_c \tag{8}$$

$$\begin{cases} u = E - i_a R_w \\ u = u_c - i_c ESR \end{cases} \tag{9}$$

Figure 20. The substitute circuit of a battery's ultra capacitor system

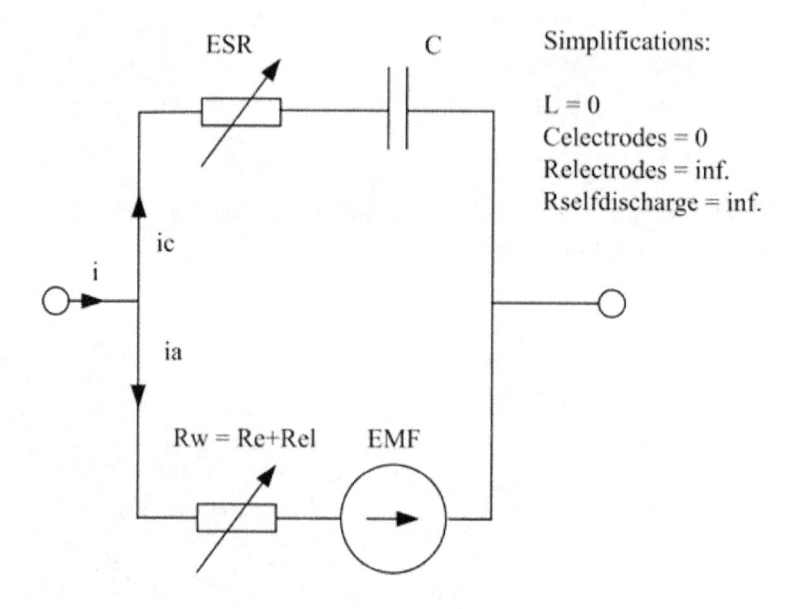

Simplifications:

L = 0
Celectrodes = 0
Relectrodes = inf.
Rselfdischarge = inf.

hence,

$$u = E - R_w(i - i_c) = E - R_w i + R_w i_c \qquad (10)$$

hence,

$$u - R_w i_c = E - R_w i \qquad (11)$$

As

$$\begin{cases} i_c = C \dfrac{u_c}{t} \\ u_c = u + i_c ESR \end{cases} \qquad (12)$$

hence,

$$i_c = C\left(\dfrac{u}{t} + \dfrac{i_c}{t} ESR\right) \qquad (13)$$

So, Equation (13) can be written in the following form:

$$u - R_w C \dfrac{u}{t} + C R_w ESR \dfrac{i_c}{t} = E - R_w i \qquad (14)$$

In relation to the differential form of this equation, it is possible to write:

$$R_w \cdot C \cdot ESR \dfrac{di_c}{dt} - R_w \cdot C \dfrac{du}{dt} + u = E - i R_w \qquad (15)$$

where: $ESR \cdot C$ – ultra capacitor time constant.

It is a kind of 1st order inertial system, where the response of the voltage drop is delayed, which is caused by a loading current, according to the time constant $R_w C$ value. After a few $R_w C$ times, the influence of the battery increases – and the battery ultra capacitor system is still able to deliver power. The most important issue is to determine proper time constant $R_w C$ and $R_w C ESR$ (ESR is the equivalent series resistance) values, which should be correlated with the power variation duration time.

Due to the nonlinearity of the system, the easiest way to resolve the worked out equations is by numerical methods.

2.2.3. Simulation Results

The battery ultra capacitor system operation was analyzed in simulation studies. Some unimportant parameters were duly neglected. The main parameters were determined as characteristics based on laboratory tests. The simulations were conducted by using the MatlabSimulink computation program environment, where all required equations, characteristics, and parameters were implemented (Blanchard, Gaignerot, Hemeyer, & Rigobert, 2002). The system was tested as the ultra capacitor secondary power source in the series hybrid drive of a passenger car, operating in ECE+EUDC driving cycles.

Simulation results were compared with a lone battery and a lone capacitor as a secondary power source in the analyzed hybrid power train. Exemplary simulation results are presented in the Figure 21.

The Internal Combustion Engine (ICE) connected with the PM permanent magnet generator (series power train) was used as the primary source for the analyzed hybrid drive. In simulation, the primary energy source generated constant power in two different value levels, separately for the ECE vehicle driving cycle (maximum vehicle speed 50km/h) and for the EUCD extended vehicle driving cycle (maximum vehicle speed 120 km/h).

As shown in Figure 21, the battery ultra capacitor system can provide the smoothest voltage alteration. In the case of a lone battery, voltage drops are bigger and faster; whilst in a lone ultra capacitor, the relative small energy capacity causes very quick discharge. The only solution is to use a higher capacity unit.

Table 3. The main parameters of the vehicle and its energy storage unit used in the simulation studies

Vehicle	
Total mass	1400kg
Rolling resistance coefficient.	0.012
Aerodynamic coefficient.	0.335
Front are	2 m^2
Wheel dynamic radius	0.304m
Total mechanical transm. ratio wheel/traction motor	3.0
Auxiliary constant load	350W

Basic Design Requirements of an Energy Storage Unit Equipped with Battery

Figure 21. The comparison of terminal voltages of battery, ultra capacitor and battery ultra capacitor set, according to the analyzed vehicle driving cycle

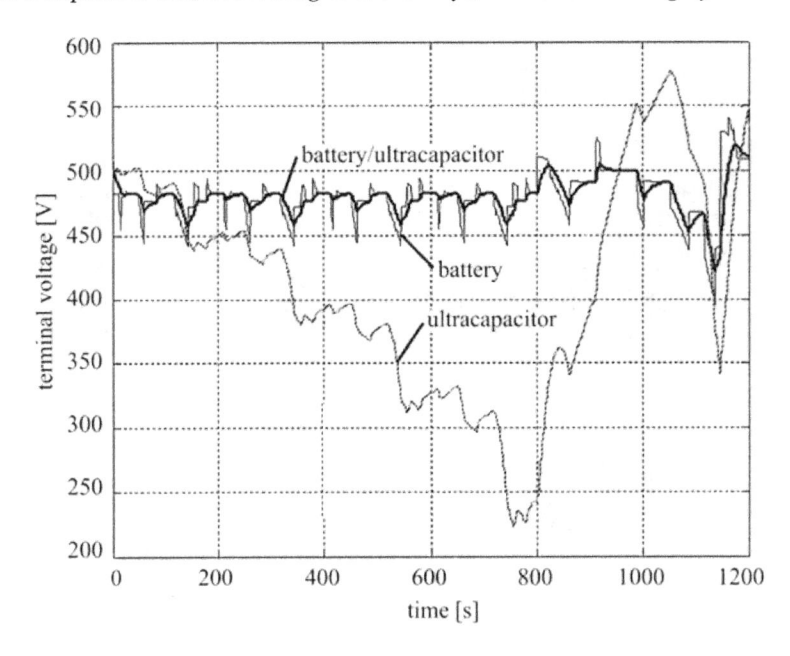

As shown in Figure 22 and Figure 23, the battery ultra capacitor set enables current smoothing and limitation. For this reason, efficiency of power flow is higher (lower ohmic losses). In the case of the battery, the current is a little higher than the resultant current of the battery ultra capacitor set. As for the ultra capacitor, big voltage variation also causes current variation. Again, the higher capacity unit is necessary to limit voltage and current variations. The other solution is to add a small battery and use it as a battery ultra capacitor set.

The above-presented modeling and simulation show some of the main advantages of a coupled battery ultra capacitor set caused by proper adjustment of the inertia factors of both set units, which means the equalization of both *RC* time-constants. The main advantages of a battery ultra capacitor set are:

- Reduced currents.
- Smoothed and averaged voltage drops.
- High-energy density together with high-power density.
- Good performance even in low temperatures.

Certainly, there are some disadvantages with the battery ultra capacitor set:

Basic Design Requirements of an Energy Storage Unit Equipped with Battery

Figure 22. The comparison of currents of battery, ultra capacitor, and battery ultra capacitor set, according to the analyzed vehicle driving cycle (positive values mean charging)

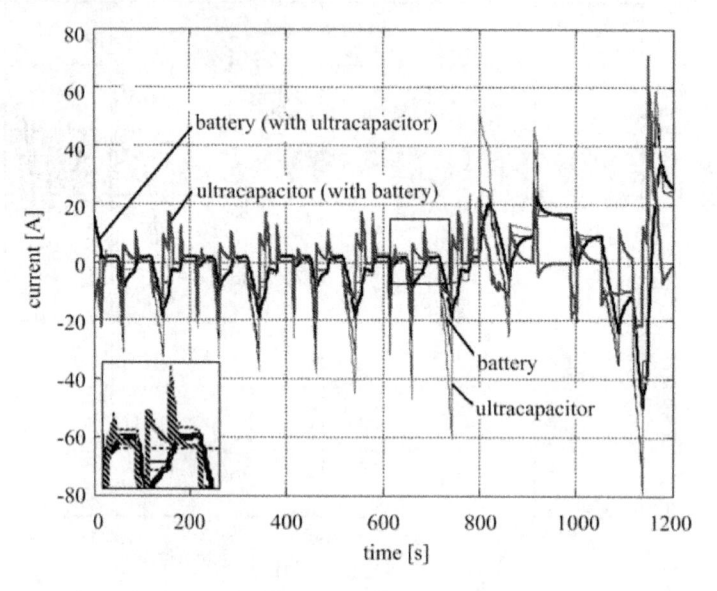

Figure 23. The comparison of currents of battery, ultra capacitor and battery ultra capacitor set in the zoom of a 200s time-window, according to the analyzed vehicle driving cycle (positive values mean charging)

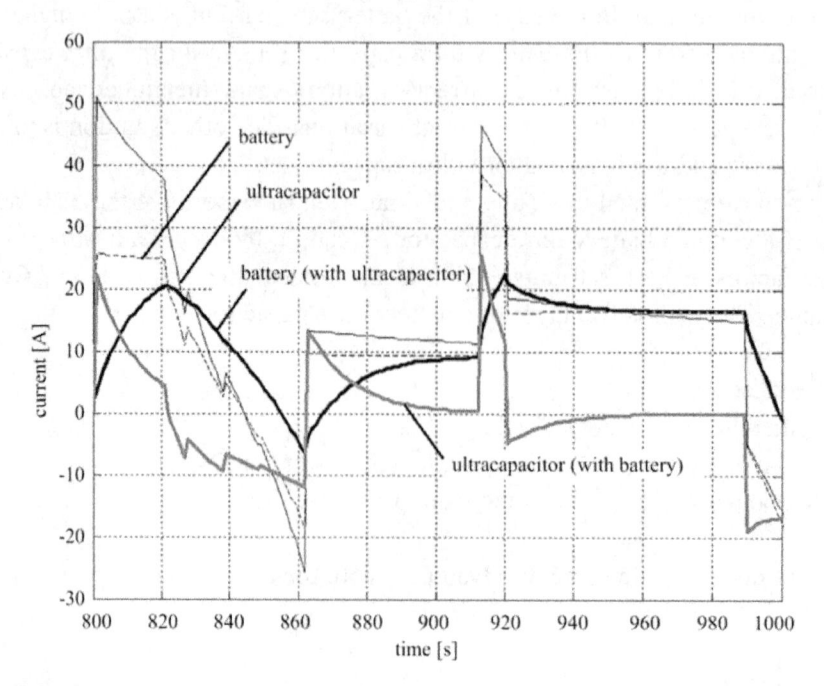

Basic Design Requirements of an Energy Storage Unit Equipped with Battery

- Higher costs.
- Additional weight and volume requirement.
- Additional monitoring and voltage balance system requirement.

The impact of a battery ultra capacitor set on entire current limitation is shown in Figures 22 and 23, where the current's histogram is presented. In the small value of the current range, the battery ultra capacitor system is better, and in the higher current range, the battery and capacitor decoupled system dominates.

3. INFLUENCE OF TEMPERATURE ON BATTERY AND SUPER CAPACITOR'S VOLTAGE EQUALIZATION

As is well known, temperature influences battery performance. The impact of temperature alteration in constant external heat conditions, is a changeable transitory battery current. When you analyze one battery cell from a whole package, its current alteration not only depends on external load, but also on internal cell resistance. As cells are connected in series, load current has to be the same for every single cell. If one of them has an internal resistance different than the other, the same current causes different terminal voltage and temperature value in this cell. This could be the reason behind battery event explosion. This impact is especially dangerous in classic Li-ion batteries, and above all, super capacitors. The repeatable quality cell manufacturing is the background to avoiding voltage equalization devices. This is complicated and costly in practical application. In the last part of this chapter, this problem is outlined.

3.1. Battery's Temperature Distribution

For Li-ion battery management, the temperature must be monitored and adjusted to the optimum operating range. In order to know how battery temperature changes, according to different discharge current, and what the difference of temperature is in different places of the battery, this impact was tested in a laboratory. During the test, at the same time, three thermal sensors are used to measure the temperature of the room, electrode and one point of the exterior body of the cell. The room temperature remains about 23~24°C during the test. Figures 24 and 25 show for instance, the temperature change trend of the battery's exterior and electrode, according to different discharge current and time (Van Mierlo, Maggetto, & Van den Bossche, 2003).

From Figures 24 and 25, it is possible to see that there is a difference between the temperatures of electrode and the exterior body of the battery. The biggest difference is about 35°C, when the discharge current is 14A. When discharging current and time increases, the increasing speed of the temperature of the electrode is faster than the exterior.

When the battery management system is designed, knowledge about the temperature distribution in the battery's different places is necessary. It is better to set the thermal sensor directly to the battery's electrode, because the electrode's temperature is close to inner battery temperature, which strongly influences battery performance.

4. VOLTAGE EQUALIZATION

The battery's thermal management is very important, which in previous paragraphs of this chapter was proved. If one cell or one module of the battery pack is discharged more than another, its internal resistance increases. Then, the drop of voltage on these cell terminals grows in the case of the same current, as in other cell series connected. It means, the power of the invalid cell terminals is increasing and heat emission is growing. This impact causes critical conditions for the battery in its unbalanced state.

Contemporary technology of cell manufacturing is mature enough and permits us to obtain the same quality of the whole production. So, the battery's integration

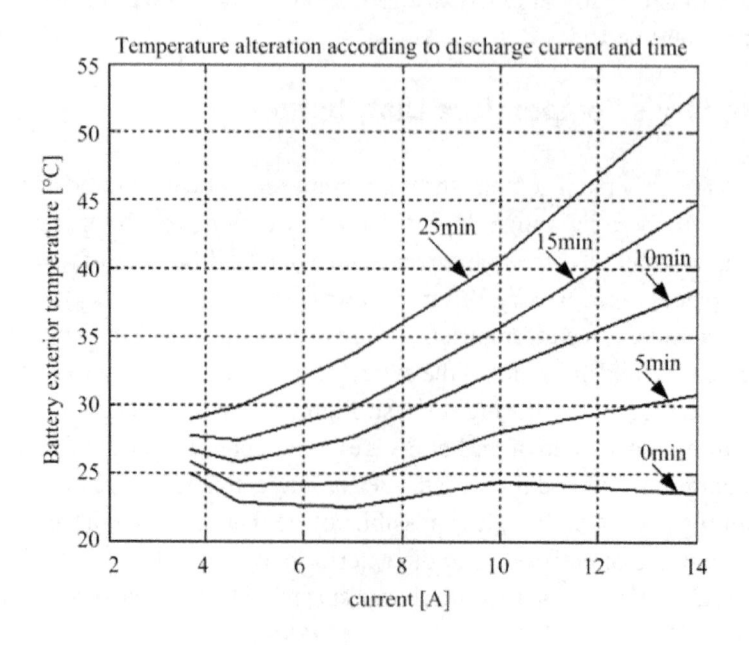

Figure 24. The temperature change trend of a battery's exterior, according to different discharge current and time (Chang, 2005)

Basic Design Requirements of an Energy Storage Unit Equipped with Battery

Figure 25. The temperature change trend of the battery's electrode, according to different discharge current and time (Chang, 2005)

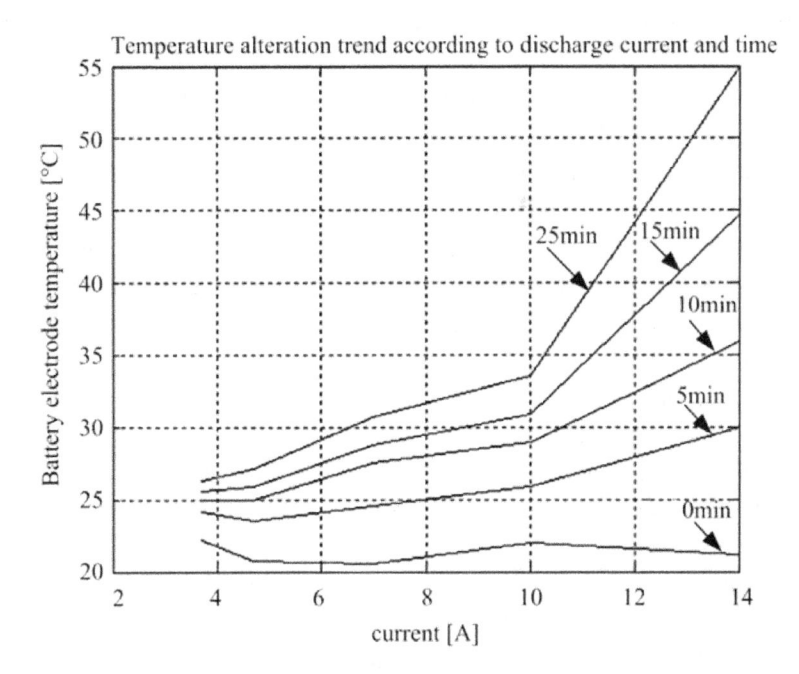

can include parallel and series cell connection, which is especially important in the case of the battery's high Coulomb capacity.

Anyway, for middle and low battery capacity, as in the case of high-power HP batteries applied in hybrid power trains, the consideration of a cell balancing mechanism is advisable.

Charging shuttling cell balancing mechanisms consists of a device that removes the charge from a selected cell, stores that charge, and then delivers it to another cell. The following 'flying capacitor' (Figure 26), is a typical charging shuttling method. It is necessary to emphasize that in ultra capacitors' voltage equalization, the same system of devices can be used. In this last case, the ultra capacitors have to replace the battery's cells, according to schemes in the following figures.

The control electronics close the proper switches to charge capacitor C across cell B_1. Once the capacitor is charged, the switches are opened. The switches are then closed to connect capacitor C across cell B_2. The capacitor then delivers a charge to B_2 based on the differential of voltage between B_1 and B_2.

The capacitor is then connected in the same manner across B_3, B_4,...B_n, B_1,... The highest charged cells will charge C, and the lowest charged cells will take a

Basic Design Requirements of an Energy Storage Unit Equipped with Battery

charge from C. In this way, the charge of the most-charged cells is distributed to the least-charged cells. The only electronic controls needed for this method is a fixed switching sequence to open and close the proper switches.

Batteries usually offer a relatively flat cell terminal voltage across a broad range of battery state-of-charge, SOC, from 40%~80%. The batteries, in charging to sustain hybrid electric vehicles power trains, HEV, operate in the mid-range of their state-of-charge values, and this is where the cell-to-cell voltage differentials are the smallest, thus limiting the usefulness of charge shuttling techniques.

Charging shuttling techniques are useful for electric vehicle power trains, EV, and charging- depleting hybrid electric vehicle power trains, HEV, applications. As the batteries in these applications can be fully charged, the voltage differential between a fully-charged cell and a less-charged cell is greater near the ends of the voltage curve. This increases the effectiveness of the technique.

This method shunts selected cells, with high value resistors, to remove the charge from the highest cells, until they match the charge of the lowest cells (see Figure 27). This circuit is the simplest and cheapest cell balancing implementation. If the resistor value is chosen properly so that it is small, the physical resistor size and switch rating can be small. This method can be operated continuously, with the resistors turning on and off as required. The efficiency of the dissipative technique can be improved by the application of adaptive and learning control algorithms. This method is suitable for charge-sustaining hybrid electric vehicle power train, HEV, applications.

Figure 26. The flying capacitor charge shuttling method of battery cells or ultra capacitor voltage equalization

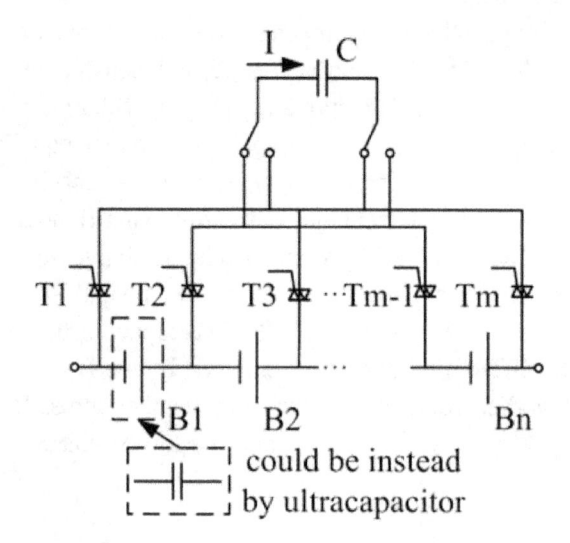

Basic Design Requirements of an Energy Storage Unit Equipped with Battery

Figure 27. The dissipative method of battery cells or ultracapaitor voltage equalization

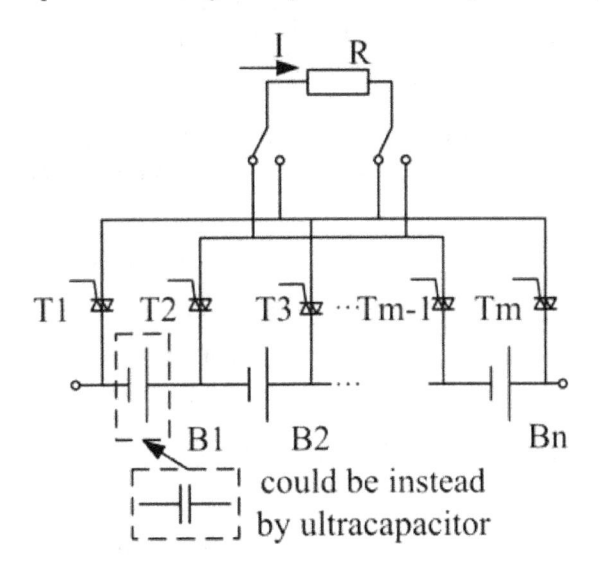

Charge-sustaining hybrid electric vehicle power train, HEV, applications, typically feature regenerative braking, battery charging and electric motoring. These features put high demands on the battery pack for both charging and discharging. The battery pack is usually not kept in a fully-charged condition, and it is marginally-charged, leaving room at the top for charge acceptance. Thus, charging shunting is not an applicable solution.

Since the charge-sustaining hybrid electric vehicle power trains, HEV designs, feature battery packs significantly smaller than their charging-depleting HEV or EV power train counterparts, charge shuttling methods become more attractive with smaller peak switch currents. However, the amount of energy dissipated in the capacitor and switching losses may not justify the increased complexity and expense. The dissipative method is effective without the complexity and expense. However, the algorithm development is significantly more involved.

The voltage equalization system developed in the author's laboratory applies the flying capacitor charge shuttling (see Figure 26).

REFERENCES

Ayad, M. Y., Rael, S., & Davat, B. (2003). Hybrid power source using supercapacitors and batteries. In *Proceedings of 10th European Conference on Power Electronics and Applications (EPE2003)*. EPE.

Barsaq, F., Blanchard, P., Broussely, M., & Sarre, G. (2004). Application of li-ion battery technology to hybrid vehicles. In *Proceedings of ELE European Drive Transportation Conference*. Estorial, Portugal: ELE.

Bartley, T. (2005). Ultra capacitors and batteries for energy storage in heavy-duty hybrid-electric vehicles. In *Proceedings of 22nd International Battery Seminar and Exhibit*. Fort Lauderdale, FL: IEEE.

Blanchard, P., Gaignerot, L., Hemeyer, S., & Rigobert, G. (2002). Progress in SAFT li-ion cells and batteries for automotive application. In *Proceedings of EVS19*. EVS.

Burke, A. (2002). Cost-effective combinations of ultra capacitors and batteries for vehicle application. In *Proceedings of AABC*. AABC.

Burke, A., & Miller, M. (2003). Ultra capacitor and fuel cell applications. In *Proceedings of EVS20*. EVS.

Cegnar, E. J., Hess, H. L., & Johnson, B. K. (2004). A purely ultra capacitor energy storage system hybrid electric vehicles utilizing a based DC-DC boost converter. In *Proceedings of IEEE Applied Power Electronics Conference APEC'04*, (vol. 2, pp. 1160 – 1164). IEEE.

Chang, Y. (2005). *Battery modeling for HEV and battery parameters adjustment for series hybrid bus by simulation*. (MSc thesis). Warsaw University of Technology. Warsaw, Poland.

Chu, A. (2007). Nanophosphate li-ion technology for transportation application. In *Proceedings of EVS 23*. EVS.

Chu, A., & Braatz, P. (2002). Comparison of commercial supercapacitors and high-power lithium-ion batteries for power-assist applications in hybrid electric vehicles. *Journal of Power Sources, 112*, 236–240. doi:10.1016/S0378-7753(02)00364-6

He, Z., Zhang, C., & Sun, F. (2002). Design of EV BMS. In *Proceedings of EVS19*. EVS.

Kalman, P. G. (2002). Filter SOC estimation for Li PB HEV cells. In *Proceedings of EVS19*. EVS.

Kelly, K. (2007). Li-ion batteries in EV/HEV application. In *Proceedings of EVS 23*. EVS.

Kim, J., Lee, S., & Cho, B. H. (2010). SOH prediction of li-ion battery based on hamming network using two patterns recognition. In *Proceedings of EVS 25*. EVS.

Kuhn, B., Pitel, G., & Krein, P. (2005). Electrical properties and equalization of li-ion cells in automotive application. In *Proceedings of VPPC*. IEEE.

Lecout, B., & Liska, I. (2004). NiMH advanced technologies batteries for hybrid public transportation system. In *Proceedings of ELE European Drive Transportation Conference*. ELE.

Lukic, S. M., Wirasingha, S. G., Rodriguez, F., Cao, J., & Emadi, A. (2006). Power management of an ultra capacitor/battery hybrid energy storage system in an HEV. In *Proceedings of IEEE Power and Propulsion Conference*. IEEE.

Miller, J. M., McCleer, P. J., & Everett, M. (2005). Comparative assessment of ultracapacitors and advanced battery energy storage systems in power split electronic-CVT vehicle power trains. In *Proceedings of IEEE International Electric Machines and Drives Conference IEMDC2005*. IEEE.

Miller, J. M., McCleer, P. J., Everett, M., & Strangas, E. (2005). Ultra capacitor plus battery energy storage system sizing methodology for HEV power split electronic CVT's. In *Proceedings of IEEE International Symposium on Industrial Electronics*. IEEE.

Piórkowski, P. (2004). *Study of energy's accumulation efficiency in hybrid drives of vehicles*. (Ph.D. thesis). Warsaw University of Technology. Warsaw, Poland.

Rsekranz, C. (2007). Modern battery systems for HEV. In *Proceedings of EVS 23*. EVS.

Rutquist, P. (2002). Optimal control for the energy storage in HEV. In *Proceedings of EVS19*. EVS.

Schupbach, R. M., & Balda, J. C. (2003). The role of ultra capacitors in an energy storage unit for vehicle power management. In *Proceedings of IEEE 58th Vehicular Technology Conference*. IEEE.

Schupbach, R. M., Balda, J. C., Zolot, M., & Kramer, B. (2003). Design methodology of a combined battery-ultra capacitor energy storage unit for vehicle power management. In *Proceedings of IEEE Power Electronics Specialists Conference*. IEEE.

Stienecker, A. W. (2005). A combined ultra capacitor – Lead acid battery energy storage system for mild hybrid electric vehicles. In *Proceedings of IEEE VPPC*. IEEE.

Strabnick, R., Naunin, D., & Freger, D. (2004). Online SOC determination and forecast for EV by use of different battery models. In *Proceedings of ELE European Drive Transportation Conference*. ELE.

Szumanowski, A. (2001). *Fundamentals of hybrid drives design*. Warsaw: ITE Press.

Szumanowski, A. (2006). *Hybrid electric vehicle drive design based on urban buses*. Warsaw: ITE Press.

Szumanowski, A., Chang, Y., & Piórkowski, P. (2005). Method of battery adjustment for hybrid drive by modeling and simulation. In *Proceedings of IEEE Vehicle Power and Propulsion (VPP) Conference*. IEEE.

Szumanowski, A., Chang, Y., & Piórkowski, P. (2006). Battery parameters adjustment for series hybrid bus by simulation. *Electrotechnical Review*, 2, 139.

Szumanowski, A., Nguyen, K. V., & Piórkowski, P. (2000). Analysis of charging-discharging of nickel metal hydride (NiMH) battery and its influence on the fuel consumption of advanced hybrid drives. In *Proceedings of GPC*. GPC.

Szumanowski, A., Piorkowski, P., & Chang, Y. (2007). Batteries and ultra capacitors set in hybrid propulsion systems. power engineering, energy and electrical drives. In *Proceedings of Powereng*. IEEE.

Timmermans, J.-M., Zadora, P., Cheng, Y., Van Mierlo, J., & Lataire, P. (2005). Modeling and design of super capacitors as a peak power unit for hybrid electric vehicles. In *Proceedings of IEEE 2005 VPPC*. IEEE. doi:10.1109/VPPC.2005.1554635

Van Mierlo, J., Maggetto, G., & Van den Bossche, P. (2003). Models of energy sources for EV and HEV: Fuel cells, batteries, ultra-capacitors, flywheels and engine-generators. *Journal of Power Sources*, *128*(1), 76–89. doi:10.1016/j.jpowsour.2003.09.048

Zolot, M. D., & Kramer, B. (2002). *Hybrid energy storage studies using batteries and ultra capacitors for advanced vehicles*. Paper presented at the 12th International Seminar on Double Layer Capacitors and Similar Energy Storage Devices. Deerfield Beach, FL.

Chapter 7
Basic Hybrid Power Trains Modeling and Simulation

ABSTRACT

Chapter 7 is devoted to the basic and existing in present-day vehicles, power train modeling, and simulation. Generally, there are series and parallel hybrid power trains. In both cases, the role of the internal combustion engine and its dynamic modeling is significant. The two aspects of modeling should be considered. The one devoted to the energy distribution, the second to the local internal combustion engine's control. For the Internal Combustion Engine (ICE) the dynamic modeling method is proposed. Using the simulation of the well-determined map of the ICE can be accepted. In the practical application of a series power train, it is necessary to consider different control strategies of the internal combustion engine's operation. The most significant are the "constant torque" and the "constant speed" control method. The other important problem, because the Internal Combustion Engine's (ICE) generator unit is a strong nonlinear object, is the modeling of the permanent magnet generator, connected by the shaft with the ICE. As for the common parallel hybrid power train, two of its types were, in dynamic modeling, tested by simulation. One of them is the hybrid power train equipped with an automatic (robotized) transmission. Generally, it is possible to state that this transmission can be used as the Automatic Manual Transmission (AMT) or the Dual Clutch. The second one is the split sectional hybrid power train and is the most simple solution. The Hybrid Split Sectional Drive (HSSD) applied in an urban bus is also presented.

DOI: 10.4018/978-1-4666-4042-9.ch007

INTRODUCTION

The series and parallel drives belong to most common hybrid power trains. This chapter is focused on analysis of these most typical drives, especially emphasizing the role of the internal combustion engine in power train architecture and its influence on the world of energy's economy.

1. THE INTERNAL COMBUSTION ENGINE AS A PRIMARY ENERGY SOURCE: DYNAMIC MODELING

Modeling of the Internal Combustion Engine (ICE) is generally a difficult task. The best solution for hybrid drive design is using engine-operating maps, which are possible to obtain after special laboratory bench-tests. The typical (simplified) engine-operating map is shown in Figure 1 (Szumanowski, 2006; Szumanowski, & Hajduga, 2006; Szumanowski, Hajduga, & Piórkowski, 1998b).

It is clear that the Internal Combustion Engine's (ICE) operating points in its hybrid drive design should be in the area of its lowest fuel consumption. The proper design process of the hybrid power train should include simulation of the internal combustion engine's operation related to its map. The result of the simulation can be illustrated by internal combustion engine operating points on the map. However, it is not complete information because only the location range of internal combustion engine operating points is known, and the time frequency of these

Figure 1. The illustration of the internal combustion engine ICE operating map in the shape of static engine characteristics - its output shaft torque versus shaft angular velocity

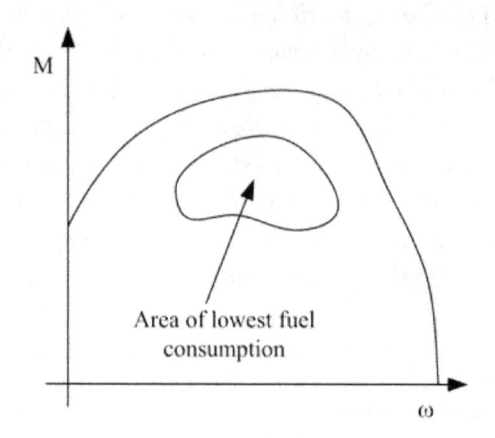

Basic Hybrid Power Trains Modeling and Simulation

operating points when appearing, is unknown. However, it is possible to indicate it, using only dynamic Internal Combustion Engine (ICE) modeling and simulation.

The alterations of an internal combustion engine's momentary power, torque and angular velocity during vehicle driving should be determined. Certainly, the power of load shaft (N_{ICE}) can be computed, according to torque (M_{ICE}) and angular velocity (ω_{ICE}) values:

$$N_{ICE} = f(M_{ICE}, \omega_{ICE})$$

The mathematical models worked out in previous chapters are necessary for this computation. These models need adequate configuration, according to analyzed power train architecture. Additionally, according to block equations set, for example using the Matlab Simulink program, the control functions of the hybrid power train should be considered. This means that only by simulation of the whole power train, can the Internal Combustion Engine (ICE) operating points be designed.

The engine's downsizing means that internal combustion engine power, in the case of the hybrid power train, can be decreased, during downtown or urban traffic. Two types of internal combustion engine should be analyzed - gasoline and diesel. The comparable emissions of the gasoline internal combustion engine are higher (especially CO_2) than a diesel engine, but the gasoline engine does not emit constant particles. Of course, the low power of the gasoline engine can limit the value of 'on line' emissions. The other difference is in the moment of inertia values. The gasoline engine's inertia is lower than in comparable diesel engines, and is more sensitive to the load dynamic changes of torque at output shafts.

The modeling of the thermal combustion engine is very complicated, because the object is strongly nonlinear. For this reason, the approximation functions depicted by the high stage of polynomial or by a set of 'spline' functions are practically useless for simulating the drive system, consisting of a few mechanical-electrical components.

The based-on real data, obtained from laboratory tests, generic, dynamic, internal combustion engine ICE modeling approach is, for example, proposed (Haltori, Aoyama, Kitada, Matsuo, & Hamai, 2011; Zhong, 2007). The basis for dynamic engine modeling is experimental data in the form of static, internal combustion engine torque versus its output shaft angular velocity characteristics, as shown in Figure 2.

It is very important to note that the torque curves for different throttle valve positions (angle α_{te}) are asymmetric and alternated in an individual way. For this reason, it is very difficult to approximate these curves in a proper way. After analysis, the depiction of the above-mentioned curves by a square-powered multinomial is correct. Directly using this function is not correct, because the thermodynamic

Figure 2. The exemplary engine's experimental torque plots a). illustration of transformation M_{te} versus ω_{te} co-ordinates using a square-powered multinomial as the approximation function of a measured internal combustion engine characteristic; b). maximal engine torque points plot; c). throttling valve - means injection - open-angled plot obtained as a set of angles between tangents in maximum torque points and α_{te} axis (differentials in points)

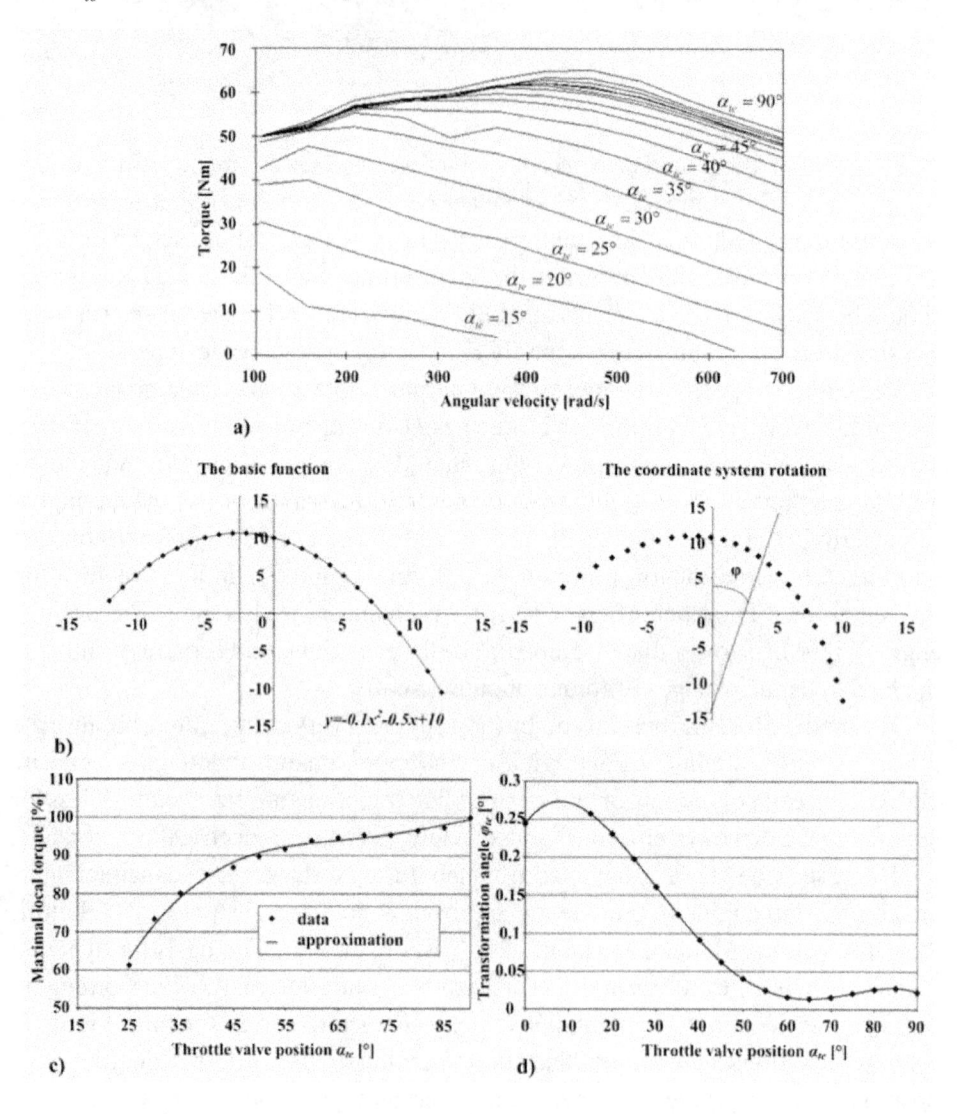

engine's behavior is quite different. However, the increasing power factor of the multinomial is unacceptable, due to computing reasons. In the case of simulation hybrid drives, the method of determining the mathematical model of the internal combustion engine is described as follows:

Basic Hybrid Power Trains Modeling and Simulation

It is necessary to rotate basic co-ordinates $\omega_{te} - M_{te}$ (ω --engine angular velocity; M -- engine torque) into angle ϕ_{te}. Then, new co-ordinates $\omega_{te}' - M_{te}'$ (transformed engine angular velocity and torque respectively) can be obtained, according to the general equation:

$$x = x'\cos\phi - y'\sin\phi$$
$$y = x'\sin\phi + y'\cos\phi \tag{1}$$

where:

- x, y : Basic co-ordinates;
- x', y' : The transformed co-ordinates;
- ϕ : General angle of transformation.

The following equations determine the result of the above-mentioned transformed angle:

$$\phi_{te} \rightarrow f(M_{\max}(\alpha_{te}), \alpha_{te}) \tag{2}$$

or

$$\phi_{te} \rightarrow f(M_{\max}(\alpha_{te}), \omega) \tag{3}$$

From the upper equation, we can get the following result:

$$\phi_{te}(\alpha_{te}) = f\left(\frac{M_{\max}(\alpha_{te})}{M_{\max}(\alpha_{te\,\max})}\right), \quad \alpha_{te\,\max} = 90° \tag{4}$$

In Equation (4), the angle ϕ_{te} is depicted as a function of local maximum torque value (which means the maximum torque value for a certain fixed constant throttle valve position, see Figure 2) related to maximum-maximorum torque value (which means maximum torque of the engine).

Approximation of transforming the engine's torque by a square-powered polynomial is possible, if the basic set of equations is used.

The transformed torque of the engine is:

Basic Hybrid Power Trains Modeling and Simulation

$$M_{te}'(\omega_{te}') = a(\alpha_{te})\omega_{te}'^2 + b(\alpha_{te})\omega_{te}' + c(\alpha_{te}) \tag{5}$$

where $a(\alpha_{te}), b(\alpha_{te}), c(\alpha_{te})$ are coefficients of the Equation (5) obtained by the approximation method, described as follows:

$$\begin{aligned}a(\alpha_{te}) &= a_m \alpha_{te}^m + a_{m-1}\alpha_{te}^{m-1} \\ &+ \ldots + a_1 \alpha_{te} + a_0 \\ b(\alpha_{te}) &= b_n \alpha_{te}^n + b_{n-1}\alpha_{te}^{n-1} \\ &+ \ldots + b_1 \alpha_{te} + b_0 \\ c(\alpha_{te}) &= c_k \alpha_{te}^k + c_{k-1}\alpha_{te}^{k-1} \\ &+ \ldots + c_1 \alpha_{te} + c_0\end{aligned} \tag{6}$$

The transformed equations, after basic co-ordinates are rotated into ϕ_{te} angle, are as follows:

$$\begin{aligned}\omega_{te}'(\phi_{te}) &= \omega_{te} \cos\phi_{te} - M_{te} \sin\phi_{te} \\ \phi(\alpha_{te}) &= a_\alpha \alpha_{te}^m + b_\alpha \alpha_{te}^{m-1} \\ &+ c_\alpha \alpha_{te}^{m-2} + \ldots + d_\alpha \alpha_{te} + e_\alpha\end{aligned} \tag{7}$$

where:

$a_\alpha, b_\alpha, c_\alpha, d_\alpha, e_\alpha$

are coefficients necessary for determining $M_{te\max}(\alpha_{te})$ according to Equation (7).

All coefficients of Equations (6), and (7) should be determined, individually, for a concrete type of engine.

After transformation from $\omega_{te} - M_{te}$ to $\omega_{te}' - M_{te}'$, it is necessary to turn back to real approximated torque curves in following way:

The basic torque in algebraic form is:

$$M_{te} = -\omega_{te}' \sin\phi_{te} + M_{te}' \cos\phi_{te} \tag{8}$$

where:

Basic Hybrid Power Trains Modeling and Simulation

- ω_{te}' : The transformed value of angular velocity of engine shaft,
- M_{te}' : The internal combustion engine torque after transformation of the natural engine's torque, indicated by experiment,
- ϕ_{te} : The angle, after conversion determined by the realistic curve, connecting the points of the maximum, realistic, engine torque $M_{te\,\max}$ (natural engine characteristic) as the function of the engine's throttling valve open angle α_{te}, or corresponding to the fuel injection, if the engine is equipped with it.

The mathematical model of the thermal engine can be described by the following equations set:

$$\begin{cases} M_{te} = -c_\omega \omega'_{te} \sin \phi_{te} + M'_{te} \cos \phi_{te} \\ J \dfrac{d\omega_{te}}{dt} = M_{te} - M(t) \end{cases} \qquad (9)$$

where:

- $M(t)$: External load torque reduced to the engine shaft,
- $c_\omega = 1$ [Nms] : Calculated, proportionality constant.

The approximated torque characteristics are shown in Figure 2, and the mathematical model of the engine, using the presented method, is shown in Figure 3.

The maximum error of the mathematical model compared with experimental data is less than 15% for all families of curves.

The presented methodology is very useful for the complicated power train, and it can be preferred to hybrid drive modeling, in the matter of simplified, computer calculation.

After modeling, it is possible to obtain a 'map' of real ICE by modeling, based on the lab bench-tests, and after computer simulation. The 'maps' can also be used for a simulation study of the hybrid power train.

The following figures show an exemplary 'map' for two different cubic-capacity diesel engines.

The control of the internal combustion engine, ICE, is based on its output torque creation dependent on the fuel injection. In this case, generally, the output response is the angular velocity of the engine's shaft, respectively, to external load conditions.

The following internal combustion engine model should be considered in the design of the ICE angular velocity controller (Figure 6).

Figure 3. The approximated internal combustion engine torque characteristics in the shape of the static engine's characteristic - its output shaft torque versus its shaft angular velocity

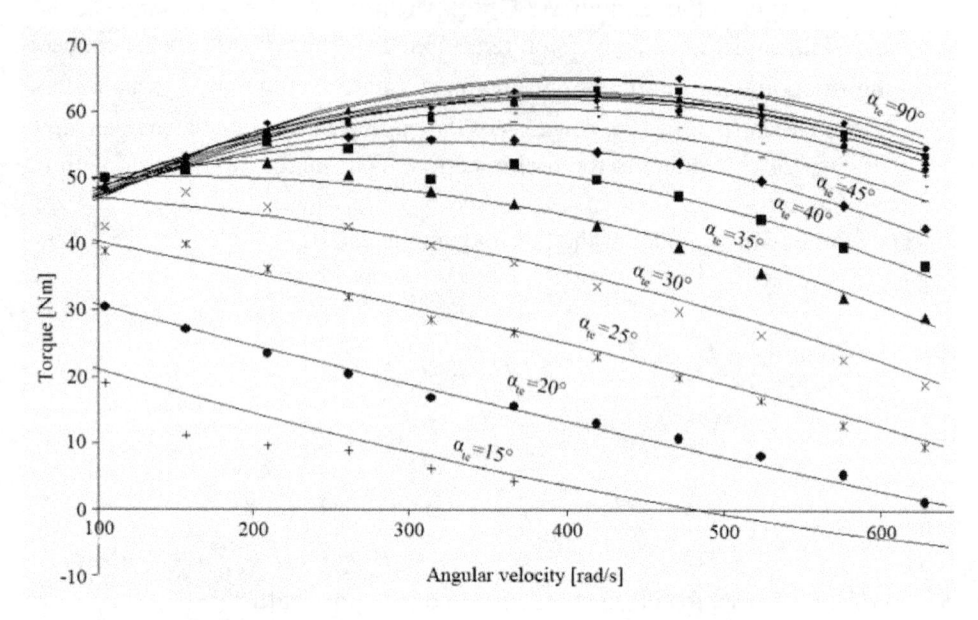

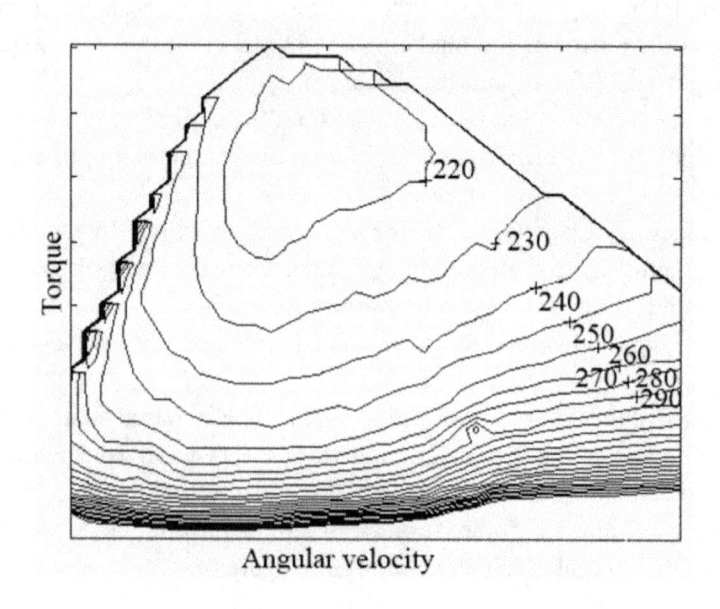

Figure 4. The map of a 1.18L internal combustion diesel engine (fuel consumption curves are indicated as g/kWh)

Basic Hybrid Power Trains Modeling and Simulation

Figure 5. The map of a 8L internal combustion diesel engine (fuel consumption curves are indicated as g/kWh)

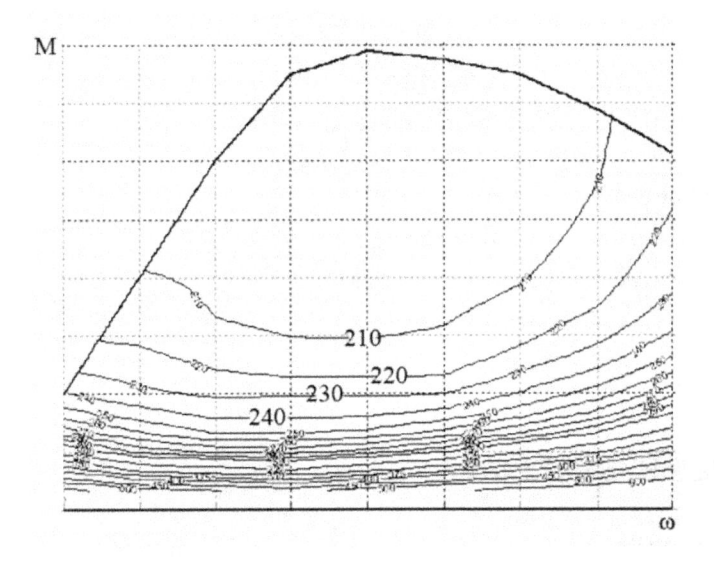

Figure 6. The simplified block layout for the internal combustion engine, ICE, equipped with angular velocity regulator

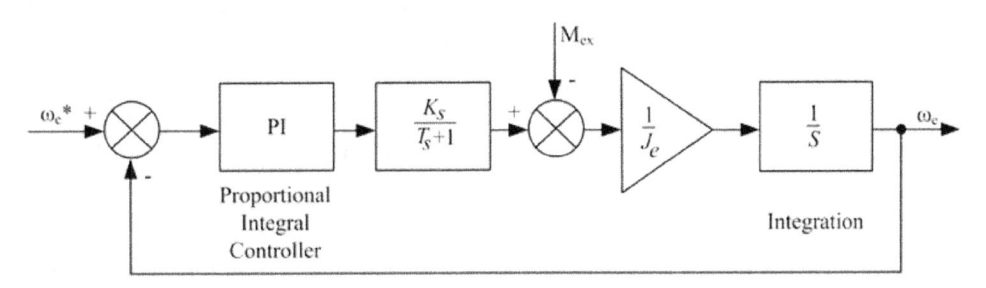

$$\begin{cases} \dfrac{dX_e}{dt} = K_{ie}\omega_e + K_{ie}\omega_e^* \\ J_{ein} = X_e - K_{pe}\omega_e + K_{pe}\omega_e^* \\ \dfrac{dM_e}{dt} = \dfrac{K_e}{T_e}J_e - \dfrac{1}{T_e}M_e \\ \dfrac{d\omega_e}{dt} = \dfrac{1}{J_e}(M_e + M_{ex}) \end{cases} \qquad (10)$$

where:

Basic Hybrid Power Trains Modeling and Simulation

- X_e: Integration block output signal;
- J_{ein}: Injection unit control signal;
- K_{ie}: Angular velocity regulator integration gain;
- K_{pe}: Angular velocity regulator proportional gain;
- ω_e: Output ICE angular velocity;
- ω_e^*: Input ICE angular velocity;
- M_e: Output ICE torque;
- M_{ex}: External ICE load torque;
- J_e: Equivalent ICE moment of inertia;
- K_s: Inertial block gain;
- T_e: Equivalent ICE time constant

2. SERIES HYBRID DRIVE

In Chapter 2, the power distribution process in the hybrid drive (including the series) was discussed. The power generated by the internal combustion engine, ICE, in the theoretical point of view, can be constant, permanently or intermittently. In practical application, it's necessary to consider different control strategies of the internal combustion engine operation. The most important factors are 'constant torque' and 'constant speed' of this engine's operation (Szumanowski, 2006; Szumanowski & Hajduga, 2006; Szumanowski, Hajduga, & Piórkowski, 1998a; 1998b).

Figure 7 shows the momentary power flow in joint-point, electrically connecting the internal combustion engine, ICE-generator unit, with the battery. Control of the

Figure 7. The momentary power–energy flowing through a junction point, in the case of the series hybrid power train

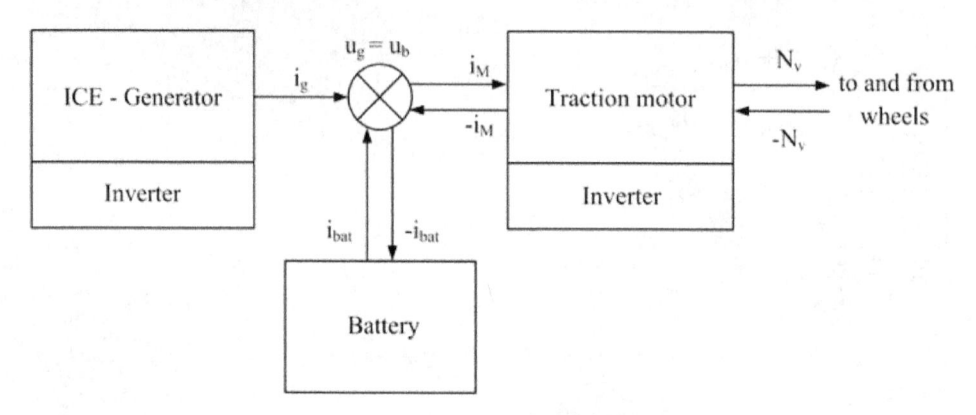

Basic Hybrid Power Trains Modeling and Simulation

series hybrid drive is focused on proper ICE-generator momentary voltage generation, in terms of which voltage should be equal to the battery voltage, at any one time.where:

- i_g, u_g : Current and voltage of generator;
- i_b, u_b : Current and voltage of battery;
- i_M, u_M : Input current and voltage of a traction motor;
- N_V : Power proportional to a traction wheel's load power.

At any one moment, the power balance at the junction point of the hybrid power train is depicted as follows:

Vehicle acceleration:

$$\left. \begin{array}{l} i_g - i_b = i_M \rightarrow \quad when \quad n_g > n_V \\ i_g + i_b = i_M \rightarrow \quad when \quad n_g < n_V \end{array} \right\}$$

Momentary power of the vehicle and generator:

$$n_V, n_g$$

Vehicle regenerative brake:

$$i_g - i_b = -i_M \quad \rightarrow \quad for \quad -n_V$$

Active vehicle stop:

$$i_g - i_b = 0 \quad if \quad i_M = 0 \quad \rightarrow \quad for \quad n_V = 0$$

Momentary generator power:

$$n_g = i_g u_g$$

The power of the generator can be expressed as follows:

$$\begin{array}{l} n_g = i_g u_g = (e_g - i_g R_g) i_g \\ e_g = f(\psi, \omega_g) \cong c\psi\omega \end{array} \tag{11}$$

Finally,

$$n_g = c\psi\omega_g i_g - i_g^2 R_g \quad \rightarrow \quad \omega_g = \frac{n_g + i_g^2 R_g}{ci_g\psi} \tag{12}$$

where:

- e_g: Momentary electromotive force, EMF, value of generator;
- c: Proportionality factor;
- ψ: Magnetic flux;
- ω_g: Angular velocity of the generator;
- R_g: Internal resistance of the generator.

The angular velocity ω_g of the generator is equal to that of the internal combustion engine (directly connected by shafts). The momentary electromagnetic torque of the generator ($m_g = c_1 i_g \psi$) is proportional to the rotational torque of internal combustion engine (m_{ICE}) and $m_g = m_{ICE}\eta$ ($\eta < 1$ because of efficiency). So, the internal combustion engine's (ICE), momentary torque, power, and angular velocity can be expressed as follows:

$$m_{ICE} = f(i_g) = c_2 i_g \Psi;$$

$$\omega_{ICE} = \omega_g = \frac{n_g + i_g^2 R_g}{ci_g \psi};$$

$$m_{ICE}\omega_{ICE} = n_{ICE}$$

If

$$\psi = \text{const} \quad \rightarrow \quad \text{PM generator,}$$

$$c_2\psi = c_3, \quad c\psi = c_4,$$

and

Basic Hybrid Power Trains Modeling and Simulation

$$n_{ICE} = c_3 i_g \left(\frac{ng + i_g^2 R_g}{c_4 i_g} \right) \quad \rightarrow$$

$$n_{ICE} = \frac{c_3}{c_4}(n_g - i_g^2 R_g) \tag{13}$$

There are three following possibilities for the internal combustion engine's (ICE) operation:

1. $n_{ICE} = \text{const}$, in this case, at any given moment, the control system should realize the condition: $n_g = i_g^2 R_g$;
2. $\omega_g = \omega_{ICE} = \text{const}$, in this case, at any given moment, the control system should realize the condition:

$$\frac{n_g}{c_4 i_g} + \frac{i_g^2 R_g}{c_4 i_g} = \text{const};$$

3. $m_{ICE} = \text{const}$, in this case, at any given moment, the control system should realize the condition:

$$c_3 i_g = \text{const} \quad \rightarrow \quad i_g = \text{const}.$$

The assumption $\psi = \text{const}$ can be realized when the permanent magnet (PM) generator is used. The efficiency of permanent magnet (PM) electric machines is the highest. For this reason, this type of generator is strongly recommended. The momentary values of the induction motor can be evaluated after similar analyses.

If the permanent magnet (PM) generator is used, and the condition $\omega_g = \text{const}$ can be fulfilled, then the result of $e_g = \text{const}$ at any one time can be obtained. In the case of other control strategies of the internal combustion engine - permanent magnet generator, the result is $e_g = \text{variable}$.

For the three above-mentioned control strategies, the internal combustion engine's operating points are shown in Figure 8.

The theoretical analysis presented above cannot be strictly realized in practice. First of all, the internal combustion engine - generator unit is a nonlinear object. Secondly, the problem is connected to the accuracy of the control. It means that in real time control, there is uncertainty regarding the internal combustion engine's (ICE) operating points. This impact will be shown in the attached simulation results.

Basic Hybrid Power Trains Modeling and Simulation

Figure 8. The internal combustion engine's (ICE) operating points for different control strategies: a) constant power; b) constant speed; c) constant torque

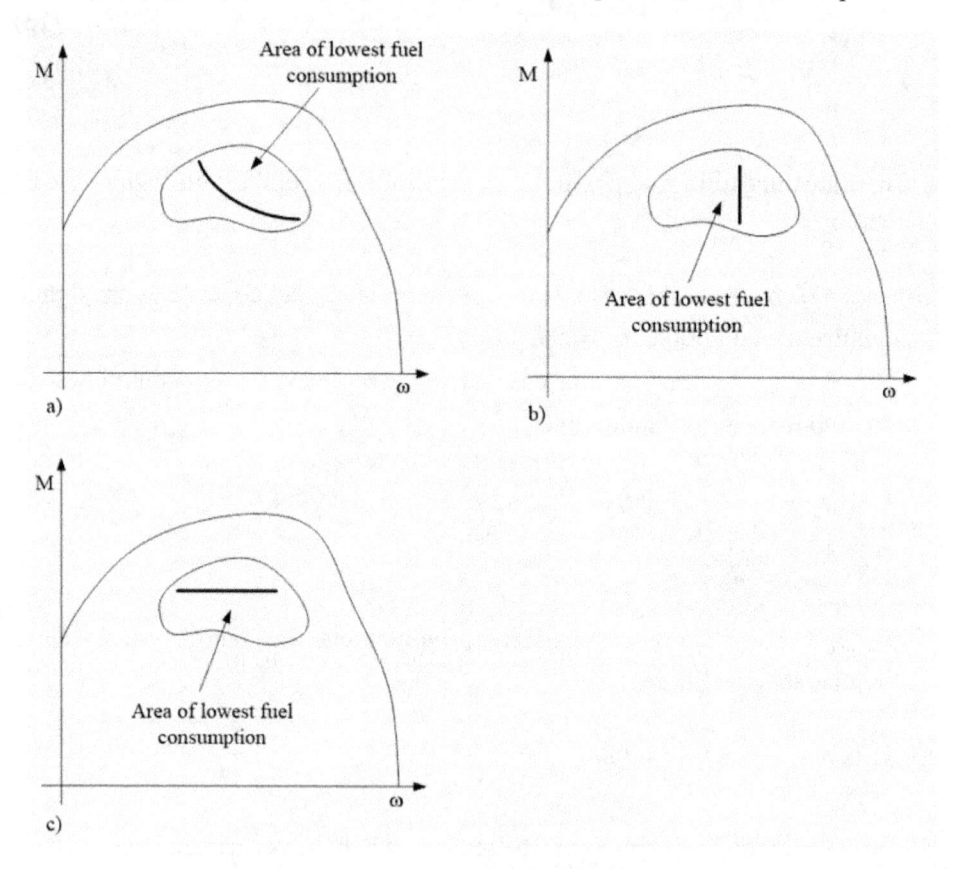

Figure 9 shows the vector graph of the permanent synchronous (PMS) generator with the construction of buried magnets (see Chapter 4).

The permanent magnet, synchronous generator electromotive force (EMF) vector, respectively, to the stator windings (d − axe), can be described as the following equation:

$$e_{gq} = u_g + jx_g i_q \tag{14}$$

After some form changes of the above equation, the power triangle can be obtained, as illustrated by Figure 10.

Further deduction is as follows:

Basic Hybrid Power Trains Modeling and Simulation

Figure 9. The vector graph for a permanent magnet, synchronous (PMS) electric machine (δ - load angle, ϕ - power factor angle)

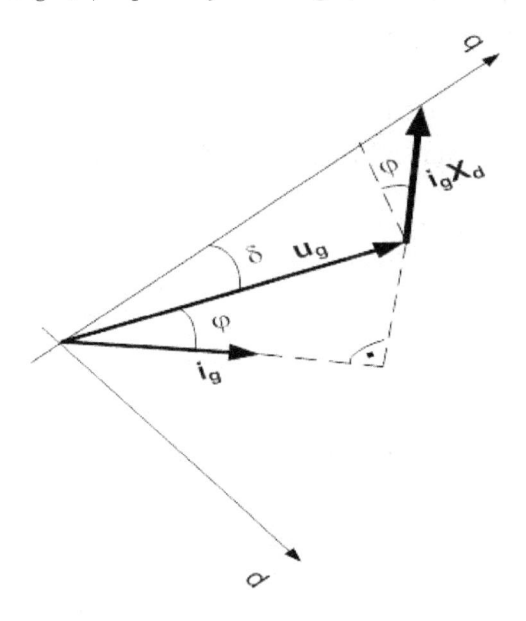

Figure 10. The power triangle related to the terminals of the permanent magnet, synchronous generator

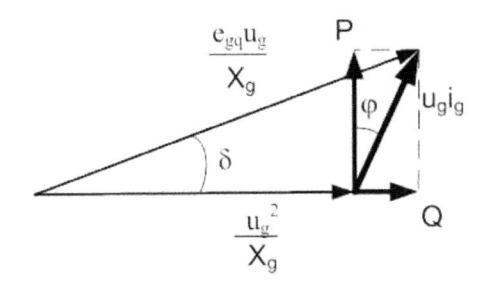

$$S = u_g i_g$$

$$P = u_g i_g \cos\phi = \frac{e_{gq} u_g}{x_g} \sin\delta \qquad (15)$$

$$Q = u_g i_g \sin\phi = \frac{e_{gq} u_g}{x_g} \cos\delta - \frac{u_g^2}{x_g}$$

where:

- S : Apparent power component;
- P : Power component;
- Q : Passive power component.

Note: $P = N_g$ or, at any moment n_g

If $\cos\phi = 1$ (for maximum efficiency of PM synchronous motor operation, see Chapter 4)

Then:

$$Q = 0 \Rightarrow \frac{e_{gq} u_g}{x_q} \cos\delta = \frac{u_g^2}{x_g} \qquad (16)$$

Finally, the following equation can be obtained:

$$e_{gq} \cos\delta = u_g$$

This is the basic control equation of the PM synchronous generator to get the condition $\cos\phi = 1$. Simultaneously, the additional equation is generated, according to the vector graph (see Figure 9):

$$x_g i_g \cos\phi = e_{gq} \sin\delta$$

For $\cos\phi = 1$, the following equation can be obtained:

$$x_g i_g = e_{gq} \sin\delta$$

and

$$e_{gq} = p\omega_g \psi$$

where:

- ψ : Constant flux of permanent magnets;
- p : Number of magnetic poles.

If $x_g = p\omega_g L_g$, then:

Basic Hybrid Power Trains Modeling and Simulation

Figure 11. Vector graph of constant flux of permanent magnets and voltage of permanent magnet, synchronous generator projection on d, q axes for $\cos\phi = 1$

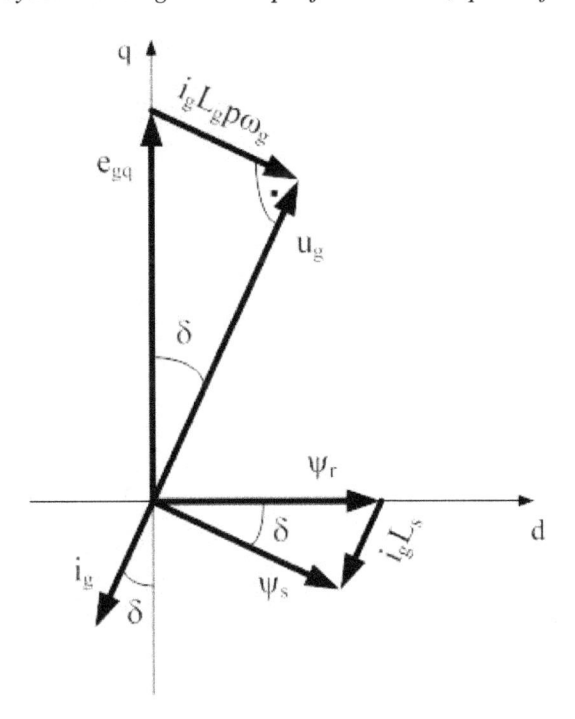

$$i_g = \frac{\psi}{L_g}\sin\delta \qquad (17)$$

Figure 11 shows the vector graph of PM generator for the condition $\cos\phi = 1$ where: ψ_s -flux generated by stator.

Note: flux ψ is rotating because the permanent magnets are fixed in the rotor.

Based on the PM synchronous motor model and the assumption $\cos\phi = 1$, the following equations can be obtained:

$$\begin{cases} i_d = i_g \sin\delta \\ i_q = i_g \cos\delta \end{cases} \qquad (18)$$

and

$$M_g = \frac{3}{2}pi_q\psi \qquad (19)$$

Basic Hybrid Power Trains Modeling and Simulation

After some transformation, the following equations can be obtained:

$$\begin{cases} u_d = e_{gq} \cos\delta \sin\delta = p\psi\omega_g \cos\delta \sin\delta \\ u_q = e_{gq} \cos^2\delta = p\psi\omega_g \cos^2\delta \end{cases} \quad (20)$$

$$\begin{cases} i_d = \dfrac{\psi}{L_g} \sin^2\delta \\ i_q = \dfrac{\psi}{L_g} \sin\delta \cos\delta \end{cases} \quad (21)$$

and

$$M_g = \frac{3}{2}\frac{p\psi}{L_g} \sin\delta \cos\delta \quad (22)$$

The equation set (20-22) is the backdrop to permanent magnet (PM), synchronous generator control for the condition $\cos\phi = 1$ in steady, operating states of this electric machine.

For $\cos\phi = 1$, the voltage of permanent magnet synchronous generator is depicted as follows:

$$u_g = p\omega_g \psi \cos\delta \quad (23)$$

During the power flow of a series hybrid drive operation, at any moment u_g should be equal to the battery's voltage u_b. So, the alterations of $\cos\delta$ offer the possibility of obtaining a constant ω_g. This control must be used in the case of engine–operating modes of 'constant power' or 'constant speed'.

In the case of 'constant torque', $\omega_g = \text{variable}$, p and ψ are always constant. Then, the product $\omega_g \cos\delta$ decides the alterations of u_g and u_b.

For $\cos\phi = 1$, the generator's torque:

$$M_g = \frac{3}{2}\frac{p\psi}{L_g} \sin\delta \cos\delta$$

can be illustrated by Figure 12.

Basic Hybrid Power Trains Modeling and Simulation

Figure 12. The loading, angular-power factor angle of permanent magnet generator versus its output torque, expressed per unit

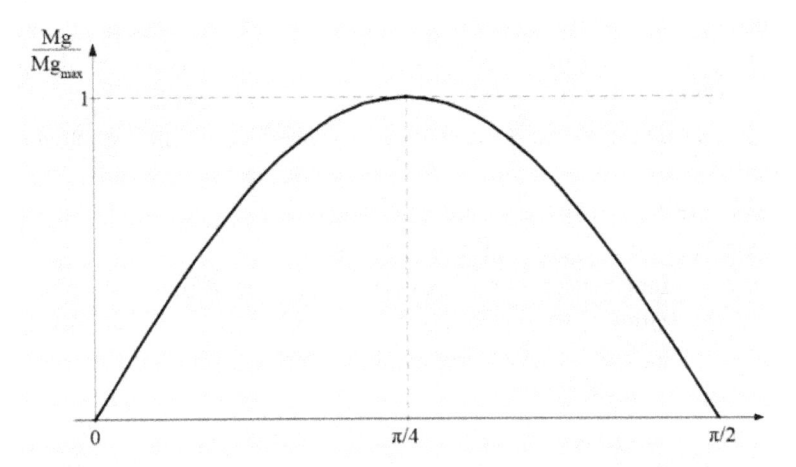

Considering the above-mentioned facts, using the engine-operating mode of 'constant torque', it is easier to obtain $\cos\phi = 1$. In this case, it's certain that ω_g should change, according to the control of the internal combustion engine's angular velocity. Alternatively, it's easier for internal combustion engine control when the engine-operating mode is 'constant speed', for this means that $\omega_g = const$ can be obtained.

The response is possible to attain after simulation tests for real drive components. It's necessary to know that in the transitory state, both ω_g and angle δ should change. The goal of designing a permanent magnet, synchronous generator, control system (inverter operating in a majority in-pulse-width-modulation mode) is to obtain the result that the loading angle alternates in a proper, narrow range near the maximum torque value ($\delta = \dfrac{\pi}{4}$). Correspondingly, the alteration of ω_g is also limited.

The accuracy of control depends on the following conditions:

- The use of an additional ICE controller to limit 'on line' changes of ω_g;
- The use of proper feedbacks of torque and speed;
- The quality of a pulse-modulation range of inverter operating in PWM mode.

It's possible to achieve all the above-mentioned conditions in practical application.

The exemplary simulations of a 15T urban bus power train are attached below. The background to the simulation tests was a typical, urban, driving cycle with a maximum vehicle speed of 50 km/h. Parameters of a series hybrid bus power train are shown in Table 1. As the target of simulations is to find the minimum fuel consumption, after many tests, proper battery parameters are adjusted. Battery 43.2 kWh was adjusted (300V and capacity 144Ah). For this reason, the full series hybrid drive is obtained. It is difficult enough to define what is 'mild' or 'full'. The difference is based on the proportion of power provided by the engine and battery. In borderline cases, 'zero' hybrid means pure engine drive, whereas 'completely full' means pure electric drive. In the first case, electricity consumption is zero: and in the second one, fuel consumption is also zero.

Anyway, the fuel consumption is lower in the case of the 'full' hybrid than in the case of the 'mild'. Certainly in the 'full' hybrid case, the operating points on the same diesel engine's map are removed in the direction of lower rpm, which is opposite to the 'mild' hybrid.

The general conclusion in analyzed cases is that change of voltage has a lower influence on fuel consumption than battery capacity, assuming similar battery energy (kWh).

Increasing the voltage is connected with increasing a battery's internal resistance and decreasing a battery's current. However, the most important is the relationship of a maximum battery current to its nominal 1C current. Two design solutions regarding nickel metal hydride NiMH battery parameters were considered: 500V/80Ah and 300V/144Ah. In analyzed cases, the voltage ratio of 500V and 300V is equal to about 1.6. The capacity ratio of 144Ah and 80 Ah is equal to about 1.8. The increasing of capacity is higher than the decreasing of voltage. This influences strongly on the decreasing internal resistance of the battery in the case of 300V, in comparison to the 500V pack. Additionally, the battery's internal resistance is in inverse ratio to its capacity. Finally, respecting the nonlinear character of a battery's internal resistance in the case of a 300V and 144 Ah pack, despite increasing discharge-charge currents (the same traction wheel power requirements), the EMF of the battery is stiffer. During driving, the cycle electromotive force of the battery (EMF_b) is higher than the electromotive force of the generator (EMF_g). For the condition of momentary equalization of both sources of voltage, the higher power is taken from the battery (according to $EMF_b/EMF_g > 1$) than from the engine-generator set. This is the reason for decreasing both fuel and electricity consumption. As the battery state- of-charge factor (k; SOC) keeps the same level at the beginning and at the end of the driving cycle, the fuel consumption decrease is absolutely more important.

The following attached figures show selected characteristics of the so-called 'full' hybrid drive for three different internal combustion engine (ICE) control strategies; 'constant speed', 'constant torque', and 'sloping speed'.

Basic Hybrid Power Trains Modeling and Simulation

Table 1. Parameters of an urban bus with a series hybrid power train

Vehicle data	Battery	Traction motor
Mass: 15000 kg Frontal area: 6.92 m^2 Coeff. of aerodynamic drag: 0.55 Coeff. of rolling friction: 0.01 Tire rolling radius: 0.51 m	NiMH Nominal capacity: 144Ah Nominal voltage: 300V	Number of poles p: 24 PM flux: 0.05775 Wb Coils inductance L 0.000076 H Coils resistance R: 0.04Ω Torque max: 275Nm Speed max: 8500rpm
	ICE	Generator
	Diesel 1180ccm JTD	Number of poles p: 24 PM flux: 0.10311 Wb Coils inductance L 0.000248 H Coils resistance R: 0.04Ω

Figure 13. The map of 'constant speed' of an internal combustion diesel engine operation controlled, according to a selected vehicle driving cycle

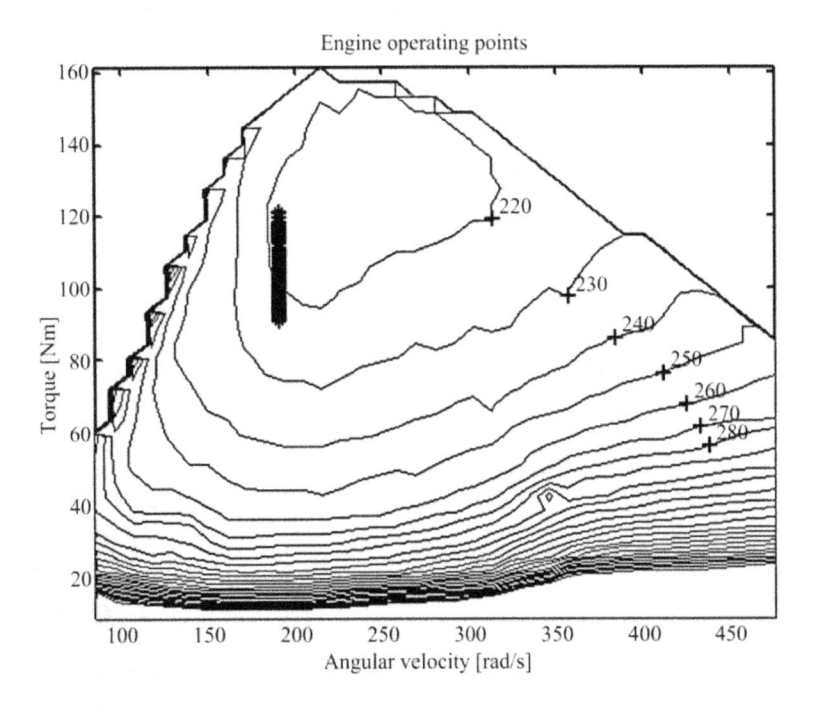

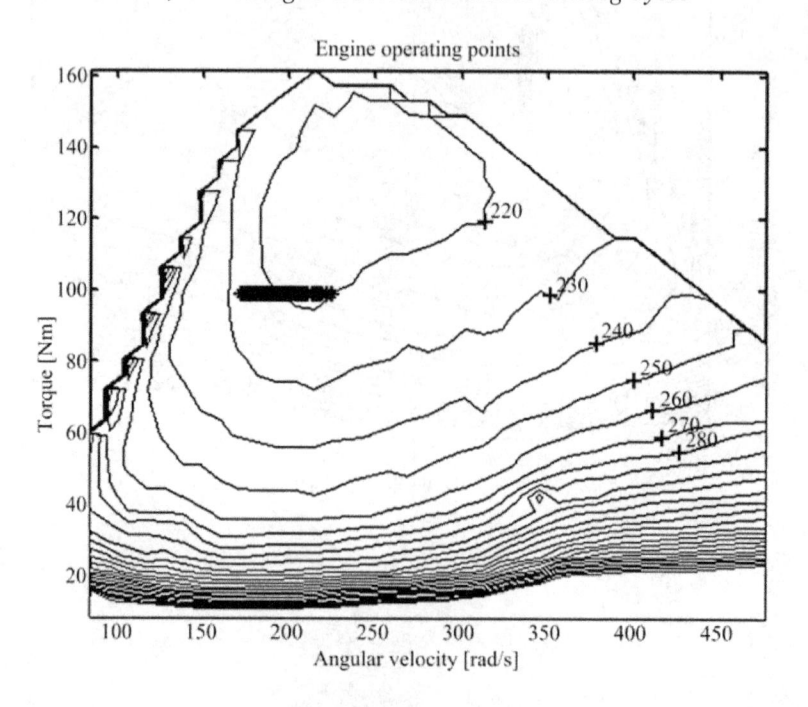

Figure 14. The battery state-of-charge factor (SOC) alterations, during a vehicle driving cycle for a 'constant speed' controlled internal combustion engine

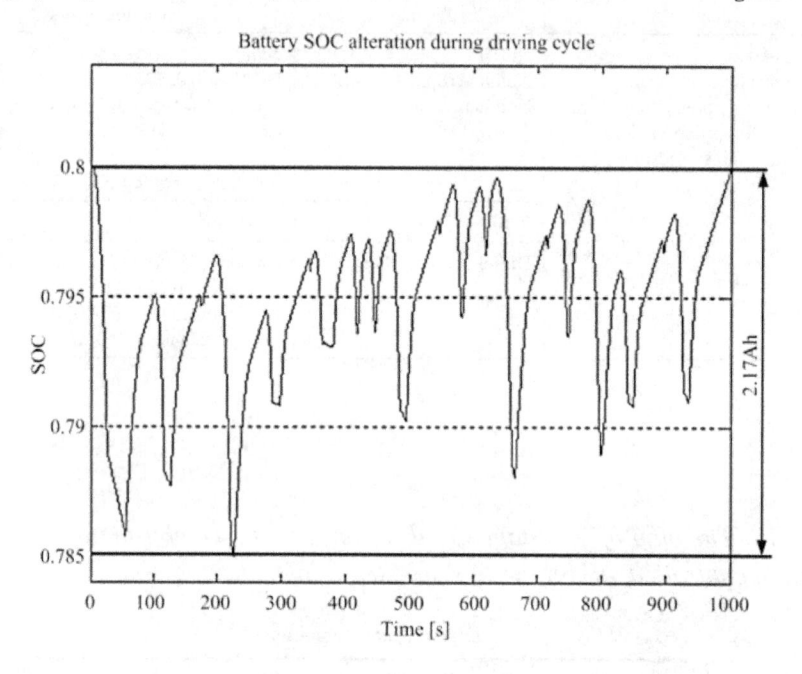

Figure 15. The map of 'constant torque' of an internal combustion diesel engine operation controlled, according to a selected vehicle driving cycle

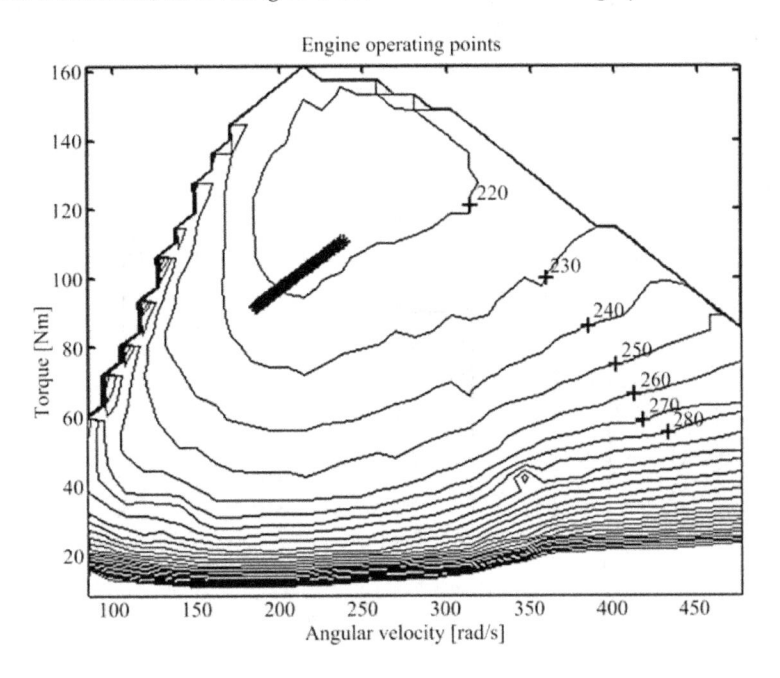

Basic Hybrid Power Trains Modeling and Simulation

Figure 16. The battery state-of-charge factor (SOC) alterations during a vehicle driving cycle for a 'constant torque' controlled internal combustion engine

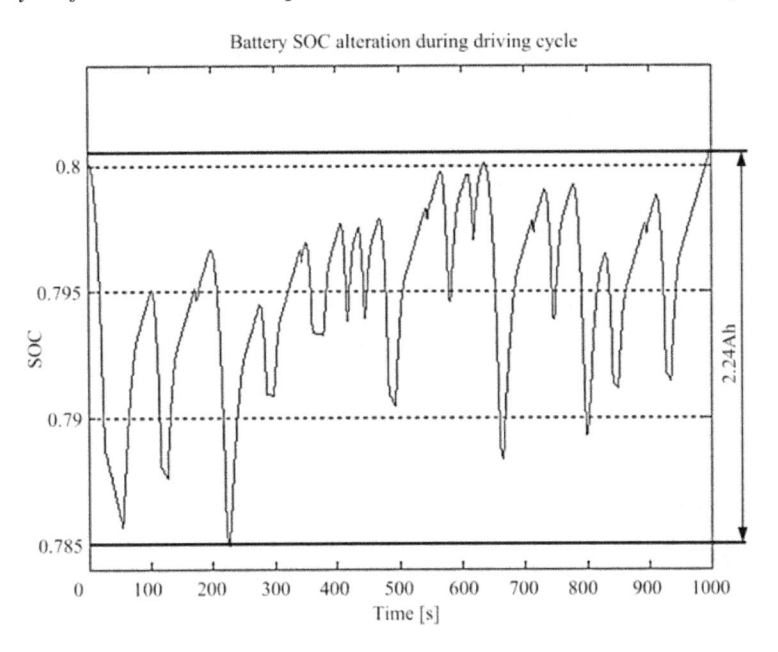

Figure 17. The map of 'sloping speed' of an internal combustion diesel engine operation controlled, according to a selected vehicle driving cycle

Figure 18. The battery state-of-charge factor (SOC) alterations during a vehicle driving cycle for a 'sloping speed' controlled internal combustion engine

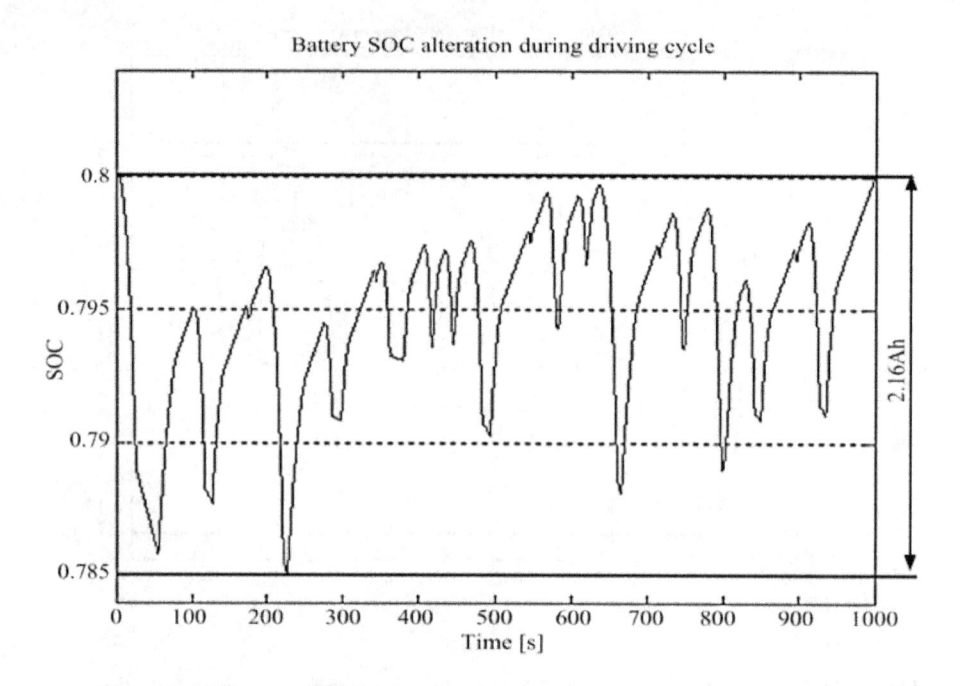

Figure 19. The fuel consumption (l/100km) comparison of three control strategies applied to the internal combustion diesel engine control unit in a series hybrid bus drive and its comparison with a conventional drive

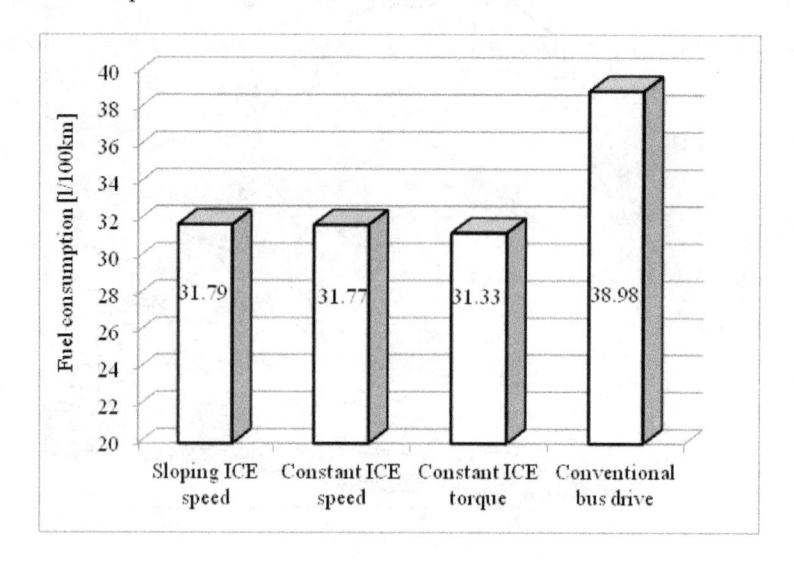

According to the simulation results, some comments are as follows:

- Assumed control based on 'constant torque' and 'constant speed' with the limitation of the generator's angular velocity permits us to obtain a proper ICE-generator set operation;
- The smallest alteration of power, torque, and velocity of the ICE-generator set confirmed the theoretical analyses;
- The presented method is necessary for design of the series hybrid drive;
- It is interesting to note (Figures 14, 16, 18) that the battery charge's alteration is only a little more than 2Ah for a 15-ton bus, referring to its one driving cycle. This alteration means the difference between the maximum and the minimum values of the battery state-of-charge factor (SOC), respectively, to the analyzed driving cycle (see Chapter 2). Certainly, the transitory battery's load power and its current require a high and proper battery Coulomb capacity.
- A diesel engine tested in presented simulations for 'mild' hybrid drive, is significantly less profitable. The attached simulations show only results, when the hybrid drive is not optimized (minimum fuel consumption). Much better is the case of the 'full' series hybrid. The fuel saving in comparison to the conventional diesel drive is about 7 l/100km;

In the case of the gasoline engine used in a 'full' series hybrid drive, the operating points of the internal combustion engine, ICE, are located in the maximum efficiency area, in terms of high torque and small angular velocity (Figures 13, 15 and 17). As for the small diesel engine ('mild' hybrid), its operating points are located in a similar torque range, but at the higher angular velocity in the most efficient area. For this reason, fuel consumption is also higher than in the case of the properly-adjusted diesel engine. For the presented method, used for a hybrid drive design, this solution only shows the need for proper usage of tools for modeling and simulation.

3. DRIVE ARCHITECTURE EQUIPPED WITH AN AUTOMATIC (ROBOTIZED) TRANSMISSION

Another configuration of parallel hybrid drive systems is that they are equipped with a robotized, as well as an Automatic Manual Transmission (AMT) or Dual Clutch. The generic scheme referring to this type of hybrid power train is shown in Figure 20.

Basic Hybrid Power Trains Modeling and Simulation

In the robotized gearbox, the shifting is performed automatically, and respectively, to the output's rotational speed defined by the nominal of this automatic gearbox's output shaft angular velocity value ω_{nom}. The Total Transient Time (TTT) of gearshift change is described by the following equation (Ippolito, & Rovera, 1996):

TTT = TRT + TWT + TIT

in which:

TWT = TDR + TSR + TSS + TER

where:

- **TRT:** Time of traction reduction,
- **TWT:** Time without traction,
- **TIT:** Time of traction increasing,
- **TDR:** Time to disengage the previous ratio,
- **TSR:** Time to select the new ratio,
- **TSS:** Time to synchronize the shafts,
- **TER:** Time to engage the new ratio.

In order to reduce as much as possible the total transient times, a small hydraulic actuator is used, capable of quick and accurate shifts, controlled by an electronic control system. The time has been reduced to about 1s for each gearshift change. The automatic change of mechanical ratios is adjusted for the necessity of limiting the internal combustion engine's operation. It is done in both directions: - decreasing and increasing ratios. During regenerative braking, the values of ratios, in particular, are increased, keeping a higher speed of the permanent magnet machine, operating as a generator.

Computer simulation analysis is conducted with the following data (see Figure 20 and Table 2). In proper computations, the watt efficiency of the drive is included (the efficiency of the hydraulic unit is neglected).

Note:

- Fuel consumption by IC engine ~3.7 l/100 km (according to assumed data).
- All values of speed and torque are reduced to the shaft of the gearbox, where speed and torque of the internal combustion engine and the electric machine are added.

Basic Hybrid Power Trains Modeling and Simulation

Table 2. Data used in analysis, according to the scheme presented in Figure 20.

Vehicle	
Total mass	1300 (kg)
Traction wheels inertial torque	1,7 (kgm^2)
Tire dynamic radius	0,263 (m)
Rolling resistance factor	0,0008
Flat surface of vehicle	1,6 (m^2)
Aerodynamic drag coefficient	0,33
IC engine ~ 900 cc	
Mechanical ratio between engine shaft and main shaft: 0,583	
PM motor	
Mechanical ratio between PM motor shaft and main shaft: 1	
Rated power	18,9 (kW)
Rated speed	7500 (rpm)
Voltage	195 (V)
Battery	
Capacity	~ 10 (kWh)
Nominal voltage	216 (V)
Nominal current	50 (A)
Main differential gear	**4,923**
6-speed gear-box: Data of FIAT Punto	
i_I	3,545
i_{II}	2,157
i_{III}	1,480
i_{IV}	1,121
i_V	0,902
i_{VI}	0,744

Compared to planetary transmission (see Chapter 8, according to the scheme from Figure 8.36) the fuel consumption for the same driving and engine condition is about 0.5l/100 km higher.

In the case of the above-presented simulation results (Figure 21), the internal combustion (ICE) engine is switched off during the regenerative braking, typical for pure electric vehicles, by disconnecting clutch C2. Certainly, another option for operation of the internal combustion engine is possible. It can be switched off, additionally, during the vehicle's steady-speeds. This means that only the electric

Figure 20. The analyzed parallel hybrid power train equipped with automatic transmission

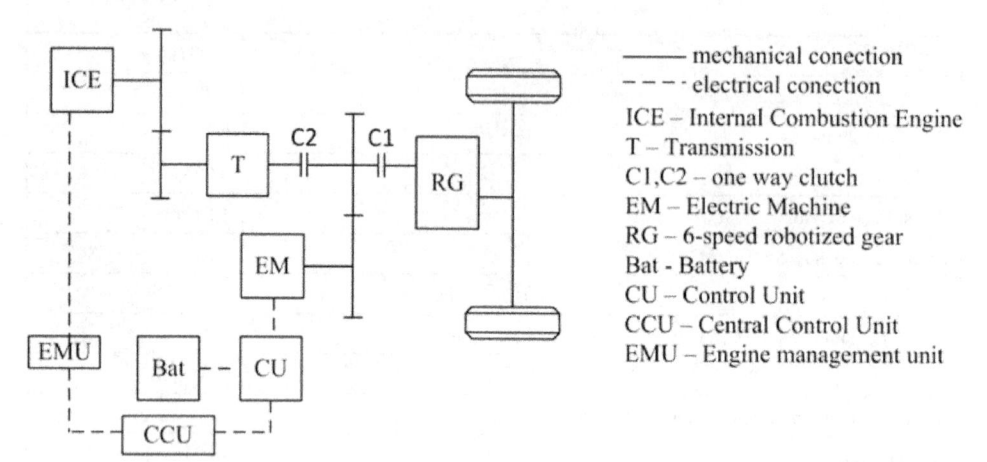

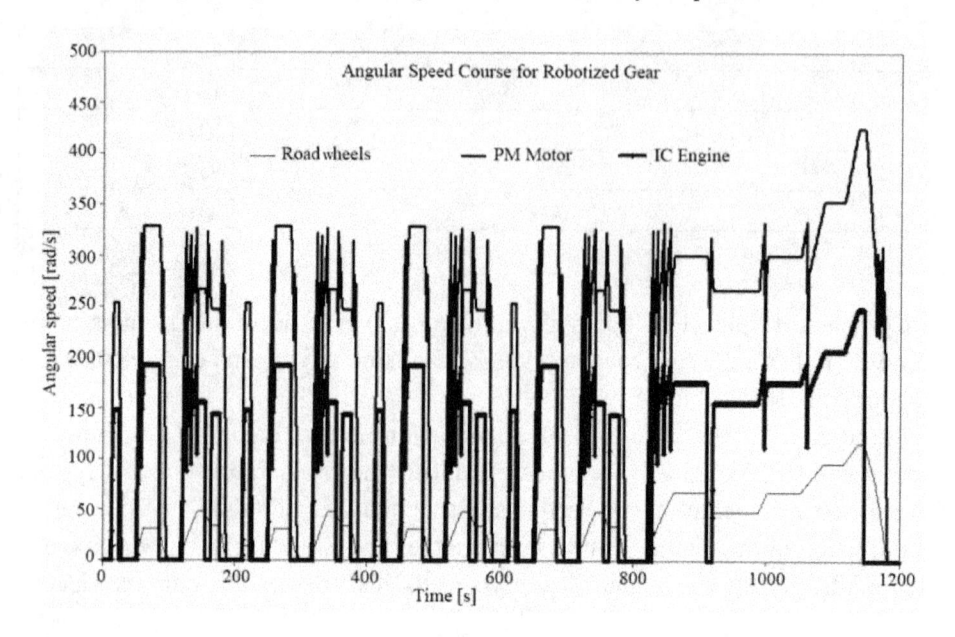

———— mechanical conection
- - - - electrical conection
ICE – Internal Combustion Engine
T – Transmission
C1,C2 – one way clutch
EM – Electric Machine
RG – 6-speed robotized gear
Bat - Battery
CU – Control Unit
CCU – Central Control Unit
EMU – Engine management unit

Figure 21. The angular speeds alteration of traction vehicle wheels, output shaft of permanent magnet, motor, internal combustion engine, according to the analyzed vehicle driving cycle for robotized gear transmission of the power train

Basic Hybrid Power Trains Modeling and Simulation

Figure 22. The torque's alteration of traction vehicle wheels, output shaft of permanent magnet, motor, internal combustion engine, according to an analyzed vehicle driving cycle for robotized gear transmission of the power train

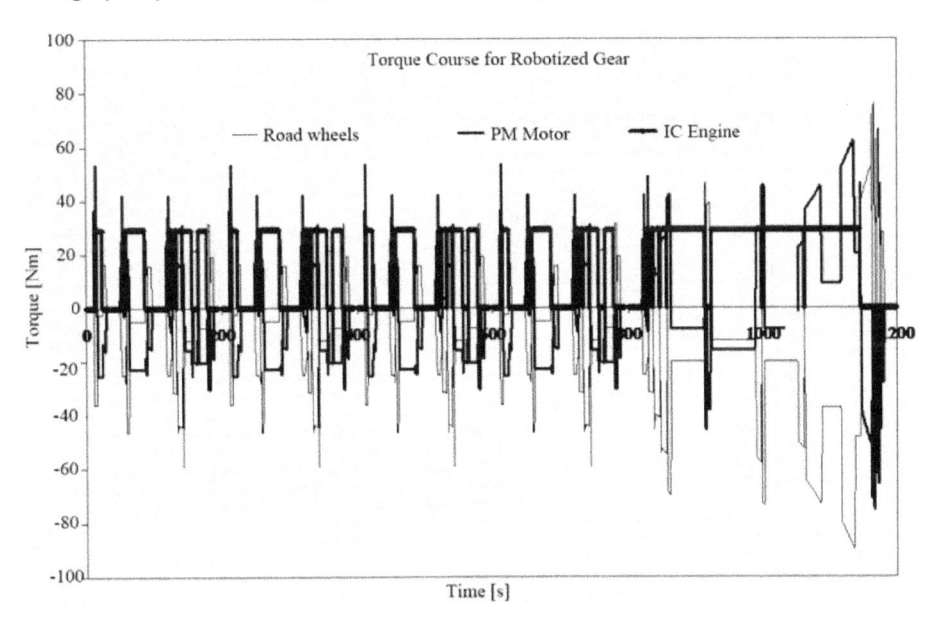

Figure 23. The powers alteration of traction vehicle wheels, output shaft of a permanent magnet, motor, internal combustion engine, according to an analyzed vehicle driving cycle for a robotized gear transmission of the power train

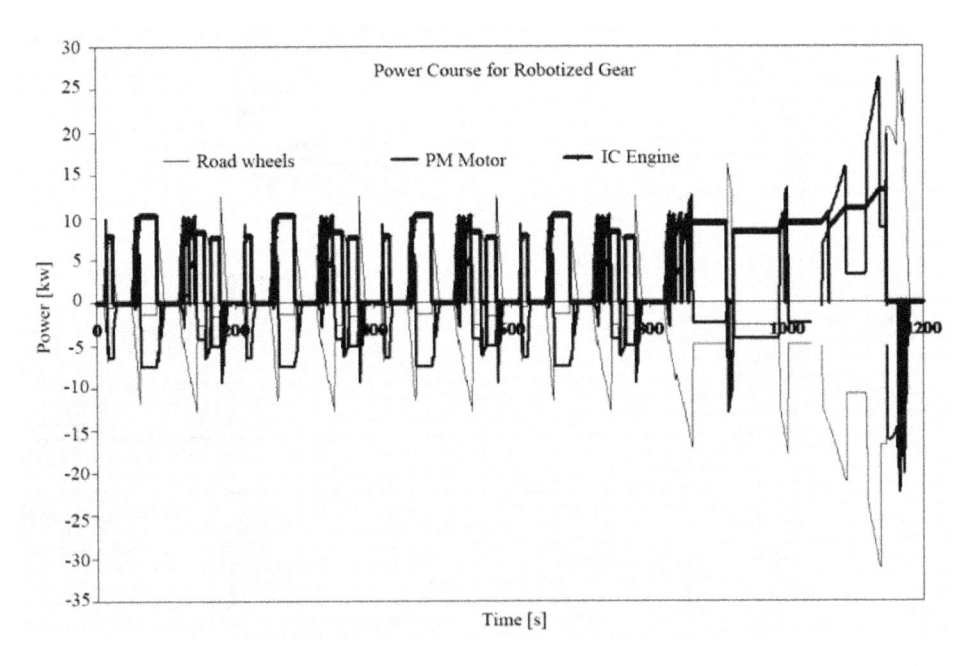

Figure 24. The battery voltage and current alteration, according to an analyzed vehicle driving cycle for a robotized gear transmission of the power train

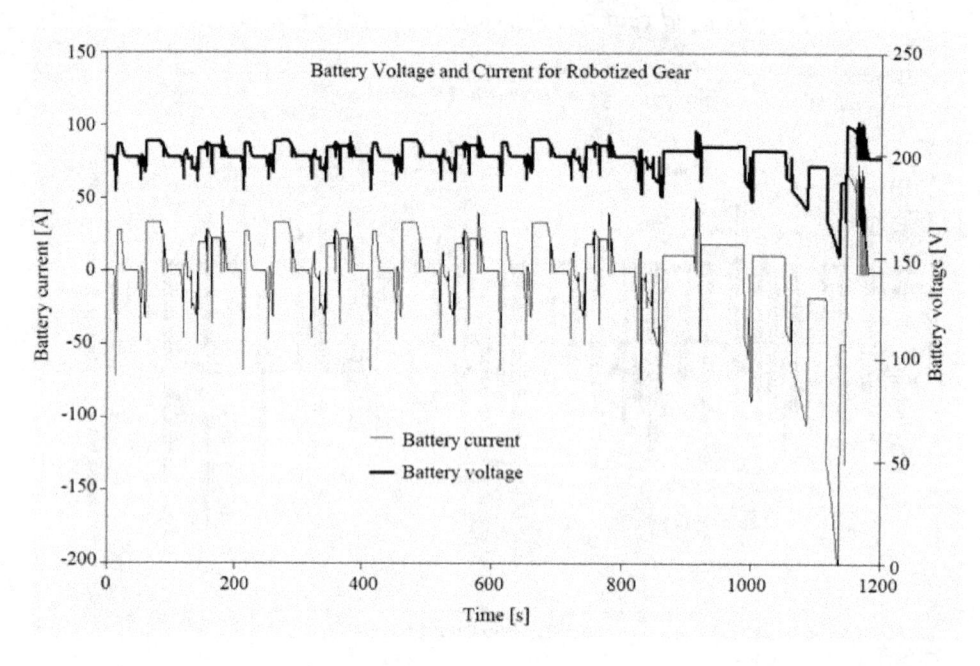

Figure 25. The battery's state-of-charge (SOC) factor value alteration, according to an analyzed vehicle driving cycle for a robotized gear transmission of the power train

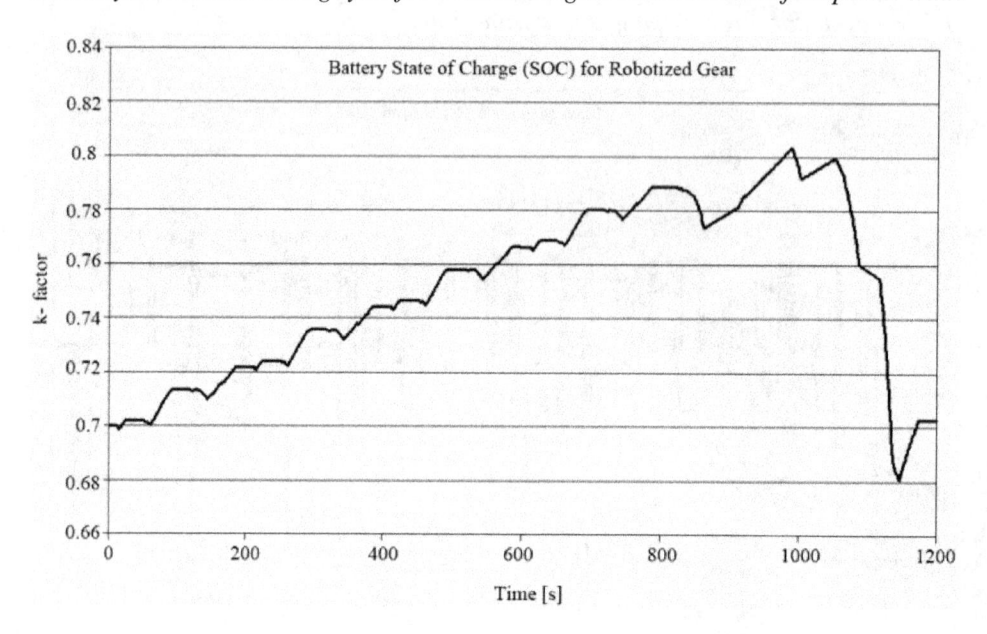

Basic Hybrid Power Trains Modeling and Simulation

Figure 26. The internal combustion gasoline engine fuel consumption alteration, according to an analyzed vehicle driving cycle for a robotized gear transmission of the power train

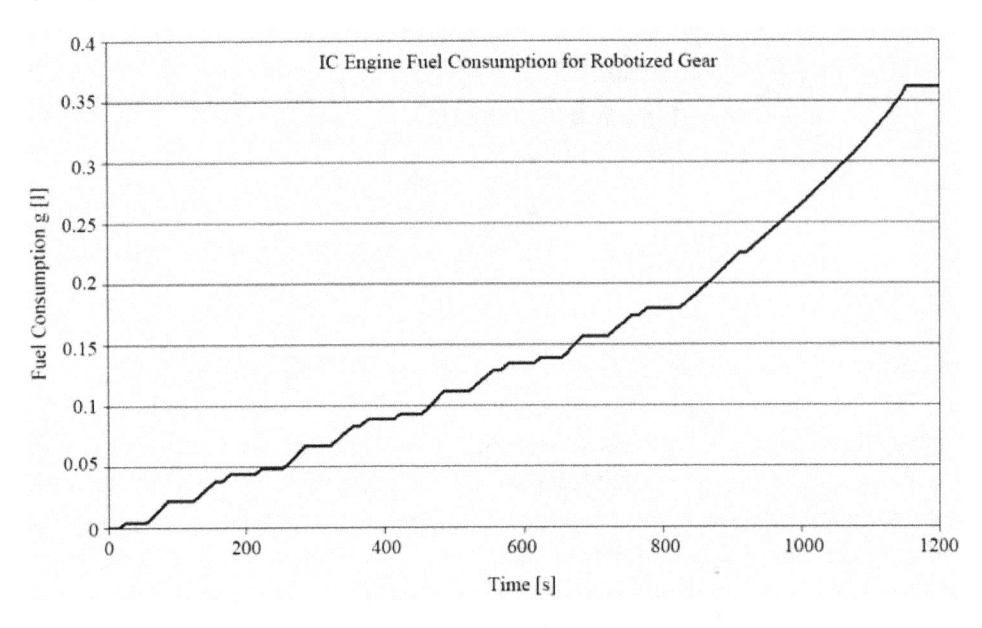

Figure 27. The internal combustion gasoline engine's operating points, located at its map according to an analyzed vehicle driving cycle for a robotized gearbox transmission of the power train

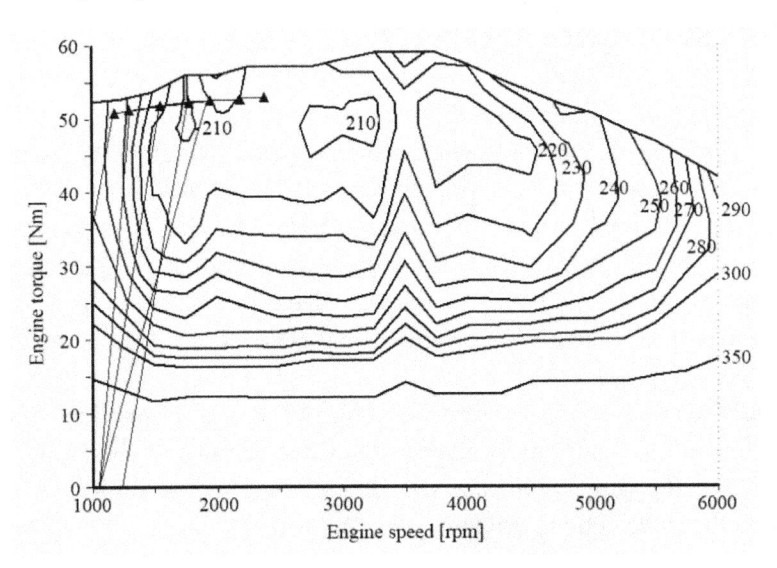

Basic Hybrid Power Trains Modeling and Simulation

Figure 28. The urban bus propelled by a hybrid power train (hybrid split sectional drive, HSSB)

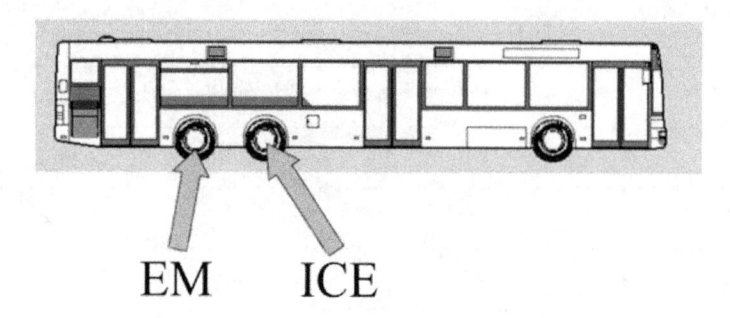

motor propels the vehicle at these time periods (pure electric drive). Additionally, the electric machine, propelled by an internal combustion engine, can charge the battery (clutch C1 is disconnected) during standstill or 'off-road' periods.

The presented drive system integrated with a robotized gearbox enables the internal combustion engine to operate at its best points on the fuel consumption map. This is achieved by selecting, at any time, the proper gearbox ratio, which maximizes operation efficiency, simultaneously, preserving the level of acceleration expected by the driver. It also enables us to avoid the idle condition of the engine which, particularly in urban driving, is responsible for the overall fuel consumption.

4. SPLIT SECTIONAL DRIVE

The hybrid split sectional drive (HSSD) exemplary applied in an urban bus is shown in Figure 28. Only taking into consideration the energy analysis of this bus with a hybrid drive, it's easy to note that the hybrid power train, HSSD, has the typical features of the regular parallel system and in general, its construction is the most simple.

In the case of this hybrid power train, one axis is propelled by the internal combustion diesel engine (ICE), (exactly the same drive construction as in a classic pure engine power train), and the second one is propelled by the electric traction motor, EM.

In order to compare and analyze the impact of both configurations of CHPTD, compact hybrid planetary transmission (see Chapter 8) and hybrid split sectional drive, HSSD, the same engine and the same electric motor, similar control algorithm and the same vehicle driving cycle (Warsaw Driving Cycle) are used in the simulation study.

Basic Hybrid Power Trains Modeling and Simulation

It is necessary to assume the following boundary terms:

- Friction factors between all traction wheels of bus and road are the same;
- The control system ignores the bus's dynamic behavior during its turning on curves, which means the split power between two axes is determined only by energy distribution conditions;
- The bus's start is realized only by the electric motor;
- The internal combustion diesel engine (ICE) is started by the bus's inertia torque;
- Regenerative braking takes place when the internal combustion diesel engine, ICE, is 'switched off' and the electric traction motor is operating in generator mode.

Table 3. The main parameters of an urban bus and its power train

Mass of bus	15000 kg
ICE	Diesel 180kW
Electric motor	PM 49kW
Battery	NiMH 150Ah, 240V
Main gear ratio	4.63
Gearbox ratios*	1.36/1.84

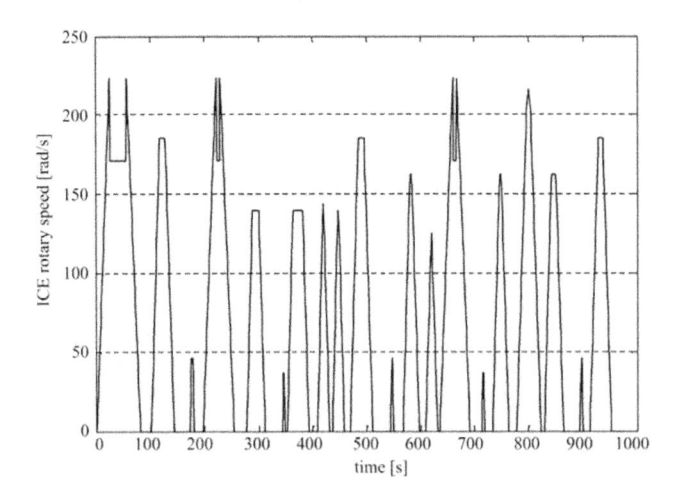

Figure 29. The alterations of the internal combustion diesel engine's (ICE) rotational speed during a vehicle urban driving cycle

Figure 30. The alterations of an internal combustion diesel engine's (ICE) output shaft torque during a vehicle urban driving cycle

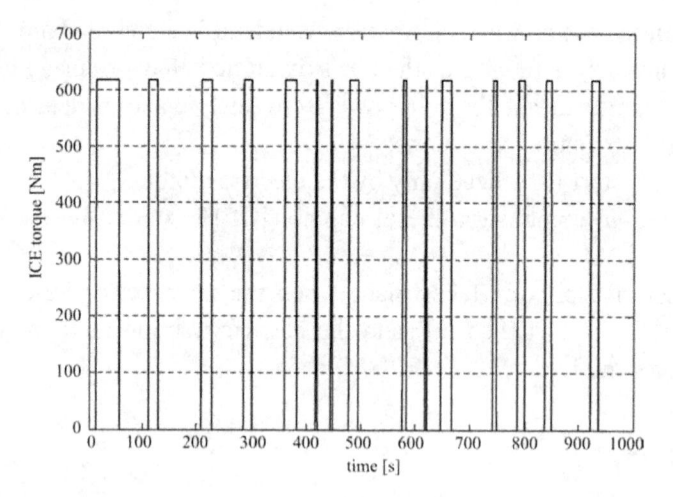

Figure 31. The power requirements and distributions in relation to vehicle traction wheels, internal combustion diesel engine output shaft, and permanent magnet, traction motor during a vehicle urban driving cycle

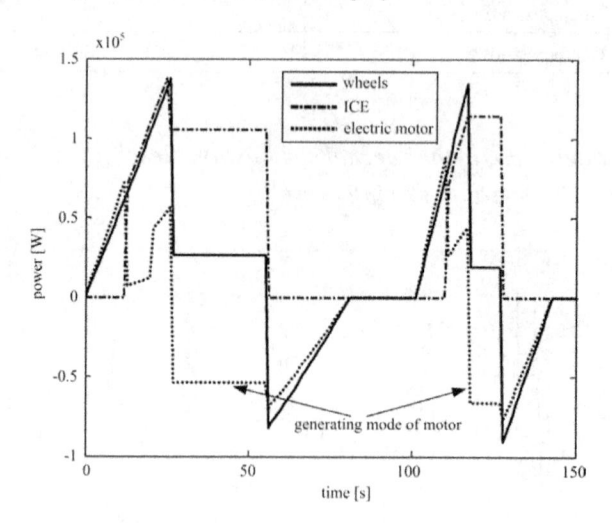

Basic Hybrid Power Trains Modeling and Simulation

Figure 32. The alterations of battery pack state-of-charge factor (SOC), and indication of its maximal Coulomb capacity change during a vehicle urban driving cycle

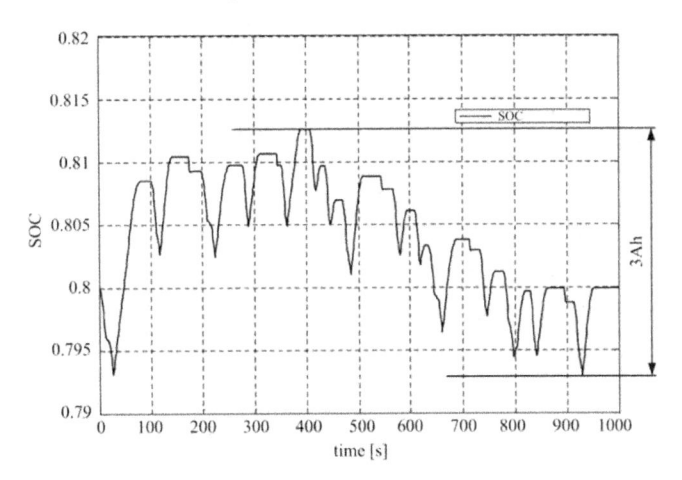

Figure 33. The fuel consumption alteration of an internal combustion diesel engine (ICE) during a vehicle urban driving cycle

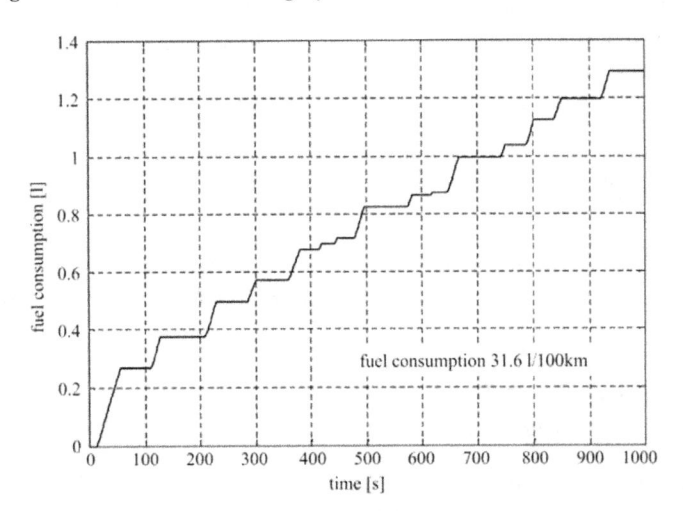

Figure 34. The internal combustion diesel engine's (ICE) operating points[1] located on its map

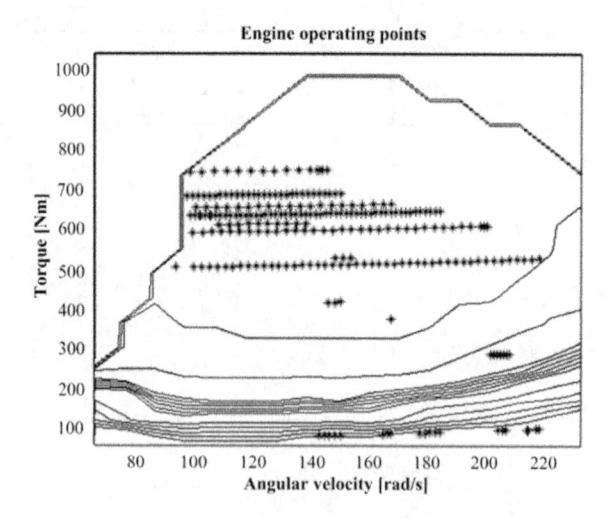

This approach adopts the typical energy analysis of hybrid drives depicted in previous chapters.

Similar to the compact hybrid planetary transmission drive, CHPTD, (see Chapter 8), the form of bus starting is 'pure electric acceleration'.

Some simulation results of the hybrid split sectional drive (HSSD) urban bus are as shown in Figures 29, 30, 31, 32, 33, and 34. (Zhong, 2007).

REFERENCES

Antoniou, A. Komythy, Brench, J., & Emadi, A. (2005). Modeling and simulation of various hybrid electric configurations of the HMMWV. In *Proceedings of VPPC*. IEEE.

Bullock, K. J., & Hollis, P. G. (1998). Energy storage elements in hybrid bus applications. In *Proceedings of EVS15*. EVS.

Burke, A. F. (1992). *Development of test procedures for hybrid electric vehicles*. INEL US Department of Energy INEL Field Office.

Eifert, M. (2005). Alternator control algorithm to minimize fuel consumption. In *Proceedings of VPPC*. IEEE.

Haltori, N., Aoyama, S., Kitada, S., Matsuo, I., & Hamai, K. (2011). *Configuration and operation of a newly-developed parallel hybrid propulsion system*. Nissan Motor Co Technical Papers.

Kruger, M., Cornetti, G., Greis, A., Weidmann, U., Schumacher, H., Gerhard, J., & Leonhard, R. (2010). Operational strategy of a diesel HEV with focus on the combustion engine. In *Proceedings of Aachen Colloquium*. Aachen Colloquium.

Neuman, A. (2004). Hybrid electric power train. In *Proceedings of ELE*. ELE.

Noil, M. (2007). Simulation and optimization of a full HEV. In *Proceedings of EVS 23*. EVS.

Rutquist, P. (2002). Optimal control for the energy storage in a HEV. In *Proceedings of EVS19*. EVS.

Schofield, N. (2006). Hybrid PM generators for EV application. In *Proceedings of VPPC*. IEEE.

Schussler, M. (2007). Predictive control for HEV – Development optimization and evaluation. In *Proceedings of ELE European Conference*. ELE.

Shimizu, K., & Semya, S. (2002). Fuel consumption test procedure for HEV. In *Proceedings of EVS19*. EVS.

Szumanowski, A. (1994). Simulation study of two and three-source hybrid drives. In *Proceedings of EVS12*. EVS.

Szumanowski, A. (1997). Advanced more efficient compact hybrid drive. In *Proceedings of EVS14*. EVS.

Szumanowski, A. (1999). *Evolution of two steps of freedom planetary transmission in hybrid vehicle application*. Global Power Train Congress.

Szumanowski, A. (2006). *Hybrid electric drives design*.

Szumanowski, A., & Hajduga, A. (2006). Optimization series HEV drive using modeling and simulation methods. In *Proceedings of VPPC*. IEEE.

Szumanowski, A., Hajduga, A., Chang, Y., & Piórkowski, P. (2007). Hybrid drive for ultralight city cars. In *Proceedings of ELE European Conference*. ELE.

Szumanowski, A., Hajduga, A., & Piórkowski, P. (1998a). Evaluation of efficiency alterations in hybrid and electric vehicles drives. In *Proceedings of Advanced Propulsion Systems*. GPC.

Szumanowski, A., Hajduga, A., & Piórkowski, P. (1998b). Proper adjustment of combustion engine and induction motor in HV drives. In *Proceedings of EVS15*. EVS.

Szumanowski, A., Hajduga, A., Piórkowski, P., & Brusaglino, G. (2002). Dynamic torque speed distribution modeling for hybrid drives design. In *Proceedings of EVS19*. EVS.

Szumanowski, A., Hajduga, A., Piórkowski, P., Stefanakos, E., Moore, G., & Buckle, K. (1999). Hybrid drive structure and power train analysis for florida shuttle buses. In *Proceedings of EVS 16*. EVS.

Szumanowski, A., & Krasucki, J. (1993). *Simulation study of battery engine hybrid drive*. Paper presented at the 2nd Polish-Italian Seminar Politecnico di Torino. Turin, Italy.

Szumanowski, A., & Nguyen, V. K. (1999). *Comparison of energetic properties of different two-source hybrid drive architectures*. Global Power Train Congress.

Szumanowski, A., & Piórkowski, P. (2004). Ultralight small hybrid vehicles – Why Not? In *Proceedings of ELE*. ELE.

Tamburro, A., Mesiti, D., Ravello, V., Pesch, M., Schenk, R., & Glauning, J. (1997). *An intergrated motor generator development for an effective drive train re-engineering*. CRF Technical Papers.

Vaccaro, A., & Villaci, D. (2004). Prototyping a fussy-based energy manager for parallel HEV. In *Proceedings of ELE*. ELE.

Zhong, J. (2007). Regenerative braking system for a series hybrid electric city bus. In *Proceedings of EVS 23*. EVS.

ENDNOTES

[1] This bus drive is equipped with an ordinary gearbox. It means speed can be changed manually or automatically. In the case of a Warsaw driving cycle of an urban bus, only two speeds (ratios 1.36 and 1.84) are used. Speed changes in order to keep the ICE's operating points in an optimal area.

Chapter 8
Fundamentals of Hybrid Power Trains Equipped with Planetary Transmission

ABSTRACT

Chapter 8 describes the most advanced hybrid power trains, which were generally depicted in Chapter 1. The presented figures consist of the two degrees of freedom planetary gears. It seems to be the best system of energy, split between the Internal Combustion Engine (ICE), the battery, and the electric motor, but unfortunately, it is also the most costly solution for its manufacture. This type of hybrid power train should be preferred as the best drive architecture composition from the technical point of view. For this reason, this chapter, in a detailed way, describes the features and the modeling approach to the planetary hybrid power train. Certainly, most attention is paid to the planetary two degrees of freedom gears, yet not only to them. Cooperating with the planetary gears, additional and necessary devices are considered. The role and modeling auxiliary drive components, such as the automatic clutch-brake device and mechanical reducers are discussed in this chapter. The design of electromechanical drives related to the planetary gear of two degrees of freedom controlled by the electric motor can be transformed to the purely electromagnetic solution. An example of the mentioned gear is given in the chapter. It is a complicated construction with the rotating stator of a complex, electrical machine requiring multiple electronic controllers. The increasing output torque of the electromechanical converter and its connection with the mechanical two degrees of freedom planetary gears are depicted as well.

DOI: 10.4018/978-1-4666-4042-9.ch008

Fundamentals of Hybrid Power Trains Equipped with Planetary Transmission

INTRODUCTION

One of the most attractive power trains is based on planetary transmission as summing or differing the ICE and battery (via electric motor) energy flows. Certainly, this planetary gear is specially designed as two degrees of freedom, and additionally, equipped with a system of clutch/brake units. This chapter and the following one are dedicated to planetary power trains, because its energy economy can be most effective among all-known hybrid drive architectures, certainly, if it is properly designed.

1. PLANETARY GEAR POWER MODELING

The scheme of the planetary gear is shown in Figure 1. As it is exemplary, the sun wheel can be connected with an Internal Combustion Engine (ICE), through auxiliary transmission, whilst the ring is connected with the motor shaft and carrier through the drive reducer, which is connected with the axles of road wheels. Angular velocities of gear shafts, according to the assumed descriptions, fulfill the constraint equation:

$$\omega_1 + k_p \omega_2 - \left(1 + k_p\right)\omega_3 = 0 \tag{1}$$

where:

- $k_p = \dfrac{z_2}{z_1}$: The base gear ratio,

Figure 1. The kinematic scheme of planetary gear

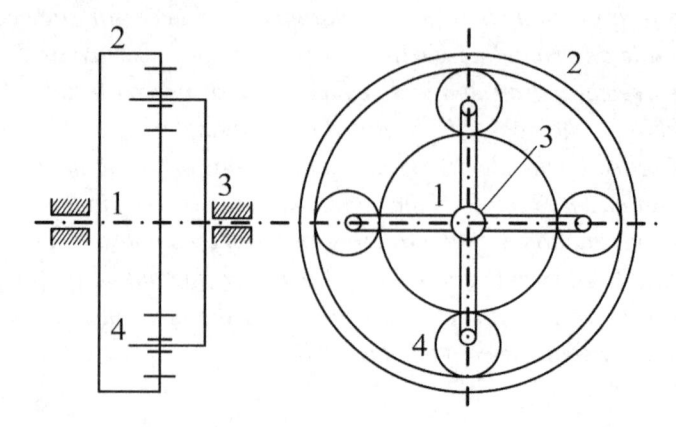

1. sun wheel   2. ring (crown) wheel   3. cage (carrier, yoke)   4. planet wheel.

Fundamentals of Hybrid Power Trains Equipped with Planetary Transmission

- z_1: Number of teeth of the sun wheel,
- z_2: Number of teeth of the crown wheel,
- $\omega_1, \omega_2, \omega_3$: Angular velocity of sun, crown and yoke wheels, respectively.

The motion equation has the following form:

$$J_1\dot{\omega}_1 = \eta_1 M_1 - \frac{1}{k_p}\eta_2 M_2$$
$$J_3\dot{\omega}_3 = M_3 + \frac{k_p+1}{k_p}\eta_3 M_2 \qquad (2)$$

where:

- J_1: Total moment of inertia of sun wheel and connecting elements reduced to sun shaft;
- J_3: Total inertial torque obtained from a reduction of the vehicle mass, road wheels and gears reducer, and inertial torques to the carrier shaft;
- M_1: External torque acting on the sun shaft;
- M_2: External torque acting on the ring shaft;
- M_3: External torque acting on the carrier and corresponding to the vehicle motion resistance reduced to the appropriate shaft;
- η_1, η_2, η_3: Substitute coefficients of internal power losses.

Internal power losses in the planetary gear (in practice to about 98%) can be neglected and the Equation (2) is simplified to the form:

$$J_1\dot{\omega}_1 = M_1 - \frac{1}{k_p}M_2$$
$$J_3\dot{\omega}_3 = M_3 + \frac{k_p+1}{k_p}M_2 \qquad (3)$$

If torque from inertial forces is considered as external torque, it is possible to write:

$$M_{1T} = M_1 - J_1\dot{\omega}_1$$
$$M_{3T} = M_3 - J_3\dot{\omega}_3 \qquad (4)$$

For this reason, the model changes form:

$$M_{1T} = \frac{1}{k_p} M_2$$

$$M_{3T} = -\frac{k_p + 1}{k_p} M_2 \qquad (5)$$

The torque and power equilibrium equations refer to shafts as follows:

$$M_{1T} + M_2 + M_{3T} = 0$$
$$M_{1T}\omega_1 + M_2\omega_2 + M_{3T}\omega_3 = 0 \qquad (6)$$

In case it is necessary to consider the efficiency of mechanical transmission, the power Equation (6) can be approximately written in a form, respectively, to input and output torques.

$$M_{1T}\omega_1 + \eta_{PG}\left(M_2\omega_2 + M_{3T}\omega_3\right) = 0$$

or

$$\eta_{PG}\left(M_{1T}\omega_1 + M_{3T}\omega_3\right) + M_2\omega_2 = 0 \qquad (7)$$

or

$$\eta_{PG}\left(M_{1T}\omega_1 + M_2\omega_2\right) + M_{3T}\omega_3 = 0$$

where:

$$\eta_{PG} = \frac{1 - \mu^* k_{p1}}{1 + \mu^* k_{p2}}$$

$$k_{p1} = \frac{k_p + 1}{k_p - 1};$$

$$k_2 = \left|\frac{3 - k_p}{k_p(k_p - 1)}\right|;$$

$$\mu^* = \frac{8\mu}{nz \sin^2 2\alpha}$$

Fundamentals of Hybrid Power Trains Equipped with Planetary Transmission

- μ: The extended coefficient of the friction between gear teeth;
- z: Tooth number of gear for condition $\omega_3 = 0$;
- n: Factor depends on co-operating numbers of gear (most often for $\omega_3 = 0$ - $n = 2$);
- α: The pressure angle between co-operated teeth.

In the differential gear box modeling, it is possible to use the above-mentioned mathematical tools, taking into consideration the symmetrical operation of external gear shafts connected to traction wheels.

In another case, when the mechanical transmission is reduced to one degree of freedom – it means a reducer or a multiplicator gear – the set of Equations (7) changes to the following form:

e.g. $M_3 = 0$, $\omega_3 = 0$

$$J_1 \dot{\omega}_1 = \eta_1 M_1 - \frac{1}{k_g} \eta_2 M_2 \tag{8}$$

where $k_g = k_p$ – gear ratio.

2. DESIGN OF THE PLANETARY GEAR WITH TWO DEGREES OF FREEDOM APPLIED TO THE TWO-SOURCE HYBRID ELECTRIC DRIVE SYSTEMS

However, it is clear that the method of summing of power in two-source hybrid power trains with an active receiver using planetary gear with two degrees of freedom is the most efficient method, with a wide-enough change of kinematic ratio. Furthermore, in case of great power transferred, the planetary gear is small and compact in construction.

A planetary gear with two degrees of freedom changes the angular velocity of output shaft with a constant ratio of input and output torques. Therefore, it is not a classic torque, but velocity continuous ratio transmission (continuous variable transmission CVT). To get the Continuous Variable Transmission (CVT) function, the planetary gear has to be torque-controlled. The best torque converter for controlling the planetary gear is the electric motor operating in 4Q (controlled in 4 quarters of the co-ordinate system).

One of the solutions is a power train system designed by FEV (see Chapter 1, Figure 15).

Fundamentals of Hybrid Power Trains Equipped with Planetary Transmission

The transmission shown in Figure 15 from Chapter 1 has features as follows:

- Three shaft layout; compact construction
- Seven ratios for combustion engine (ICE) operation, spread 6.53
- Four ratios for EM (electric motor) operation, spread 2.91
- 200Nm internal combustion engine (ICE) torque (possibility to install)
- Electric motor EM integrated into the housing; 20kW continuous power, 38kWpeak power and 143Nm continuous torque, 250Nm in overload peak
- Automatic power shift
- Hybrid operation, pure electric driving, regenerative braking, internal combustion engine start (ICE) and restart by EM (electric motor).

The first prototype of this transmission was equipped with electro-hydraulic actuator controlling clutches and brakes engagement-disengagement. Certainly, the development of electro-mechanical actuator technology is a more efficient and perspective solution (see Chapter 10). The other exemplary one from the family of hybrid power trains equipped with planetary transmission is CHPTD (Compact Hybrid Planetary Transmission Drive) which was tested in the author's laboratory (see Chapter 1, Figure 16). As of 1994, it has been improved, until it attained the form patented in 2008 The patent from the year, 1994, was published the year before Toyota patented its Planetary Transmission Hybrid power train, later to be used in the drive system of the Toyota Prius vehicle. This does not mean, of course, that both these power trains are exactly the same as in particular construction and power distribution control. An analogous system with three planetary gears was developed at the turn of the century by Allison (GM).

The above-mentioned compact hybrid planetary transmission drive (CHPTD) power train, equipped with only one electric machine, is a useful example for explanation modeling and the simulation process in the case of the planetary transmission hybrid power trains, and ultimately, this one is used as an exemplary propulsion system base for modeling and simulation studies presented in this book.

At this point, our role is not to judge which of these drive systems is the best. It is possible to only mention that numerous tests show the superiority of the compact planetary transmission power train, particularly for applications in 'plug-in hybrid' drives, behind which, unfortunately, no world-class company stands. Further explanation will be carried out, taking into consideration, the basic option of a compact hybrid planetary transmission drive shown in Figure 11, dedicated to the design of a 'planetary gear with two degrees of freedom'. The basic constraints are given, and the mathematical model necessary for drive design and mechanical gear selection done by computer simulation.

Fundamentals of Hybrid Power Trains Equipped with Planetary Transmission

3. PLANETARY GEARS POSSIBLE FOR APPLICATION IN HYBRID POWER TRAINS

Only the gears with a negative internal ratio should be taken into consideration. Two gears of this kind are shown in Figure 2.

For solutions shown above, the base internal ratio is defined by the proportions:
Chain A:

$$i_{12} = -\frac{Z_2}{Z_1} = k_p$$

Chain B:

$$i_{12} = -\frac{Z_3}{Z_1}\frac{Z_2}{Z_{31}} = k_p$$

where:

- z: Appropriate number of teeth;
- k_p: Base ratio for braked wheel 3.

Figure 2. The schemes of planetary chains with negative internal ratios

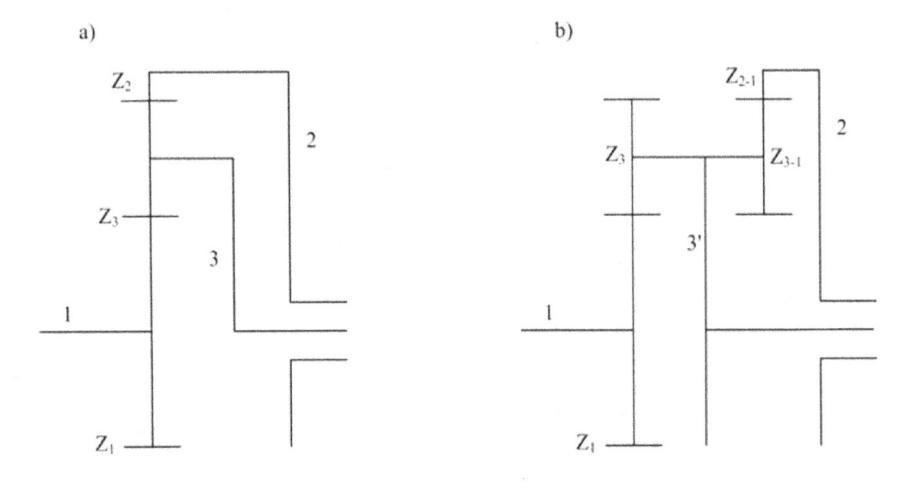

1. - sun gear; 2. - crown (ring) gear; 3. - carrier (yoke).

In hybrid drives, case a) from Figure 2 is most frequently used. Construction and assembly constraints are so large, that for assembly of electric motor-planetary gear, ratios possible to achieve are sufficient. For standard realizations, these ratios should be in the range of:

$$i_{12} = k_p \in \langle -1,5; -4 \rangle \qquad (9)$$

Standard realization constraints, in case of a small difference between the number of sun gear and satellite gear teeth refers to the possibility of interference between teeth. Therefore, a well-known condition should be fulfilled:

$$Z_2 - Z_3 \geq 10 \rightarrow \alpha = 20° \rightarrow f = 1$$
$$Z_2 - Z_3 \geq 12 \rightarrow \alpha = 15° \rightarrow f = 1$$
$$Z_2 - Z_3 \geq 7 \rightarrow \alpha = 20° \rightarrow f = 0,8$$
$$Z_2 - Z_3 \geq 9 \rightarrow \alpha = 15° \rightarrow f = 0,8$$

where:

- **α:** Angle of action;
- **f:** Addendum coefficient.

The standard assembly condition is:

$$\frac{Z_1 + Z_2}{j} = X \qquad (10)$$

where:

- **j:** Number of satellites;
- **X:** Have to be integer

Further to this and the following chapters, exemplary calculations will be provided, along with constraints related to torques on the wheels of the gear and their angular velocities.

Fundamentals of Hybrid Power Trains Equipped with Planetary Transmission

Figure 3. The scheme of a planetary gear with two degrees of freedom used in hybrid power trains

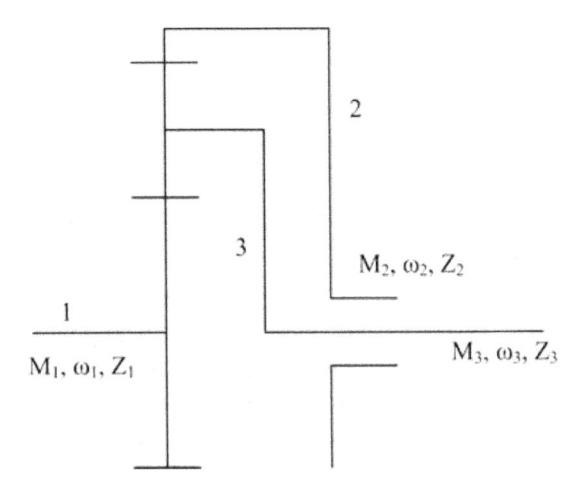

M_1, ω_1, Z_1
M_2, ω_2, Z_2
M_3, ω_3, Z_3

3.1. Summing of Power by a Single Planetary Chain with Two Degrees of Freedom

Scheme of that kind of chain is shown in Figure 3.
Power summing takes place when:

1. Power is transmitted to shaft 1 and 3 and transmitted from shaft 2.
2. Power is transmitted to shaft 1 and 2 and transmitted from shaft 3.
3. Power is transmitted to shaft 2 and 3 and transmitted from shaft 1.

Power subtraction takes place when:

1. Power is transmitted to shaft 2 and transmitted from shafts 1 and 3.
2. Power is transmitted to shaft 1 and transmitted from shafts 2 and 3.
3. Power is transmitted to shaft 3 and transmitted from shafts 1 and 2.

Directions of the power flow through the planetary gear with two degrees of freedom depend on the directions of torques, as well as directions and values of shaft angular velocities. The instantaneous ratio (expressed by the quotient) ω_1/ω_2 expresses a dynamic ratio i_d, by which we can determine the flow directions of respective powers.

In hybrid drive systems with an internal combustion engine and planetary gear with two degrees of freedom - this is, in fact, the case we consider (see Figure 3), the following cases of power distribution take place:

- **Summing:** Case 2. hereinafter designated as A.
- **Subtraction:** Case 2*. hereinafter designated as B.

In case A, the internal combustion engine with constant direction of angular velocity and torque is attached to the sun wheel 1. In case B, – the internal combustion engine (ICE) is transmitted (providing) power to the ring (crown) gear 2, where the electric motor (EM), with variable direction of angular velocity and torque (4Q) is attached, at the same time, receiving energy, whilst the generator mode and recharging batteries and the shaft carrier (yoke) are transmitting power to the traction wheels of the vehicle. Obviously, the vehicle has a variable torque direction in relation to the shaft of the carrier 3, while maintaining a constant direction of the shaft 3 angular velocity. Direction of the shaft 3 torque is positive (sign$\omega_3 = +1$), when the vehicle has been braked, and its kinetic energy, during the regenerative braking, is converted into the energy stored in the battery of accumulators.

Moreover, there are two modes of hybrid power train at work:

- **C:** Purely electric propulsion when the vehicle starts. One of the degrees of freedom is constrained. The Sun gear shaft has the brake applied and the engine is disconnected. Therefore, the clutch-brake constructions (see Figure 12) are required. In addition, case C takes place during regenerative braking.
- **D:** Driving only by the means of the internal combustion engine. In this case, the ring (crown) wheel shaft is braked, and the electric motor is disconnected from the battery's voltage by an electronic controller - inverter.

3.2. Power Summing by a Planetary Gear in the Hybrid Drive System

Case A takes place for

$$0 \leq i_d \leq +\infty$$

or

$$0 \underset{(\omega_2 \to 0)}{\leq} \frac{\omega_2}{\omega_1} \underset{(\omega_1 \to 0)}{\leq} +\infty$$

Fundamentals of Hybrid Power Trains Equipped with Planetary Transmission

Figure 4. The directions of torques and angular velocities as well as respective power flows for the case A ($0 < i_d < +\infty$)

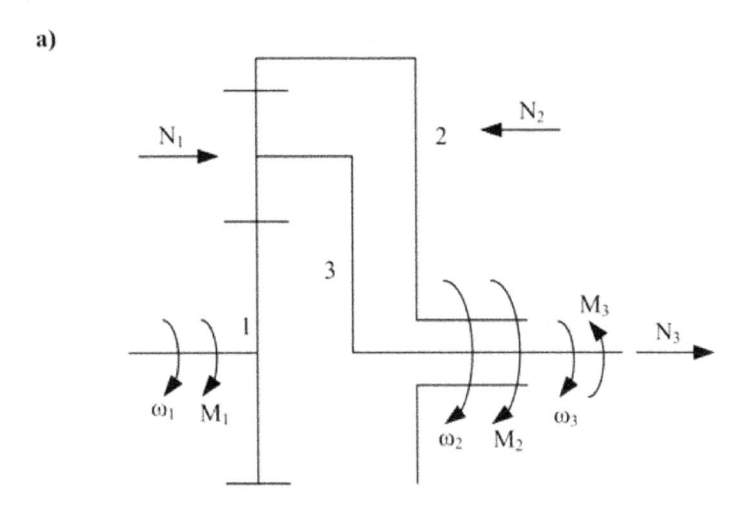

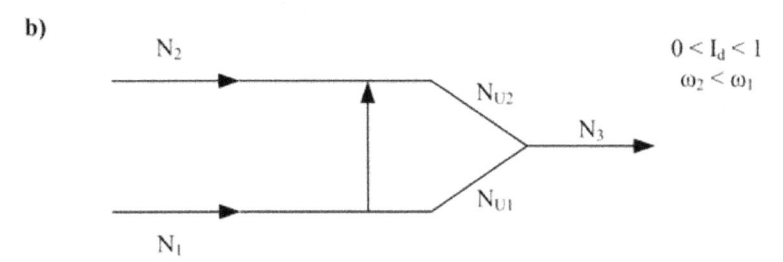

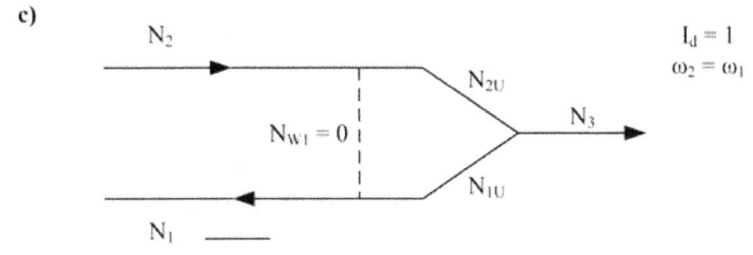

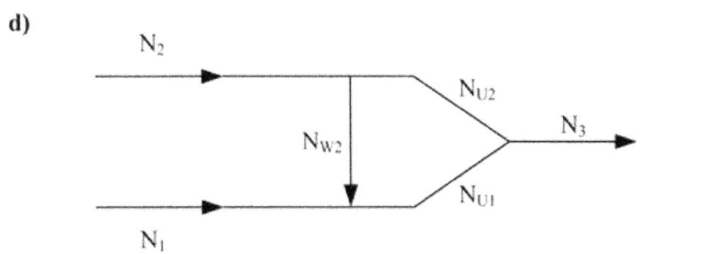

where:
- **N_W**: Power transmitted by the relative motion in relation to the appropriate shaft;
- **N_U**: Power transmitted by the relative motion through the appropriate shaft.

Fundamentals of Hybrid Power Trains Equipped with Planetary Transmission

During the normal operation of the hybrid drive, always $\omega_2>0$ and $\omega_1>0$, that is (i.e.). $0<\omega_2/\omega_1<+\infty$. Directions of torques and angular velocities of planetary gear shafts 1, 2, 3, and power flows are depicted in the Figure 4.

3.3. Power Subtraction by a Planetary Gear in the Hybrid Drive System

Case B takes place for $1/i_{1a} \le i_d \le 0$ or $1/k_p \le i_d \le 0$. Directions of torques and angular velocities, as well as power flows, are depicted in Figure 5.

Apart from the above-mentioned general cases, there are states (modes) of planetary gear operation where there are special modes, satisfying the requirements of the hybrid drive.

Figure 5. The directions of torques and angular velocities and respective power flows for the case B ($1/k_p \le i_d \le 0$)

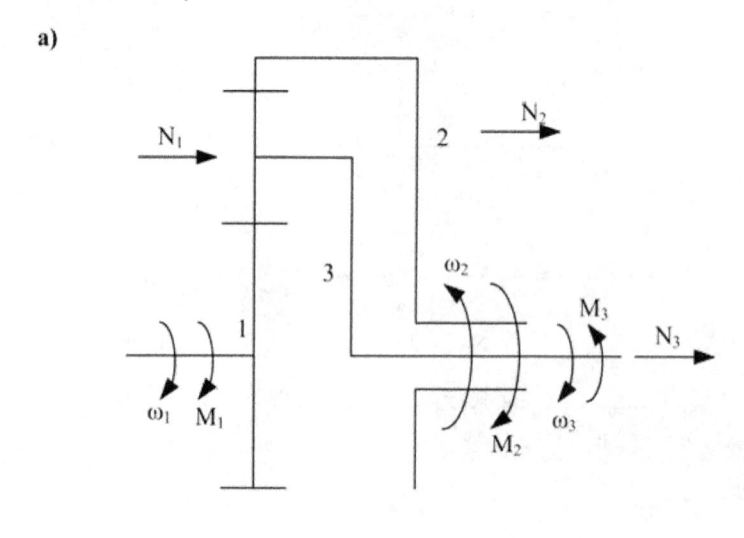

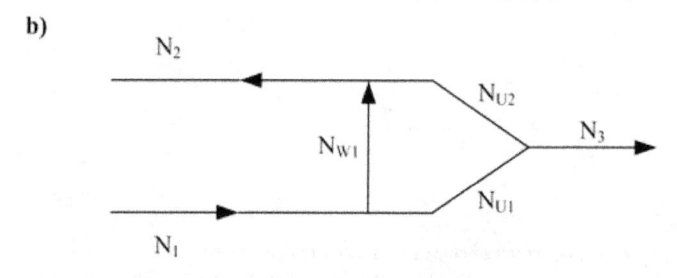

3.4. Power Summing by a Planetary Gear in the Hybrid Drive System: Pure Electric Driving

We will label this case A_1. In this scenario (see Figure 6), the clutch connecting the internal combustion engine with the sun gear's shaft 1 is decoupled, and shaft 1 is braked, that is (i.e.) ω1= 0 (id = +∞).

3.5. Power Summing by a Planetary Gear in the Hybrid Drive System: Driving Only by the Means of the Internal Combustion Engine

We will label this case A_2. In this scenario (see Figure 7), the electric motor is disconnected from the battery of accumulator voltage and its shaft is braked. The internal combustion engine, through the clutch, propels the sun gear 1; $\omega_2 = 0$, $i_d = 0$.

Figure 6. The directions of torques and angular velocities, as well as respective power flows for the case A_1 ($i_d = +\infty$)

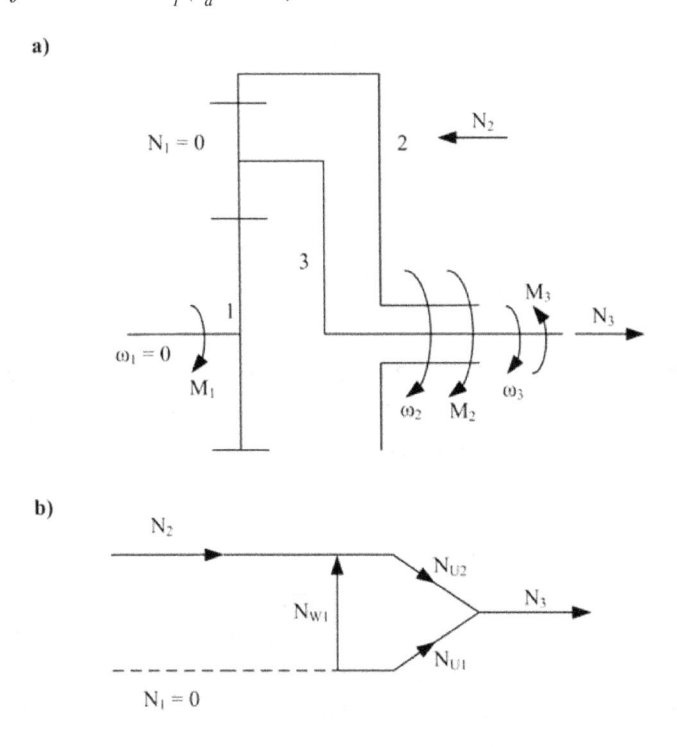

M_1 – reaction torque on the braked shaft 1.

279

Figure 7. The directions of torques and angular velocities, as well as respective power flows for the case ($i_d = 0$)

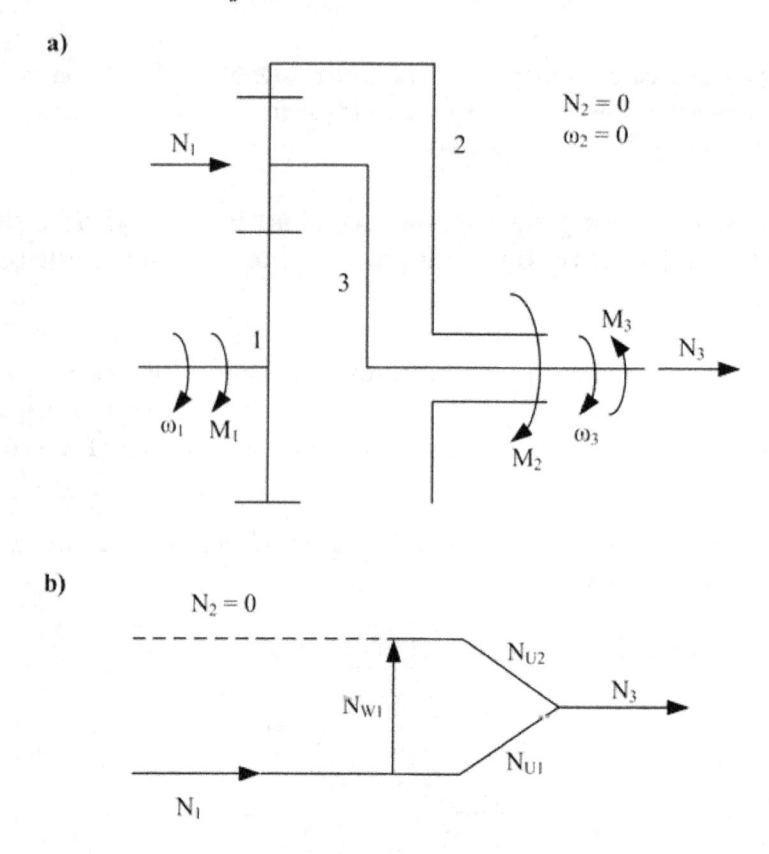

a)

b)

M_2 – reaction torque on the braked shaft 2.

3.6. Power Subtraction by a Planetary Gear in the Hybrid Drive System: Charging the Battery of Accumulators by the Internal Combustion Engine Traction Electric Motor in Generator Mode Operation

We will label this case B_1. This scenario takes place, when the vehicle wheels are braked – the carrier (yoke) shaft 3 (see Figure 8) is blocked. The internal combustion engine, through the clutch, propels the sun gear's input shaft 1, through the ring (crown) wheel output shaft 2 and electric motor operating as a generator. The electric motor, through the inverter, charges the battery of accumulators; $i_d = 1/k_p$, $\omega_2 = k_p \omega_1$.

Figure 8. The directions of torques and angular velocities, as well as respective power, flows for the case B_1

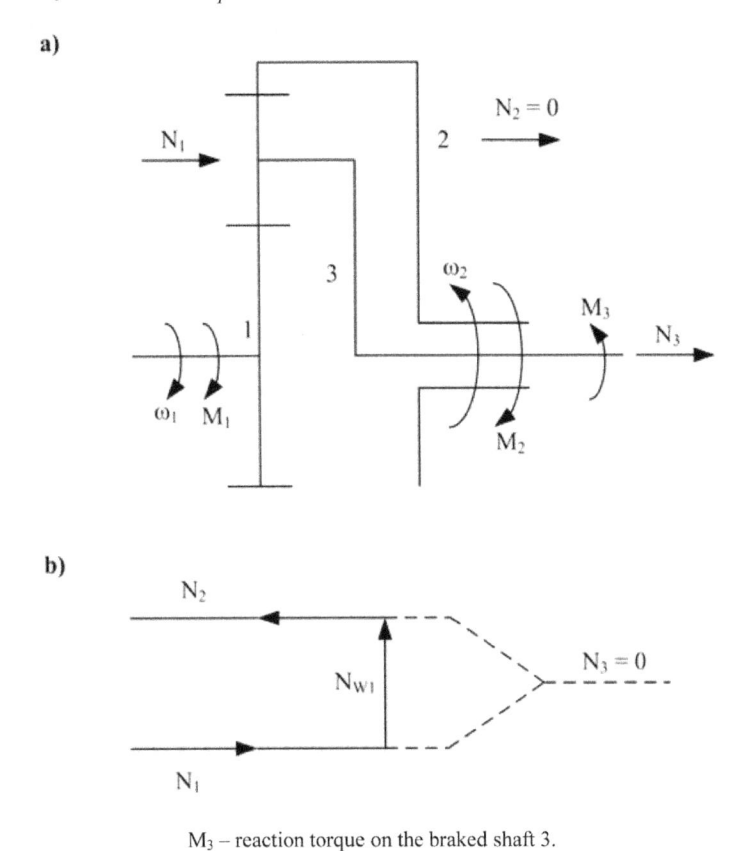

M_3 – reaction torque on the braked shaft 3.

3.7. Power Subtraction by a Planetary Gear in the Hybrid Drive System: Internal Combustion Engine Start-Up, Using the Vehicle's Kinetic Energy, and by the Transient Braking of a Ring (Crown) Gear Shaft, as a Result of Electric Motor Generator Mode

We will label this case B_2. This scenario takes place, when a purely electric operation changes in its hybrid operation. The internal combustion engine is started by the transient braking of the ring (crown) gear shaft 2, by the electric motor transitory operating in the generator mode.

Before that, coupling of the clutch (see Figure 9), connecting the internal combustion (thermal) engine with the sun gear shaft 1, and the unblocking of shaft 1, have to be done; $0 \leq i_d \leq +\infty$.

281

Figure 9. Directions of torques and angular velocities, as well as respective power flows for the case B_2

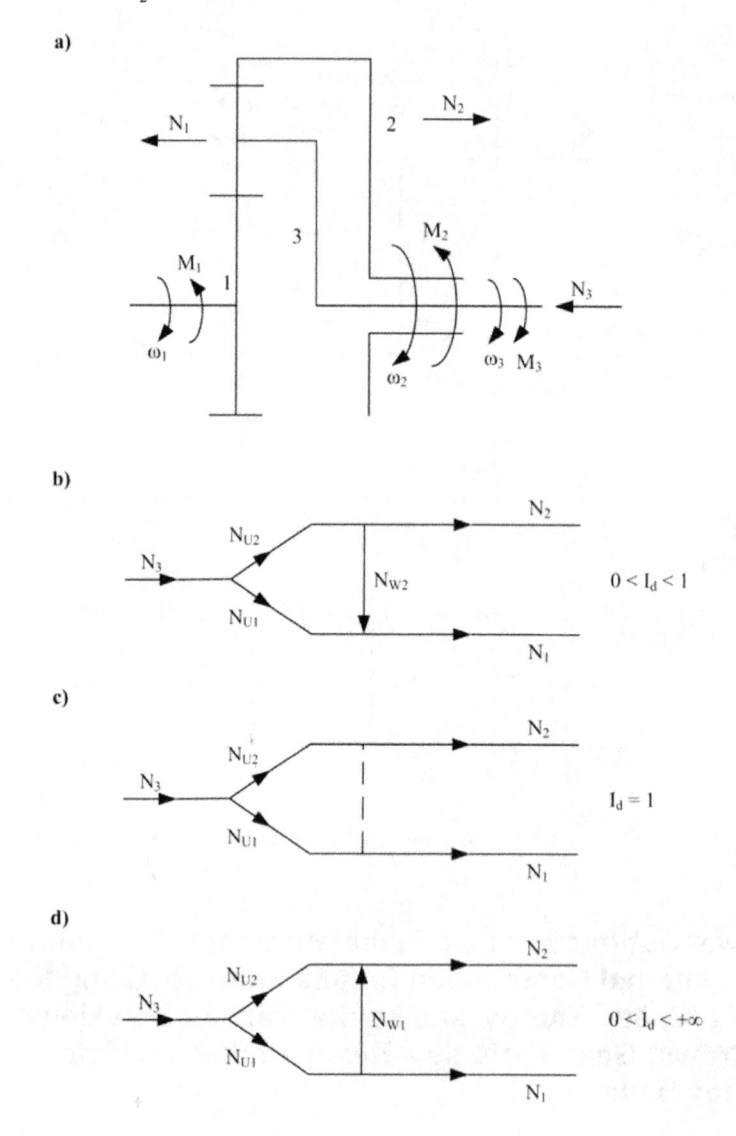

3.8. Power Subtraction by a Planetary Gear in the Hybrid Drive System: Vehicle Regenerative Braking

We will label this case B_3. This scenario takes place, when the internal combustion engine is disconnected from sun gear shaft 1 through the clutch (see Figure 10), and shaft 1 is blocked. The kinetic energy of the vehicle through shafts 3 and 2,

Figure 10. Directions of torques and angular velocities, as well as respective power flows for the case B_3

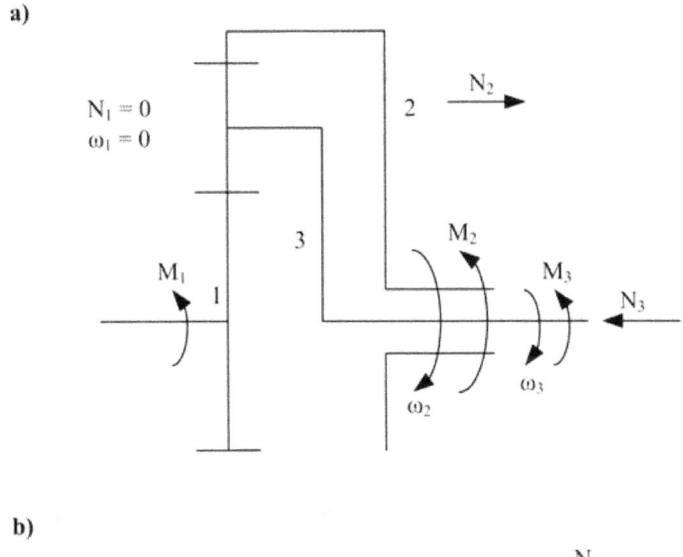

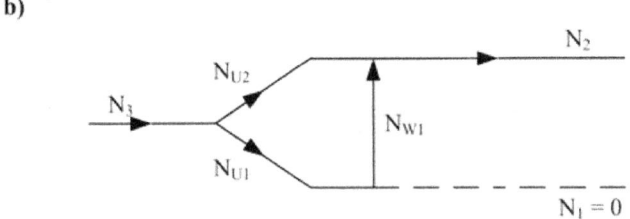

M_1 – reaction torque on the braked shaft 3.

propels the electric motor operating as a generator and charging (moment of vehicle braking, which is shown in the figure below) the battery of accumulators; $i_d = +\infty$.

4. COMPACT HYBRID PLANETARY TRANSMISSION DRIVE (CHPTD)

The layout of the basic scheme of compact planetary transmission drive, CHPTD, is shown in Figure 11. Figure 12 presents the control net necessary for proper basic compact planetary transmission drive operation.

This drive architecture is characterized by the following shaft connections: The internal combustion engine, ICE, via a mechanical reducer and clutch-brake system is linked to the planetary sun wheel (wheel 1 of Figure 13). The electric machine is connected to the crown wheel (wheel 2). The planetary yoke wheel (wheel 3)

Fundamentals of Hybrid Power Trains Equipped with Planetary Transmission

Figure 11. Layout of the compact hybrid drive with planetary transmission-block scheme

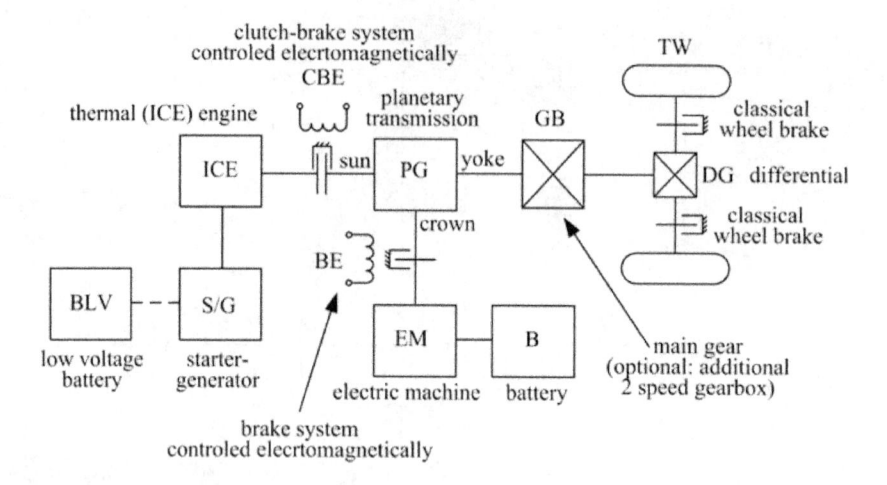

Figure 12. Layout of compact hybrid drive with planetary transmission-control system block

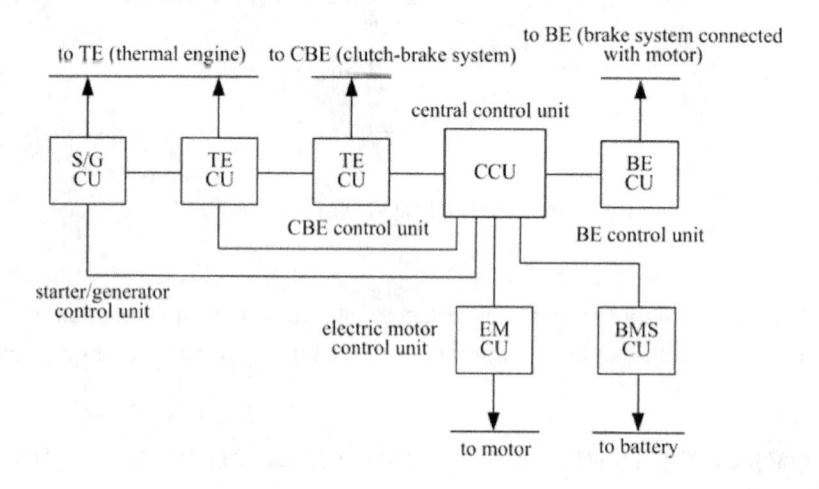

transmits the sum (with a positive or negative sign, depending on the drive operating mode) of power generated by the engine and motor through the main and differential gear set to traction wheels.

The drive applied in the vehicle can operate in two modes of power train operation during vehicle starting (see Figure 15):

- Hybrid acceleration
- Pure electric acceleration

Fundamentals of Hybrid Power Trains Equipped with Planetary Transmission

Figure 13. Compact hybrid planetary transmission (CHPTD) power train: basic connections scheme

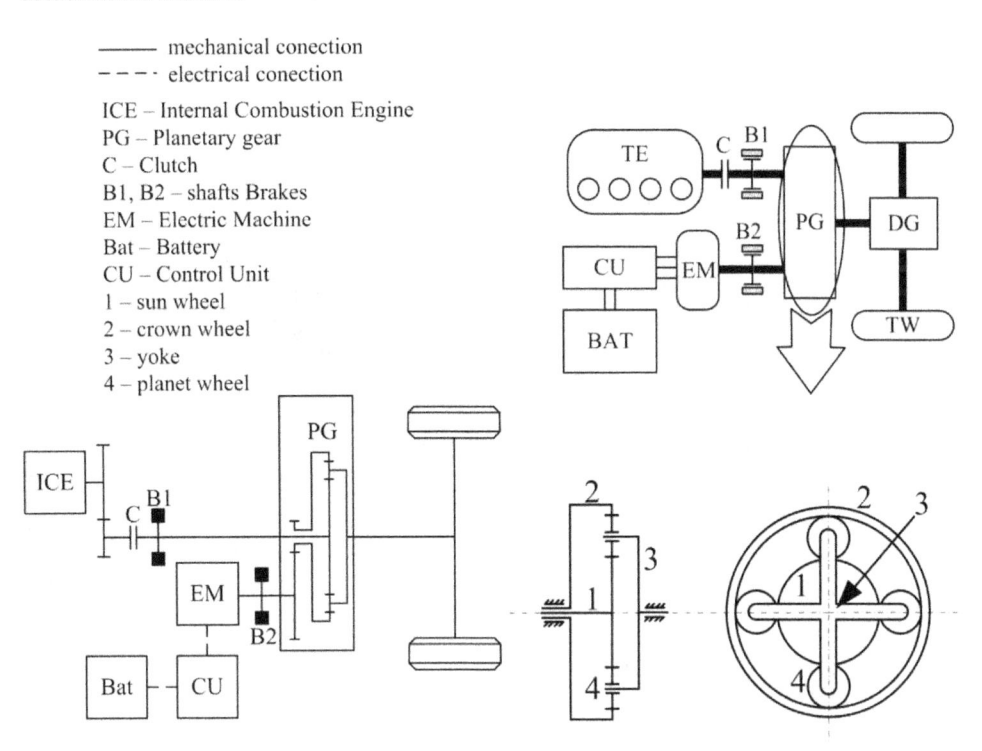

——— mechanical conection
– – – · electrical conection

ICE – Internal Combustion Engine
PG – Planetary gear
C – Clutch
B1, B2 – shafts Brakes
EM – Electric Machine
Bat – Battery
CU – Control Unit
1 – sun wheel
2 – crown wheel
3 – yoke
4 – planet wheel

In both cases, the role of the 'clutch-brake' system is very important. The internal combustion engine, ICE, is joined with one shaft of the clutch (see Figures 11 and 19). The other shaft of the clutch is linked with the sun wheel. This shaft is additionally equipped with the brake system, simultaneously operating with the clutch during its engagement and disengagement. When the clutch is disengaged, the sun wheel brakes. During the engagement of the clutch, the brake is released. The second brake is installed in the motor shaft connected with the crown wheel. When the torque of the motor is nearly zero (the voltage of the motor achieves zero value), the crown wheel can be braked. The brake system permits us to reduce the number of planetary gear degrees of freedom from 2 to 1. When the planetary gear operates in two degrees of freedom, the hybrid operation can be obtained. When the planetary gear is one degree of freedom, pure engine or pure electric operating mode is possible. In comparison with the series or Toyota parallel-series hybrid drive (applied in Prius), the compact hybrid planetary transmission CHPTD power train is more flexible for providing the diversity of propulsion ways. Both CHPTD operating modes - hybrid and pure engine - require the start of the engine.

Fundamentals of Hybrid Power Trains Equipped with Planetary Transmission

Figure 14. The velocity changing relationship of planetary gearbox shafts for a) pure electric power train operation mode, and b) hybrid power train operation mode during vehicle starting

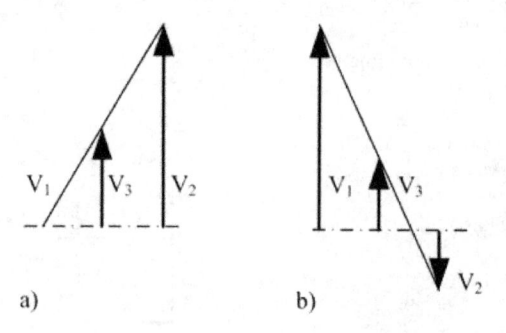

Figure 15. The simplified illustration of the compact hybrid planetary transmission drive power train (CHPTD) operation, emphasizing the internal combustion engine's start during vehicle driving

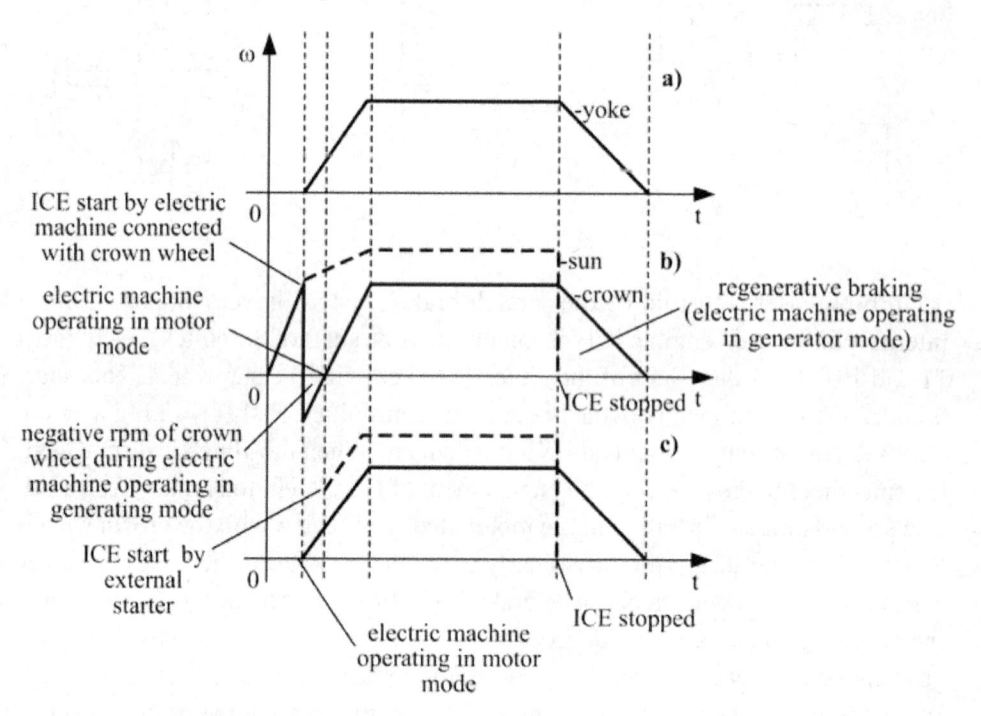

a) Angular velocity of yoke wheel - proportional to vehicle speed;
b) Angular velocity of sun wheel - proportional to internal combustion engine speed and crown wheel - proportional to the speed of electric motor-hybrid acceleration;
c) Angular velocity of sun wheel - proportional to engine speed and crown wheel - proportional to the speed of the electric motor - pure electric acceleration.

Fundamentals of Hybrid Power Trains Equipped with Planetary Transmission

Figure 16. The speed vector relations, respectively, with three rotating shafts of planetary gearbox (Szumanowski, Piórkowski, Hajduga, & Ngueyen, 2000)

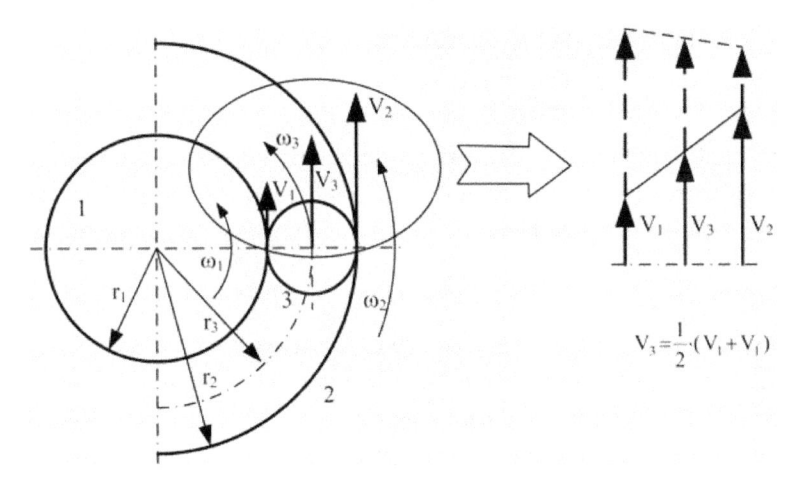

Some details will be depicted later in this chapter. Generally, internal combustion engine ICE start uses additional low voltage starter-generator. It's necessary to mention the engagement and disengagement of clutch and brake system, ICE 'switch on' and 'switch off' can be carried out by electronic signals and adequate electromagnetic actuators.

For both operating modes of CHPTD, when the ICE is 'switched off', the corresponding clutch disengaged and sun wheel braked, it is possible to reach regenerative braking. During regenerative braking, the kinetic energy of the vehicle is transferred from traction wheels, via the yoke wheel and crown wheel to the motor, and the motor operates in generator mode to charge the battery. At this time, planetary transmission operates as an ordinary reducer (one degree of freedom).

The possible speed vector relations, respectively, with three rotating shafts of planetary gearbox are shown in Figure 16.

The output power (N_3) of the drive can be described by the Equation (11):

$$N_3 = \frac{M_3}{k+1} \omega_1 + \frac{M_3 k}{k+1} \omega_2 \tag{11}$$

$$N_i = M_i \omega_i \tag{12}$$

where:

Fundamentals of Hybrid Power Trains Equipped with Planetary Transmission

- N_i: Mechanical power of ith element;
- M_i: Torque on the shaft of ith element;
- ω_i: Rotational speed of ith element;
- k: Basic ratio of planetary gearbox.

According to Equation (11), the expected value of the output power (N_3) can be achieved by proper rotational speed operation (ω_1 and ω_2). This also allows us to choose an internal combustion engine's (ICE) and permanent magnet motor's (PM)

Figure 17. The compact hybrid planetary transmission drive operation and its energy flow corresponding to: a) power train pure electric mode – the sun wheel connected with the internal combustion engine is braked by an electromagnetically controlled clutch-brake system, b) power train pure internal combustion engine mode – the crown wheel connected with permanent magnet motor is braked by an electromagnetically controlled clutch, and c) power train hybrid operation mode – all planetary gears are rotating (Szumanowski, 2006; Szumanowski et al., 2000)

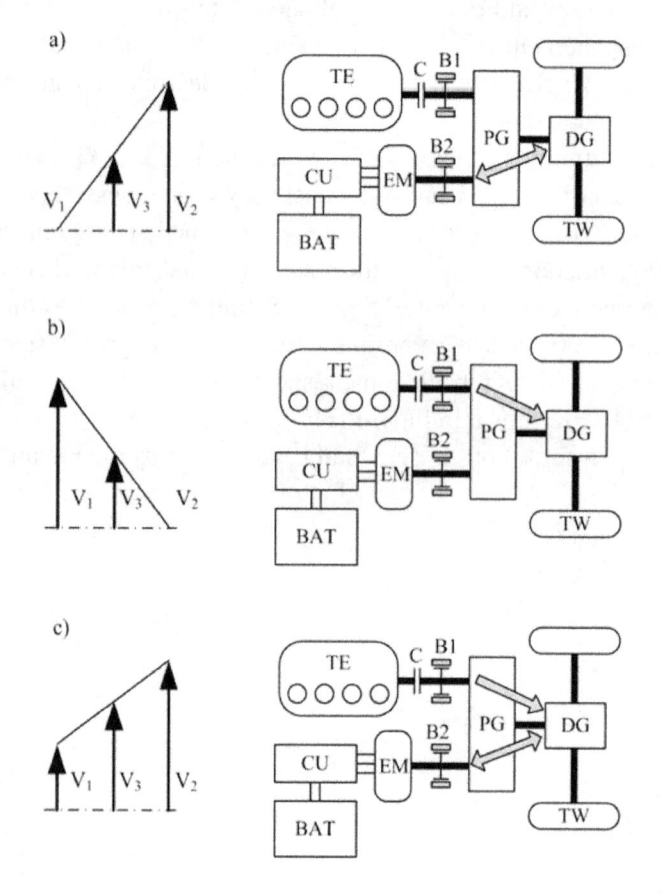

288

Fundamentals of Hybrid Power Trains Equipped with Planetary Transmission

operating points during the vehicle driving cycle, which significantly influences the minimization of the losses of energy and finally, the fuel consumption.

The internal combustion engine (ICE) starts using the traction electric motor or external starter in relation to power train operation conditions and energy flow requirements, corresponding to pure electric, pure internal combustion engine or hybrid operation modes, which are shown in Figures 17(a, b, c).

As is easy to note in Figure 15, the start of the internal combustion engine is when the vehicle accelerates, and the permanent magnet motor is operating in generator mode, and for this reason, the crown wheel shaft of the planetary gear has to be, for the proper moment in time, braked. Additionally, it is also easy to note the direction of the permanent magnet motor angular velocities (or rpm speed) as this process is changed, and in comparison to previous motor speed, this direction is negative. Of course, the planetary crown wheel shaft also changes its angular velocity direction. This is not a positive feature of the power train operation. For the smooth operation of the planetary transmission power train, it is necessary to avoid this impact. This is possible to obtain by proper adjustment of the proportion of planetary gear shaft speeds and adapt to these conditions, the torque-speed control of the permanent magnet motor. The determination of the above-mentioned conditions of proper planetary gears' speed distribution can be done in the following way. In the case of 'pure electric acceleration' of the vehicle and after properly-adjusting the vehicle speed (planetary gear shaft speeds), the internal combustion engine (ICE) starts. In order to avoid the negative angular velocity of the crown wheel occurring at this moment, the following conditions must be fulfilled:

For

$$\omega_3 = \omega_3,$$

$$\omega_1 > 0, \text{ and } \omega_2 > 0$$

$\omega_3 *$ is the minimum-adjusting of the speed of the vehicle when the ICE is started, and then, the hybrid drive mode of the vehicle will begin.

For the planetary gearbox, the following equation is fitting:

$$\omega_1 + k\omega_2 - (1+k)\omega_3 = 0$$

According to the equation above, the following cases can be obtained:
For

$$\omega_3 = 0 \quad \rightarrow \quad \omega_1 = -k\omega_2 ;$$

For

$$\omega_2 = 0 \quad \rightarrow \quad \omega_1 = (1+k)\omega_3 ;$$

For

$$\omega_1 = 0 \quad \rightarrow \quad \omega_2 = \frac{1+k}{k}\omega_3$$

If

$$\omega_1 > 0 , \quad \text{sign}\,\omega_1 = +1$$

ω_3 and ω_1 are impossible to relate to negative values.

Figure 19 illustrates this analysis, and shows the considering relations in 3D form. It is easy to note the proper area for the internal combustion engine's (ICE) start. When ω_1 is below the value $(1+k)\omega_3{}'$, ω_2 is a negative value. In this case, it's possible to accelerate the vehicle, but it's necessary to change the direction of the crown wheel speed. The internal combustion engine should start, when the vehicle stops. This kind of vehicle acceleration during its start, is called hybrid mode acceleration.

Here are some explanations for Figure 18:

1. In Figure 18:

$$\omega_2 = \omega_2{}' \quad \text{for } \omega_3 = 0$$
$$\omega_3 = \omega_3{}' \quad \text{for } \omega_2 = 0$$
$$\omega_3 = \omega_3{}^* \quad \text{for } \omega_2 > 0; \; \omega_1 > 0;$$

2. $\omega_1 = f(\omega_2)$ is for the condition $\frac{d\omega_3}{dt} = const$, and the line $\omega_1 = f(\omega_3)$ is for the condition $\frac{d\omega_3}{dt} = const$;

3. At the crossing point of three axes,

$$\omega_2 = 0, \omega_3 = 0, \omega_1 = -k\omega_2{}'$$

Fundamentals of Hybrid Power Trains Equipped with Planetary Transmission

Figure 18. The illustration of proper adjustment of the internal combustion engine's start (ICE start) conditions by proper adjustment of the planetary gear shafts angular velocity distribution in the case of the power train operating in a pure electric mode, during the vehicle's start and acceleration. Note: When the internal combustion engine (ICE) is started, then the hybrid power train mode of operation will begin.

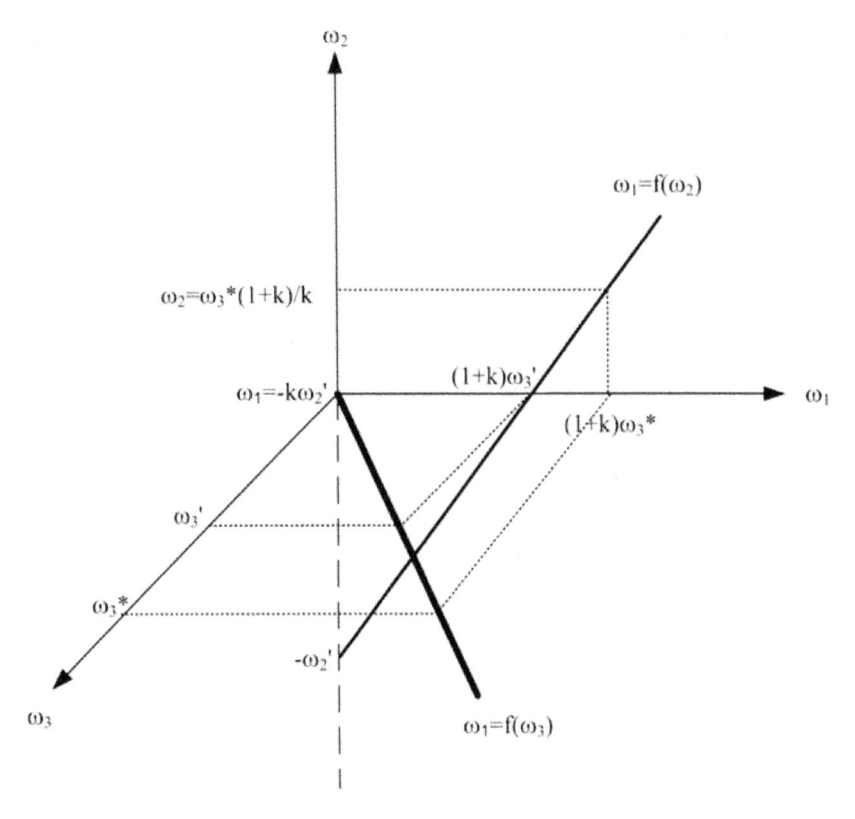

Figure 19. The layout scheme of the clutch-brake system at the sun wheel of planetary transmission

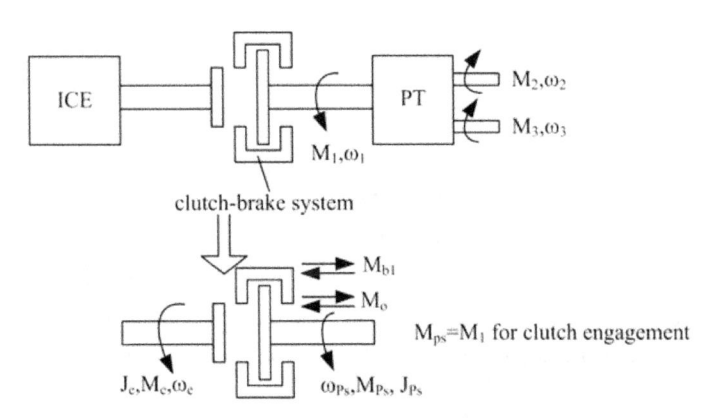

$M_{ps} = M_1$ for clutch engagement

291

Considering the basic equations set describing planetary transmission, ignoring the power efficiency of the gear, it is possible to work out the presented analysis, according to Figure 19.

As the background to general analysis, the use of an external ICE starter is considered.

Basic equations of planetary transmission for efficiency $\eta = 1$ can be expressed as follows:

$$J_1 \frac{d\omega_1}{dt} = M_1 - \frac{1}{k} M_2$$

$$J_3 \frac{d\omega_3}{dt} = M_3 + \frac{1+k}{k} M_2 \qquad (13)$$

$$\omega_1 + k\omega_2 - (1+k)\omega_3 = 0$$

Figure 19 shows the basic connection of the sun wheel with the engine through the clutch-brake system and the new further depiction of the indicated date.

For the starting engagement of clutch and brake release in a general case, it's possible to write:

$$J_{PS} \frac{d\omega_{PS}}{dt} = M_{PS} + M_t \qquad (14)$$

$$J_e \frac{d\omega_e}{dt} = -M_t + M_e \qquad (15)$$

where:

- M_t: Extended torque of clutch friction;
- $|M_{PS}| = |M_1| = |M_e|$: Torque of the output shaft of internal combustion engine ICE for clutch engagement;
- M_{PS}: Torque reduced on the planetary sun wheel shaft;
- ω_{PS}: Angular velocity of sun wheel;
- ω_e: Angular velocity of the ICE, corresponding to angular velocity of the sun wheel;
- J_{PS}: Moment of inertia of planetary shafts reduced on the sun wheel;

Fundamentals of Hybrid Power Trains Equipped with Planetary Transmission

- J_e: Moment of inertia of the ICE.

In special cases, such as engagement of the clutch and the release shafts are braked, or in opposite ways – 'disengagement and brake' - all these processes are necessarily taken under consideration.

Ignoring the energy modeling of the clutch-brake system applied in the hybrid drive, as well as torque pulsations, torsion torques, vibrations, etc., two states of clutch-brake system operation are considered:

1. The start-clutch is disconnected and the sun wheel shaft braked.

The boundary conditions for the steady state are as follows:

$$\omega_{PS} = 0, \quad \omega_e = \omega_{e\min}$$

corresponding to the idle speed of the engine (In this case, the start of ICE by an external starter is analyzed); M_e - ICE torque corresponding to the engine idle operator; $M_{PS} = M_{b1}$ - brake torque produced by external electromagnetic actuator; M_{el} - internal combustion engine (ICE) loss torque.

$$J_e \frac{d\omega_e}{dt} = M_e - M_{el}$$

$$0 = M_{b1} - \frac{1}{k} M_2 \quad \rightarrow \quad M_{b1} = \frac{1}{k} M_2 \tag{16}$$

$$J_3 \frac{d\omega_3}{dt} = M_3 + \frac{k+1}{k} M_2$$

$$k\omega_2 - (1+k)\omega_3 = 0$$

2. Steady state-clutch engaged, the brake is released.

The boundary conditions are as follows:

$$M_e(t) = M_1(t) = M_{PS}(t),$$

$$\omega_{PS}(t) = \omega_e(t) = \omega_1(t),$$

$M_{b1} = 0$, J_1 – inertial torque of the sun wheel.

Fundamentals of Hybrid Power Trains Equipped with Planetary Transmission

Hence:

$$(J_e + J_1)\frac{d\omega_1(t)}{dt} = M_e(t) - \frac{1}{k}M_2(t)$$
$$J_3\frac{d\omega_3(t)}{dt} = M_3(t) + \frac{k+1}{k}M_2(t) \qquad (17)$$
$$\omega_1(t) + k\omega_2(t) - (1+k)\omega_3(t) = 0$$

$J_1 << J_e$, so, the simplified equation can be obtained:

$$J_e\frac{d\omega_1(t)}{dt} = M_e(t) - \frac{1}{k}M_2(t) \qquad (18)$$

During the process after the start and before the steady state, the set of equations is as follows:

$$J_{PS}\frac{d\omega_{PS}(t)}{dt} = M_{PS}(t) + M_t(t) + M_{b1}(t)$$
$$J_e\frac{d\omega_e(t)}{dt} = -M_t(t) + M_e(t) - M_{el}(t) \qquad (19)$$
$$M_t(t) = aM_o(t)$$

where $M_o(t)$ - clutch engagement (or disengagement) torque produced by external electromagnetic actuator, a - parameters of the equation.

For example, M_t, M_{b1} versus adequate angular velocity characteristics can be illustrated by Figure 20.

The planetary transmission operating in hybrid mode, and also the brake of the crown wheel (electric motor shaft) is required. In this case, the only engine propels the vehicle. The motor shaft can only be equipped with the brake system (see Figure 21).

According to Figure 21 and considering the brake of the shaft, the following equations can be obtained for the required condition $M_2(t) = M_{b2}$:

Fundamentals of Hybrid Power Trains Equipped with Planetary Transmission

Figure 20. The torques M_t, M_{b1} crossing point indication in relation to the angular velocity of the sun wheel

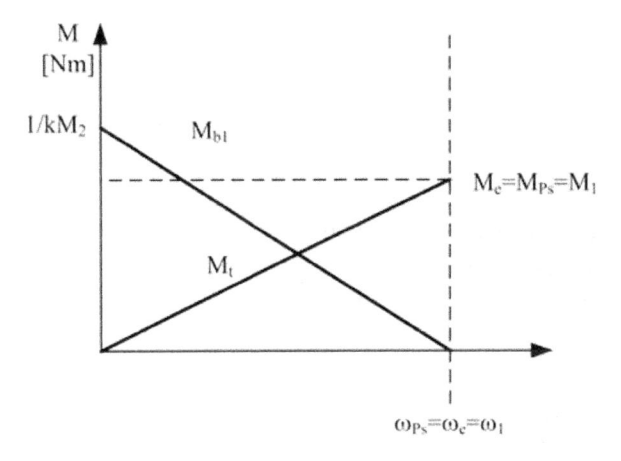

Figure 21. The scheme of the brake unit on the crown shaft connected with the permanent magnet motor

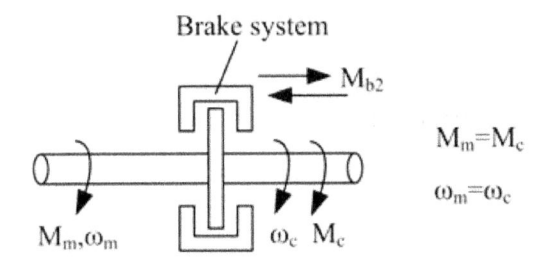

M_{b2} is produced by external electromagnetic actuator

Where:

M_m - Torque of the electric motor;

ω_m - Angular velocity of the electric motor;

M_c - Torque of the crown wheel;

ω_c - Angular velocity of the crown wheel.

$$\begin{cases} J_e \dfrac{d\omega_1(t)}{dt} = M_e(t) - \dfrac{1}{k} M_{b2} \\ J_3 \dfrac{d\omega_3(t)}{dt} = M_3(t) + \dfrac{1+k}{k} M_{b2} \end{cases}$$
$$\omega_1(t) - (1+k)\omega_3(t) = 0 \qquad (20)$$
$$\omega_2 = 0$$

Note: During this time, the electric motor is switched off, which means its electromagnetic torque is equal to zero.

From the above equations, the following equation can be obtained:

$$M_{b2} = -(M_e(t) + M_3(t))$$
$$+ J_e \dfrac{d\omega_1(t)}{dt} + J_3 \dfrac{d\omega_3(t)}{dt} \qquad (21)$$

Hence, for

$$J_e = \text{const}, \quad J_3 = \text{const}$$

and according to Willis equations for $\omega_2 = 0$

$$\omega_3 = \dfrac{1}{1+k} \omega_1 \qquad (22)$$

Hence,

$$(J_e + J_3 \dfrac{1}{1+k}) \dfrac{d\omega_1(t)}{dt}$$
$$= A \dfrac{d\omega_1(t)}{dt} \qquad (23)$$

where,

$$A = J_e + J_3 \dfrac{1}{1+k} \qquad (24)$$

Finally, the following equation can be obtained:

Fundamentals of Hybrid Power Trains Equipped with Planetary Transmission

$$M_{b2} = -M_e(t) - M_3(t) + A\frac{d\omega_1(t)}{dt} \quad \text{or}$$
$$M_{b2} = -M_e(t) - M_3(t) + A\frac{d\omega_e(t)}{dt} \tag{25}$$

When the vehicle drives in pure electric mode, there are three ways for the engine start to change operating modes from an electric drive to the hybrid drive:

1. External starter is used, which has been depicted before;
2. Non-electric brake of the crown wheel is used;
3. Electric brake of the crown wheel is used during motor operation as a generator.

In cases 2 and 3, the same impact can be obtained. It means both methods use the brake of the crown wheel (engine start). The clutch-brake system of the crown shaft should operate in the same way as described before, in both cases.

The energy losses of the clutch-brake system referring to power flow distribution in a compact hybrid planetary transmission power train, CHPTD, can be ignored. The clutch-brake system operation is quick (less than half of a second), and controlled by electromagnetic actuators. This kind of control permits us, simultaneously, to change 'pull-push' phases of the clutch, the engagement or disengagement of shaft brakes, and the internal combustion engine's ICE start (see Chapter 10).

Zero energy consumption in a steady state of the shaft, during clutch-brake system operation (the action of engagement and disengagement) is the target of design (see Chapter 10). This solution was developed by research engineering centers, as well as the Fiat Research Center (Ecodriver Hybrid System) or LUK companies, and also the concept of this kind of electromagnetic controlled clutch-brake unit was worked out in the authors' laboratory.

It's also important to note the electromagnetic actuators of the 'clutch-brake' system should be fed by low voltage (12~24V).

4.1. The Start of Internal Combustion Engine

The Internal Combustion Engine (ICE) can start in two ways, which are discussed here:

- Short-time electro-dynamic braking of permanent magnet motor (PM) shaft (generator mode);
- Short-time mechanical braking of the permanent magnet motor (PM) shaft (only mechanical brake).

The short-time electro-dynamic braking of the permanent motor (PM) shaft means its generator mode operation, and this time, the electric energy is transferred to a battery, which needs the determination of a properly-controlled algorithm in practice, realized by the motor inverter. This looks complicated as a complex solution, but frankly speaking, it is effectively possible by programming the microprocessor controller, which was proved by laboratory tests, in dynamic conditions, for planetary hybrid power train operation. For this reason, and taken under consideration significant energy effectiveness, the internal combustion engine's start by short-time braking its shaft, during motor generator mode operation, is recommended. Of course, the mechanical braking of the permanent magnet motor is also possible. Anyway, this solution does not permit dynamic control of the braking torque in an effective way, so the additional control unit connected with the mechanical brake device is necessary.

The following figures show the proper distribution of torques and speeds during the internal combustion's start, when the vehicle is driven and accelerating.

When the Internal Combustion Engine (ICE) starts, inverter sets of the permanent magnet motor (PM) are in an idle state for the initial period of the braking process. At this moment, the electromagnetic torque of the permanent magnet PM motor is equal to zero (Figure 22).

Figure 22. The torques of the internal combustion engine (ICE) and permanent magnet motor (PM) distribution during the internal combustion engine's start when the vehicle is driving and accelerating (the case of a compact hybrid planetary transmission drive (CHPTD) power train applied in a 15-ton urban bus)

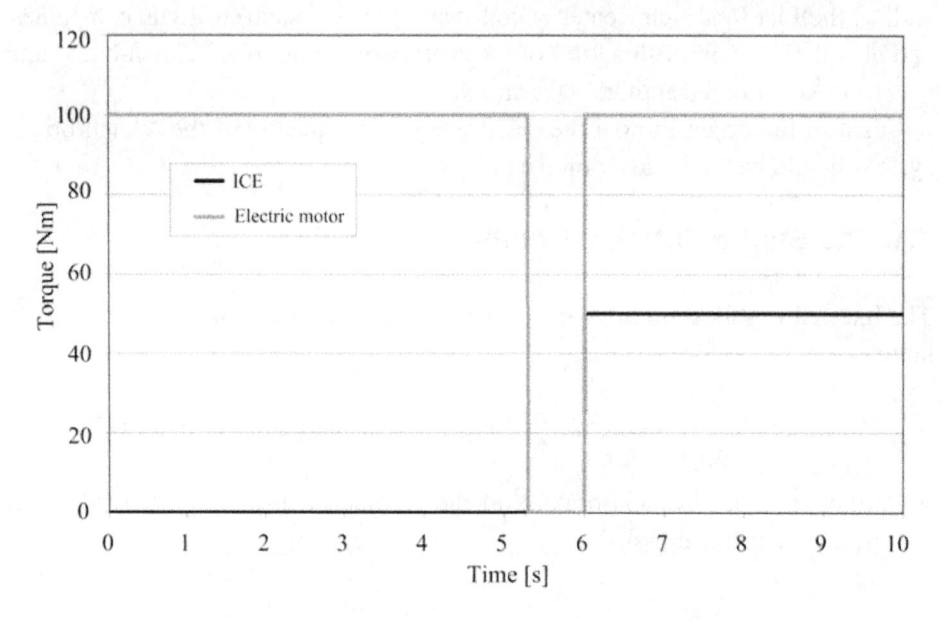

Fundamentals of Hybrid Power Trains Equipped with Planetary Transmission

Figure 23. The angular velocity distribution of the planetary gear shafts referring its connections with the internal combustion engine (ICE), electric permanent magnet motor and vehicle traction wheels (the case CHPTD power train applied in a15-ton urban bus)

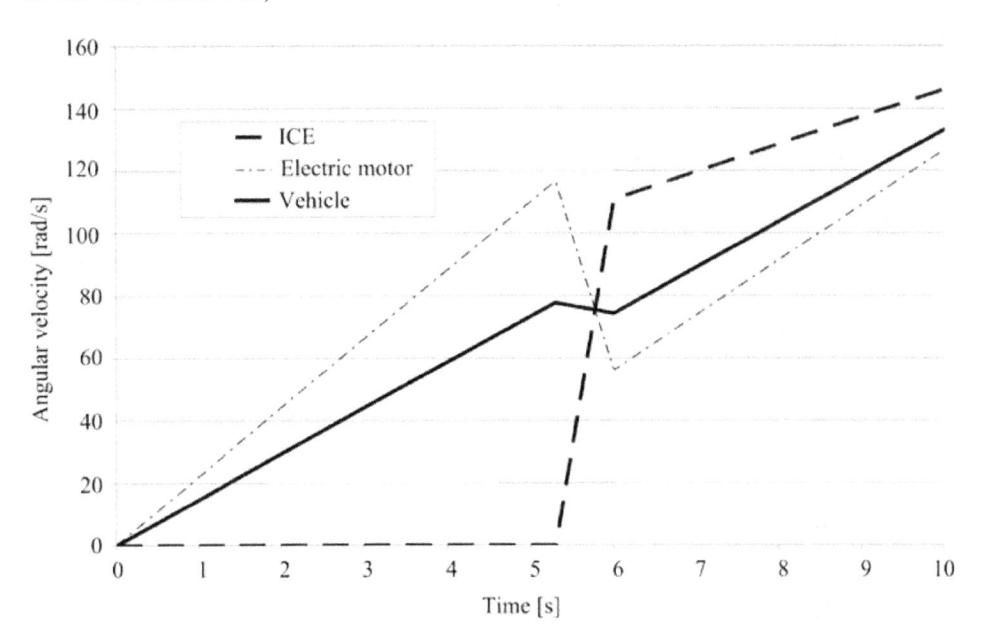

When the permanent magnet motor (PM) is in an idle state and its electromagnetic torque is equal to zero, then brake B1 (see Figure 17 – the internal combustion engine output shaft is connected to the planetary sun wheel, B1) is released, and brake B2 (on the shaft of the permanent magnet motor connected with the crown wheel) is engaged for a short time (Figure 23).

The engagement and disengagement sequence of the above-mentioned brakes make it possible to start the internal combustion engine (ICE) quickly, without sensitive disturbance of the vehicle's speed (see Figure 23).

In case of the 'hybrid acceleration' of the vehicle, the internal combustion engine (ICE) can be started in two ways:

1. By external additional starter;
2. By traction motor when the vehicle stops.

According to the first way, the starter is well adjusted to various vehicles' driving places. The internal combustion engine's (ICE) start in the second way is more complicated, especially when the vehicle is running. In this case, the way the internal combustion engine starts, is similar to vehicle acceleration in the 'pure electric acceleration' mode depicted in detail above.

The contemporary ICE is equipped with a starter-generator and low voltage system necessary for lights, displays, engine sensors, etc., which is typical equipment for a conventional vehicle, so this unit can be conveniently applied in a hybrid vehicle. It is necessary to state that the additional starter also can be used in a CHPTD power train. When the regular alternator (starter-generator) feeds low voltage, it enables us, every time, during vehicle driving, to start the internal combustion engine. When the vehicle brakes, the internal combustion engine has to be switched off. When the vehicle accelerates, the internal combustion engine should start in a very short time (less than half of a second). This is not so difficult when the temperature of ICE is high enough. For this reason, the energy used for an ICE start, including clutch-brake action, can be ignored in a modeling and simulation study of a compact hybrid planetary transmission drive. The error is estimated at lower than one percent.

The electric vehicle's acceleration causes discharge of the battery at the beginning of the driving cycle. This process generates the higher battery discharging current, so it is generating higher energy losses which should be covered by the internal combustion engine (ICE) and also by the kinetic energy of the vehicle, ac-

Figure 24. The permanent magnet (PM) motor's efficiency map (275Nm, 8500rpm) for its output shaft, various angular velocities, and torque direction (4-quadrant map)

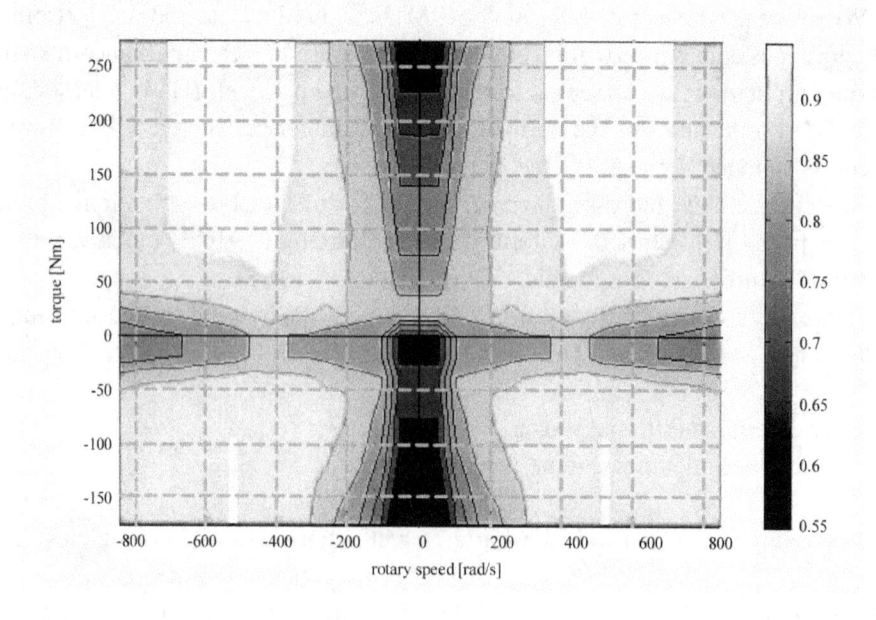

Fundamentals of Hybrid Power Trains Equipped with Planetary Transmission

cumulating in the battery during the vehicle regenerative braking process, to obtain the balance of the battery state-of-charge (SOC) at the beginning and at the end of driving cycle. The electric start, however, eliminates unnecessary ICE operation for low vehicle driving speed – e.g. in traffic (zero emission, lowest noise emission).

The hybrid vehicle's acceleration causes the battery to be charged at the beginning of the acceleration phase (inversely to the electric vehicle's acceleration). In general, the average current (but not short peaks) is lower in this mode and the battery operates with higher efficiency. The disadvantage of this mode is that it is necessary to change the permanent motor's (PM) rotary speed direction. The change of the permanent magnet motor's angular velocity direction requires:

- More advanced inverter technology (4-quadrant inverter);
- Temporary operation of (PM) motor with speeds close to zero (very low efficiency).

Both of the vehicle acceleration modes have advantages and disadvantages, and it is difficult to decide which one is better only on the basis of theoretical considerations – for both mentioned cases, a simulation study is applied. The pure electric vehicle's start has more advantages, and for this reason, it should be recommended in hybrid power train applications.

For further consideration of the analysis of the compact hybrid planetary transmission power train's (CHPTD) drive schemes (as a consequence of relating to proper mechanical ratio adjustment) are shown in Figure 25.

The hybrid architectures shown in Figure 25 are necessary for fulfilling the driving requirements, especially for high speeds, of more than 100 or 120 km/h. These requirements are also included in the extended ECE (NEDC) cycle (see Figure 26). According to the requirements of car-makers, the following list of performance targets should be achieved by hybrid vehicles:

- Accelerations- 0-100 km/h < 15 s; 80-120 km/h <20 s;
- Maximum continuous speed (without battery supporting) >130 km/h;
- Maximum continuous uphill grading (without battery supporting) >25%.

There are very critical conditions for the proper operation of a hybrid power train's configuration. First of all, the internal combustion engine should be adjusted for maximum speed and uphill grading of the vehicle. Also, mechanical ratios should be adequate to these speeds and torques (see Figure 25a – gear G2). However, for comparison of different drive architectures, still more important is the analysis, according to basic statistical duty driving cycles. For this reason, in this chapter, only the vehicle statistic driving cycles are considered.

Figure 25. The schemes of a compact planetary transmission drive power train equipped with additional systems of electromagnetic clutches, causing the changing mechanical ratios between: the yoke wheel shaft of planetary gear connected to vehicle traction wheels (G2); crown wheel shaft of planetary gear connected with the electric machine (G1): a) double system, b) single system

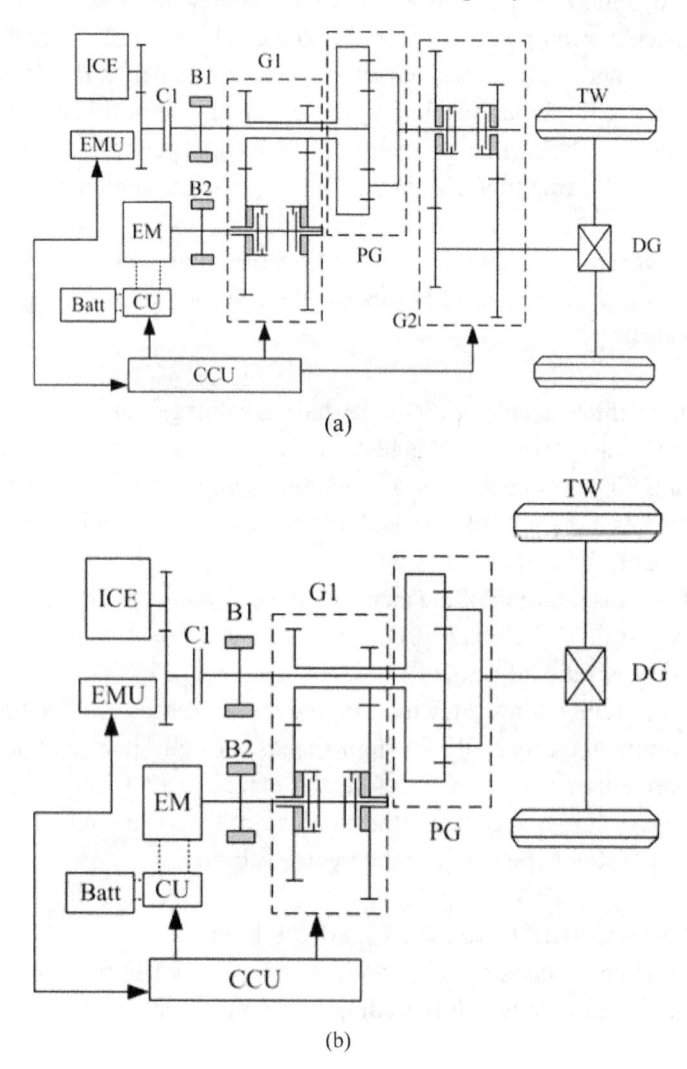

According to the power train scheme which is shown in Figure 25, the simulation results of the compact hybrid planetary transmission drive are included below. The data taken for the simulation studies is presented in Table 1.

Note: Fuel consumption by the internal combustion engine is 3.2 l/100 km (according to assumed data). Energy balance for the ECE driving cycle: aerodynamic

Fundamentals of Hybrid Power Trains Equipped with Planetary Transmission

Figure 26. The European statistic driving cycle, NEDC, for road vehicles

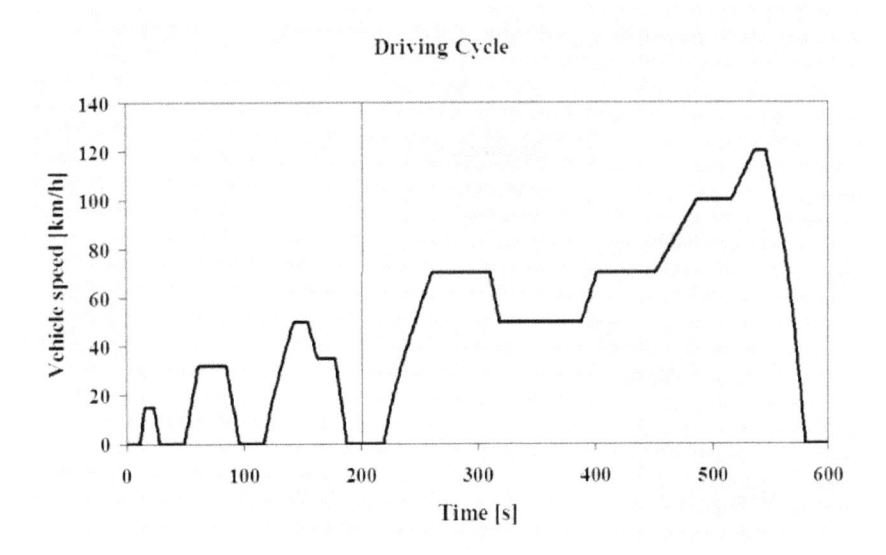

Table 1. The power train data used for simulations, according to the drive scheme presented in Figure 25b

Vehicle	
Total mass	1300 (kg)
Traction wheels inertia	1,7 (kgm^2)
Tire dynamic radius	0,263 (m)
Rolling resistance factor	0,0008
Frontal area of the vehicle	1,6 (m^2)
Aerodynamic drag coefficient c_x	0,33

drag – 1185,9 (kJ); rolling resistance – 1016,8 (kJ); kinetic energy – 1053,2 (kJ); regenerative braking energy – 559 (kJ) (efficiency of energy recuperation η_r=53%); internal combustion engine energy – 2452,8 (kJ).

It is easy to note that the changing of the ratio between the planetary gear and permanent magnet motor (PM) causes an increase of the internal combustion engine speed for faster driving. This is the easiest way, from the practical point of view, to control the internal combustion engine's rotational speed. The one strategy of the internal combustion engine's (ICE) control is to stabilize its rotational speed at different levels, as shown in Figure 28.

Fundamentals of Hybrid Power Trains Equipped with Planetary Transmission

Figure 27. The permanent magnet motor connected with the planetary crown wheel shaft and the planetary yoke wheel shaft, connected to the vehicle traction wheels rotational speeds alteration, for changeable, additional, transmission sets

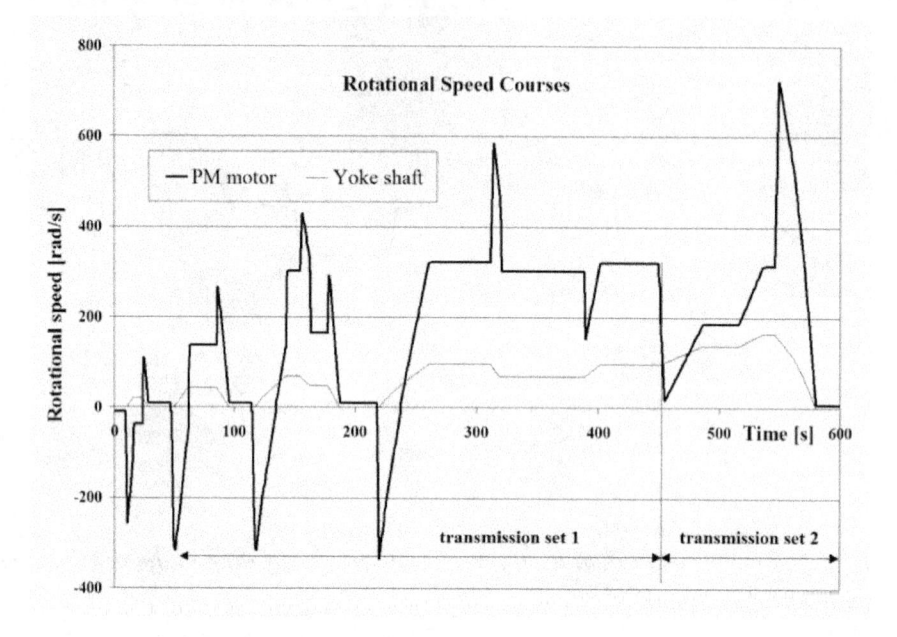

Figure 28. The internal combustion engine and the planetary yoke wheel shaft connected to the vehicle traction wheels rotational speeds alteration, for changeable, additional, transmission sets

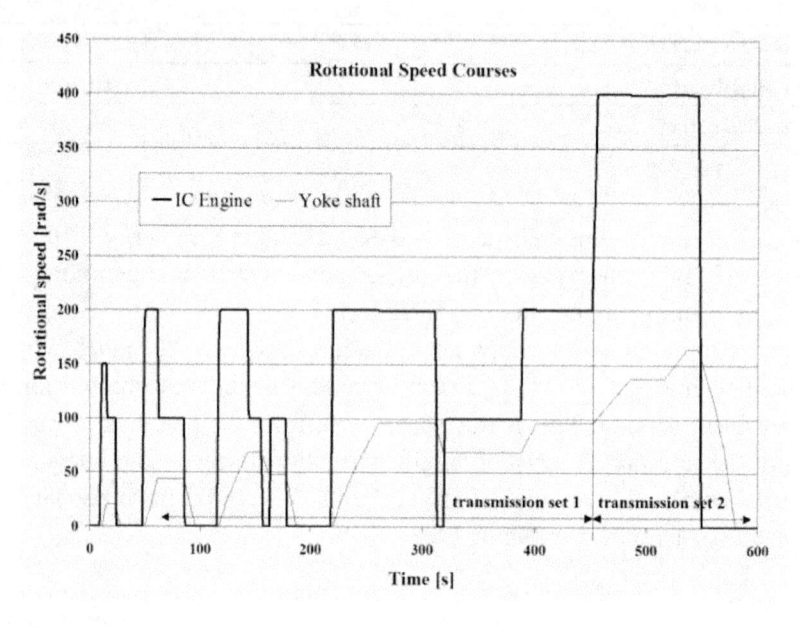

Fundamentals of Hybrid Power Trains Equipped with Planetary Transmission

Figure 29. The permanent magnet motor and the planetary yoke wheel shaft connected to the vehicle traction wheels rotational torques alteration, for changeable, additional, transmission sets

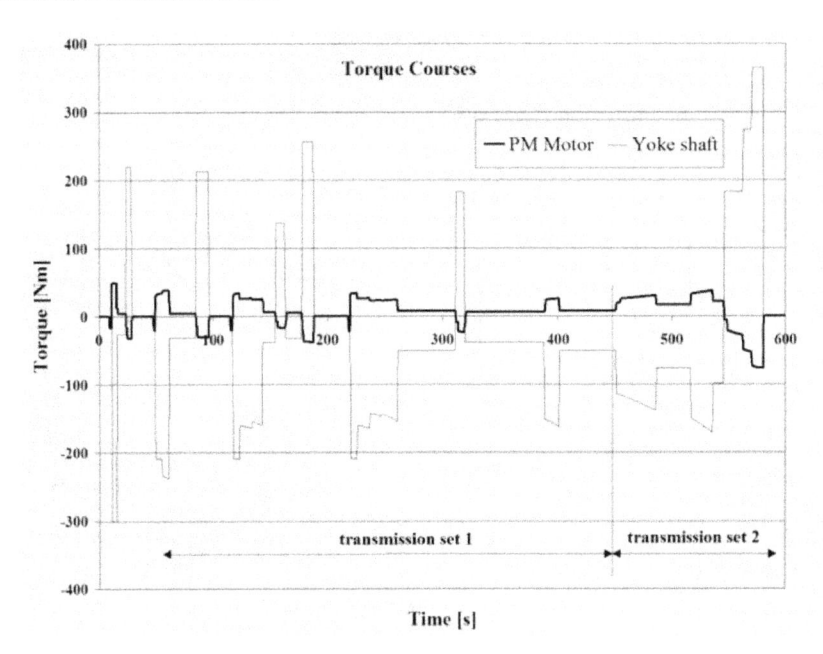

Figure 30. The internal combustion engine and the additional brake device torques alteration, for changeable, additional, transmission sets

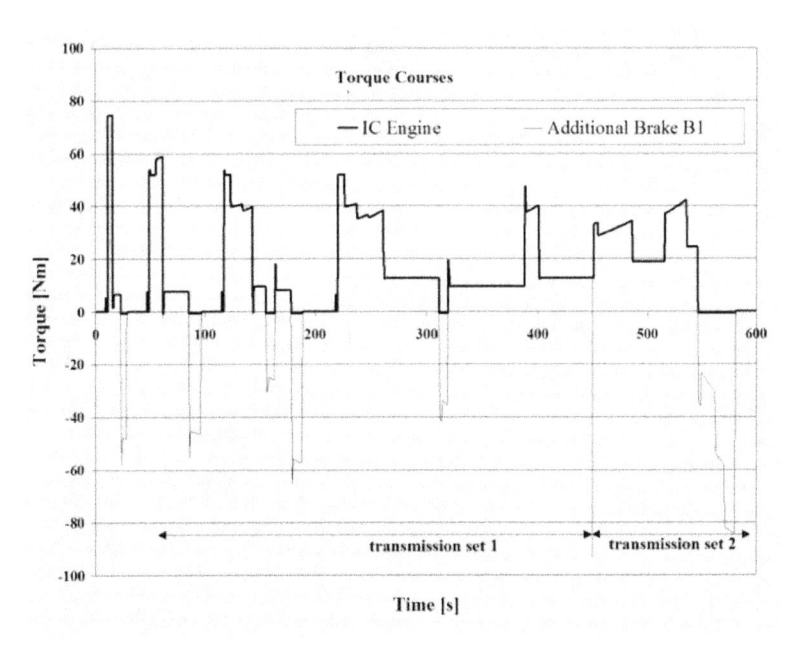

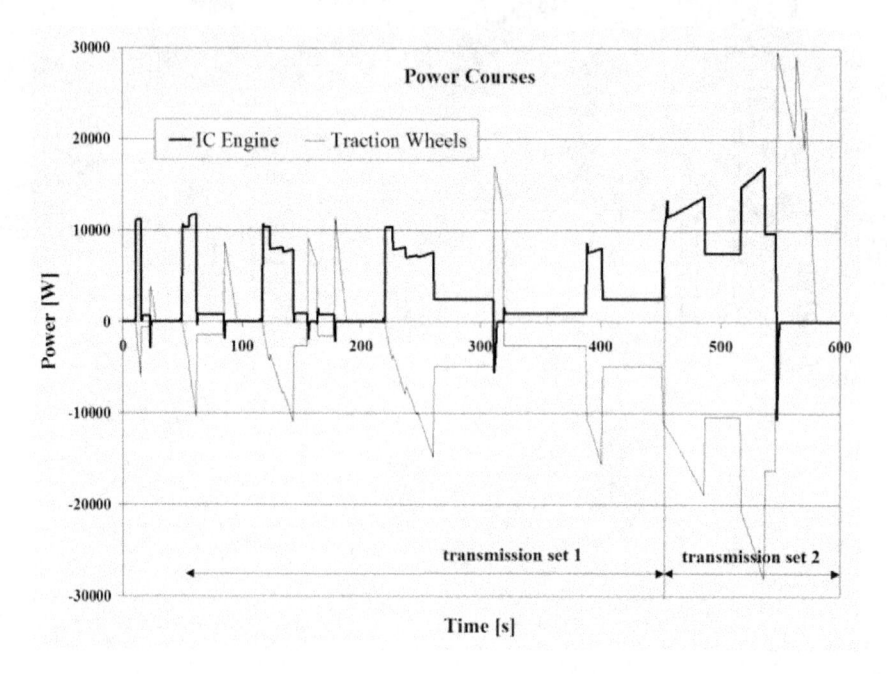

Fundamentals of Hybrid Power Trains Equipped with Planetary Transmission

Figure 31. The permanent magnet motor and vehicle traction wheels power distribution for changeable, additional, transmission sets

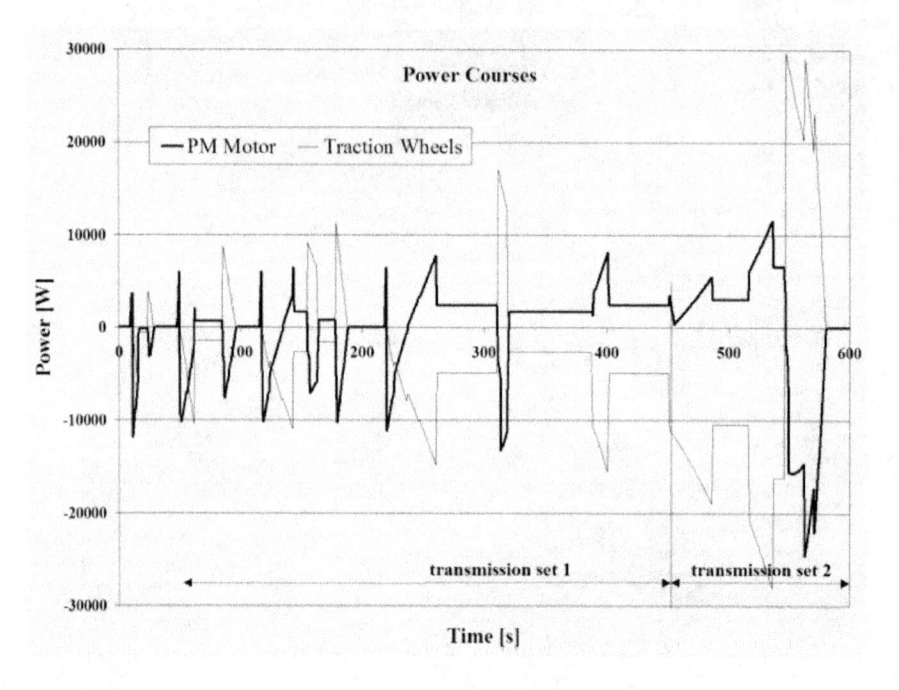

Figure 32. The internal combustion engine and the vehicle traction wheels power distribution for changeable, additional, transmission sets

Figure 33. The permanent magnet motor current and its loading angel alteration for changeable, additional, transmission sets

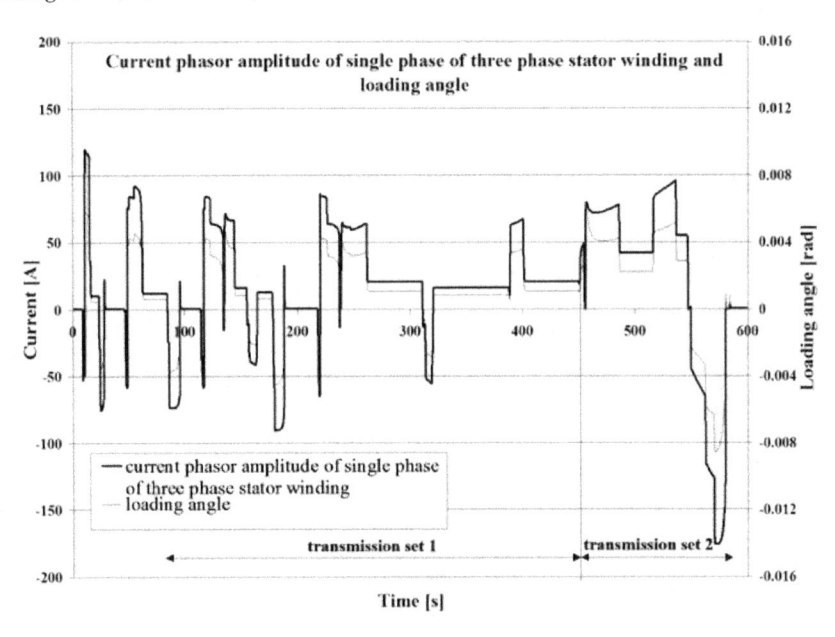

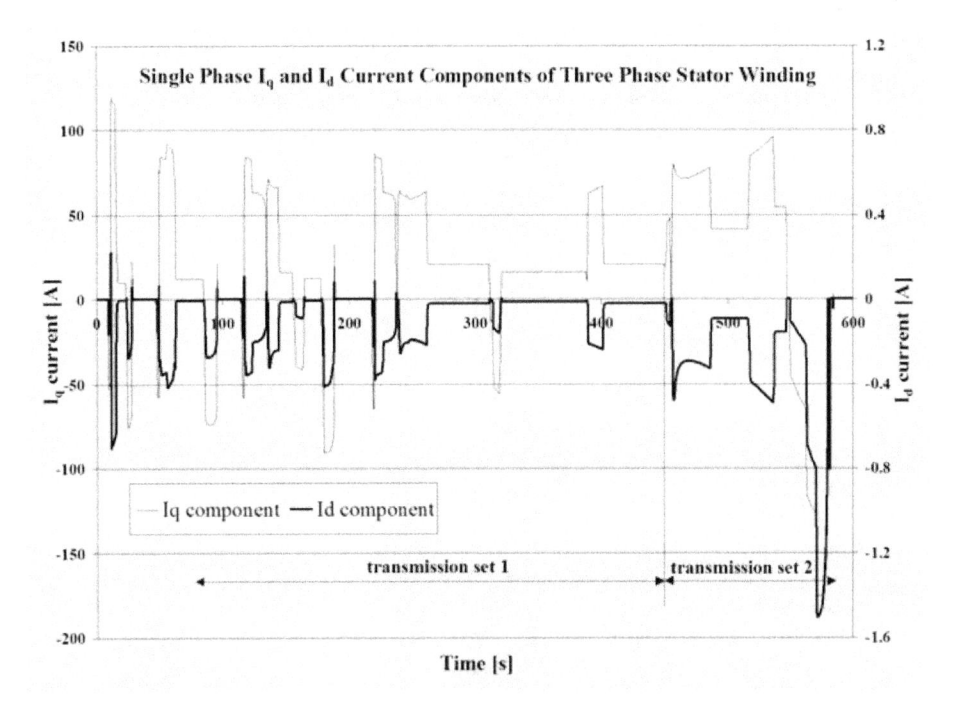

Figure 34. The permanent magnet motor current components alteration for changeable, additional, transmission sets

Fundamentals of Hybrid Power Trains Equipped with Planetary Transmission

Figure 35. The permanent magnet motor voltage components alteration for changeable, additional, transmission sets

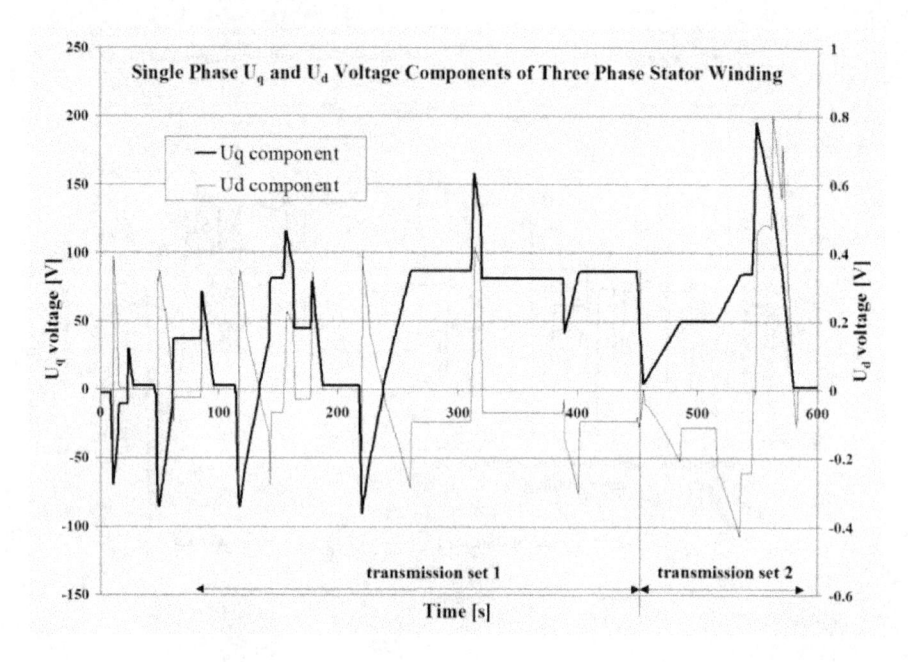

Figure 36. The battery state-of-charge factor and its internal resistance alteration for changeable, additional, transmission sets

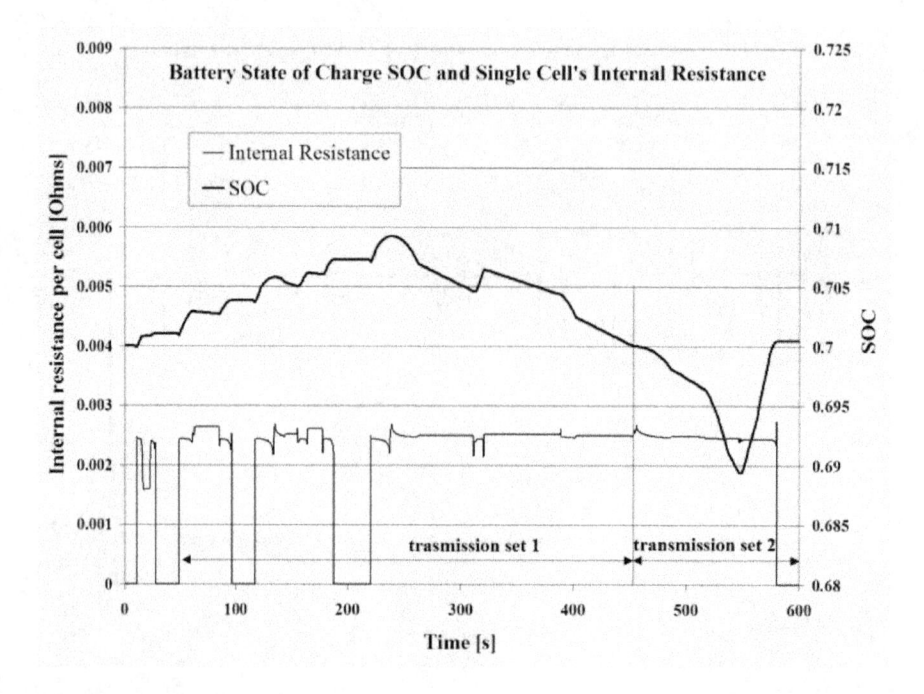

Fundamentals of Hybrid Power Trains Equipped with Planetary Transmission

Figure 37. Two of the electromagnetic solutions, representing a system with a planetary gear with two degrees of freedom (the electromagnetic hybrid power train system of the type as shown in Figure 37b was developed by the FIAT Research Center – CRF – Hybrid EMCVT): a) planetary hybrid electromechanical drive, substituted by the hybrid electromagnetic system; b) construction and control principles of the hybrid electromagnetic power train

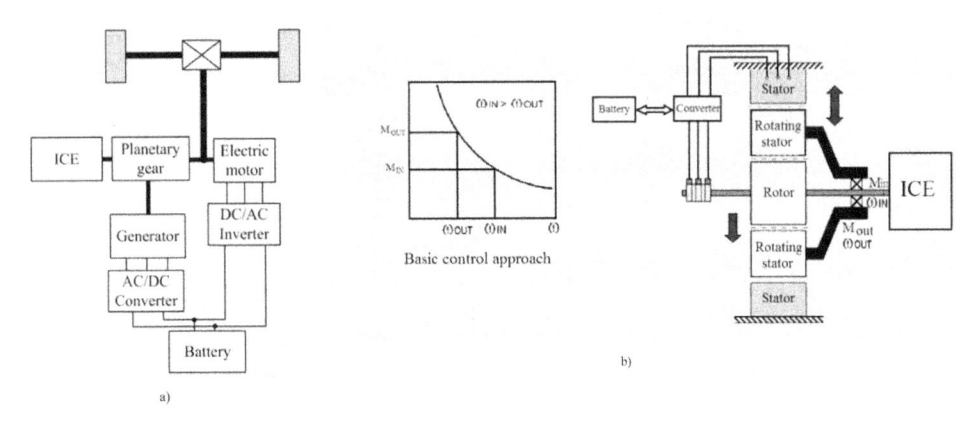

5. DESIGN OF POWER-SUMMING ELECTROMECHANICAL CONVERTERS

Design of electromechanical drives related to the planetary gear of two degrees of freedom controlled by the electric motor, can be transformed into a purely electromagnetic solution. An example of the mentioned solution is depicted in Figure 37. It is a complicated construction with the rotating stator of a complex electrical machine requiring multiple electronic controllers.

The essence of this electromagnetic power train design (as in the classic mechanical planetary gear with two degrees of freedom) is the summing of mechanical and electrical power. Of course, the angular velocities of the electromagnetic power train rotating system components are continuously changing. There is a similar impact as in two degrees of freedom planetary transmission controlled by an electric motor. The complexity of the drive depicted in Figure 37b, provides an output torque multiplication, proportionally, to input torques (in constant proportions), similarly in a planetary gear. Although in drives, e.g. of hybrid vehicles, the summing of power from the internal combustion engine and the electrochemical batteries supplying the electric motor seems attractive, but still, the drive depicted in Figure 37b, should be considered experimental.

However, there is a simpler solution, unfortunately, without the function of torque multiplication (Figure 38a). If it is added to this planetary gear (Figure 38b),

Figure 38. The permanent magnet motor with a rotating stator summing mechanical power from the internal combustion engine and electrical power from the batteries (a); and the entire drive system with the planetary gear (b)

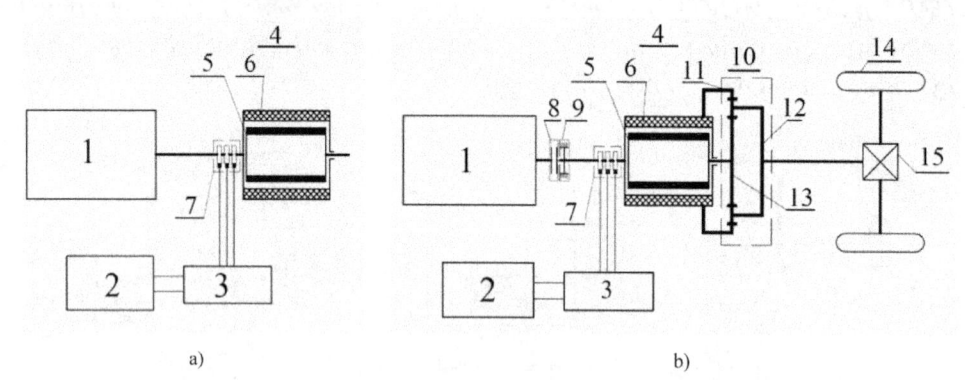

1 – internal combustion engine; 2 – battery of electrochemical accumulators;
3 – control unit; 4 – electric motor; 5- rotor; 6 – stator; 7- slip rings and brushes;
8 – clutch; 9 – brake; 10 – planetary gear with two degrees of freedom;
11 – ring (crown) wheel; 12 – cage (carrier); 13 – sun wheel; 14 – traction wheels;
15 – main gear.

it is possible to obtain, although complicated, the multiplication of output torque, achieved, despite everything, in a less complicated way than in the case of the solution from Figure 37b.

The design of the drive system from Figure 38 is based on analysis of the operation of the electromechanical converter in the form of a synchronous, permanent magnet motor.

5.1. Analysis of Operation of an Ideal Synchronous Motor with Rotating Stator

Let the ideal synchronous motor be the basis of consideration. The ideal means neglecting internal, electromagnetic losses (iron and leakage losses) and mechanical losses (aerodynamic loss, friction loss in bearings and in the assembly of brushes and slip rings supplying the permanent magnet (PM) electrical machine rotating stator controlled by an inverter).

Figure 38 shows the scheme of power distribution in the ideal motor with a rotating stator.

- **Assumption:** Efficiency - $\eta=1$, $\phi = $ const.
- **Power Balance:** $M_2\omega_2 + M_3\omega_3 + M_1\omega_1 = 0$ (26)

Fundamentals of Hybrid Power Trains Equipped with Planetary Transmission

Figure 39. Power distribution in the permanent magnet motor (PM) with a rotating stator (ω_n – respective angular velocities)

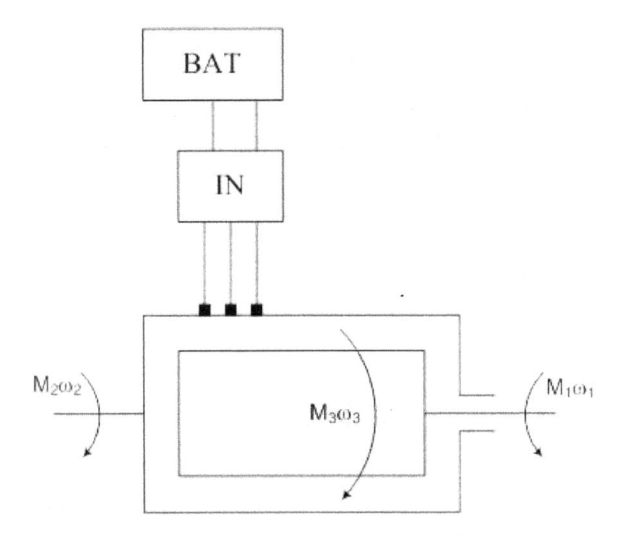

$M_2\omega_2$ - mechanical power supplied to rotating stator
$M_3\omega_3$ - electric power supplied to the stator via brushes with voltage frequency (through the inverter from a battery of accumulators); $\text{sign}\,\omega_3 = \pm 1$
$M_1\omega_1$ - power on the rotor shaft; $\text{sign}\,M_1 = \pm 1$

Electromagnetic torque M_3 is a function of the stator current I_s and flux Φ. In the case of the synchronous motor, for $\cos\varphi \approx 1$ electric power absorbed by the stator $N_s = EI_s$, for stationary conditions $N_s = c\Phi f I_s$ or $N_s = c\Phi\omega_3 I_s$, because $\omega_3 = f =$ inverter frequency. Thus, $M_s = c\Phi I_s$, c – constant depending on construction, and $\Phi = $ const.

5.1.1. Fundamental Kinematic Equations for the ideal Synchronous Motor with Rotary Stator

When $M_2 = 0$ and $\omega_2 = 0$, that $N_s \approx M_1\omega_1$, i.e. this means $c\Phi I_s\omega_3 = M_3\omega_3 = M_1\omega_1$.
Hence:

$$\omega_1 = \omega_3 = \frac{Af}{p} \tag{27}$$

where:

- A: Coefficient of the proportionality;
- f: Inverter output (motor input) voltage frequency;

- **_p_:** Number of pair of poles ($p = 1$).

As ω_3 is, in given conditions, proportional to the stator synchronous frequency and equal to ω_1 ($\omega_3 = \omega_1$), thus $M_3 \approx M_1$. This means that the electromagnetic torque – when neglecting losses – is equal to the rotor torque, and the electromagnetic field power is transferred to the synchronous motor shaft. It is described by a system of Equations (28):

$$\begin{cases} M_3 = M_1 \\ \omega_3 = \omega_1 \end{cases} \qquad (28)$$

When $M2 \neq 0$ and $\omega2 > 0$; $\operatorname{sign}\omega2 = +1 = \text{const.}$, this means that the stator magnetic field rotates in relation to rotor magnets with angular velocity greater than the inverter's voltage frequency f, and if the sign of ω3 is positive in relation to ω1 - lower velocity, then the sign of ω3 is negative. Hence, the new synchronous angular velocity of filed rotation is expressed by a formula:

$$\omega_3{}^* = \omega_2 \pm \omega_3 \text{ and } \omega_3{}^* = Bf^* \qquad (29)$$

and

$$\omega_1 = \frac{Bf^*}{p}, \; p = 1$$

where:

- **_f*_:** Is a new stator electromagnetic field frequency;
- **_B_:** Is a new coefficient of proportionality.

Thus, field rotation velocity in new conditions equals:

$$\omega_1 = \omega_2 \pm \omega_3 \qquad (30)$$

Thus,

$$\omega_2 \pm \omega_3 - \omega_1 = 0 \qquad (31)$$

Fundamentals of Hybrid Power Trains Equipped with Planetary Transmission

Figure 40. The distribution of angular velocities and torques of the permanent magnet, synchronous motor equipped with a rotating stator

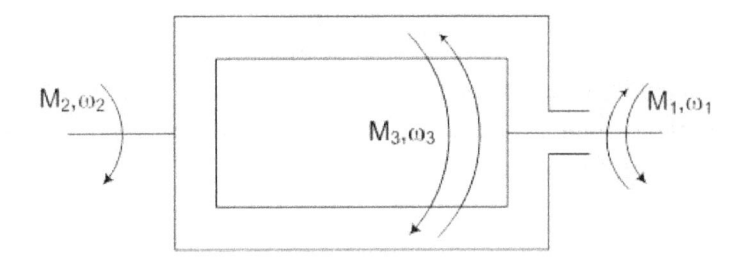

In this simple way, a fundamental kinematic Formula (31) for an ideal synchronous motor with rotating stator has been obtained.

Consider the case, when $\omega_1 = 0$, thus $\omega_2 = \omega_3$, sign $\omega_3 = -1$, which means, that the stator's field angular velocity, in relation to its motionless state, is equal, and in opposite direction to the angular velocity of the stator. Then, $M_3 \omega_3 = M_2 \omega_2$; i.e.:

$$\begin{cases} M_3 = M_2 \\ \omega_3 = \omega_2 \end{cases} \quad (32)$$

This means that the resultant field rotation velocity, in relation to the rotor rotation equals zero, and thus, it is motionless.

The external drive torque equals, with an opposite sign, the electromagnetic torque. It is obvious, that this torque physically exists, because the stator windings are rotating with a given angular velocity ω_2 around motionless rotor magnets, generating a constant flux Φ. Then, in the consequence of power balance, the power of the propelled stator is transferred to e.g. a battery of electro-chemical accumulators.

5.1.2. Torque Equations for the Ideal Synchronous Motor with Rotary Stator

Assuming no loss, the electromagnetic torque M_3 always equals the torque M_1, because Φ = const., and stator current I_s determines both torque values at the same time.

$$M_3 = M_1 \quad (33)$$

The above-mentioned equation is valid for any values of stator angular velocity ω2, including ω2 = 0.

It follows from this proposal that a resultant magnetic field in the gap between rotating stator and rotating rotor serves as a coupling element for a rotating stator and rotor, using the electromechanical analogy.

Consider, therefore, once again the case when $\omega_2 = 0$ and for $M_3\omega_3 + M_2\omega_2 = M_1\omega_1$ is possible to obtain:

$$M_3\,\omega_3 = M_1\,\omega_1; \; \omega_1 = \omega_3$$

thus,

$$M_3 = M_1$$

or

$$M_3 / M_1 = 1 \tag{34}$$

The Equation (34) defines the torque ratio between the electromagnetic field and the torque of the ideal, synchronous motor with one pair of poles.

Another equivalent case for $\omega_1 = 0$ has the form of:

$$M_3\,\omega_3 = M_2\,\omega_2\,;\; \omega_3 = \omega_2$$

thus,

$$M_3 / M_2 = -1 \tag{35}$$

Equation (35) defines the torque ratio between the electromagnetic field and the stator external drive torque for the same conditions, as in the case of the Equation (34).

In this case, when $\omega_1 \neq 0$ and $\omega_2 \neq 0$ the following system of equations is valid:

$$\begin{cases} M_3 = M_1 \\ M_3\omega_3 + M_2\omega_2 - M_1\omega_1 = 0 \\ \omega_3 + \omega_2 - \omega_1 = 0 \end{cases} \tag{36}$$

where:

- $M_3 = c\,\Phi\,I_s$;

Fundamentals of Hybrid Power Trains Equipped with Planetary Transmission

- $\Phi = const$: Electromagnetic torque (I_s - stator winding current);
- ω_3: Angular velocity of field rotation in stator windings proportional to the inverter frequency;
- M_2: Stator drive torque;
- ω_2: Angular velocity of stator rotation;
- M_1: Torque on the synchronous motor shaft;
- ω_1: Synchronous motor rotor's angular velocity.

Solution to the system of Equations (36) is, as follows:

$$M_3 \omega_3 + M_2 \omega_2 - M_1(\omega_2 + \omega_3) = 0$$

$$M_3 \omega_3 + M_2 \omega_2 - M_3 \omega_2 - M_3 \omega_3 = 0$$

$$M_2 \omega_2 - M_3 \omega_3 = 0$$

$$M_2 = M_3 \tag{37}$$

or

$$M_1 \omega_3 + M_2 \omega_2 - M_1(\omega_2 + \omega_3) = 0$$

$$M_1 \omega_3 + M_2 \omega_2 - M_1 \omega_2 - M_1 \omega_3 = 0$$

$$M_2 \omega_2 - M_1 \omega_2 = 0$$

$$M_1 = M_2 \tag{38}$$

The above-mentioned case described by the Equations (36) takes place when the power of rotating stator and electromagnetic field is transmitted to the rotor shaft; the motor mode of the synchronous motor.

If the rotor of a synchronous motor is driven by an external torque (regenerative braking), then, the following system of equations is valid:

$$\begin{cases} M_3 = M_1 \\ M_2 \omega_2 - M_3 \omega_3 + M_1 \omega_1 = 0 \\ \omega_3 + \omega_2 - \omega_1 = 0 \end{cases} \tag{39}$$

Assuming, that sign $\omega_2 = +1$ and $M_2 > 0$ or $M_2 = 0$, which means that the direction of torque and angular velocity ω_2 is always the same or both quantities are equal to zero. This case takes place when, for example, the stator is driven by the internal combustion engine, which has constant angular velocity direction.

The method of solving Equations (39) is analogous to the case of Equations (36):

$$M_2 \omega_2 - M_3 \omega_3 + M_3(\omega_2 + \omega_3) = 0$$

$$M_2 \omega_2 - M_3 \omega_3 + M_3 \omega_2 - M_3 \omega_3 = 0$$

$$M_2 \omega_2 + M_3 \omega_2 = 0$$

$$M_2 = -M_3 \tag{40}$$

or

$$M_2 \omega_2 - M_1 \omega_3 + M_1(\omega_2 + \omega_3) = 0$$

$$M_2 \omega_2 - M_1 \omega_3 + M_1 \omega_2 - M_1 \omega_3 = 0$$

$$M_2 \omega_2 + M_1 \omega_2 = 0$$

$$M_2 = -M_1 \tag{41}$$

The next case is the change of direction of the field in air-gap rotation. It means that sign $\omega_3 = -1$ and sign $M_3 = -1$; when $\omega_2 - \omega_3 > 0$, sign$\omega_1 = +1$. This causes a power transfer from the rotating stator to both the rotor and – through stator windings – to, e.g. electro-chemical accumulators. The analyzed system takes the form:

$$\begin{cases} M_3 = M_1 \\ M_2 \omega_2 - M_3 \omega_3 - M_1 \omega_1 = 0 \\ \omega_3 - \omega_2 - \omega_1 = 0 \end{cases} \tag{42}$$

i.e.

$$M_2 \omega_2 - M_3 \omega_3 - M_3(\omega_2 - \omega_3) = 0$$

$$M_2 \omega_2 - M_3 \omega_3 - M_3 \omega_2 + M_3 \omega_3 = 0$$

Fundamentals of Hybrid Power Trains Equipped with Planetary Transmission

$M_2 \omega_2 - M_3 \omega_2 = 0$

$M_2 = M_3$ (43)

or

$M_2 \omega_2 - M_1 \omega_3 + M_1(\omega_2 - \omega_3) = 0$

$M_2 \omega_2 - M_1 \omega_3 - M_1 \omega_2 + M_1 \omega_3 = 0$

$M_2 \omega_2 - M_1 \omega_2 = 0$

$M_2 = M_1$ (44)

5.2. Basic Simulation

Summing of power from two sources, the internal combustion engine ICE and the battery, by a permanent magnet, synchronous, motor (PM) is simple and efficient, if there are continuous adjustments of angular velocity of the motor output shaft,

Figure 41. The angular velocities of permanent magnet motor stator ω_2 and its rotating magnetic field ω_3

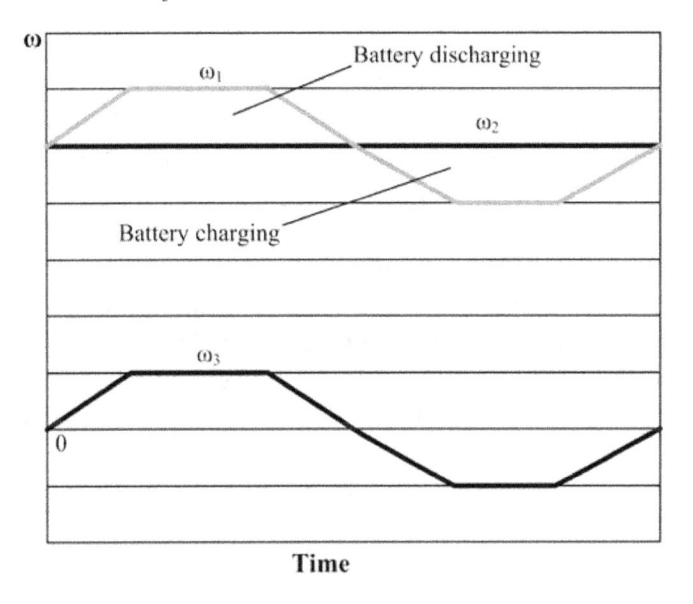

under and above the transitory angular velocity of electromagnetic field in the stator. This effect can be achieved at a constant, permanent magnet, motor (PM) stator angular velocity, which is connected with the constant power operation of the internal combustion engine (ICE), propelling the permanent magnet, motor stator, by appropriately changing its field frequency.

Change of direction of the field in air-gap rotation (Equations 42 - 44) automatically causes the generator mode of the permanent magnet, motor, and power transfer from the rotating stator to both the permanent magnet, motor rotor and the electrochemical accumulators whose values are dependent on the resultant electromagnetic field frequency, without a change of direction, and the value of angular velocity of the externally-driven stator. In hybrid power trains, it is important because of the stabilization of the internal combustion engine operation, which achieved in this way (higher efficiency, lower fuel consumption, lower emissions), determines its angular velocity in practice. As a result of laboratory tests, the chart diagram of angular velocity and torques of permanent magnet, synchronous motor (respectively to the scheme from Figure 37) was obtained (Figure 41). The permanent magnet, synchronous motor is working in generator mode, charging batteries for $\omega_2 = const$, $M_3 = M_1$.

The equations derived from this work are clearly described before the starting courses depicted in Figure 41.

Equations introduced in this paragraph prove the equality of torques on the permanent magnet, synchronous, motor stator and rotor, during the significant and continuous changes of rotor angular velocity, both in the case of generator and motor mode of the electric, synchronous machine.

In the case of vehicle drives, obviously including hybrid power trains, proper torque distribution on the vehicle traction wheels, in both cases of accelerating and regenerative braking, is equally most important. For that reason, there must be a multiplication of the synchronous machine torque, as in the case depicted in Figure 37b. This solution can cause the following changes: change of the permanent magnet, synchronous motor, electromagnetic torque courses, the change of stator torque and the change of rotor angular, velocity and torque. The planetary gear with two degrees of freedom, properly connected to the rotating stator and internal combustion engine shaft, will provide on the output torque (traction wheel input), a summing of power from the internal combustion engine and the battery, and also torque multiplication, proportional to the planetary gear ratio, during the operation.

ACKNOWLEDGMENT

Special thanks to those companies and individuals mentioned in this chapter, whose presentations, e.g. during EVS 20 - 25 - 26, permitted me to depict their great technical achievements.

REFERENCES

Chan, C. C., & Chau, K. T. (2001). *Modern electric vehicle technology*. Oxford, UK: Oxford University Press.

Dietrich, P., Ender, M., & Wittmer, C. (1996). *Hybrid III power train update*. Electric & Hybrid Vehicle Technology.

Ehsani, M., Gao, Y., Gay, L., & Emadi, A. (2004). *Modern electronic, hybrid electric and fuel cell vehicles – Fundamentals, theory and design*. Boca Raton, FL: CRC Press. doi:10.1201/9781420037739

Eiraku, A., Abe, T., & Yamacha, M. (1998). An application of hardware in the loop simulation to HEV. In *Proceedings of EVS 15*. EVS.

Fleckner, M., Gohring, M., & Spiegel, L. (2009). New strategies for an efficiency optimized layout of an operating control for hybrid vehicles. In *Proceedings of Aachen Colloquium*. Aachen Colloquium.

Hayasaki, K., Kiyota, S., & Abe, T. (2009). The potential of the parallel hybrid system and Nissan's approach. In *Proceedings of Aachen Colloquium*. Aachen Colloquium.

He, X., Parten, M., & Maxwell, T. (2005). Energy management strategies for HEV. In *Proceedings of IEEE Vehicle Power and Propulsion Conference*. IEEE.

Hellenbrouch, G., Lefgen, W., Janssen, P., & Rosenburg, V. (2010). New planetary based hybrid automatic transmission with electric torque converter on demand actuation. In *Proceedings of Aachen Colloquium*. Aachen Colloquium.

Ippolito, L., & Rovera, G. (1996). Potential of robotized gearbox to improve fuel economy. In *Proceedings of International Symposium Power Train Technologies for a 3-Litre-Car*. Academic Press.

Killman, G. (2009). The hybrid power train of the new Toyota Prius. In *Proceedings of Aachen Colloquium*. Aachen Colloquium.

Portmann, D., & Guist, A. (2010). Electric and hybrid drive developed by Mercedes – Benz vans and technical challenges to achieve a successful market position. In *Proceedings of Aachen Colloquium*. Aachen Colloquium.

Ruschmayer, R., Shussier, B., & Biermann, J. W. (2006). Detailed aspects of HV. In *Proceedings of Aachen Colloquium*. Aachen Colloquium.

Saenger, F., Zetina, S., & Neiss, K. (2008). Control approach for comfortable power shifting in hybrid transmission – ML 450 hybrid. In *Proceedings of Aachen Colloquium*. Aachen Colloquium.

Szumanowski, A. (1993). Regenerative braking for one and two-source EV drive. In *Proceedings of ISATA 26*. ISATA.

Szumanowski, A. (1996a). Simulation testing of travel range of vehicle powered from battery under pulse load condition. In *Proceedings of EVS13*. EVS.

Szumanowski, A. (1996b). Generic method of comparative energetic analysis of HEV drive. In *Proceedings of EVS 13*. EVS.

Szumanowski, A. (2000). *Fundamentals of hybrid vehicle drives*. ISBN 83-7204-114-8

Szumanowski, A. (2006). *Hybrid electric vehicle drives design*.

Szumanowski, A., & Bramson, E. (1992). Electric vehicle drive control in constant power mode. In *Proceedings of ISATA 25*. ISATA.

Szumanowski, A., & Brusaglino, G. (1992). Analysis of the hybrid drive consisting of electro-chemical battery and flywheel. In *Proceedings of EVS 11*. EVS.

Szumanowski, A., & Hajduga, A. (1998). Energy management in HV drive advanced propulsion systems. In *Proceedings of Advanced Propulsion Systems*. GPS.

Szumanowski, A., & Hajduga, A. (1998). Energy management in hybrid vehicles drive. In *Proceedings of Advanced Propulsion Systems*. GPC.

Szumanowski, A., & Jaworowski, B. (1992). The control of the hybrid drive. In *Proceedings of EVS11*. EVS.

Szumanowski, A., Piórkowski, P., Hajduga, A., & Ngueyen, K. (2000). The approach to proper control of the hybrid drive. In *Proceedings of EVS 17*. EVS.

Takaoka, T., & Komatsu, M. (2010). Newly developed Toyota plug-in hybrid system and its vehicle performance. In *Proceedings of Aachen Colloquium*. Aachen Colloquium.

Trackenbrodt, A., & Nitz, L. (2006). Two-mode hybrids = adoption power of intelligent system. In *Proceedings of Aachen Colloquium*. Aachen Colloquium.

Vaccaro, A., & Villaci, D. (2004). Prototyping a fussy-based energy manager for parallel HEV. In *Proceedings of ELE European Drive Transportation Conference*. ELE.

Yamagouchi, K. (1996). *Advancing the hybrid system*. Electric & Hybrid Vehicle Technology.

Yamagouchi, K., Miyaishi, Y., & Kawamoto, M. (1996). Dual system – Newly-developed hybrid system. In *Proceedings of EVS 13*. EVS.

Chapter 9
Basic Simulation Study during the Process of Designing the Hybrid Power Train Equipped with Planetary Transmission

ABSTRACT

Chapter 9 is devoted to simulation research showing the influence of changes of the power train's parameters and control strategy on the vehicle's energy consumption, depending on different driving conditions. The control strategy role is to manage how much energy, frankly speaking, how much of the torque-speed relations referring to the power alteration, are flowing to or from each component. In this way, the components of the hybrid power train have to be integrated with a control strategy, and of course, with its energetic parameters to achieve the optimal design for a given set of constraints. The hybrid power train is very complex and non-linear to its every component. One effective method of system optimization is numerical computation, the simulation, as in the case of the multivalent suboptimal procedure regarding the number of electrical mechanical drive's elements, whose simultaneous operation is connected with the proper energy flow control. The minimization of a power train's internal losses is the target. The quality factor is minimal energy, as well as minimal fuel and electricity consumption. The fuel consumption by the hybrid power train has to be considered in relation to the conventional propelled vehicle. First of all, the commonly chosen statistic driving cycles should be taken into consideration. Unfortunately, this is not enough. The additional tests as for the vehicle's climbing, acceleration, and power train behavior, referring to real driving situations, are strongly recommended during the drive design process.

DOI: 10.4018/978-1-4666-4042-9.ch009

Basic Simulation Study during the Process of Designing the Hybrid Power Train

INTRODUCTION

The approach of the design hybrid drive system is based on dynamic modeling and simulation. The transient operation process can be studied in details with this dynamic model in Matlab - Simulink, and also, the control strategy can be optimized by running the simulation and monitoring the operation of each component: the operating points of ICE, fuel consumption (energy consumption), the power distribution, the torque and rotary speed of the ICE and motor, the operating efficiency of the motor, the change of battery SOC, current and voltage.

A power control strategy is needed to control the flow of power and to maintain adequate reserves of energy in the storage devices. Although this is an added complexity, not found in conventional vehicles, it allows the components to work together in an optimal manner to achieve multiple design objectives, such as high fuel economy and low emissions. In HEVs, the composing of power train components and the way they are connected, are hybrid drive system configuration (hardware), and the management of the power flow among the components is called the 'control strategy'.

The flexibility in HEV design comes from the ability of the control strategy to manage how much power is flowing to, or from, each component. This way, the components can be integrated with a control strategy, to achieve the optimal design for a given set of constraints. The hybrid power train is very complex and nonlinear respectively, with its every component. One effective method of system optimization is numerical, computation simulation, as in the case of multivalent, suboptimal procedure regarding the number of the electrical mechanical drive's elements, whose simultaneous operation is connected with the proper energy flow control. The minimization of power train internal losses is the target. The quality factor is minimum energy, as well as fuel and electricity consumption. The power train parameter optimization can be worked out, by also using the above-mentioned method. This means powers, torques, voltage, capacity, proper adjustment, etc.

Essentially, this chapter shows the required design approach based on digital simulation, whose result is an energy flow optimization (Szumanowski, 1999; Szumanowski, Hajduga, & Piórkowski, 1998a; Szumanowski & Nguyen, 1999).

The vehicle model, including a proper driving vehicle with load resistance equations reduced on its traction wheel, is depicted in Chapter 1, and identified as a block diagram, useful for computation in Figure 2.

The planetary transmission model is discussed in Chapter 8, and illustrated by Figure 3.

For the reason of strong nonlinear features of drive components optimizing the hybrid power train's parameters is possible, only by use computer simulations.

Basic Simulation Study during the Process of Designing the Hybrid Power Train

Figure 1. The simulation flow chart of a hybrid electric vehicle power train's block diagram of a dynamic model in Simulink

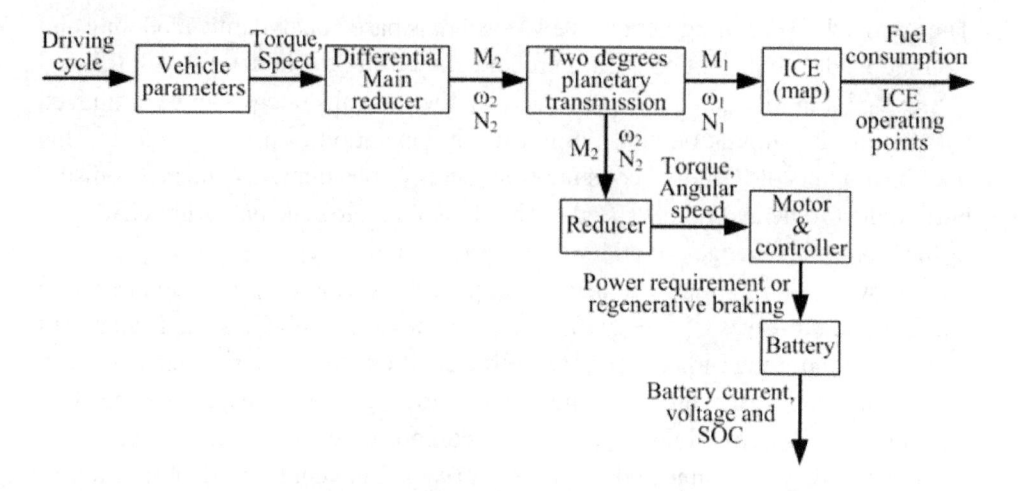

Figure 2. The vehicle external driving load: a block diagram of a dynamic model in Simulink

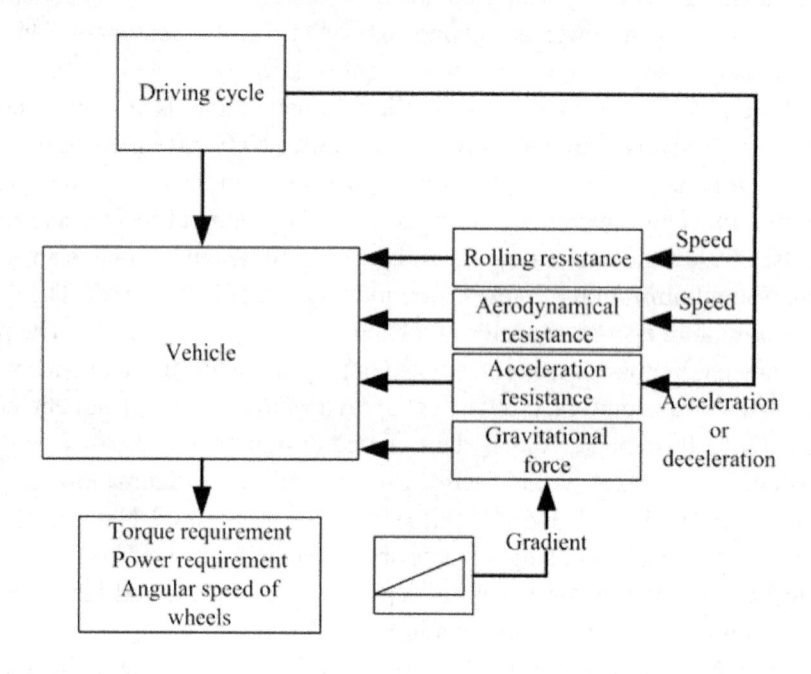

After choosing the analyzed power train architecture, the following simulations are suggested:

Basic Simulation Study during the Process of Designing the Hybrid Power Train

Figure 3. The planetary transmission modeling background used in modeling in Simulink

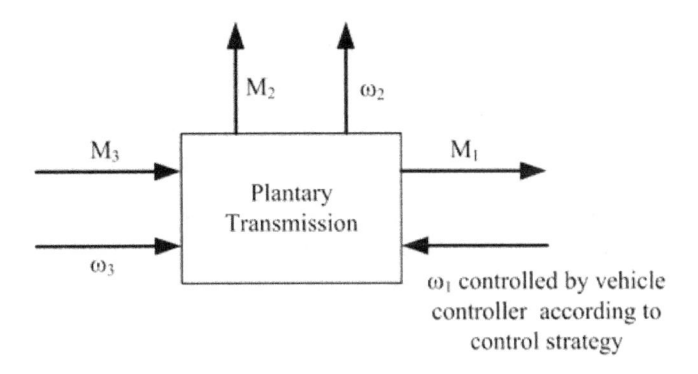

1. Comparison of the internal combustion engine fuel consumption in the case of conventional and hybrid power trains
2. Influence of power train control strategy on energy consumption
3. Comparison of a battery pack's dynamic characteristics in the case of a power train, which is equipped with different nominal voltages, capacities, or types of battery packs, assuming approximately, the same battery's nominal energy
4. Studies for power train design, based on a common statistic driving cycle
5. Test of a power train for a vehicle's maximum gradeability driving
6. Test of a power train for a vehicle's maximum driving acceleration
7. Additional simulation analysis of a power train's operation for different real vehicle driving conditions, confirmed the required operation parameters of previous power train designs

For these reasons, this chapter contains multiple simulation results, which are very useful during the hybrid electric vehicle power trains' design process. Two types of vehicles, a 15- ton urban bus and 5-ton shuttle service bus are used for the above-stated considerations.

1. SIMULATION STUDIES OF THE HYBRID ELECTRIC POWER TRAIN BASED ON AN URBAN BUS

According to the above-mentioned points, 1-3 simulation tests were done for a fifteen-ton hybrid bus (its parameters are included in the Table 1), equipped with hybrid power train architecture (CHPTD) presented in Figures 8.12, 8.13, 8.14.

Table 1. The parameters of analyzing an urban bus and its CHPTD power train components, taken under the simulation studies

Vehicle	Battery	Mechanical Transmission
Mass: 15000kg Flat surface: 6.92m^2 Coefficient of aerodynamics drag: 0.55 Coefficient of rolling friction: 0.01 Tire rolling radius: 0.51 m	Type: Ni-MH Nominal capacity: 64Ah Nominal voltage: 300V	Nominal ratio of planetary gear: 2.27 Mechanical ratio between ICE and sun wheel shaft: 1.62 Mechanical ratio between motor and crown wheel shaft: 4.98 Mechanical ratio between yoke wheel shaft and vehicle wheel: 4.98 Planetary gearbox basic ratio: 2.26
Traction motor	**Internal Combustion Engine**	
Number of poles p: 24 PM flux: 0.05775Wb Coils inductance L: 0.000076H Coils resistance R: 0.04Ω Torque max: 275Nm Speed max: 8500rpm	IVECO 7800ccm, 170 kW	

1.1. Simulation Results

In the simulation tests, a Warsaw urban bus driving cycle (Figure 4) is used. The control algorithm makes it possible to apply both vehicle start modes – hybrid and pure electric (up to 18 km/h bus speed; after achievement by the bus of this speed, the internal combustion engine ICE is started – see the previous Chapter, 8). Alterations of the battery's current, its state-of-charge factor, SOC, and the internal combustion diesel engine's fuel consumption results for both vehicle acceleration modes are shown as follows (Szumanowski, 2006; Szumanowski, Chang, & Piórkowski, 2005a):

1.2. Comparison of a Simulation Result of a Compact Hybrid Planetary Transmission Power Train, CHPTD, Equipped with 500V 80Ah Ni-MH and 300V 128Ah Ni-MH Nickel Metal Hydride Batteries

The following comparison of a simulation result is about a CHPTD power train, equipped with a 500V 80Ah Ni - MH battery and 300V 128Ah Ni - MH battery. In order to get the comparable result, to attempt to keep the same or similar main parameters of CHPTD, it was necessary to just adjust some parameters of the battery. There is no big change in the main parameters of the analyzed hybrid power train in both cases, but the nominal parameters of both battery packs are different (Chang, 2005; Szumanowski et al., 2005a).

There is no big difference in the engine's operating points, power requirement and distribution, and fuel consumption, because there is only a very small change

Figure 4. Warsaw urban bus driving cycle

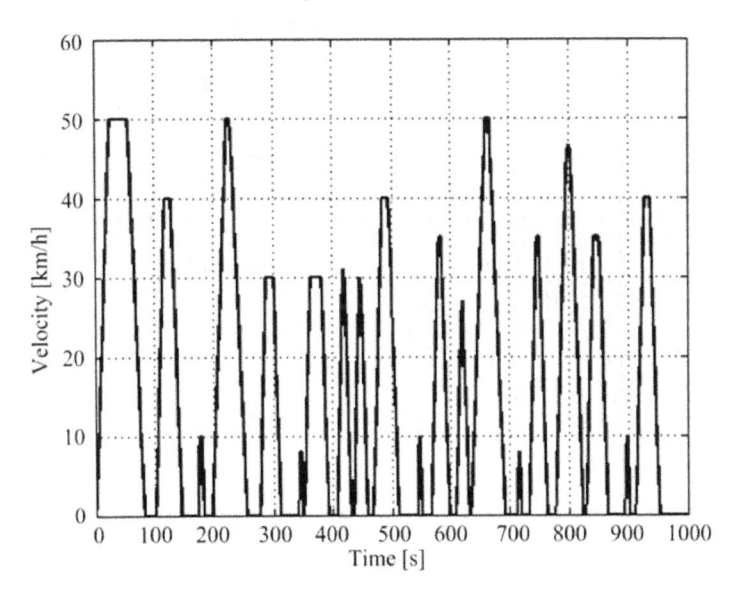

Figure 5. The battery's current alteration in the case of the pure electric urban bus's starting acceleration, according to the assumed driving cycle

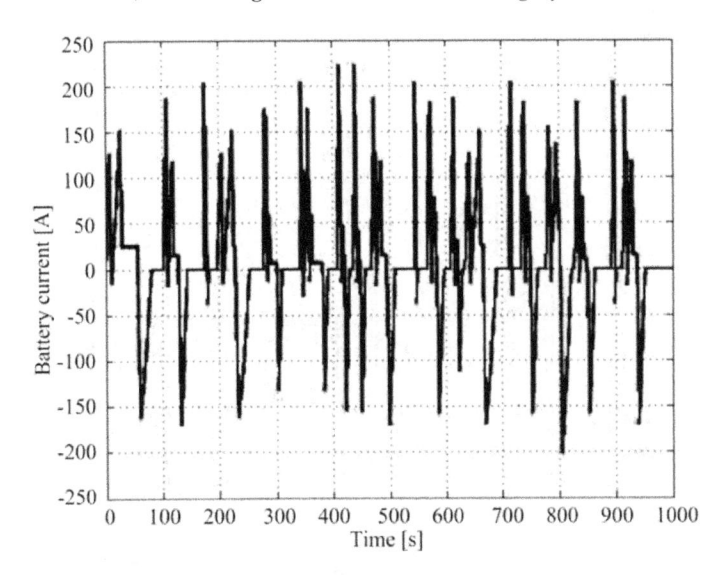

in the differential ratio of the main gearbox. As the power requirement for the battery is the same and there is a big difference in the battery's nominal capacity and electromotive force, EMF, between both Ni - MH battery packs, the battery's current alteration is a big difference. For a 500 V 80 Ah Ni - MH battery pack, the

Figure 6. The battery state-of-charge factor, SOC, alteration, in the case of a pure electric urban bus's starting acceleration, according to an assumed driving cycle

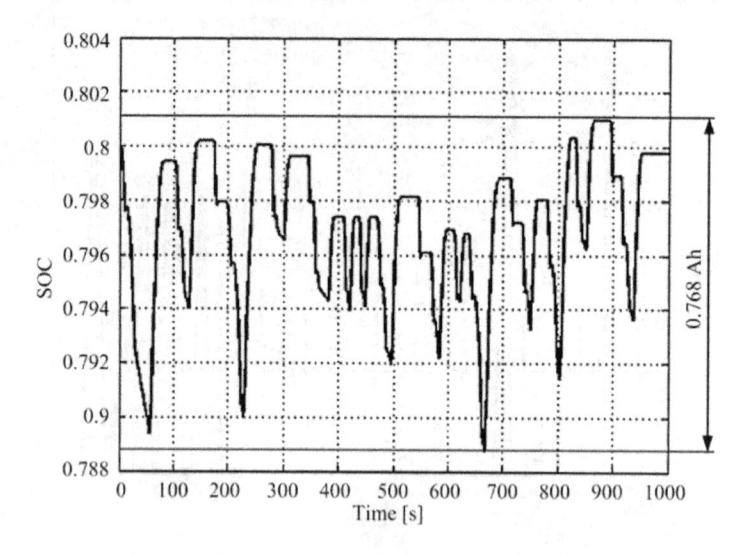

Figure 7. The battery's current alteration in the case of a hybrid urban bus's starting acceleration, according to an assumed driving cycle

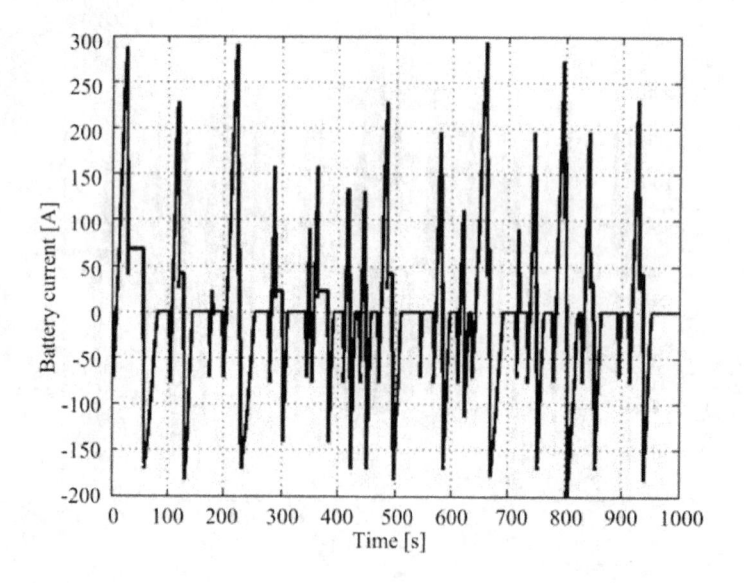

current alteration range is 127 A ~ 115 A, which corresponds to the battery's capacity alteration range of 1.59 C ~ 1.44 C. For 300 V 128 Ah Ni - MH battery pack, the current alteration range is 213 A ~ 193 A, which corresponds to the battery capacity alteration range 1.66 C ~ 1.5 C. Though the current alteration range 213

Basic Simulation Study during the Process of Designing the Hybrid Power Train

Figure 8. The battery state-of-charge factor, SOC, alteration in the case of a hybrid urban bus's starting acceleration, according to an assumed driving cycle

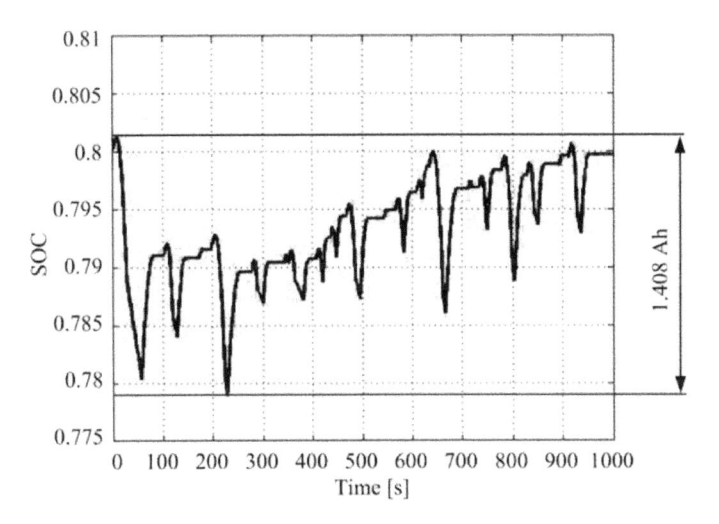

Figure 9. The internal combustion diesel engine's operating points' alteration in the case of a hybrid urban bus's starting acceleration, according to an assumed driving cycle (the specific fuel consumption is scaled in kg/Ws)

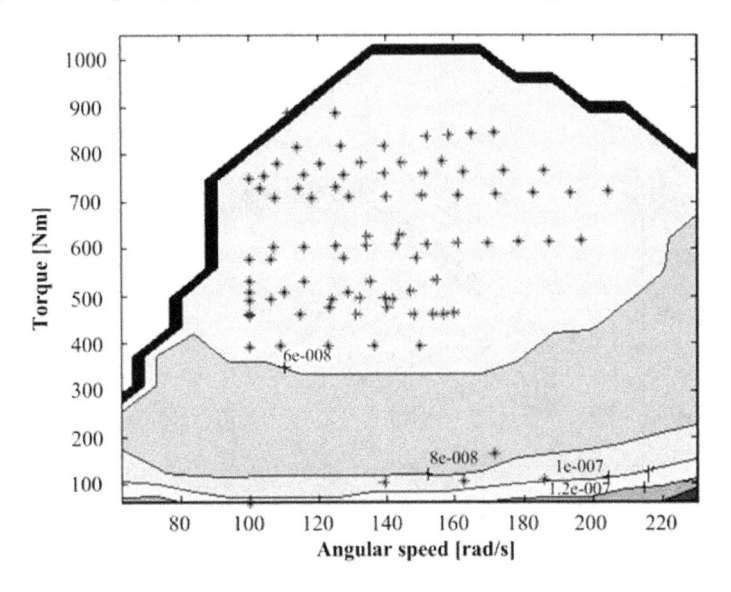

A ~ 193 A for a 300 V 128 Ah Ni - MH battery pack is bigger than the current alteration range of 127 A ~ 115 A for a 500 V 80 Ah Ni - MH battery pack, it is very similar in current density alteration for both packs.

Basic Simulation Study during the Process of Designing the Hybrid Power Train

Figure 10. The internal combustion diesel engine's operating points' alteration in the case of a pure electric urban bus's starting acceleration, according to an assumed driving cycle (the specific fuel consumption is scaled in kg/Ws)

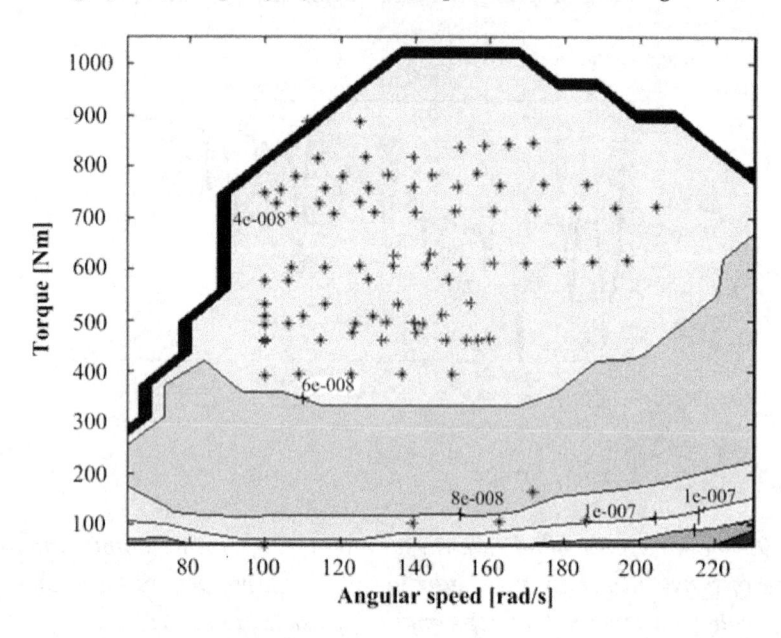

Figure 11. The internal combustion diesel engine's operating points' alteration in the case of the conventional-only diesel engine drive of the urban bus, according to an assumed driving cycle

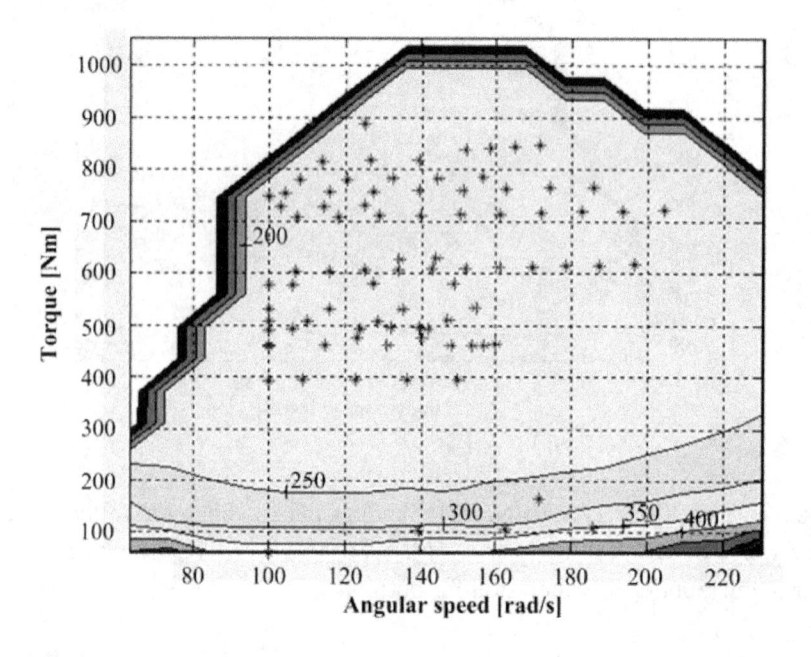

Basic Simulation Study during the Process of Designing the Hybrid Power Train

Figure 12. The comparison of the fuel consumption of a pure, electric hybrid of an urban bus's starting acceleration and conventional-only diesel engine drive of the urban bus, according to an assumed driving cycle

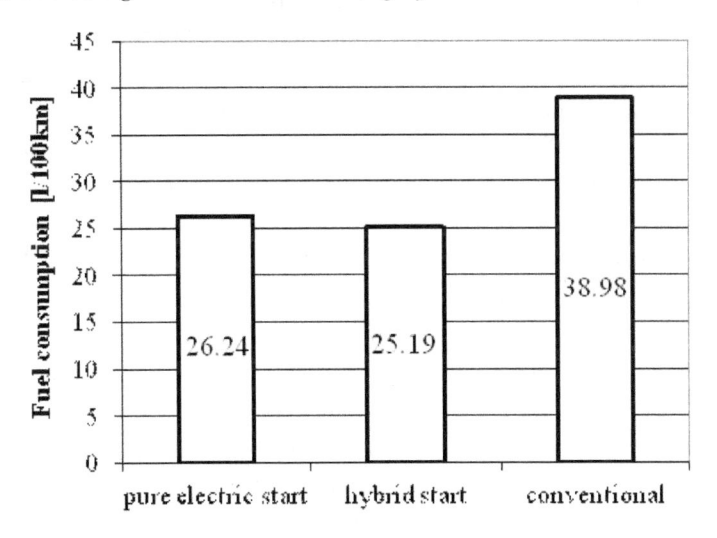

Table 2. The comparison of the main parameters of a compact planetary transmission hybrid power train, CHPTD (the parameters of the electric traction motor and the internal combustion diesel engine are the same, as in Table 1)

Battery	500V 80Ah Ni-MH battery pack	300V 128Ah Ni-MH battery pack
Nominal energy of battery pack	40 kWh	38.4 kWh
Main gear – differential ratio	4.88	4.87
The basic planetary gear ratio	2.87	2.87
Reducer ratio between PM motor and planetary gear	4.8	4.8
Reducer ratio between engine and planetary gear	1.46*1.24	1.46*1.24
Vehicle mass	15,000 kg	15,000 kg
Vehicle front area	6.92 m^2	6.92 m^2
Drag coefficient c_x	0,55	0,55
Dynamic tire radius index r_{dyn}	0,51 m	0,51 m
Nominal voltage of battery pack	500 V	300V
Nominal battery capacity	80Ah	128Ah
Fuel consumption	26.05 l/100km	26.1 l/100km
Diesel engine	7800 cc	7800 cc
Warsaw driving cycle – top speed	50km/h	50km/h
Engine control strategy parameters	(-100 0 8 9 30) (0 0 0 172 423)	(-100 0 8 9 30) (0 0 0 172 423)

The range of battery voltage alteration for a 500 V 80 Ah Ni - MH battery pack is from 487 V ~ 601 V, changing from 33 V to 61 V around 540 V (EMF). For a 300 V 128 Ah Ni-MH battery pack, the range is from 290 V ~ 362 V, changing from 34 V to 38 V around 324 V of its electromotive force's nominal value (EMF). The internal resistance of a 500 V 80 Ah Ni - MH battery is 0.475 Ω (for the battery's state-of-charge factor value, k=0. 8), which is about 2 times more than 0.178 Ω, the internal resistance of a 300 V 128 Ah Ni-MH battery pack. The battery's state-of-charge factor's, SOC, alteration for both cases is very similar.

1.3. Comparison of the Simulation Result of a CHPTD with a 500V 36Ah Li-Ion Battery and a 300V 60Ah Li-Ion Battery

The following comparison of a simulation result is about a compact hybrid planetary transmission drive, CHPTD, equipped with a 500 V 36 Ah lithium ion Li - ion battery and a 300 V 60 Ah Li - ion battery. In order to get the comparison result, the attempt is to keep the same main parameters of the CHPTD, and just adjust some parameters of the battery. There is no change in the main parameters of the hybrid drive for both cases, but nominal parameters of both battery packs are different (Szumanowski & Nguyen, 1999).

There is no difference in the engine's operating points, power requirement, and distribution and fuel consumption, because there is no change in the main parameters

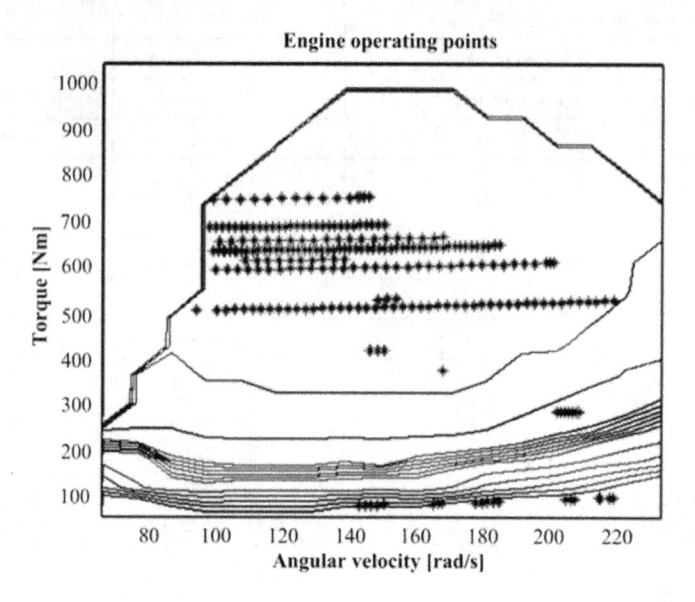

Figure 13. The internal combustion diesel engine's operating points' alteration applied in a compact hybrid planetary transmission power train, CHPTD, equipped with a 500V Ni-MH battery

Basic Simulation Study during the Process of Designing the Hybrid Power Train

Figure 14. The internal combustion diesel engine's operating points' alteration applied in a compact hybrid planetary transmission power train, CHPTD, equipped with a 300 V Ni-MH battery

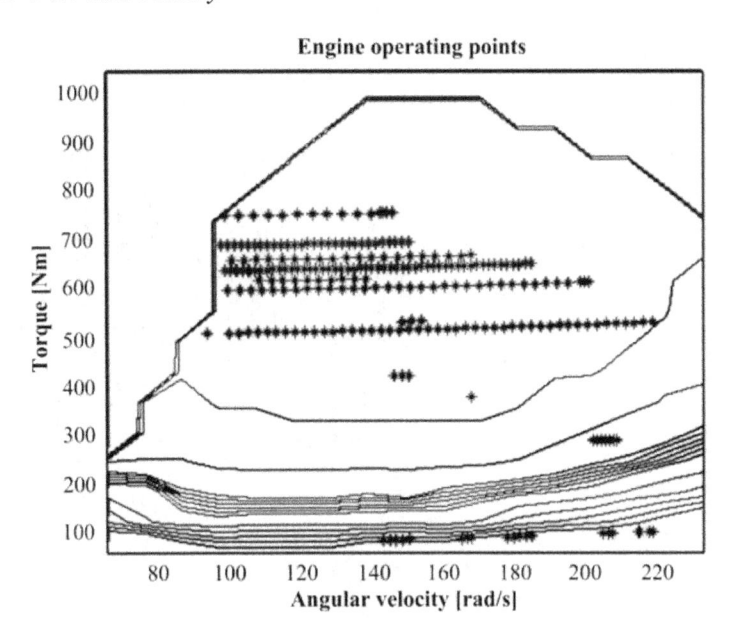

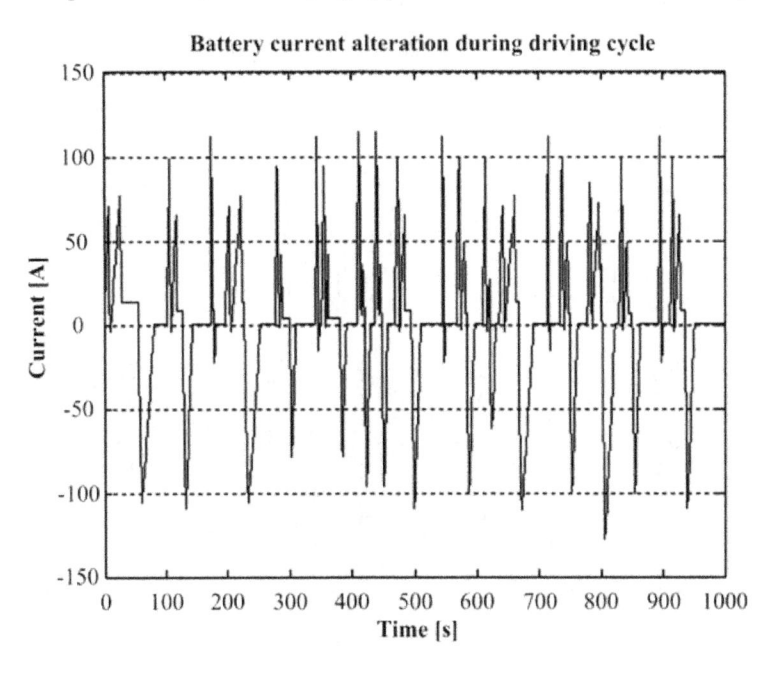

Figure 15. The current alteration of battery applied in a compact hybrid planetary transmission power train, CHPTD, equipped with a 500V Ni-MH battery

Basic Simulation Study during the Process of Designing the Hybrid Power Train

Figure 16. The current alteration of a battery applied in a compact hybrid planetary transmission power train, CHPTD, equipped with a 300V Ni-MH battery

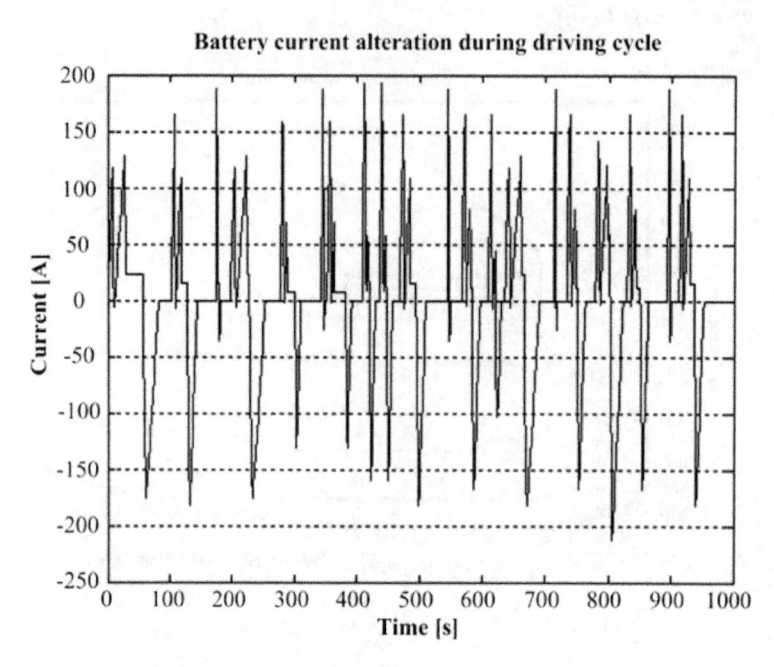

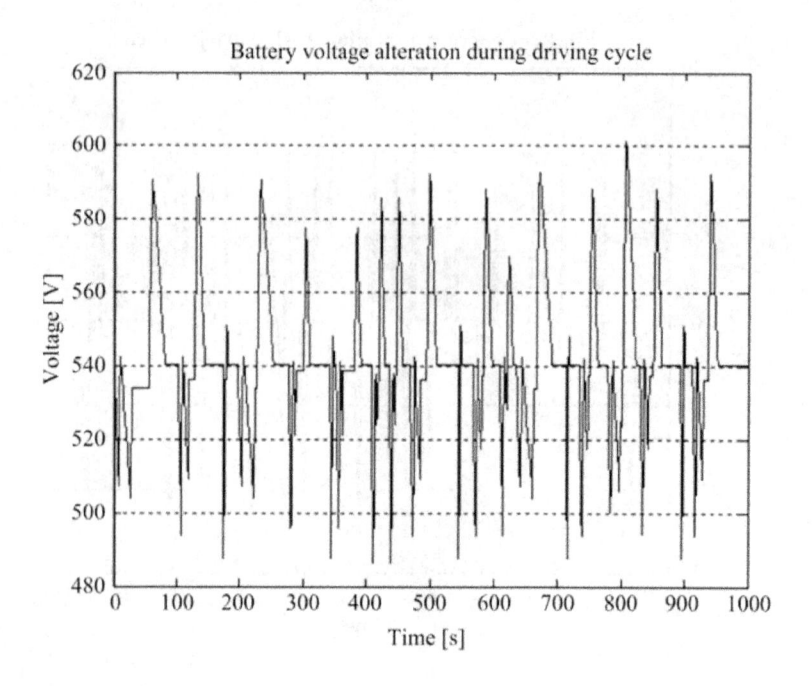

Figure 17. The voltage alteration of a battery applied in a compact hybrid planetary transmission power train, CHPTD, equipped with a 500V Ni-MH battery

Basic Simulation Study during the Process of Designing the Hybrid Power Train

Figure 18. The voltage alteration of a battery applied in a compact hybrid planetary transmission power train, CHPTD, equipped with a 300 V Ni-MH battery

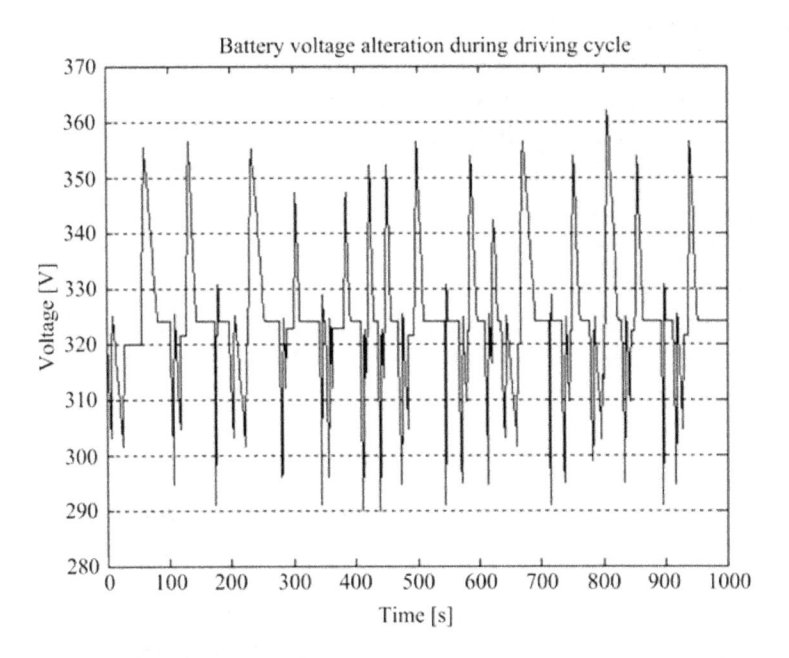

Figure 19. The comparison of internal resistance alteration: the tested 300 V and 500 V Ni-MH battery packs

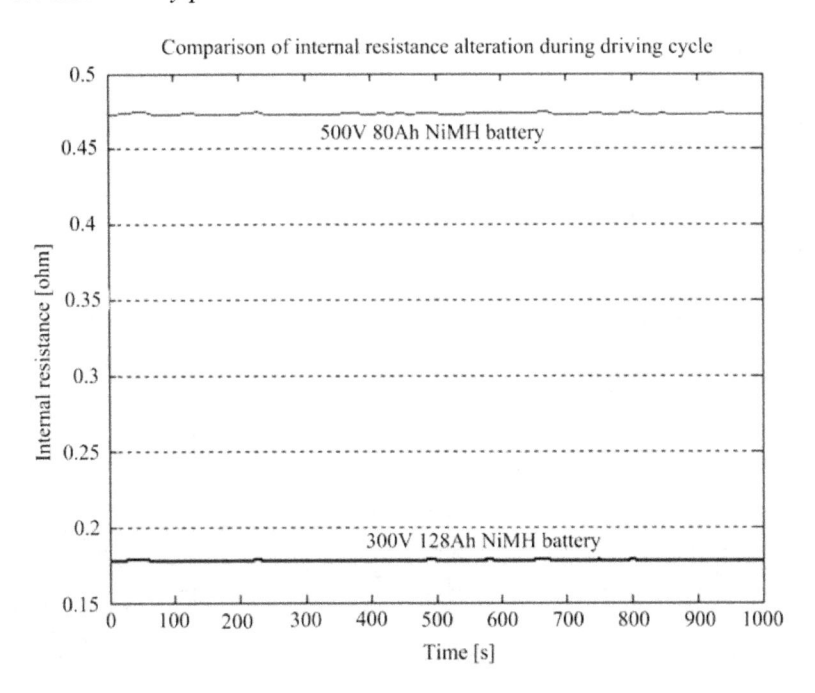

335

of the hybrid drive for both cases. Whereas the power of the battery is the same, there is a big difference of battery nominal capacity and electromotive force, EMF, for both Li-ion battery packs, and the battery's current alteration reveals a major difference. For a 500 V 36 Ah Li-ion battery pack, the current alteration is in the range135A~105A, which corresponds to the battery's capacity alteration range 3.75C~2.92C. For a 300V 60 Ah Li-ion battery pack, the current alteration is in the range 225A~175A, which corresponds to the battery's capacity alteration range 3.75C~2.92C. Though the current alteration range 225A~225A for a 300V 60 Ah Li - ion battery pack is about two times of the range 136A~132A for a 500V 36Ah Li-ion battery pack, and it is the same current density alteration for both packs.

The range of battery voltage alteration for a 500V 36 Ah Li-ion battery pack is from 534V~571V, changing around 550V (EMF) from –16V to 21V. For a 300V 60Ah Li-ion battery pack, the range is from 323V~345V, changing around 332V (EMF) from –9V to 13V. The internal resistance of a 500V 36Ah Li-ion battery is 0.158Ω (k=0. 55), which is about 3 times 0.058Ω, the internal resistance of a 300V 60Ah Li-ion battery pack. The SOC alteration for both situations is exactly the same.

2. SIMULATION STUDIES OF THE HYBRID ELECTRIC POWER TRAIN BASED ON THE SHUTTLE SERVICE BUS

2.1. Power Train and Control Strategy Depiction

In order to obtain the possibility of completely automatic control, the power train (Figures from 28 to 30) and its corresponding control strategy have been designed by simulation as follows:

Table 3. The comparison of the main parameters of a CHPTD (Parameters of vehicle, IC engine, motor and driving cycle are the same, as in Tables 1 and 3 in Chapter 8)

Battery	500V 36Ah Li-ion battery pack	300V 60Ah Li-ion –ion battery pack
Nominal energy of battery pack	18 kWh	18 kWh
Reducer ratio between PM motor and planetary gear	4.8	4.8
Nominal voltage of battery pack	500V	300 V
Nominal battery capacity	36Ah	60Ah
Fuel consumption	24.96 l/100km	24.96 l/100km
Engine control strategy parameters	(-100 0 7 8 30) (0 0 0 172 410)	(-100 0 7 8 30) (0 0 0 172 410)

Basic Simulation Study during the Process of Designing the Hybrid Power Train

Figure 20. The internal combustion diesel engine's operating points' alteration, applied in a compact hybrid planetary transmission power train, CHPTD, equipped with a 500 V Li-ion battery

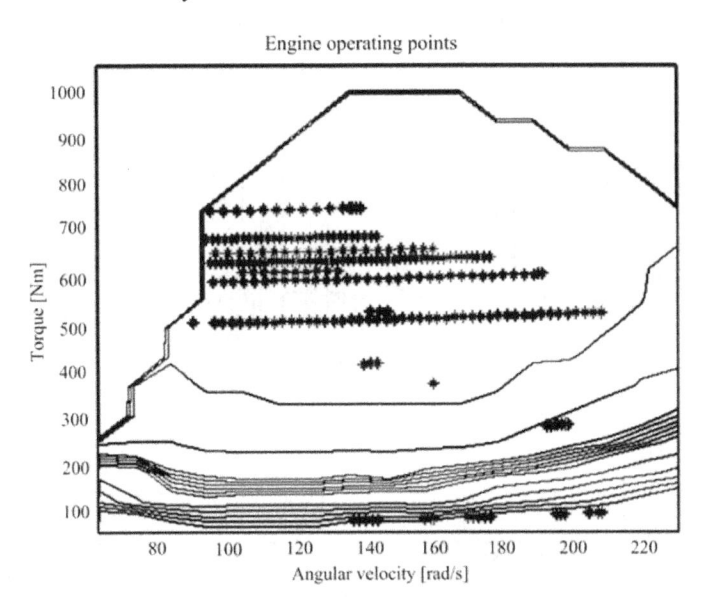

Figure 21. The internal combustion diesel engine's operating points' alteration, applied in a compact hybrid planetary transmission power train, CHPTD, equipped with a 300 V Li-ion battery

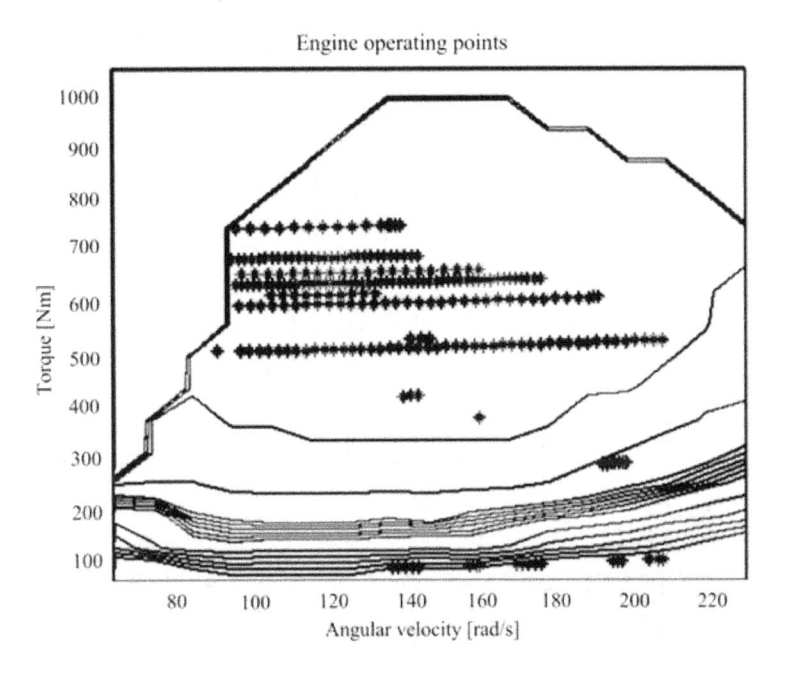

Figure 22. The current alteration of battery applied in a compact hybrid planetary transmission power train, CHPTD, equipped with a 500 V Li - ion battery

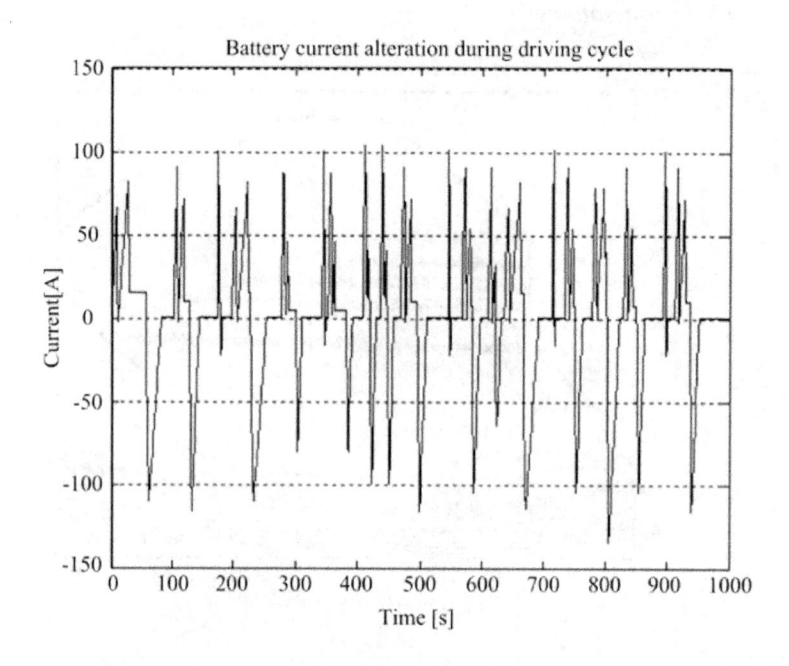

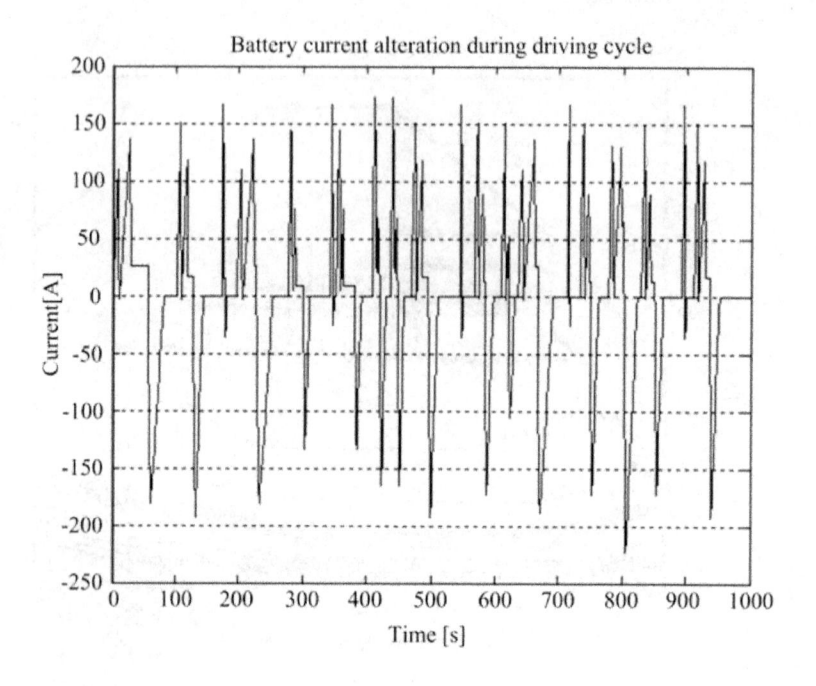

Figure 23. The current alteration of a battery applied in a compact hybrid planetary transmission power train, CHPTD, equipped with a 300 V Li-ion battery

Basic Simulation Study during the Process of Designing the Hybrid Power Train

Figure 24. The voltage alteration of a battery applied in a compact hybrid planetary transmission power train, CHPTD, equipped with a 500 V Li-ion battery

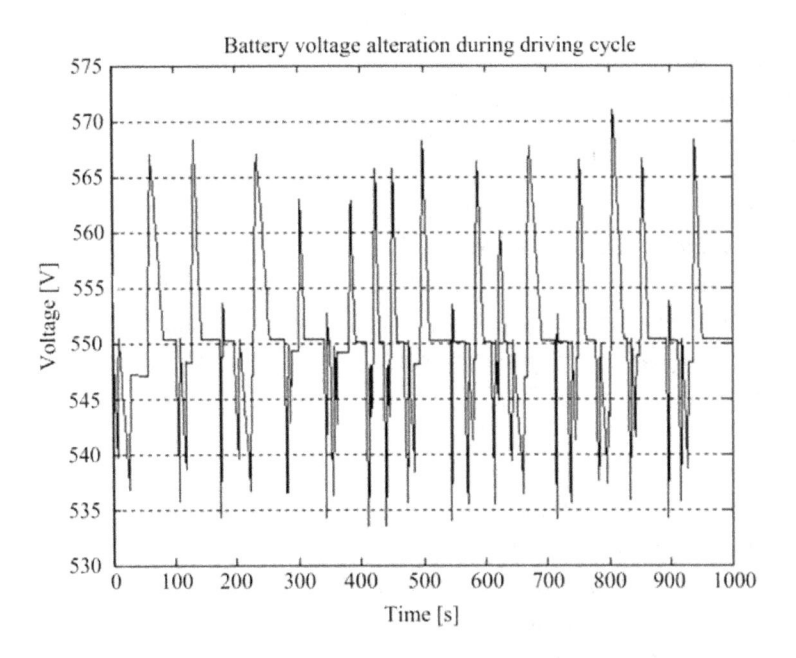

Figure 25. The voltage alteration of a battery applied in a compact hybrid planetary transmission power train, CHPTD, equipped with a 300V Li-ion battery

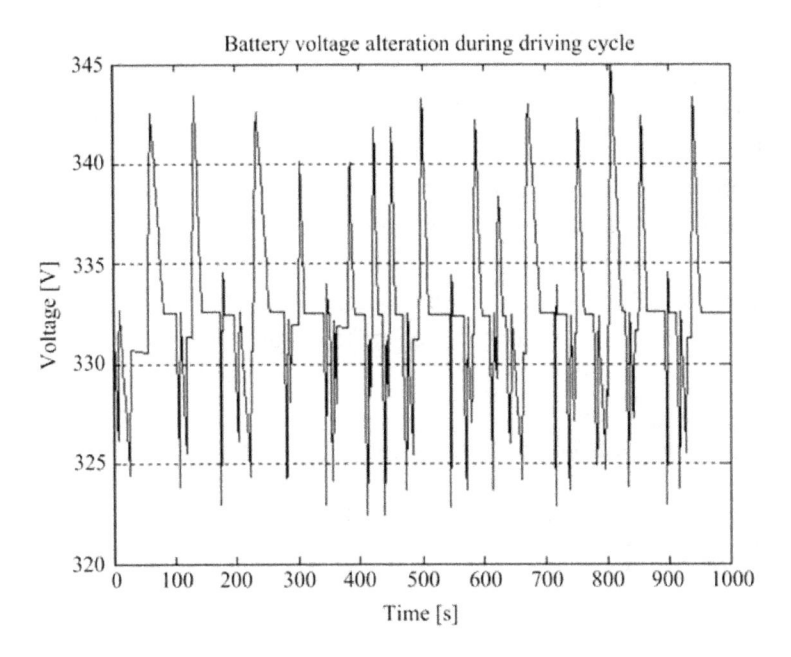

The power train is shown in Figure 31. This configuration looks more complicated than configuration 1, as shown in Figure 28, but it is easy to obtain completely automatic control of this hybrid power train system. Furthermore, if the technology of clutch-brake and planetary gear are ready for the market, the cost of configuration will be no higher than configuration 1. In order to decrease the cost of the system, the planetary reducer RP1 and RP2 are adjusted in the same parameters. The European NEDC driving cycle is used for general simulation and the top speed is maintained at 120km/h without limitation. There are also some special driving cycles used to verify the vehicle's other performance factors, such as acceleration and gradeabilty.

Table 4 shows the minibus parameters used in simulation and some simulation results such as fuel consumption, all mechanical ratios, and the parameters of the battery pack which were properly adjusted to meet the vehicle's required performance during simulation (Chang, 2005; Szumanowski et al., 2005a).

The control strategy used in the simulation is determined as the following:

1. Internal combustion diesel engine operation is still controlled, according to vehicle operating conditions - vehicle speed and vehicle external load (torque). During the vehicle's driving on normal roads, engine speed changes are limited, according to vehicle speed. The control parameters should be adjusted in order to keep the engine operating in the most efficient area. However, the factor which affects engine operating is not only vehicle speed, but also the mechanical ratio between drive wheels and engine, which means the power train's torque distribution. It's necessary to make a compromise between vehicle performance and the internal combustion engine's operating area.
2. Mechanical ratio and vehicle operating mode control- on normal roads, when vehicle speed is lower than 17 km/h, the vehicle operates in pure electric mode; when vehicle speed is 17 ~ 120km/h, the vehicle operates in hybrid mode. When vehicle speed is lower than 18 km/h, the mechanical ratio between planetary transmission (PPG) and the motor is 3.5*3.5, the ratio between differential and planetary transmission is 1. When vehicle speed is 18~120km/h, the mechanical ratio between the planetary transmission (PPG) and the motor is 3.5*1, whilst the ratio between differential and planetary transmission is 0.285.

The description of the planetary reducer operation, presented in Figure 29, is included below:

When Clutch C is connected and Brake B is free,

$$\frac{r_1}{r_3} = 0.7358$$

$$k = 2*\frac{r_3}{r_1} - 1 = 2*\frac{1}{0.7358} - 1 = 1.718$$

The ratio of planetary reducer is 0.7358.
When Clutch C is disconnected and Brake B is braked,

$$\omega_2 = \omega_1 = \omega_3$$

The ratio of the planetary reducer is 1.
The simulation results for a NEDC shuttle service bus driving cycle are shown shown in Figure 32.

2.2. Shuttle Service Bus Maximum Gradeability (30%) Simulation

- **Driving Cycle Depiction:** Vehicle start (with acceleration 0~18km/h in 61s) in 30% gradient.
- **Control Strategy:** When the vehicle's speed is lower than 8km/h, we have pure electric operating mode; when the vehicle's speed is more than 8km/h, we have pure engine operating mode.

2.3. Vehicle Ability of Acceleration (0~60km/h) Simulation (Fuel Consumption: 11.31 l/100km)

2.4. Simulation Result of Vehicle Performance

The power train configuration 2 and the corresponding control strategy are designed, based on the NEDC driving cycle. In order to check if they are designed properly for normal driving conditions, driving cycle 1 and 2 were designed for simulation with the same configuration 2, and the same control strategy.

- Driving cycle No. 1 (fuel consumption 11.07 l/100km) (see Figure 35)

The trend of the battery's state-of-charge factor value, SOC, in Driving Cycle 1 is increasing. If Driving Cycle 1 repeats five times during the simulation, the battery SOC factor value will be over 0.8 (see Figure 34). So, in the control strategy and battery management, BMS, it is necessary to limit the battery's SOC factor and change the vehicle's operating mode; more pure electric drive is necessary.

- Driving Cycle No. 2 (fuel consumption: 11.19 l/100km)

The trend of the battery's state-of-charge factor value, SOC, in Driving Cycle 2 is increasing faster than Driving Cycle 1. The battery is always charged, which is the reason why vehicle fuel consumption is higher.

2.5. The Shuttle Service Bus Simulation Result with Li-Ion Battery Pack Using NEDC

In this simulation, the vehicle parameters, the configuration of hybrid drive systems, and the control strategy are kept the same, as depicted in 1.2 and 1.3. Only a 288 V 27 Ah Ni-MH battery pack is changed into a 300 V 30 Ah Li-ion battery. As there is no change of hybrid drive configuration and control strategy, there is no change of vehicle performance. Only the battery's operation is a little changed. Battery current and voltage is in normal HP battery operating range of a hybrid electric vehicle, HEV. The battery's state-of-charge, SOC, is imbalanced, but there is no big difference compared with Ni - MH. Anyway, a 300 V 30 Ah Li - ion has a slightly bigger voltage and capacity (Ah).

3. ANALYSIS OF VEHICLE PERFORMANCE SENSITIVITY TO MECHANICAL RATIO

3.1. Basic Ratio of Planetary Transmission

The adjusted basic ratio of planetary transmission value is 2.98. If the basic ratio changes in 2.58, the simulation result is shown as in the following figures. During simulation, the control strategy and other parameters remain the same, as depicted in 1.

As we can see from the engine map, the required maximum engine torque is up to 288 Nm, and the max torque of the actual engine is 269 Nm/1900rpm. This means ICE operating points are outside of the map (see Figure 40).

If the basic ratio of planetary transmission changes from 2.98 to 3.38, the engine operating points move down a little, and fuel consumption is 10 l/100km, but the battery's SOC keeps decreasing. The battery is too much discharged.

Basic Simulation Study during the Process of Designing the Hybrid Power Train

3.2. The Reducer Ratio between Motor and Planetary Transmission

If one changes the reducer ratio between motor and planetary transmission and keep other parameters at no change, there will be no influence on the engine's operation, but it will influence on the motor and battery's operation. If the ratio product 3.5*3.5 and 3.5*1 changes into 3.0*3.5 3.0*1, the required motor torque will be up to 249 Nm, but the peak torque of the motor is only 240Nm. The battery state-of-charge factor's value, SOC, could not keep in balance.

Planetary power train parameters (1) obtained by the numerical optimization process, fulfill the requirements of minimizing energy consumption (fuel, electricity – battery SOC) and the 'designed-by-the-author' hybrid power train, as an option of CHPTD (Figure 32), has achieved the following performances:

- Vehicle acceleration time from 0 to 60 km/h 21s $\leq$25s;
- Maximum speed- 120km/h;
- Maximum gradeability- 30%;
- Fuel consumption improvement is 35.3% (for a NEDC driving cycle, with a limited top speed 100km/h, hybrid drive with configuration 2 is 10.3 l/100km; the conventional drive system is 15.93 l/100km). For a hybrid drive system, the fuel consumption means energy consumption (fuel consumption + electricity consumption), because the SOC of the battery keeps in balance at the beginning, and at the end, of the driving cycle. Electricity has been transferred into engine fuel consumption.
- The hybrid drive system can have the following operating modes: pure engine operating, pure electric operating, hybrid operating, and regenerative braking.
- This system makes it easy to realize complete automatic control.
- The configuration 2 and the corresponding control strategy, are also basically suited to the specially-designed Driving Cycle 1 and similarly typical for urban traffic.
- Battery voltage operating range (biggest) is 270 V ~370 V and the current operating range is –240 A ~ 170 A (-8 C ~ 5.67 C) during simulation, which is within the proper operating range of the hybrid electric vehicle HEV battery.

The comparison of fuel consumptions of CHPTD (configuration 2) (10.3l/100km) and the corresponding conventional drives (15.93l/100km) is as shown in Figure 45.

343

Basic Simulation Study during the Process of Designing the Hybrid Power Train

Figure 26. The comparison of internal resistance alteration between the tested 300V and 500V Li-ion battery packs

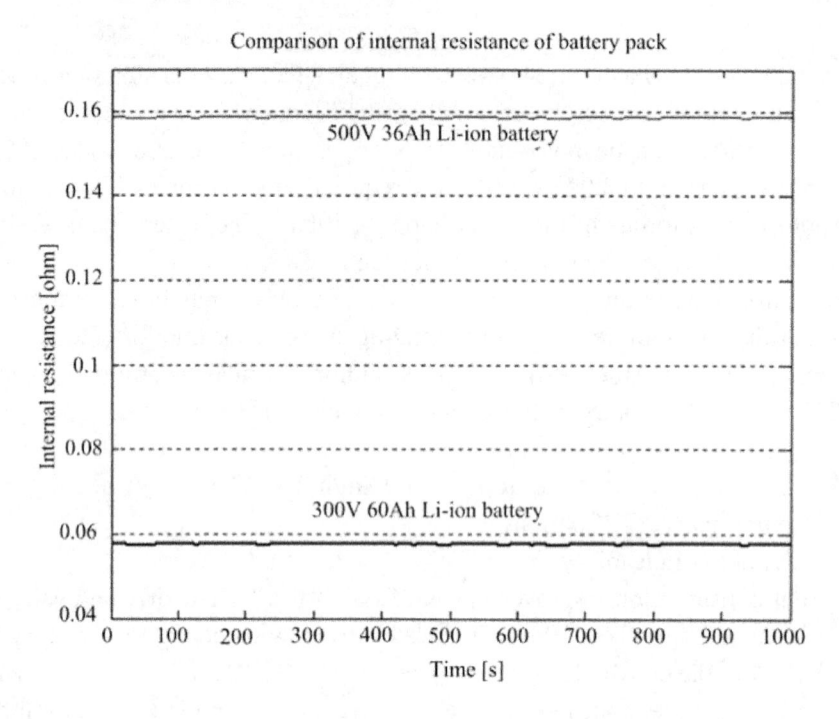

Figure 27. Designed battery's state-of-charge factor's (SOC) alteration for considering cases, in relation to the time piece of analyzing vehicle driving cycles. Note: The battery's state-of-charge factor value in the case of the full hybrid power train has the same value at the beginning, and end, of the vehicle driving cycle for all considered batteries.

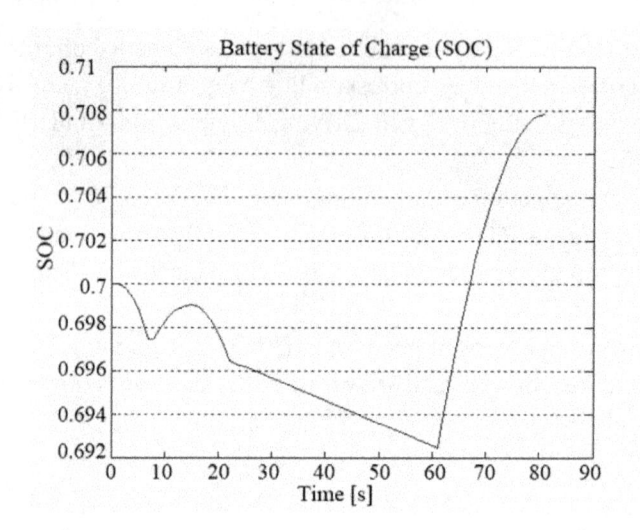

Basic Simulation Study during the Process of Designing the Hybrid Power Train

Figure 28. The configuration number 1 of an analyzing power train with optimal adjusted mechanical ratios

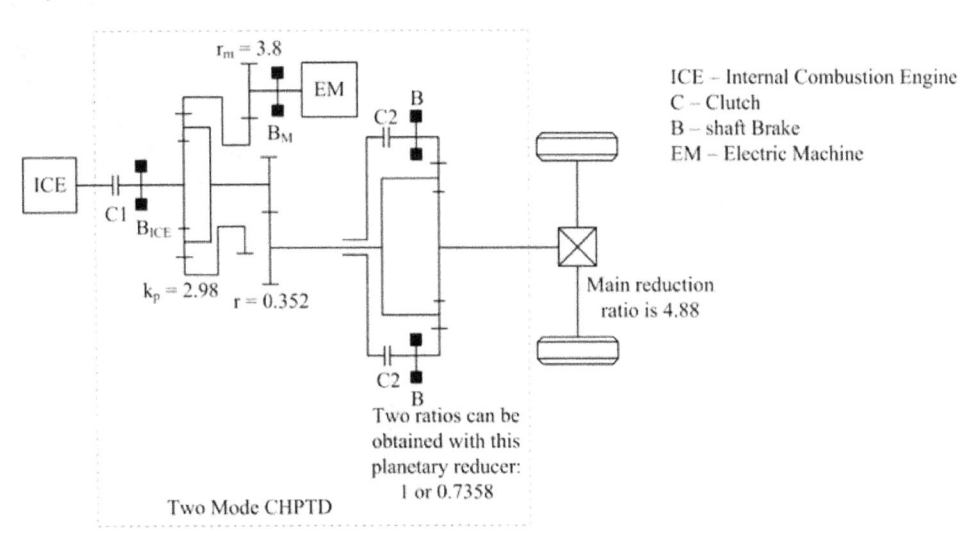

Figure 29. Two mode planetary reducer designed for configuration number 1 (Figure 28)

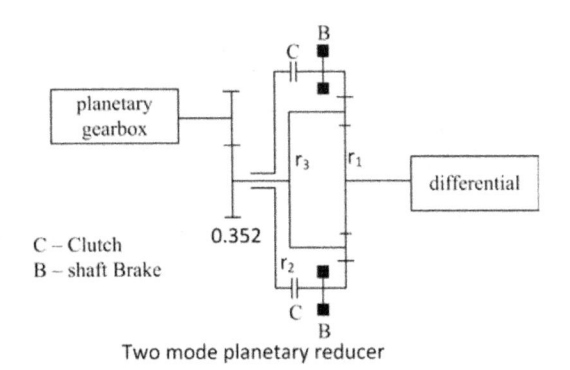

Figure 30. The layout block diagram of a considered power train

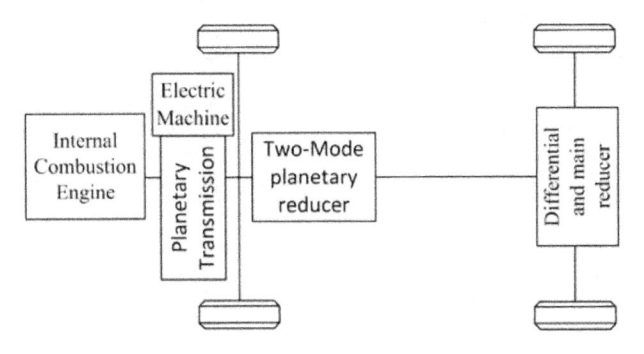

Basic Simulation Study during the Process of Designing the Hybrid Power Train

Figure 31. The configuration number 2 of a hybrid power train obtained by simulation design as the development of drive configuration, shown in Figure 28

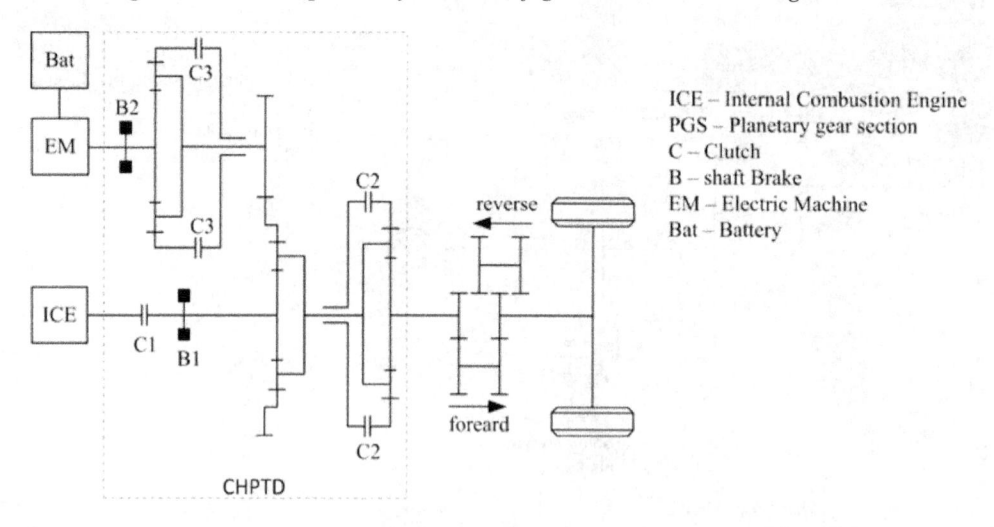

ICE – Internal Combustion Engine
PGS – Planetary gear section
C – Clutch
B – shaft Brake
EM – Electric Machine
Bat – Battery

Table 4. The parameters of analyzing a shuttle service bus and its modified CHPTD power train components, taken under simulation studies; diesel engine fuel consumption is indicated

Battery	288V27Ah Ni-MH battery pack
Main reducer ratio	5.22
The basic planetary transmission ratio	2.98
Reducer ratio between PM motor and planetary transmission	3.5*3.5 (0~18km/h) 3.5*1 (18~120km/h)
Planetary reducer between differential and planetary transmission	1 (For climbing or speed 0~18km/h) 0.285 (For speed 18~120km/h)
Vehicle mass	5,000 kg
Vehicle front area	4.64 m^2
Drag coefficient c_x	0,35
Dynamic tire radius index r_{dyn}	0,348 m
Fuel consumption	10.64 l/100km
ECE driving cycle	120km/h
Engine: Volume: 2.8 l Rotary speed ranges 850~4000 RPM; Maximum output power 87 kW/3600 rpm; Maximum torque 269 Nm/1900 rpm;	PM Motor: 240 Nm peak torque; 75kW peak, 30kW continuous power; Regenerative braking; Full power @ 250~400 VDC input

Basic Simulation Study during the Process of Designing the Hybrid Power Train

Figure 32. Attached simulation results of power train configuration: number 2. (a) Internal combustion engine characteristics: torque vs. speed, with indication of the engine's operating points. (b) Statistic new European driving cycle as a base of simulation. (c) Comparison of rotary speeds of the internal combustion engine, vehicle traction wheel and permanent magnet motor. (d) Internal combustion engine output torque vs. its speed. (e) Permanent, magnet motor output torque vs. its speed. (f) Comparison of vehicle traction wheel, internal combustion engine and permanent, magnet motor, distributed power. (g) Battery voltage alteration. (h) Battery current alteration. Note: The battery's current operating range is –240A~170V (-8C~5.67C), which is in the acceptable operating range of the battery. (i) Battery state-of-charge (SOC) alteration.

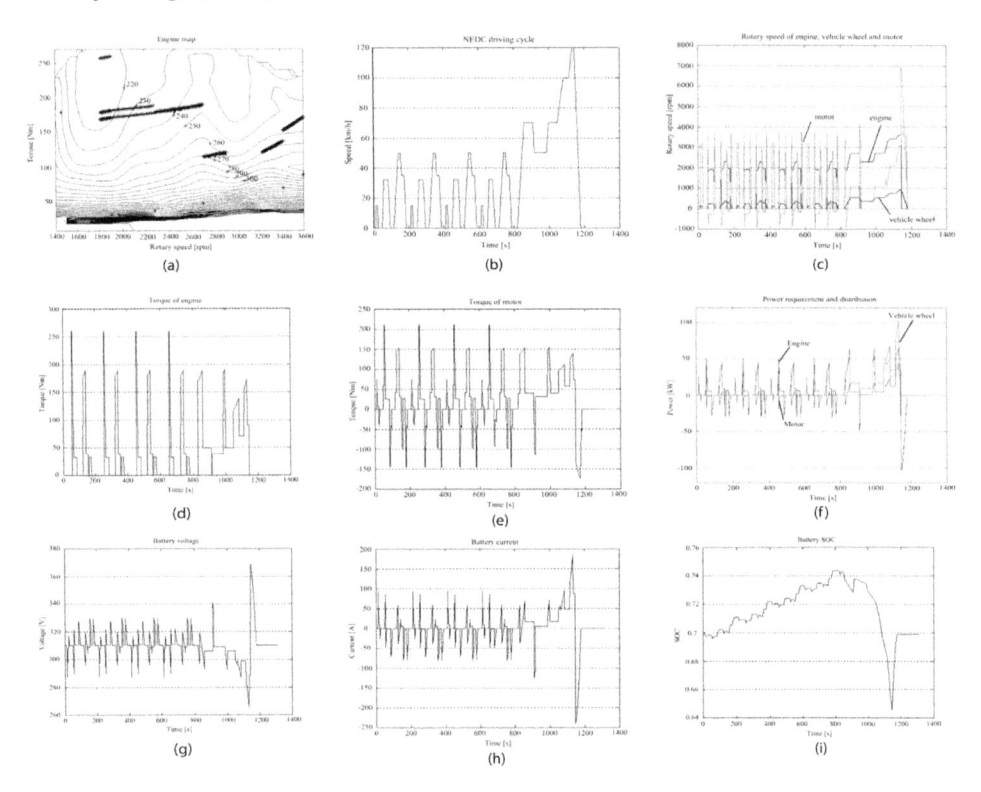

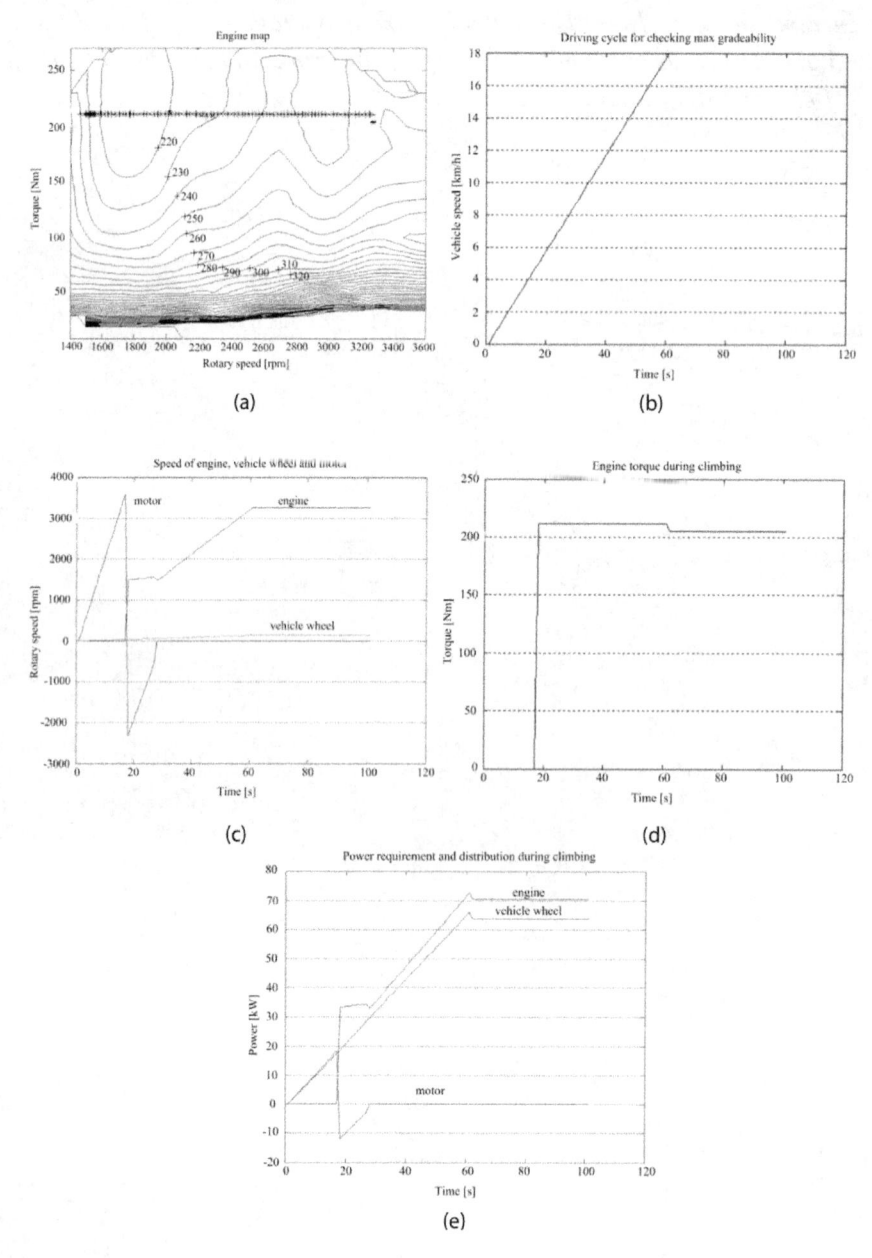

Basic Simulation Study during the Process of Designing the Hybrid Power Train

Figure 33. Attached simulation results during vehicle gradeability (power train configuration number 2) (a) Internal combustion engine characteristics: torque vs. speed, with indication of an engine's operating points.(b) Vehicle speed distribution whilst climbing. (c) Comparison of an internal combustion engine, traction wheels and permanent magnet motor speeds during a vehicle's climbing. (d) Internal combustion engine's output torque changes during a vehicle's climbing. (e) Comparison of an internal combustion engine and traction wheels' required power during a vehicle's climbing.

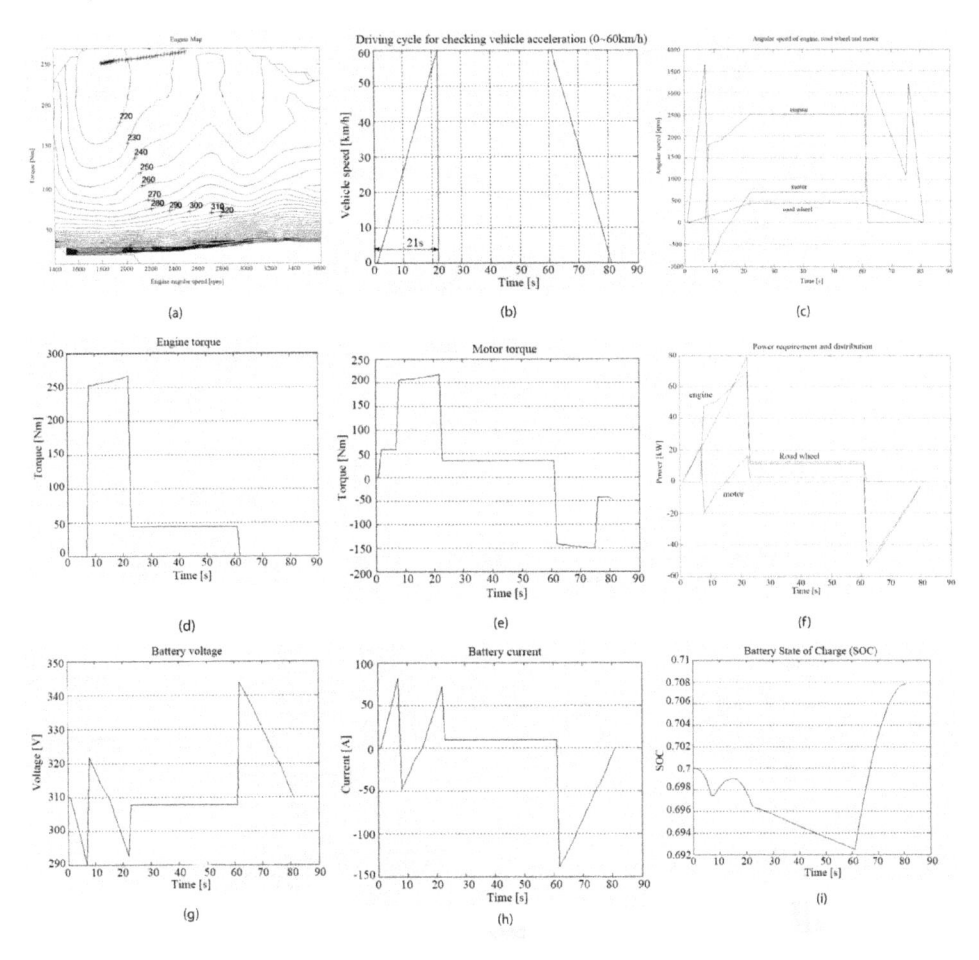

Basic Simulation Study during the Process of Designing the Hybrid Power Train

Figure 34. Attached simulation results during a vehicle's acceleration (power train configuration number 2) (a) Internal combustion engine characteristics: torque vs. speed, with indication of the engine's operating points. (b) Elementary driving cycle considered in the simulation. (c) Comparison of an internal combustion engine, traction wheels, and permanent magnet motor rotary speeds. (d) Internal combustion engine's output torque distribution. (e) Permanent, magnet motor, output torque distribution. (f) Comparison of internal combustion engine, traction wheel and permanent, magnet motor, distributed power. (g) Battery voltage alteration. (h) Battery current alteration. (i) Battery state-of-charge (SOC) alteration.

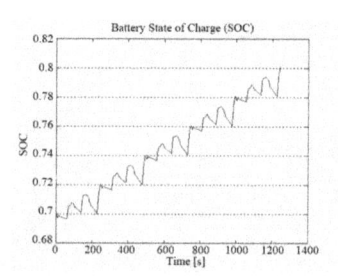

Basic Simulation Study during the Process of Designing the Hybrid Power Train

Figure 35. Attached simulation results of the Driving Cycle 1 (power train configuration 2) (a) Exemplary real driving cycle (No. 1) used for simulation. (b) Internal combustion engine characteristics: torque vs. speed, with indication of an engine's operating points. (c) Comparison of internal combustion engine, traction wheels and permanent, magnet motor, rotary speeds. (d) Internal combustion engine output torque distribution. (e) Permanent, magnet motor, output torque distribution. (f) Comparison of an internal combustion engine, traction wheel and permanent, magnet motor, distributed power. (g) Battery voltage alteration. (h) Battery current alteration. (i) Battery's state-of-charge (SOC) alteration.

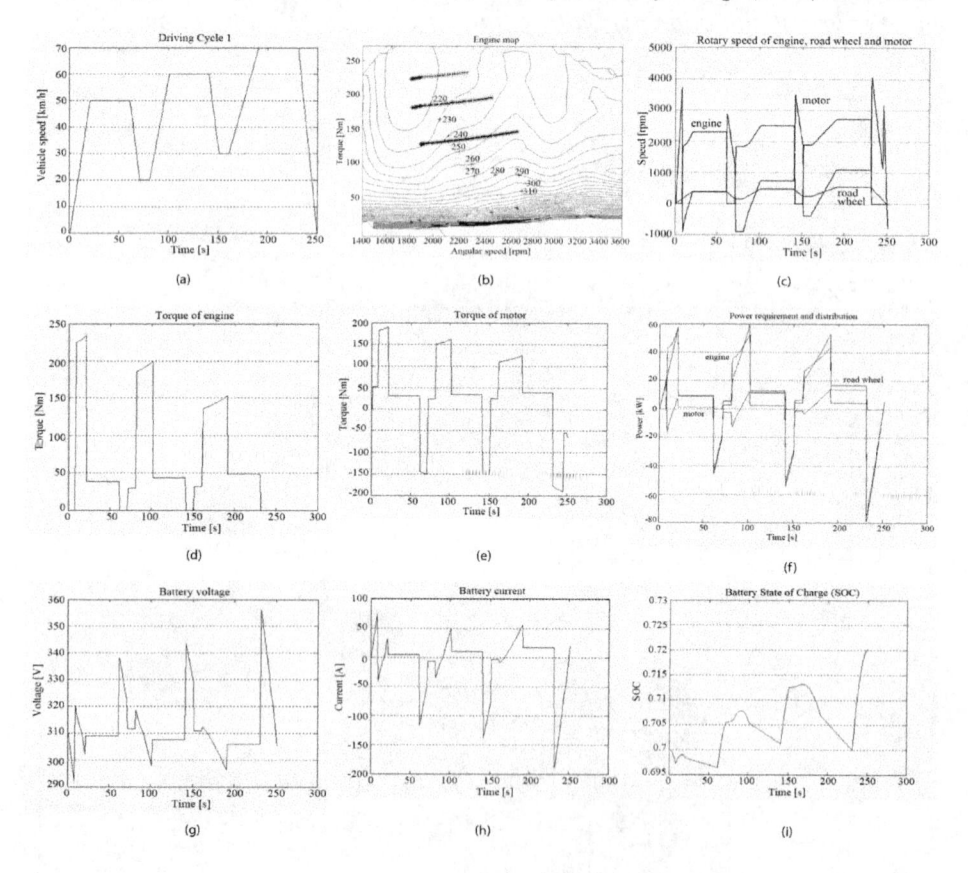

Figure 36. The battery's state-of-charge factor value, SOC, increases after five times of the driving cycle 1's repetition

350

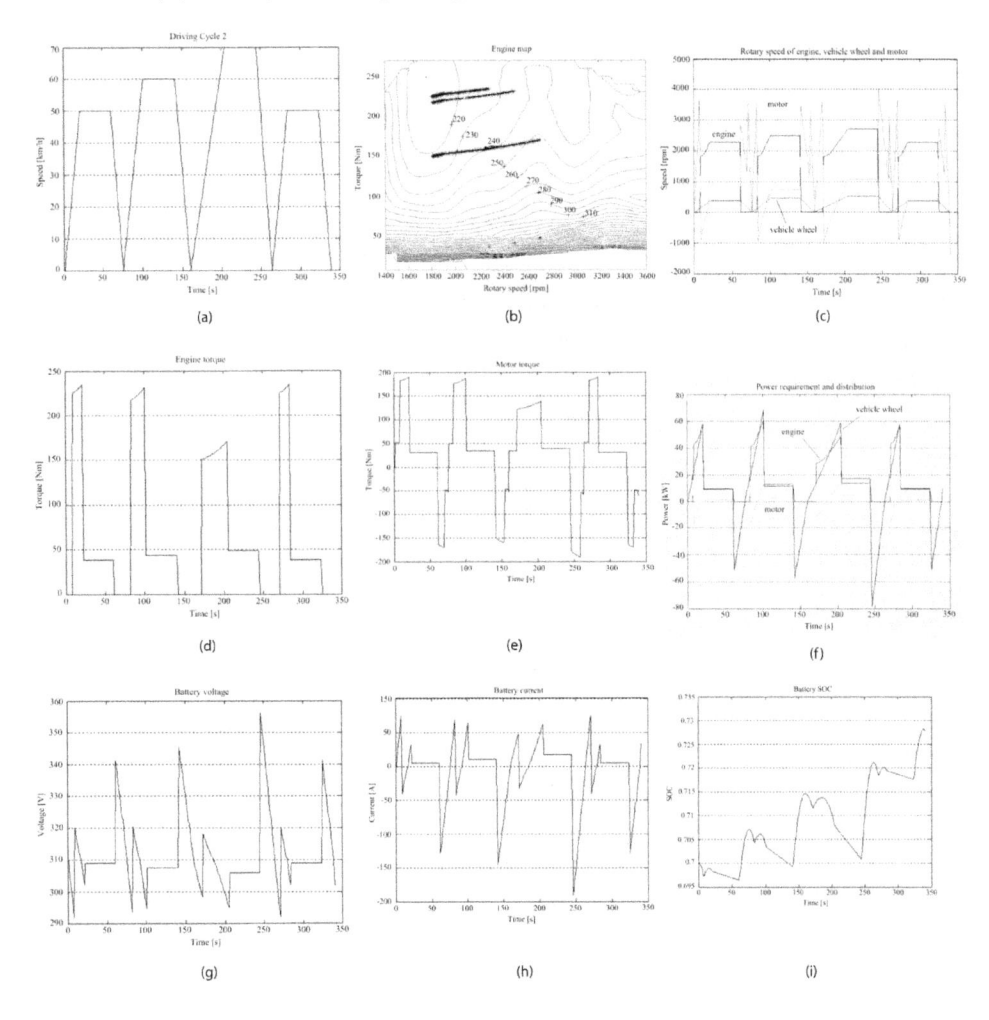

Basic Simulation Study during the Process of Designing the Hybrid Power Train

Figure 37. Attached simulation results of driving cycle 2 (power train configuration 2) (a) Exemplary real driving cycle (No. 2) used for simulation. (b) Internal combustion engine's characteristics: torque vs. speed, with indication of the engine's operating points. (c) Comparison of an internal combustion engine, traction wheels, and permanent, magnet motor, rotary speeds. (d) Internal combustion engine's output torque distribution. (e) Permanent, magnet motor, output torque distribution. (f) Comparison of an internal combustion engine, traction wheel, and permanent, magnet motor's distributed power. (g) Battery voltage alteration. (g) Battery current alteration. (h) Battery's state-of-charge (SOC) alteration.

351

Figure 38. Attached simulation results of the driving cycle 2 of power train configuration number 2, equipped with a Li-ion battery (a) Internal combustion engine characteristics: torque vs. speed, with indication of an engine's operating points. (b) Battery voltage alteration. Battery voltage around 307 V changes from 287 V ~ 333 V. (c) Battery current alteration The battery's current operating range 170 A ~ -226 A, corresponds to 5.67 C~ -8.87 C, which is located in a high-power battery; a HP battery's normal operating range.

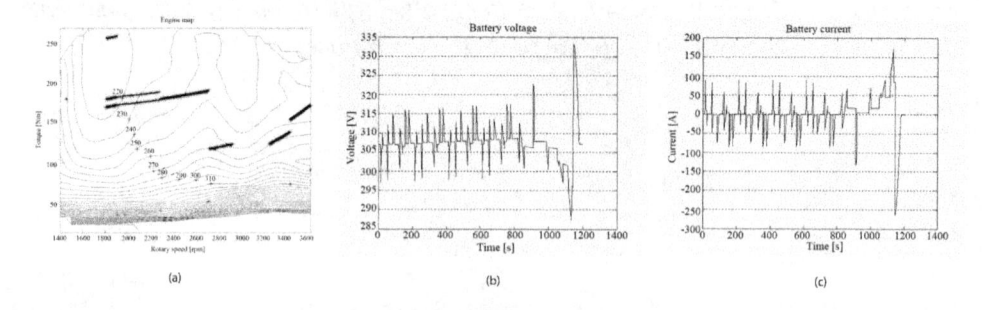

Figure 39. The Li-ion battery's state-of-charge factor, SOC, value alteration versus its analyzed vehicle driving cycle time

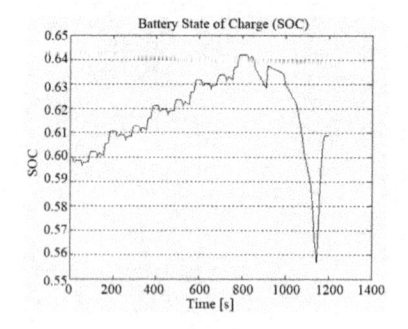

Figure 40. The illustration failure adjusted to the basic planetary transmission ratio (power train configuration number 2) in relation to the internal combustion map

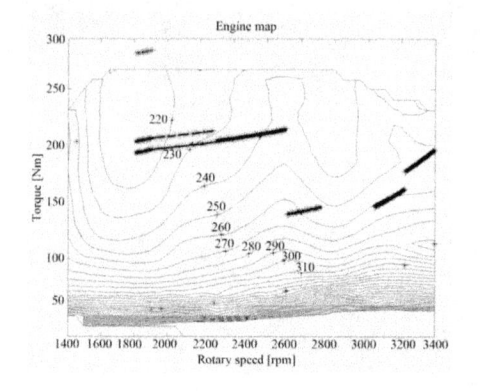

Basic Simulation Study during the Process of Designing the Hybrid Power Train

Figure 41. The internal combustion diesel engine's (ICE) operating points for planetary transmission: basic ratio 3.38 (power train configuration number 2)

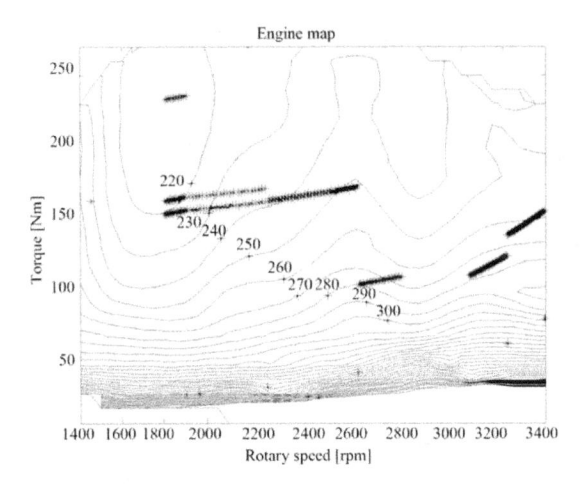

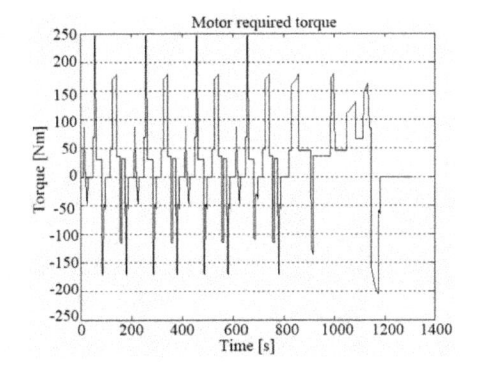

Figure 42. The state-of-charge battery factor's value (SOC) alteration, respectively, to data from Figure 41

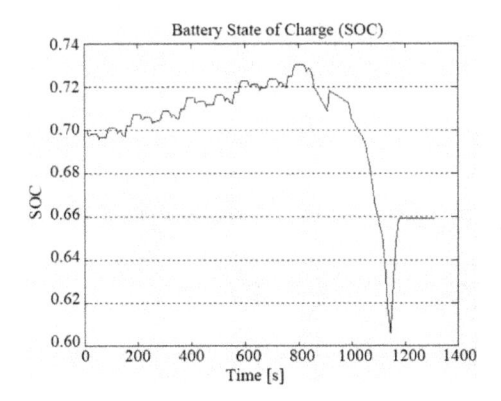

Figure 43. The permanent magnet, (PM) motor torque alteration requirements for new reducer ratio (power train configuration number 2)

Basic Simulation Study during the Process of Designing the Hybrid Power Train

Figure 44. The battery state-of-charge factor's value (SOC) alteration respectively, to data from Figure 43

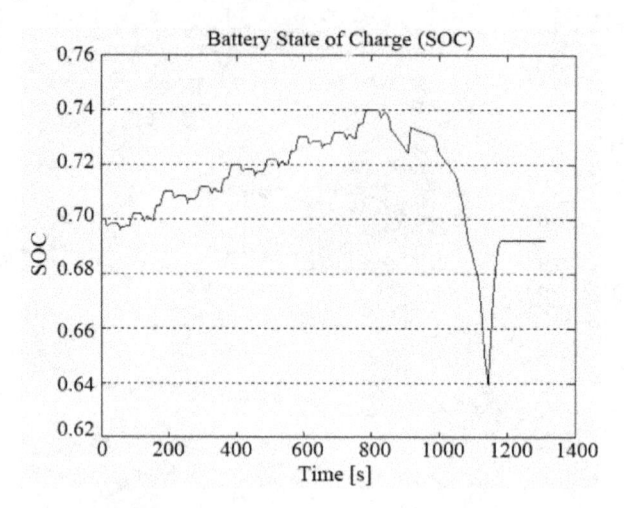

Figure 45. Comparison of the fuel consumption of the internal combustion diesel engine in conventional drive and the internal combustion diesel engine applied in different configurations of analyzing, upper, hybrid power trains (Chang, 2005; Szumanowski, Chang, & Piórkowski, 2005b)

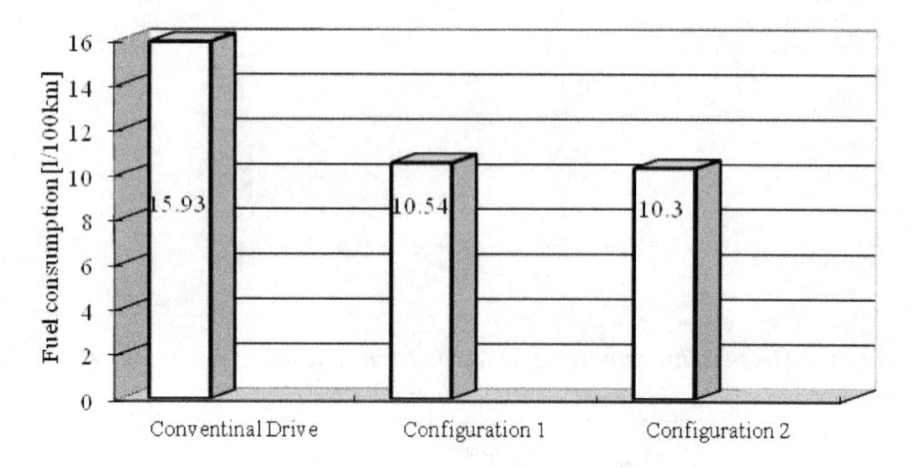

REFERENCES

Bullok, K. J., & Hollis, P. G. (1998). Energy storage elements in hybrid bus applications. In *Proceedings of EVS15*. EVS.

Burke, A. F. (1992). *Development of test procedures for hybrid electric vehicles*. INEL US Depertment of Energy INEL Field Office.

Chang, Y. (2005). *Battery modeling for HEV and battery parameters adjustment for series hybrid bus by simulation*. (MSc thesis). Warsaw University of Technology. Warsaw, Poland.

Gear, C.W. (1971). Simulations: Numerical solution of differential algebraic equations. *IEEE Transactions, 18*.

Gear, C. W. (1972). DIFSUB for the solution of ordinary differential equations. *Communications of the ACM, 14*.

Hajduga, A. (2005). *Electrical-mechanical parameters adjustment of ICE hybrid drive using simulation method*. (PhD thesis). Warsaw University of Technology. Warsaw, Poland.

Haltori, N., Aoyama, S., Kitada, S., Matsuo, I., & Hamai, K. (1998). *Configuration and operation of a newly-developed parallel hybrid propulsion system*. Nissan Motor Co..

Hecke, R., & Plouman, S. (2009). Increase of recuperation energy in hybrid vehicles. In *Proceedings of Aachen Colloquium*. Aachen Colloquium.

Hellenbraich, G., & Rosenburg, V. (2009). FEV's new parallel hybrid transmission with single dry clutch and electric support. In *Proceedings of Aachen Colloquium*. Aachen Colloquium.

Heywood, J. B. (1989). *Internal combustion engine fundamentals*. London: McGraw-Hill Company.

Ippolito, L., & Rovera, G. (1996). *Potential of robotized gearbox to improve fuel economy. Power Train Technologies for a 3-litre-car*. Italy: Siena.

Kelly, K. (2007). Li-ion batteries in EV/HEV application. In *Proceedings of EVS 23*. EVS.

Kitada, S., Aoyama, S., Haltori, N., Maeda, H., & Matsuo, I. (1998). Development of parallel hybrid vehicles system with CVT. In *Proceedings of EVS15*. EVS.

Morano, T. (2007). Pro-post hybrid parallel hybrid drive system. In *Proceedings of EVS 23*. EVS.

Samper, Z. S., & Neiss, K. (2008). Control approach for comfortable powershifting in hybrid transmission. In *Proceedings of Aachen Colloquium*. Aachen Colloquium.

Szumanowski, A. (1994). Simulation study of two and three-source hybrid drives. In *Proceedings of EVS12*. EVS.

Szumanowski, A. (1997). Advanced more efficient compact hybrid drive. In *Proceedings of EVS14*. EVS.

Szumanowski, A. (1999). *Evolution of two steps of freedom planetary transmission in hybrid vehicle application*. Global Powertrain Congress.

Szumanowski, A. (2006). *Hybrid electric vehicle drive design*. Warsaw: ITE Press.

Szumanowski, A., Chang, Y., & Piórkowski, P. (2005a). Analysis of different control strategies and operating modes of a compact hybrid planetary transmission drive. In *Proceedings of IEEE Vehicle Power and Propulsion (VPP) Conference*. IEEE.

Szumanowski, A., Chang, Y., & Piórkowski, P. (2005b). Method of battery adjustment for a hybrid drive by modeling and simulation. In *Proceedings of IEEE Vehicle Power and Propulsion (VPP) Conference*. IEEE.

Szumanowski, A., Hajduga, A., & Piórkowski, P. (1998a). Evaluation of efficiency alterations in hybrid and electric vehicles drives. In *Proceedings of Advanced Propulsion Systems*. GPC.

Szumanowski, A., Hajduga, A., & Piórkowski, P. (1998b). Proper adjustment of combustion engine and induction motor in hybrid vehicles drive. In *Proceedings of EVS15*. EVS.

Szumanowski, A., Hajduga, A., Piórkowski, P., Stefanakos, E., Moore, G., & Buckle, K. (1999). Hybrid drive structure and powertrain analysis for florida shuttle buses. In *Proceedings of EVS 16*. EVS.

Szumanowski, A., & Krasucki, J. (1993). *Simulation study of battery – engine hybrid drive*. Paper presented at the 2nd Polish –Italian Seminar Politechnico di Torino. Turin, Italy.

Szumanowski, A., & Nguyen, V. K. (1999). *Comparison of energetic properties of different two-source hybrid drive architectures*. Global Power Train Congress.

Szumanowski, A., Piórkowski, P., Sun, F., & Chang, Y. (2005). Control strategy choice: Influences on effectiveness of a HEV drive. In *Proceedings of EVS 21*. EVS.

Zhou, J. (2007). Regenerative braking system for series hybrid electric city bus. In *Proceedings of EVS 23*. EVS.

Chapter 10
Plug-In Hybrid Power Train Engineering, Modeling, and Simulation

ABSTRACT

Chapter 10 presents the principles of the plug-in hybrid power train (PHEV) operation. The power trains of the battery-powered vehicle (BEV – pure electric) are close to the plug in hybrid drives. For this reason, the pure electric mode of operation of the plug in hybrid power train is very important. The vehicle's range of driving autonomy must be extended. It means the design process has to be focused on energy economy, emphasizing electricity consumption. Simultaneously, the increasing of the battery's capacity causes its mass and volume also to increase. Generally, it is not recommended. After many tests, one can observe the strong dependence between the proper multiple gear speed, the proper mechanical transmission adjustment, and the vehicle's driving range, which in the case of the plug-in hybrid power train means long distance of a drive using the majority the battery's energy. The mechanical ratio's proper adjustment and its influence on the vehicle's driving range autonomy is discussed in the chapter. Three types of the automatic mechanical transmission are depicted: the toothed gear (ball), the belt's continuously variable transmission, and the planetary transmission system called the "Compact Hybrid Planetary Transmission Drive," equipped additionally with tooth gear reducers, connected or disconnected by the specially constructed electromagnetic clutches. The number of mechanical ratios—gear speeds—depends on the vehicle's size, mass, and function, which in the majority of cases means the maximal speed value.

DOI: 10.4018/978-1-4666-4042-9.ch010

INTRODUCTION

As was explained in Chapter 1, 'plug-in hybrid' (PHEV) power trains, considering their drive architecture, are the same (see and compare Figure 10 and Figure 11 in Chapter 1) as 'full hybrid'(HV). The difference is in control strategy, based on an extended range of the pure battery drive. This causes the state-of-charge, SOC, battery's balance not to be obtained, as in the case of the full hybrid power train. This power train operation takes place when the battery is not so deeply discharged. In this case, the hybrid electric vehicle drives in pure electric mode. If the SOC decreases below the assumed value, the PHEV power train operates with advantage hybrid or pure Internal Combustion Engine (ICE) mode, which means the power train operates as a full hybrid system. This enables us to stabilize the battery's SOC on a low, but approximately constant level, and continuing further driving similar to the full hybrid power train's HV case. Of course, the capacity and accumulated energy of batteries are higher. The PHEV power train is the next step in hybrid technology development.

Figure 1 shows the difference between full hybrid HV and plug-in hybrid PHEV power train operation (Duvall, 2005; Mitsutani, Yamamoto, & Takaoka, 2010).

The plug-in hybrid power train, PHEV, operation is possible to split in three zones:

- In the first, (1) the externally supplied electrical energy recharges the battery
- In the second, (2) the PHEV operates as a battery-powered electric vehicle EV and energy is supplied from the battery as a first priority. Anyway, conventional full hybrid HV power train operation is realized, when rapid acceleration or passing some steep terrain is required. The PHEV operates like a HV, which has a long driving range and lower fuel consumption. For this

Figure 1. The comparison of a full hybrid HV power train (a) and a plug-in hybrid PHEV power train (b), as it's the battery's state-of-charge (SOC) alteration versus time of vehicle driving cycle

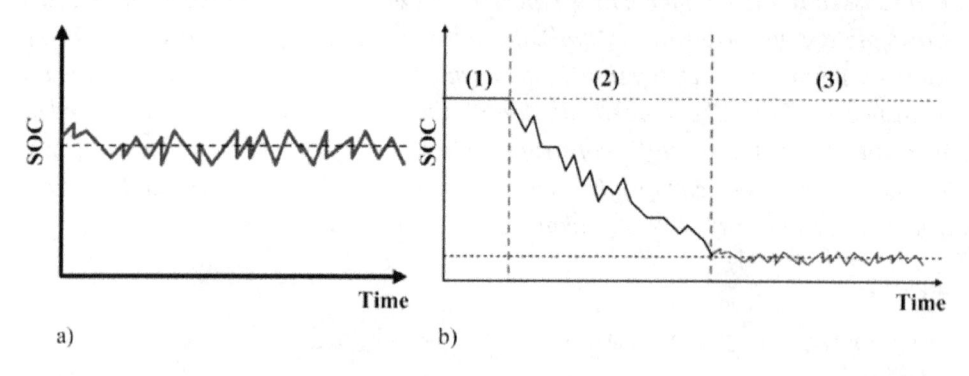

reason, the effective design of the PHEV power train can be based on HV construction, where the advantages of EV and HV can be combined by only expanding the capacity of the battery, and adding a battery charger connected to the electric grid, by socket and plug-in. Certainly, the special battery's SOC and entire power train control strategy is necessary.
- In the third, (3) the PHEV power train is controlled as a full hybrid (HV), as was mentioned previously, keeping the battery's SOC balance value on a properly-adjusted level, lower than in zone (2).

Despite depicting before, that the PHEV's operation is possible, using pure electric drive (as was mentioned), is based on using only battery power, as long as externally charged energy is remaining. When the battery is discharged enough, the internal combustion engine starts to charge the battery or/and propels the traction wheels. This kind of PHEV power train control is not as effective as the previous one. It can be accepted, when a vehicle drives mainly in downtown restricted areas. In this case, generally, a 20-kilometers distance pure electric drive is required as a maximal condition of proper value, stored, electro-chemical energy adjustment, which is directly connected with battery volume and mass.

1. PURE BATTERY MODE POWER TRAIN OPERATION

The pure electric drive is a significant phase of a plug-in hybrid power train PHEV action. In this case, the vehicle can be treated as a classic EV. Some small differences are connected with relatively low battery energy capacity and control, when the battery's SOC value alteration causes switch on/off hybrid or internal combustion engine (ICE) operation modes.

Considering the power distribution efficiency, it is easy to note that detailed analysis of a power train's losses is only one method of measuring a power train's effectiveness and efficiency improvement. The motor, its inverter and mechanical transmission create losses whose value is possible to decrease by proper adjustment of mechanical gear ratios and motor - inverter operation limitation to its best available efficiency area. This is clear; too big losses generated by the motor-transmission unit, consume additional energy from the battery, decreasing, ultimately, the EV's driving range.

The exemplary calculated results of the electric motor and entire power train's internal losses analysis are shown in Figures 2 and 3.

The power train equipped with multi-speed transmission is more efficient, especially when vehicle driving characterizes frequent accelerations and regenerative braking.

Figure 2. The torque (M) alteration curve of a controlled by pulse-width-modulation (PWM) method, permanent magnet motor (PM) as indicated on its map, and versus vehicle driving speed: a) without mechanical transmission; b) with a 4-speed gearbox

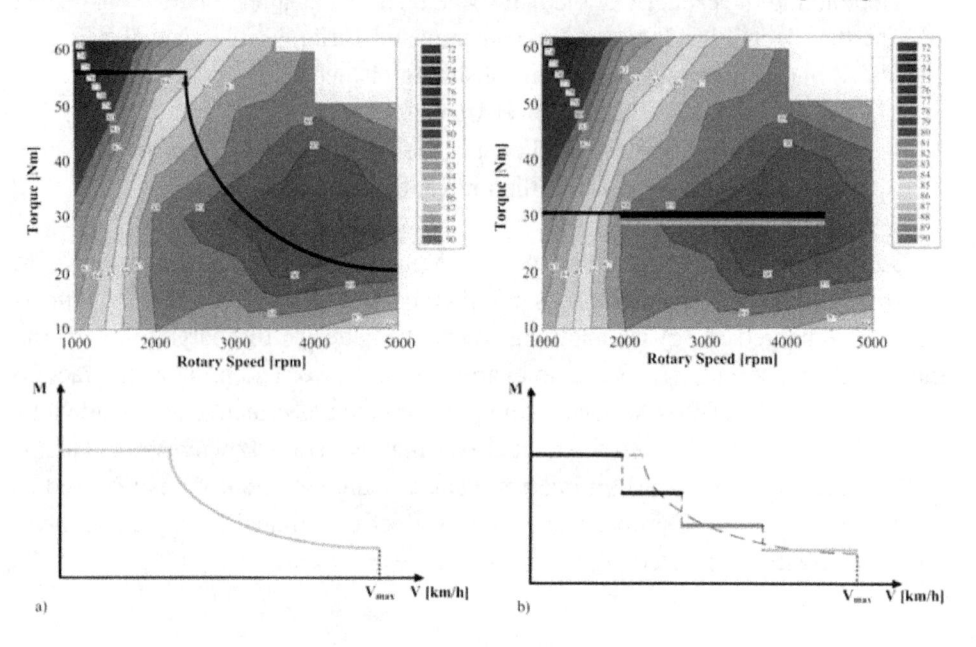

Figure 3. The entire power train's efficiency alteration comparison in the cases: motor without gearbox, and with a four-speed gearbox (see Figure 2)

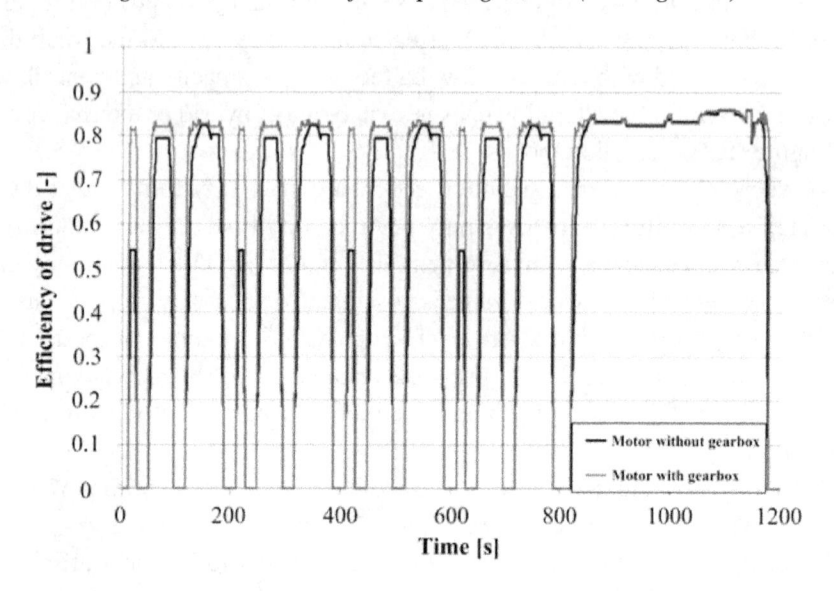

Figure 4. The electric battery-powered vehicle, EV, driving difference range comparison for the vehicle's different driving conditions, in cases when the power train is equipped with only constant ratio gear, and when it is equipped with three-speed automatic transmission. Note: Analysis was done for a downtown driving cycle with maximal speed 50km/h.

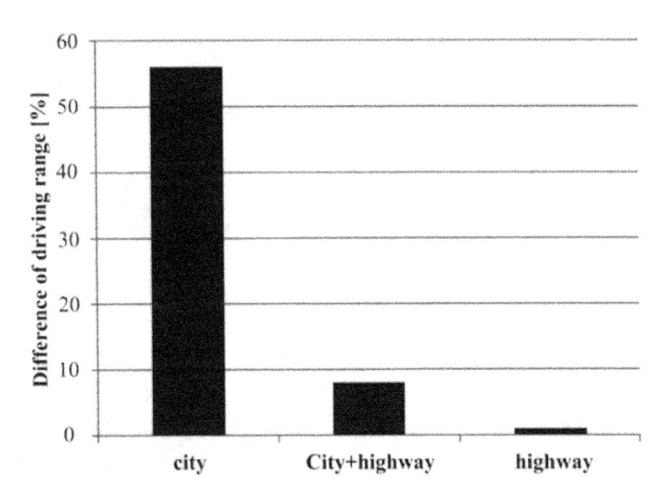

If a vehicle drives with a high-speed longer time travelling range, the difference is only about 10 percent in comparison, and about 50 percent in the case of dynamic city driving, which is most important for the electric battery-powered vehicle, EV, as well as the plug-in hybrid vehicle, PHEV.

The question is: what transmission is necessary to design characterizing low cost and mass, easy control, synchronized with a motor inverter and battery management system (BMS).

2. MECHANICAL TRANSMISSION CONCEPT PROPOSALS

There are two suggestions for light EV and PHEV: three speed 'ball' and CVT 'belt' transmissions. In both cases, the basic condition of ratio or speed changes is zero voltage on the motor's terminals (zero active motor's torque) during shifting. As the moment of inertia of a traction motor is very low, compared to a whole vehicle using the clutch, it is not necessary for a pure electric drive to refer to acceleration and regenerative braking (which means the gear ratio is changed in both modes; motor - generator). This type of control is also easy applied in a PHEV series power train architecture (see Figure 7 in Chapter 1), where the ICE's generator is used as a 'range extender', and generally, this unit has no mechanical connection to traction wheels. This exemplary solution is shown in Figure 5.

Plug-In Hybrid Power Train Engineering, Modeling, and Simulation

Figure 5. Series plug-in hybrid PHEV power train equipped with 'ball' or 'belt' transmission

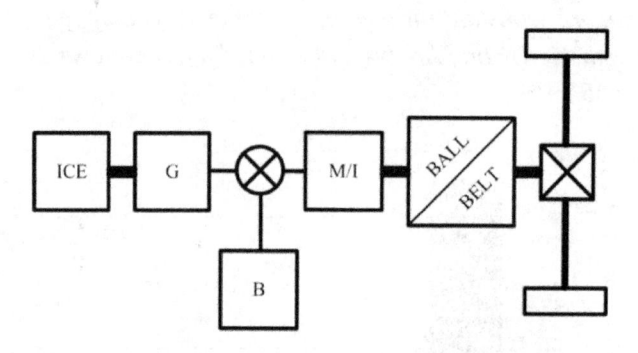

As presented in Figures 6 and 7, the charger-control unit can be used for fulfilling two functions. When the power train operates in hybrid mode, this unit controls the engine– generator co-operating with the battery. During vehicle battery charging, this same device is used as a simple charger. In a detailed way, the internal combustion engine's ICE-generator control unit is used in a hybrid mode power train operation. At this time, the electric energy generated by the ICE-generator unit, flows via a controlled junction point to the battery or to the motor, depending on traction power demands (see Chapter 7). When the battery is transitory, with too much load or discharge, the ICE - generator starts with delivery of additional energy to the power train, properly controlled, and based on feedback signals. The interrupted ICE-generator as a primary energy source operation is proposed (see Chapter 2, Figure 1). This controlled ICE-generator system is a typical 'range extender' and can be applied in an EV, correctly called PHEV, so surely a proper control strategy has to be adopted.

The second function of this unit's operation is a charging battery using external energy (from an electrical grid). This is a function of a simple charger.

Figure 6. The block scheme of charger or ICE's generator control (both functions are separately available) unit applied in a series hybrid drive (relevant to Figure 5)

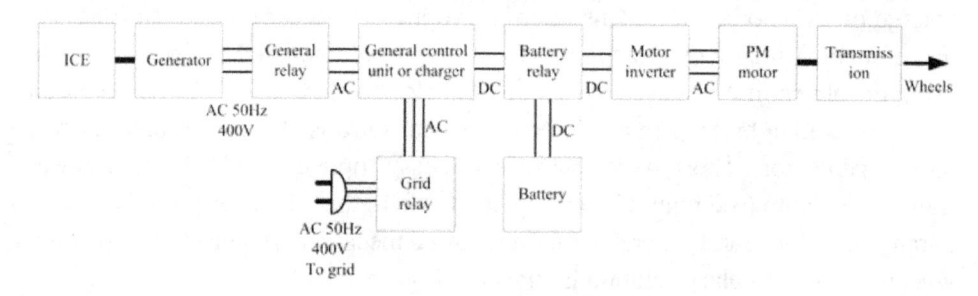

Figure 7. The on-board battery charger or internal combustion engine–generator unit controller for a plug-in hybrid series power train (author's lab)

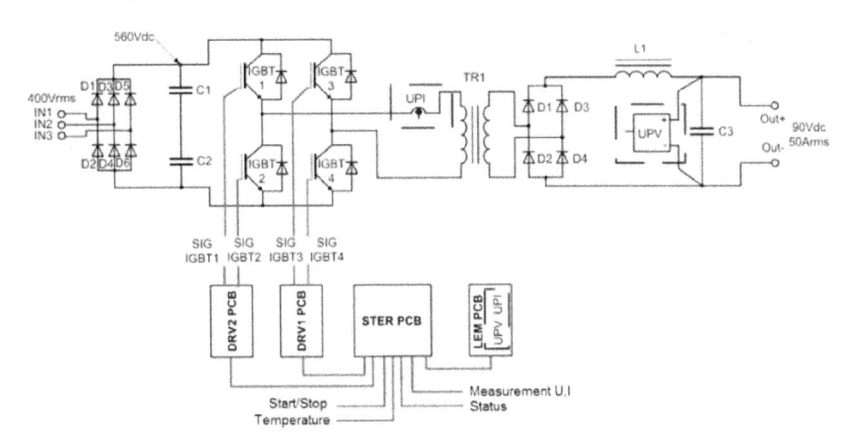

The three-or-two phase grid can be plugged-in. The experimental prototype unit (see Figures 6 and 7) was designed for a three-phase grid connection and three-phase generator output (see Figure 6). A charger control unit frequency was adjusted as 20kHz. The prototype's device mass is 15kg, and its maximal output, direct current, DC power is 4.5kW. The constant speed of the ICE/generator is controlled. The output maximum current value is 50A for a nominal constant voltage of 90V DC. The magnetic flux of the generator is assumed as constant.

The above-presented concept causes decreased costs of the PHEV power train, because the additional charger is not necessary. However, there is one condition that the output parameters of a synchronous generator have to have the same as the grid's parameters, which means input frequency and voltage. Whilst the engine-generator units as grid substitutes are in mass production, their application in a vehicle demands special ICE construction, e.g. two cylinders, liquid cooling, fast easy start, very low noise, etc. It is the last time, that the small Wankel engine which is directly connected with a PM generator, is considered. Another promising construction is the new generation of a Free Piston engine, also directly connected, but with the linear PM generator. Constant speed and torque operation is the best condition for this type of internal combustion engine (ICE) application.

The way to decrease the motor current for this same vehicle's traction wheels power load, is to use the boost converter. In this case, the battery's voltage is smaller than the motor's voltage. The power of the motor is the same when the power train is not equipped with a boost converter, but its current is smaller, because the voltage is higher. Unfortunately, the motor's output torque is also smaller.

For this reason, the limitation of the internal losses of the power train should be based on two of the best solutions; applied to the boost converter and automatic three or four speed mechanical transmission. In the case of EV and HEV, the number of a gear's speeds is significantly lower than for only an ICE power train, because the motor has a sufficient enough torque/speed converter. It means the automatic transmission for an EV and HEV should be designed in a simple way and with much lower costs than e.g. automatic manual transmission AMT power shift contemporary mass, as applied in conventional vehicles. Further exemplary solutions of a boost converter and the above-mentioned transmissions will be discussed.

The typical construction of the boost converter, operating as a step-up mode for vehicle acceleration and step-down mode during regenerative braking when the battery is charged, is shown in Figure 8.

2.1. The Concept of the 'Ball' Transmission

The idea of this automatic gearbox is based on simple ball mechanism.

It is applied in a power train, enabling us to obtain the most efficient motor operation, as well as during vehicle acceleration, and especially regenerative braking, increasing the kinetic energy recovery. Selection of construction parameters and a control system which allows us to reduce motor currents, leads to reducing copper losses, and to increasing the motor's and its inverter's operation efficiency. It all depends on the proper transmission ratio adjustment of particular gears, as well as an adequate control system, using motor-vehicle angular velocity and torque feedbacks. The target was to design a compact, light, and simple, structured gearbox. It is important because it allows making more room space for passengers and reducing the mass of a car. While the simple structure allows the cutting down of costs of production, simultaneously, the construction permits us, with slight modifications, to use this gearbox in a hybrid drive system, especially in the case of light vehicles.

'Ball gearbox' controlled by a stepper motor (Figure 9) is a variation of a three-speed gearbox. The main difference between this transmission and the normal, manual gearbox is the method of throwing particular gears. Gears on the main shaft are mounted for good, while on the input shaft, they are mounted through ball bearings. The input shaft is a hollow shaft. Inside it, there is a grading shaft which controls throwing of the gears. The gear is thrown when a cam on the control shaft, pushes balls which are in the holes in the input shaft, (balls which are used for throwing gears are different from the one chosen, and are inside the input shaft beyond the contours of the responding holes) in the seats in the space between the bearings, and inside the hub of the chosen gear. To disconnect, the gear control shaft moves until there is a pocket under the ball. Then, balls are pushed out of the seat by a simple pusher installed inside the hub (Figure 10) and moved inside the input shaft beyond the contour of its holes. Moving the control shaft to the left or right, causes the throwing of the following gears. It is controlled by a stepper motor (Figure 10). This way of changing gears, forces the

Figure 8. The exemplary boost converter design with step–down voltage ratio 043 – 0.57 between the motor's and battery's voltage, used during vehicle regenerative braking, a step–up voltage ratio 1.76 – 2.31 between the battery's and motor's voltage during a vehicle's acceleration or steady speed

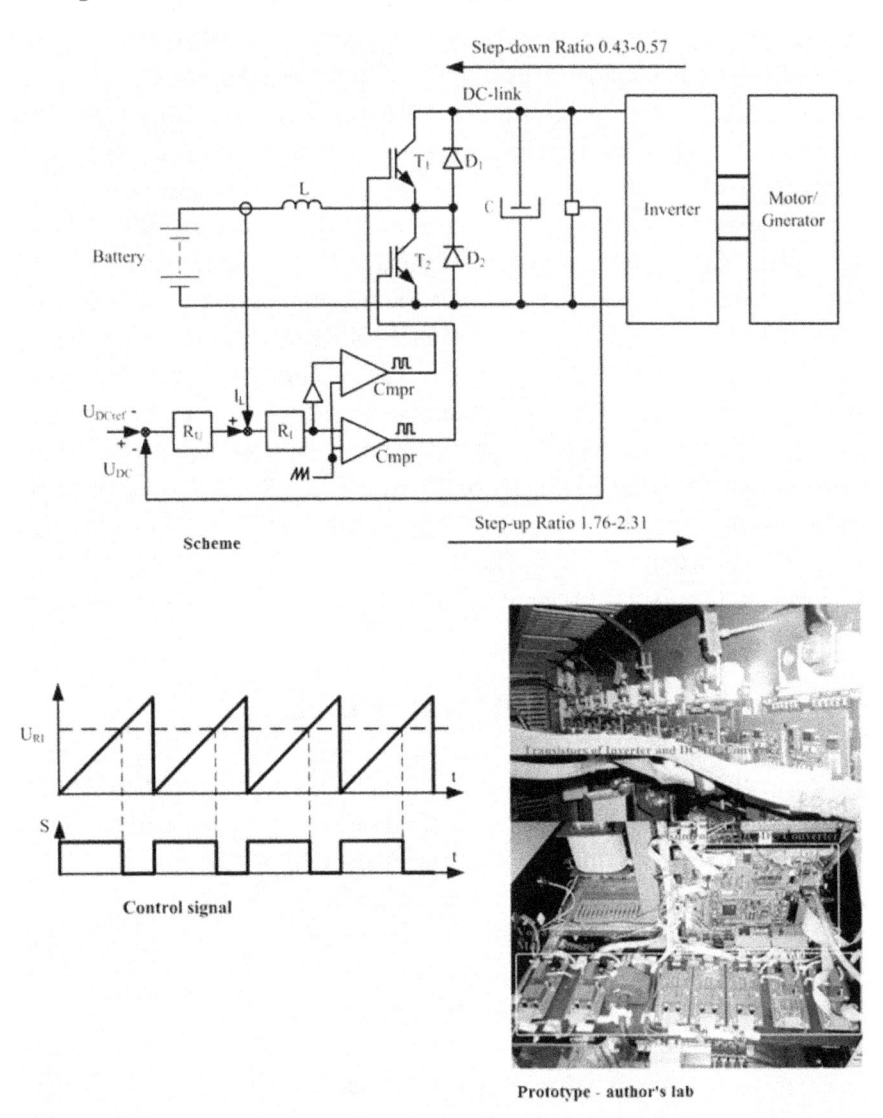

throwing of the gears sequence, which means 1=>2=>3 or 3=>2=>1. There is no possibility of changing gear from 1 to 3, or inversely, without throwing second gear. Simultaneously, the lack of clutch which forces the usage of the electric motor, allows us to make a gearbox without a neutral gear (there is always one connected to one of the gears). Proper control shaft profile, pitch of thread on set screw, rotational speed of the stepper motor and a proper control system allow us to shift gears fast. The time of a gear change won't be longer than in a traditional manual gearbox, so it will transmit the torque almost continuously, both whilst changing gear 'up' on accelerating and while changing it 'down' on braking. Changing gear down on braking, is used to generate greater braking torque and increasing the angular velocity of the motor/generator shaft and to obtain, as a result, greater efficiency of energy recuperation. When the vehicle is braking and finally stopped, the shifting mechanism secures first gear engagement. It permits vehicle start again, because it is impossible to change gear when the car stops and shafts are not turning. There can take place the situation whereby particular parts are set up in a way that balls don't hit the seats, and lock the movement of the control shaft. That's why the design process has to take this situation into consideration, a control system, and make sure that only when the vehicle is moving, is there the possibility of automatically changing the gears. It is necessary to remember that the gearbox can't transmit the torque while the gears are shifting, because the motor's feeding voltage of inverter must be reduced down to zero value. That makes the force which normally acts between the ball and its seat, and between the ball and shaft decrease, permit the ball to push out from its seat, by a special spring and fix in the pocket of the control shaft.

The reverse gear is not necessary, because it will be realized by changing the direction of motor rotation.

2.2. Concept of "Hybrid Belt" Transmission

A hybrid, continuous, variable transmission equipped with a new type of belt is a kind of CVT. This is called a hybrid belt, because power is transferred by a friction (conical wheel) and teeth (toothed wheel) connection.

There are two essential features contained within:

- Possibility of variable gear changes
- Better efficiency by using a standard cogged belt surface

This type of transmission is supposed to be used in very lightweight electric city vehicles, to reduce energy consumption and make energy recuperation more efficient.

The gear change mechanism operates as follows (Figure 11):

Nut (1) driven by a stepping motor is moving two screws in opposite directions. One screw is connected with the conical wheel by axial bearing, and the second one is moving other conical wheels by two couplers and an axial bearing. This mechanism allows for a changing ratio by sliding two conical wheels, working with a V - shaped surface of the hybrid belt.

Figure 9. The basic construction of 'ball' gear: a) general cross section, b) details (Mitsutani et al., 2010; Sekrecki, 2010)

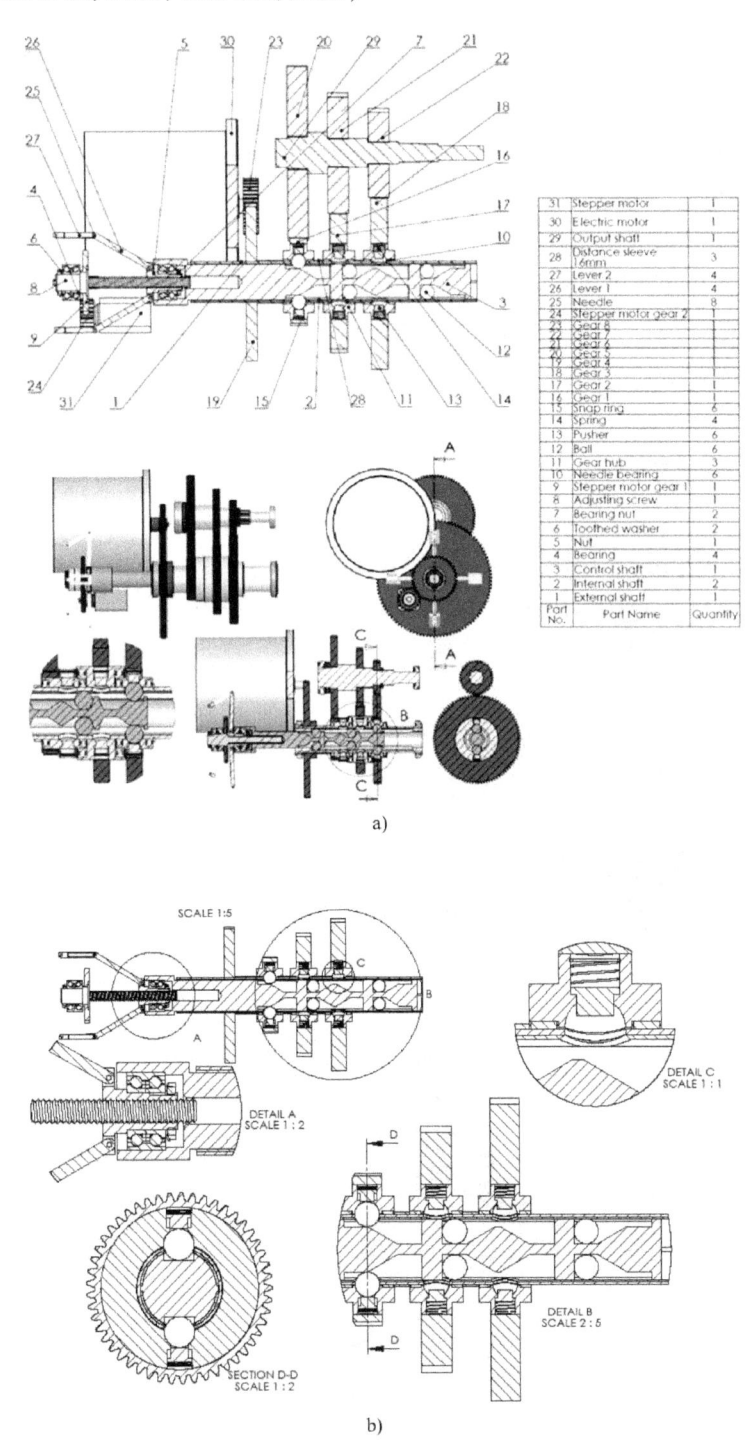

Part No.	Part Name	Quantity
31	Stepper motor	1
30	Electric motor	1
29	Output shaft	1
28	Distance sleeve 16mm	3
27	Lever 2	4
26	Lever 1	4
25	Needle	8
24	Stepper motor gear 2	1
23	Gear 8	1
22	Gear 7	1
21	Gear 6	1
20	Gear 5	1
19	Gear 4	1
18	Gear 3	1
17	Gear 2	1
16	Gear 1	1
15	Snap ring	6
14	Spring	4
13	Pusher	6
12	Ball	6
11	Gear hub	3
10	Needle bearing	6
9	Stepper motor gear 1	1
8	Adjusting screw	1
7	Bearing nut	2
6	Toothed washer	2
5	Nut	1
4	Bearing	4
3	Control shaft	1
2	Internal shaft	2
1	External shaft	1

a)

b)

Plug-In Hybrid Power Train Engineering, Modeling, and Simulation

Figure 10. The illustration of connecting and disconnecting particular gears from the first, up to the third

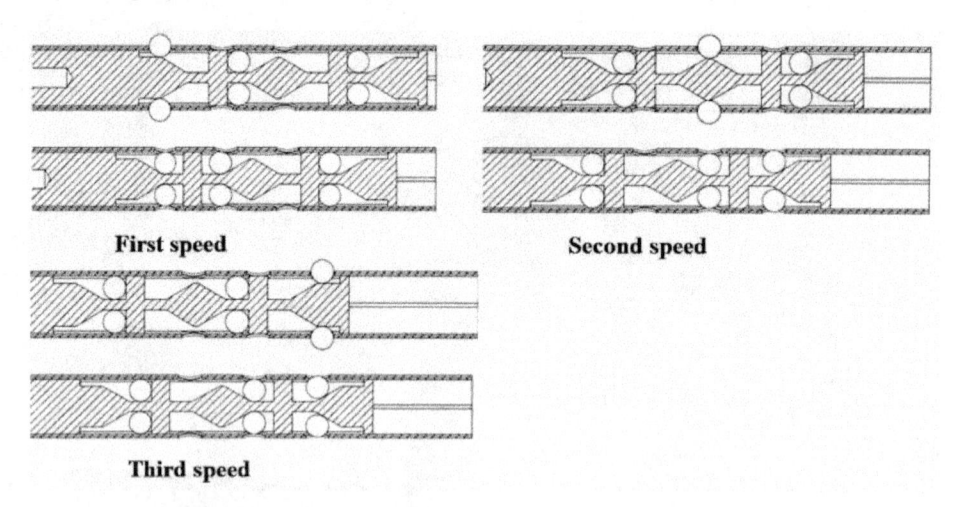

First speed

Second speed

Third speed

Figure 11. Belt transmission's basic mechanisms: 1) nut propelled by stepper motor; 2) input torque (traction motor connection); 3) output torque (traction wheels via differential connection)

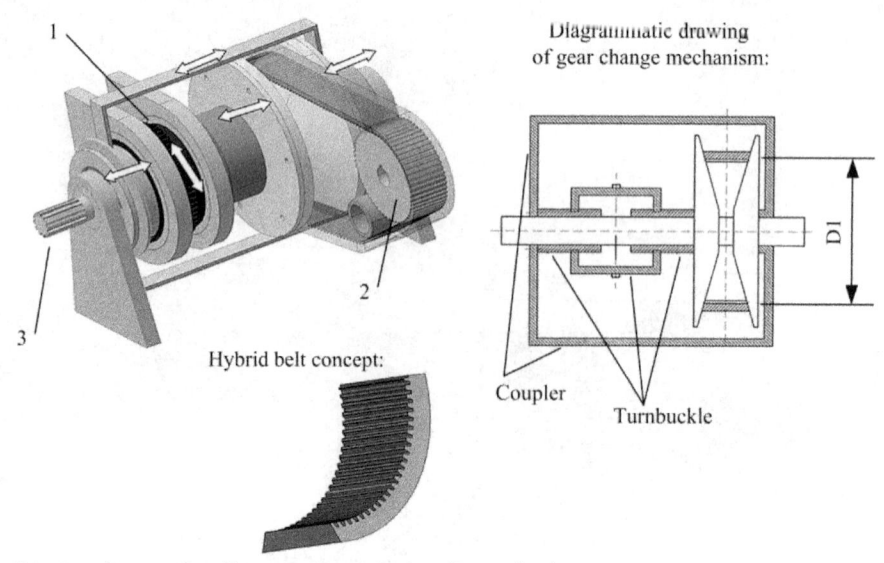

Diagrammatic drawing of gear change mechanism:

Hybrid belt concept:

Coupler

Turnbuckle

D1 - transitory active diameter of a conical variator wheel

Figure 12. The belt transmission's efficiency during the acceleration phase

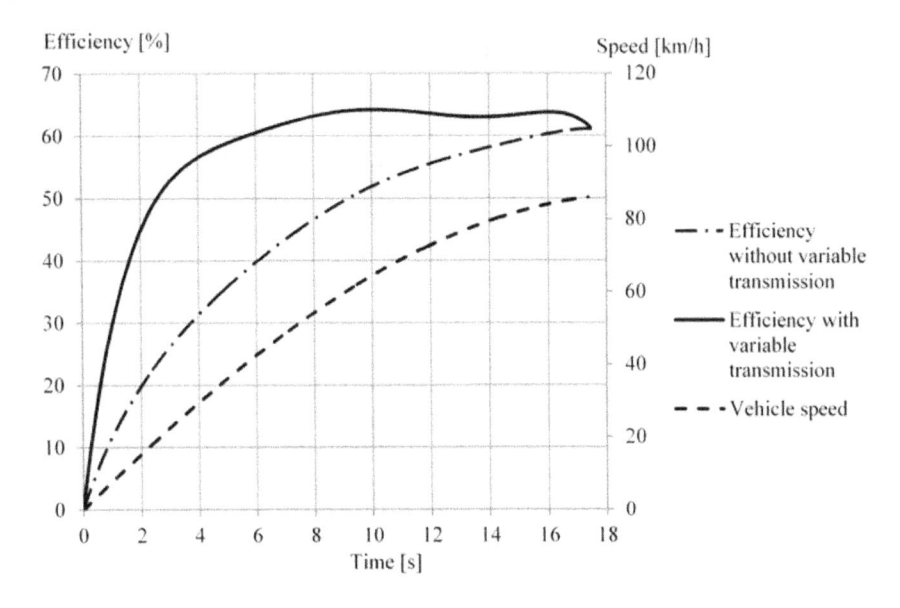

Figure 13. The exemplary characteristics of the belt's transmission:a) transmission ratio versus the belt stretcher wheel's position; b) variator active diameter versus belt stretcher wheel's position (Krawczyk, 2010; 2011)

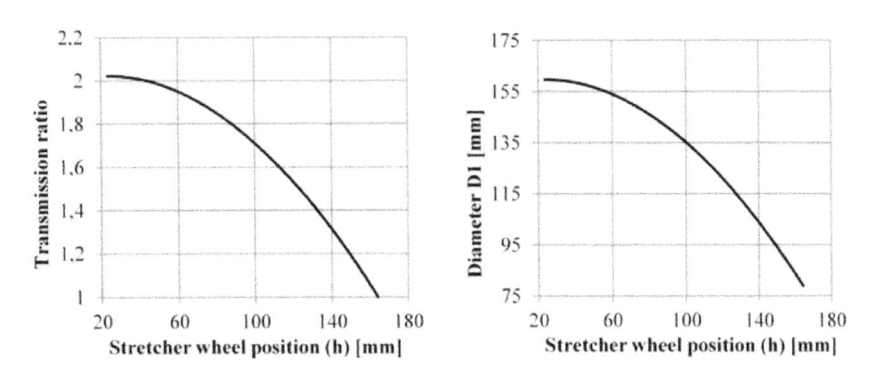

2.3. An Exemplary Simulation Study of a Power Train Equipped with Belt and Automatic Gear Transmission Controlled by Stepper Motors

In the previous section of this chapter, mechanical automatic transmission directly connected with the traction motor was discussed. This assembling is possible only in a series hybrid configuration case. If depicted automatic transmission has to be

Figure 14. The angle of contact on a conical wheel in the function of the belt stretcher wheel's position: a) scheme of transmission; b) angle of belt friction contact versus the change position of the stretcher wheel

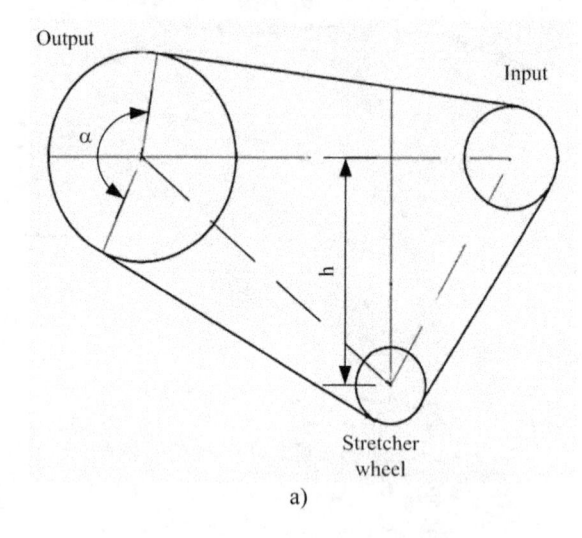

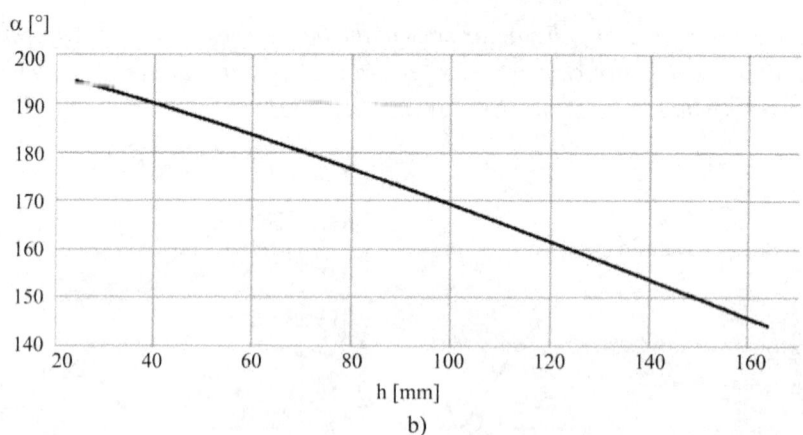

installed in a parallel or planetary hybrid power train, the proper clutch system is necessary. For control reasons, the best solution would be an electromagnetic one. Figure 15 shows the power train's simplest parallel configuration with a system consisting of a clutch/brake, whose operation is electromagnetically controlled. It is important to emphasize the clutch/brake unit should not consume energy during steady state operations. The clutches controlled by the stepper motor are also considered.

Plug-In Hybrid Power Train Engineering, Modeling, and Simulation

Figure 15. The simplest hybrid power train equipped with transmission, automatically controlled by a stepper motor. 1. Battery 2. Motor/generator 3. Internal combustion engine (ICE) 4.The electromagnetic controlled clutch-brake unit 5. Reducer (power summing joint) 6. Automatic transmission belt or ball controlled by stepper motor 7. Differential main gear 8. Traction wheels. Note: Index for number 4: (0) clutch disengaged; (1) clutch engaged. Arrows indicate energy flow.

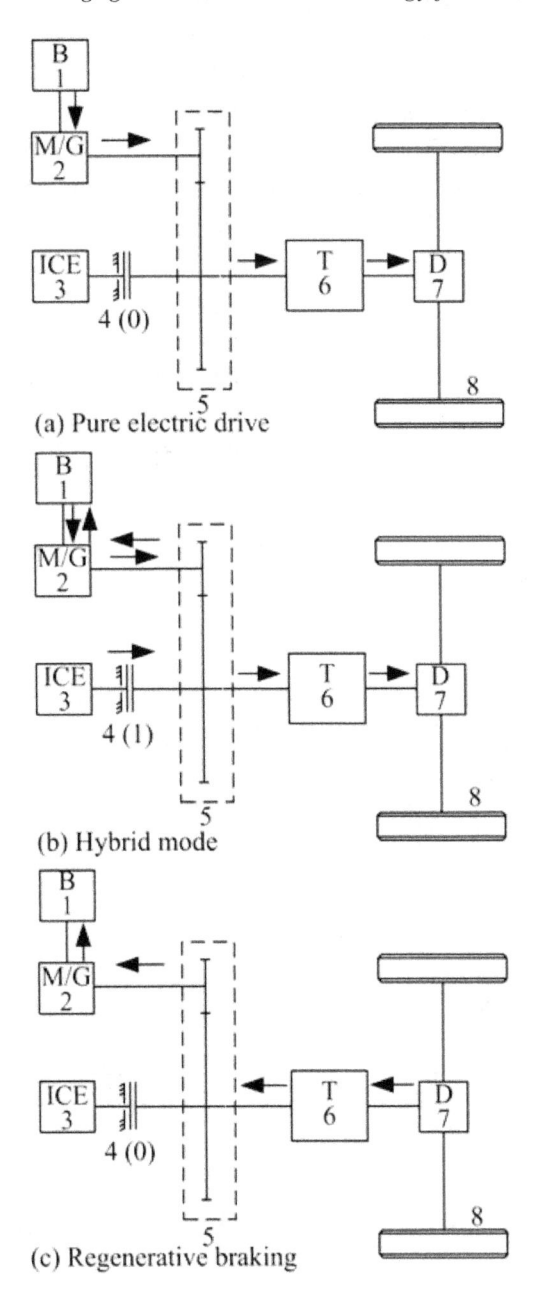

(a) Pure electric drive

(b) Hybrid mode

(c) Regenerative braking

Plug-In Hybrid Power Train Engineering, Modeling, and Simulation

The new concept of the dry, electromagnetic clutches designed as saving energy devices, possibly applied in hybrid electric power trains of passenger cars is depicted further (see Figures 16, 19, 20). This type of clutches can be used in planetary gear transmission, as is shown in Figures 15 and 16 in Chapter 1. The clutches controlled by a stepper motor and electromagnetically, respectively, are applied.

For instance, one of the proposed power train solutions, equipped with gearbox and belt transmission, controlled by a stepper motor is dicussed in detail. (Szumanowski, Chang, Hajduga, & Piorkowski, 2007).

The experimental concept of a hybrid drive is dedicated, only to the next expected generation of urban cars. The mass reduction of vehicles – without decreasing the safety impact – is one of the significant targets of car bodies. The concept is to design a stiff frame made with the composition of light materials. This approach is especially dedicated to electric and hybrid electric vehicles, where decreasing energy consumption has the most important meaning. The ultralight city car, designed as the simplest cage-tube frame construction (see Figure 42), with altitude, width and length, approximately: 1.65m, 1.4m and 3.1m was taken under consideration. The weight of this frame is only about 75kg, and total vehicle mass, including nominal payload (passengers and fully loaded trunk), is 980kg. The battery voltage is 72V. The

Figure 16. The scheme of analyzing power trains equipped with automatic clutch and belt continuous, variable transmission (CVT) controlled by a stepper motor

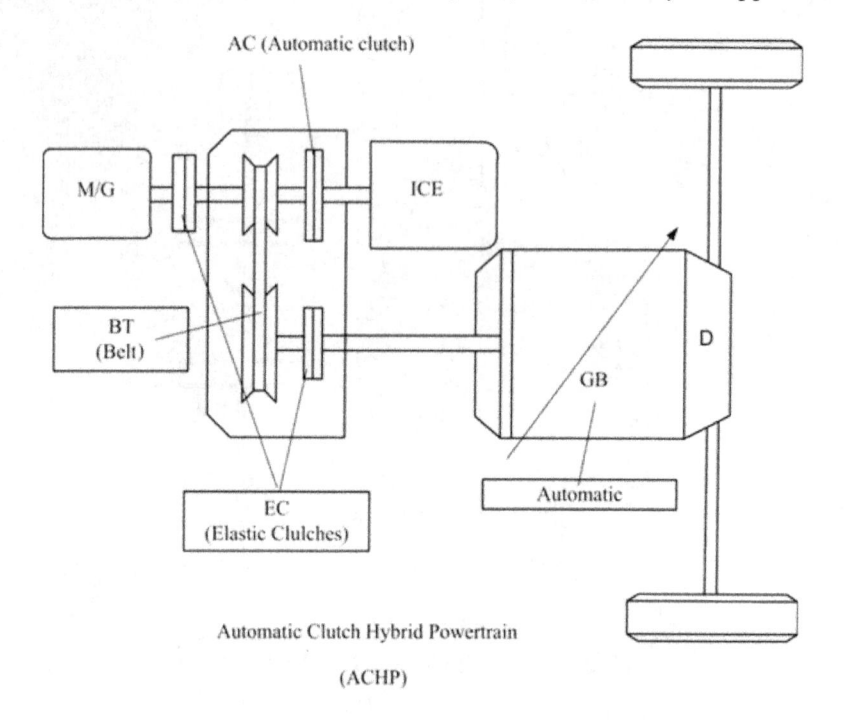

Automatic Clutch Hybrid Powertrain

(ACHP)

idea is to equip this car with a hybrid drive system, which operates between 'plug-in hybrid' and 'full hybrid'. What does it mean? The target is to obtain a 300~400 km driving range for one charge of battery, using its state-of-charge (SOC) alteration range between 40%~80%. It is clear that the SOC's balance of battery charge/discharge during real driving is possible, but not absolutely required. Special attention was put in a hybrid operating mode-power distribution, according to assumed control strategy. All considerations were conducted in a New European Driving Cycle (NEDC), with a limited top speed 95km/h for case 1 and 80km/h for case 2.

The mentioned hybrid drive is characterized by low power components, automatic changing mechanical ratios during a car's acceleration and deceleration, which can offer lower energy consumption and smoother driving. An automatic clutch system connecting the motor's and internal combustion engine's (ICE) shafts through one wheel of belt transmission, offers the possibility of a 'smart switching on-off' ICE. Figure 16 shows the hybrid power train with an automatic clutch system. This hybrid drive consists of a gearbox, controlled by a stepper motor (e.g. depicted before the ball transmission) whose speeds can be changed, without the additional clutch in conditions of automatic clutch disconnection, and the voltage value on motor terminals behind the inverter being momentarily zero. In practice, in the case of city traffic driving, for reaching max speeds of 95km/h or 80km/h, only 2 speeds of the automatic gears are necessary. It is possible to obtain by belt transmission, the connection of gearbox shafts, of the motor/generator and the internal combustion engine (ICE), through elastic clutches and automatic clutches. This last one should be controlled in electromagnetic ways (see Figure 38), or by using a more costly system equipped with stepper motors (see Figure 39).

The belt transmission ratio is changed by the stepper motor (depiction is included in paragraph 2.2 of this chapter). The ratio of belt transmission is changed automatically, during acceleration, and also deceleration, depending on the angular velocity of the automatic gearbox output shaft (vehicle speed) and required power reduced on vehicle traction (road) wheels. This last case caused an increasing regenerative braking—generator operating mode of the motor.

The automatic clutch system consists of classic, dry, dual clutches and two diaphragm springs, specially connected and controlled by an electromagnetic actuator. Electric energy is necessary only during the connecting and disconnecting time of automatic clutch operation (transitory states). In steady state, the automatic clutch is not validated by energy (see paragraph 3.1 of this chapter). This means the automatic clutch characterizes zero energy consumption.

The following parameters of the hybrid power train were considered: the internal combustion engine's (ICE) power is 12 kW (gasoline), the rated power of the permanent motor (PM) is 8 kW, the range of the belt's continuous variable transmission (CVT) is 1-2.5, battery pack voltage is 72 V and its capacity is 50 Ah.

Plug-In Hybrid Power Train Engineering, Modeling, and Simulation

The analyzed experimental power train system was designed by computer simulations, based on mathematical models presented in this book.

The following figures show power/torque flows, depending on different operating modes of the power train. These modes are obtained by implementing control algorithm.

When the vehicle accelerates or operates at a constant speed (when the angular velocity of the belt's continuous transmission (CVT) input shaft is less than 200 rad/s), the hybrid drive system operates in a pure electric mode.

When a vehicle accelerating or operating at constant speed (when the angular velocity of a belt's continuous variable transmission (CVT) input shaft is equal or more than 200 rad/s) or the belt's continuous variable transmission (CVT) output torque is less than the minimal internal combustion engine's (ICE) torque (limitation of engine operating area), the vehicle operates in the pure engine's (ICE) operating mode. The motor operates in generating mode (charging the battery) when $T_{ICE} > T_{CVT output}$.

Figure 17. The start of the vehicle in pure electric mode

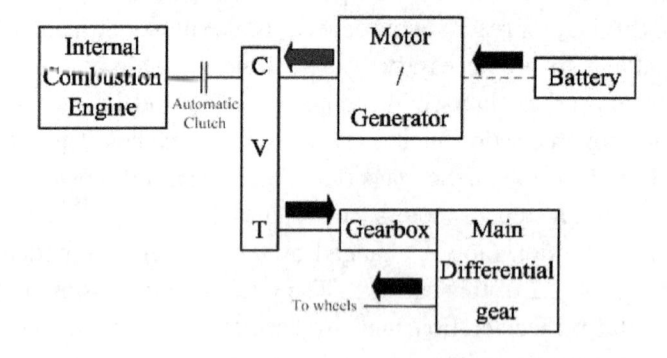

Figure 18. The start of the vehicle in the pure internal combustion engine's (ICE) operation mode

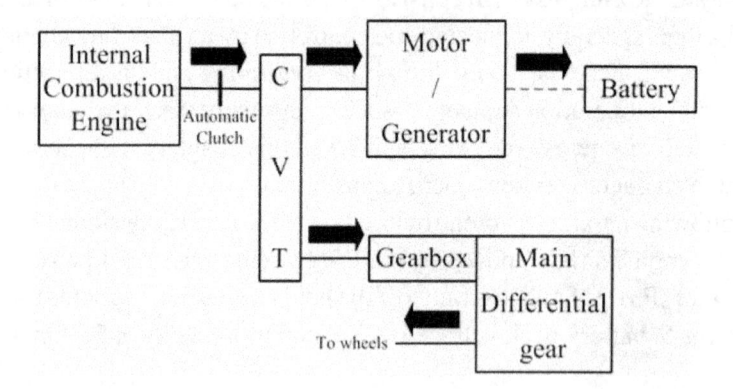

Figure 19. The power train's hybrid operation mode

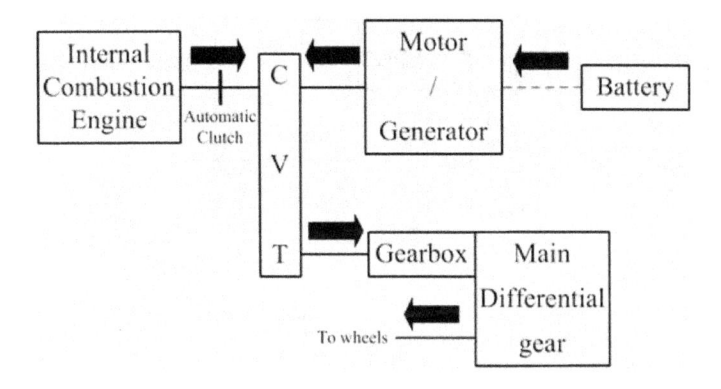

Figure 20. The first phase of vehicle regenerative braking

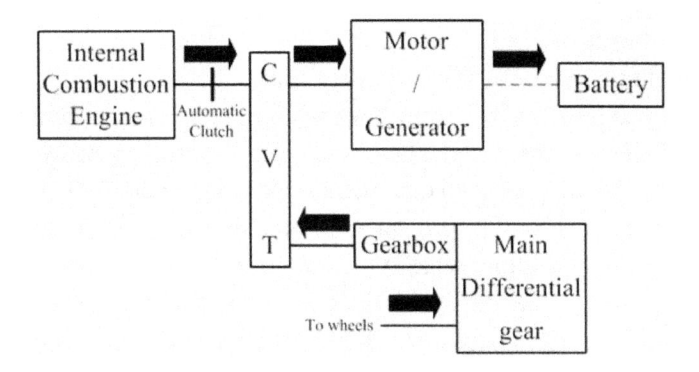

When a vehicle accelerating or operating at constant speed (when the angular velocity of a belt's continuous variable transmission (CVT) input shaft is equal or more than 200 rad/s) or the belt's continuous transmission (CVT) output torque is more than the minimal internal combustion engine's (ICE) torque, the vehicle operates in a hybrid mode. The motor operates in motoring mode when $T_{ICE} < T_{CVToutput}$.

When vehicle decelerating and the angular velocity of the belt's continuous variable transmission (CVT) input shaft is equal or more than 200 rad/s, the vehicle operates in the phase of regenerative braking.

When vehicle decelerating and the angular velocity of the belt's continuous variable transmission (CVT) input shaft is less than 200 rad/s, the vehicle operates in the second phase of regenerative braking.

Plug-In Hybrid Power Train Engineering, Modeling, and Simulation

Figure 21. The second phase of regenerative braking

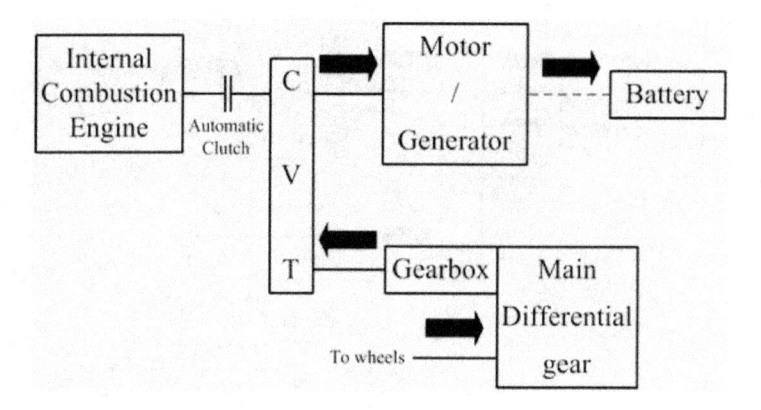

Selected results of simulations in the shape of power train dynamic characteristics of the new European driving cycle (NEDC) with a limited top speed of 95 km/h in case 1 and 80 km/h in case 2, are shown in Figures 22-32.

The developed hybrid power train for ultralight city vehicles characterizes low costs and sufficient performances. The conclusion is based on existing automotive components, which are produced on a mass scale. Using belt CVT transmission in the majority of cases of city traffic, an additional two-speed gearbox is only used. This leads to the key question: what is the most effective (also from a cost point of view) design? A new 2-speed gearbox or using an existing 5-speed gearbox operating in a limited range? It's necessary to emphasize that the speeds of gearbox are changed semi-automatically. The clutch of the gearbox does not exist. Manually changing speeds is possible when the motor voltage is zero (release the pedal of

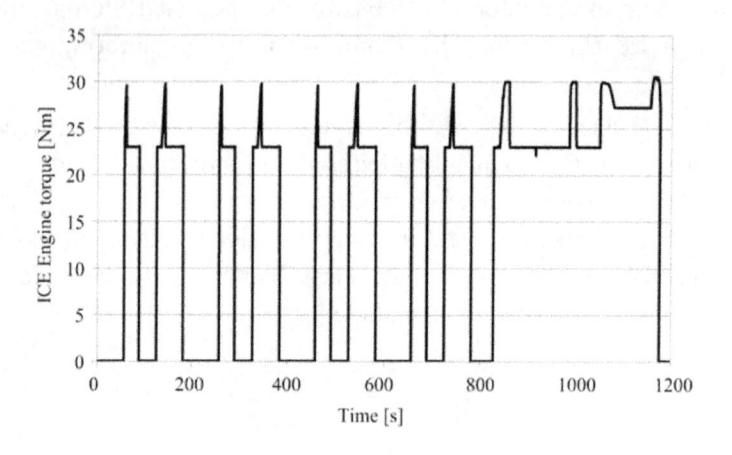

Figure 22. The torque alteration of the internal combustion engine (ICE)

Figure 23. The torque alteration of the permanent magnet (PM) motor

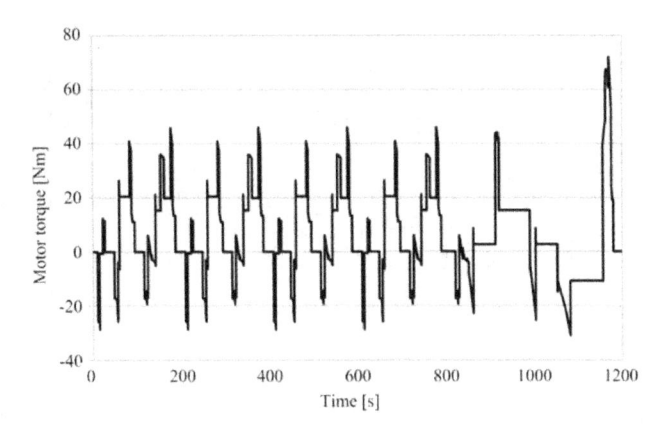

Figure 24. The angular velocity alteration of the belt's continuous variable transmission (CVT) input and output shafts

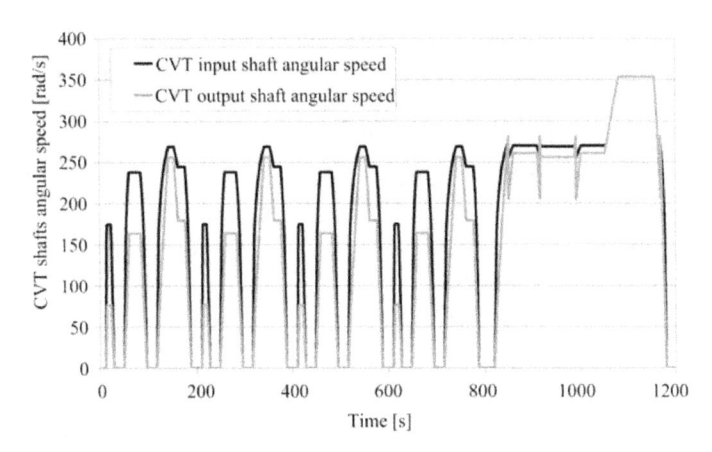

Figure 25. The alteration of the permanent magnet (PM) motor's operating efficiency

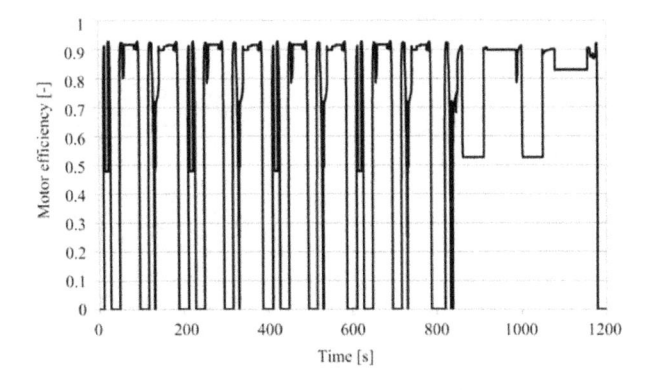

Figure 26. Permanent magnet (PM) motor's voltage and current alteration

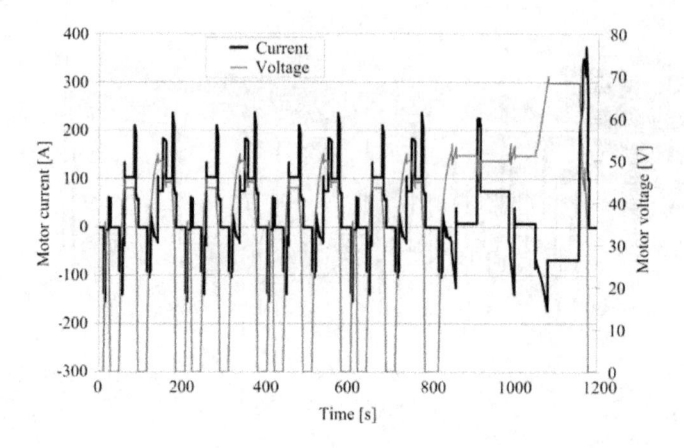

Figure 27. Battery voltage and current alteration

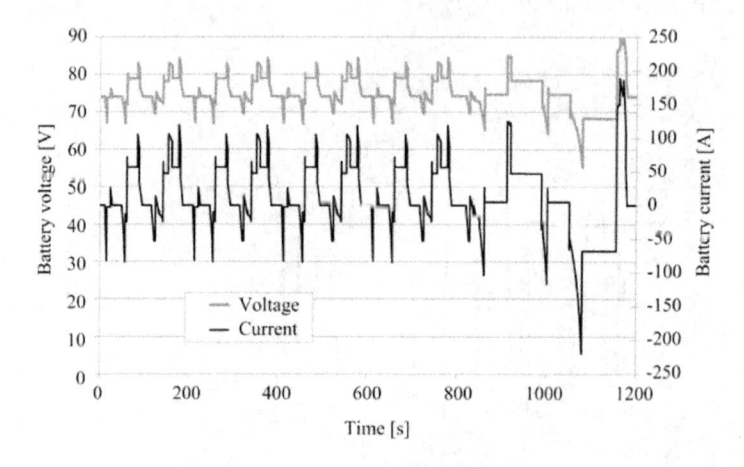

Figure 28. Automatic two-speed gearbox ratio alteration

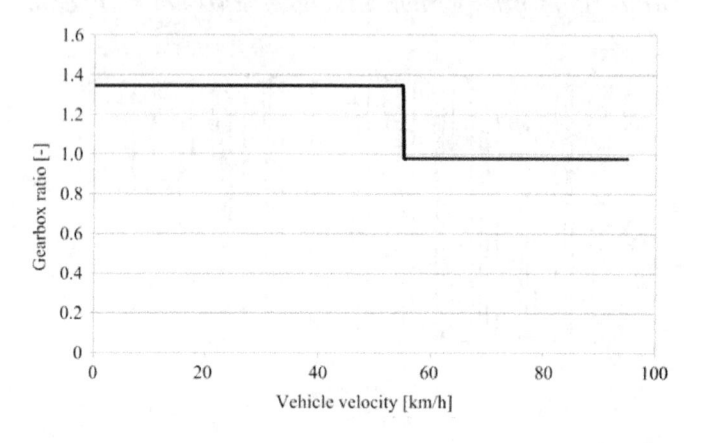

Figure 29. Belt's continuous, variable transmission (CVT) ratio alteration

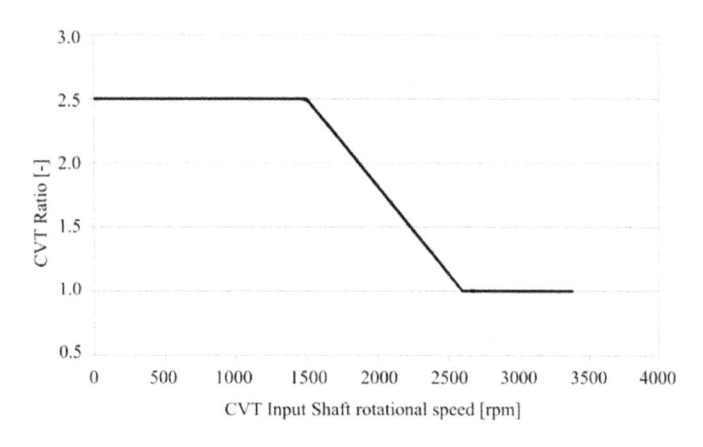

Figure 30. The operating efficiency of the power train

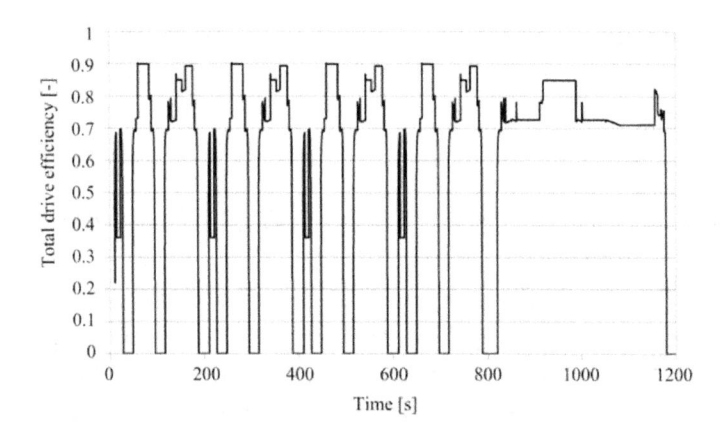

Figure 31. Battery's state-of-charge factor alteration

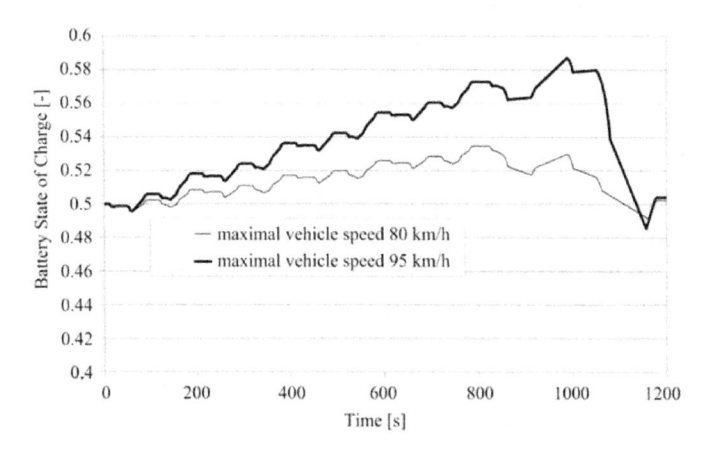

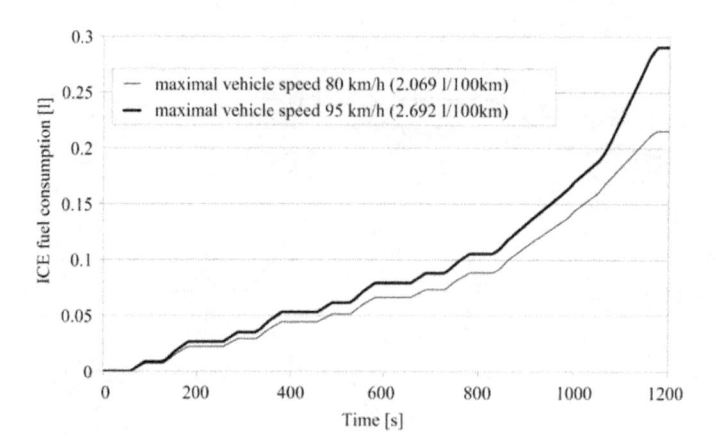

Figure 32. Internal combustion engine's (ICE) fuel consumption

acceleration), and when the automatic clutch system is in disengaged mode. The obtained results look very optimistic—especially in terms of fuel and electricity consumption, which is a very good incentive to further develop this kind of hybrid vehicle, as well as operating in a 'plug-in hybrid' or 'full hybrid' system.

3. EXEMPLARY PLUG-IN HYBRID POWER TRAIN ANALYSIS

Power train architecture in the case of the 'full hybrid' and 'plug-in' is the same. The difference is in battery pack capacity and size, and certainly its mass, volume and cost. Of course, the control strategy, as mentioned previously, is completely different.

A hybrid power train equipped with planetary transmission is an advanced solution. This is why, for an exemplary power train, the Compact Hybrid Planetary Transmission Drive (CHPTD) was chosen. The compact hybrid planetary transmission power train, CHPTD, depicted in Chapters 1, 8, and 9, is the complex system. Figure 33 shows two solutions of the new CHPTD, with an additional gearbox.

The CHPTD is a low cost solution, for it uses only one set of planetary gears and one electric motor for all operating modes, which are, namely, a pure electric mode, a pure engine mode, a hybrid mode, and lastly, a regenerative braking mode. A small internal combustion engine is employed as a range-extended power source. As a power summing unit, the planetary gearbox combines two power sources. The CHPTD could achieve higher efficiency than other existing hybrid power trains, because of its efficient power distribution via planetary transmission supported by a clutch/brake system, and additional transmission co-operating with a three- shafts clutch unit. Every one of the clutches is controlled electromagnetically, and the

Plug-In Hybrid Power Train Engineering, Modeling, and Simulation

CHPTD power train, certainly, demands a special control strategy, in case of plug-in hybrid application (Debal, Faid, & Bervoets, 2010; Smaling & Comits, 2010).

Several sets of clutch-brake system are used, together with mechanical transmission, for changing the operating modes of plug-in hybrid power trains and adjusting gear ratios. It gives more possibilities and flexibility for advanced control strategies of the plug-in hybrid power train.

The additional gearbox is necessary for adjusting the operating area of the internal combustion engine (ICE), and the electric motor in a different speed range of the vehicle's drive, which means improving the energy efficiency. Equipped with a properly shifting gearbox, the efficiency of regenerative braking can be improved as well.

3.1. The Clutch-Brake System Design

The clutch-brake system employed in the plug-in hybrid power train influences the performance of the whole system. However, the existing electromagnetic clutch-brake system consumes electric power, continuously. To minimize the energy consumption, the innovative zero steady-states electrical energy consumption clutch-brake system, according to a patented concept is selected. As the low cost solution, a dry-friction clutch is considered as the foundation of the new clutch design. The configuration of the newly-designed, clutch-brake system is shown in Figure 34 (Szumanowski, Liu, & Hajduga, 2010).

The two dual diaphragm springs, stiffly connected, and commonly moving along the x axis shaft (see Figures 34 and 35) is a key element in the design. After the movement of this shaft to a transitory final position, it pulls both springs. As one of them stays in second position by stiff connections, after a transient state, electric energy is not consumed during the steady state. Figure 35b is the characteristic of a single diaphragm spring. By connecting two diaphragm springs in the opposite direction, and setting the initial position of both diaphragms to point O on their characteristics, the characteristic of a dual diaphragm spring is obtained, as shown in Figure 35c.

The dual diaphragm spring has two steady-states, which are engagement and disengagement of the clutch (see Figure 34b). In the operation of the clutch, the dual diaphragm spring works between point B and D of its characteristics (see Figure 35c).

Two sets of electromagnetic actuator are necessary to create an actuation force in opposite directions. During engaging, the dual diaphragm spring moves from point B to point D. According to the characteristic of the dual diaphragm spring, the actuation force only applies on the actuation plate, when the dual diaphragm spring moves from point B to point C. The clutch can engage automatically, after the spring crosses the critical point C, whilst the energy consumption during the steady state is zero. For disengaging, the dual diaphragm spring works in the opposite direction to engaging.

The characteristic of a diaphragm spring is related to its materials and dimensions. The reaction force of the diaphragm spring is the function of its axial deformation.

The basic set of equations which depict input and output shafts of this clutch torque distribution is defined by the equations from (14) up to (25) included, in Chapter 8. For the detailed analysis of the zero steady state, energy consumption, clutch operation, it is necessary to add the following mathematical model to the above-mentioned equations.

The input and output shafts of the clutch are connected by the torque transferred to the clutch's plate surface (see Equation 1).

$$\begin{cases} I_{in} \dfrac{d\omega_{in}}{dt} = T_{in} - T_{cl} - \omega_{in} b_{in} \\ I_{out} \dfrac{d\omega_{out}}{dt} = T_{cl} - T_{out} \end{cases} \quad (1)$$

where:

- I_{in}, I_{out}: The equivalent moment of inertia reduced on input and output shafts;
- T_{in}, T_{out}: The input torque and resistant torque on the output shaft;
- T_{cl}: orque transmitted through the clutch's friction plates;
- $\omega_{in}, \omega_{out}$: Angular velocities of input shaft and output shaft;
- b_{in} (kg·m/s²): The torque-speed coefficient of input shaft, referring to the propulsion torque from the power source connected to this shaft.

The torque capacity of the clutch depends on the size and the material of the friction plates (see Equation 2).

$$\begin{cases} F_f = F_n \mu S \\ T_f = F_f r \\ T_{f\max} = \dfrac{rF_f}{\pi(r_2^2 - r_1^2)} \int_{r_1}^{r_2} \int_0^{2\pi} r^2 d\theta dr = \dfrac{2}{3}\dfrac{r_2^3 - r_1^3}{r_2^2 - r_1^2} F_n \mu \end{cases} \quad (2)$$

where:

- T_{fmax}: Maximal torque transferred by clutch,
- F_f: Friction force between two clutch plates,
- F_n: Axial force applied on friction plates,
- S: Area of surface calculated for equivalent radius r,
- r: Equivalent radius of the friction plate, calculated from Equation 2,
- r_1: Inner radius of the friction plate,
- r_2: Outer radius of the friction plate,
- μ: Friction coefficient.

Tf_{max} is used as a threshold to judge that the clutch is in a slipping or lock-up condition, when maximal torque is transferred. This means that for the input torque, higher than the maximal friction torque, the clutch plates will be continually slipping and it is not possible for the clutch to be fully engaged. Below the value T_{fmax}, the clutch can be normally engaged. For T_{fmax}, clutch plates, slip boundary conditions are defined.

When the clutch is slipping, the kinetic friction coefficient is used to calculate the torque transferred by the clutch, and the torque transmitted through the clutch plates, equates to the full capacity. The direction of the torque depends on the relative velocity of the two plates (see Equation 3).

$$\begin{cases} T_{f\max k} = RF_n \mu_k \\ T_{cl} = \text{sgn}(\omega_e - \omega_v) T_{f\max k} \end{cases} \quad (3)$$

k indicates the kinetic condition.

When the clutch is locked, the two plates are coupled and have the same angular velocity. Then, by solving Equation 1, the result is the torque transmitted by the clutch (see Equation 4).

$$\begin{cases} \omega_{in} = \omega_{out} = \omega \\ T_{cl} = \dfrac{I_{in} T_{out} + I_{out}(T_{in} - b_{in}\omega)}{I_{in} + I_{out}} \end{cases} \quad (4)$$

During engaging,

$$F_n = F_s - F_e \quad (5)$$

In Equation 5, the reaction force of the dual diaphragm spring (F_s) is constant when the spring is at point D.

In Equation 5, F_s is the reaction force of the dual diaphragm spring. As the reaction force of the diaphragm spring is a function of its elastic deformation, F_s are constant when the diaphragm is at point D in Figure 3c. According to experience concerning the classic dry friction clutch, the time of transient process (diaphragm spring moving between point B and D) is much shorter than the time of the clutch plates' angular velocity synchronization ($|\omega_{in} - \omega_{out}| \to 0$). The behavior of the clutch in the transient process, which is not relevant in this case, is ignored.

The looked-for actuation force of the electromagnetic actuator (F_e) is the function of the current in an electromagnet coil. Then, the behavior of clutch engagement, which means the angular velocity of the output shaft, is controlled by the current in the electromagnet coil. Thus,

$$\omega_{out} = f(i, T_{in}) \qquad (6)$$

where: i - the current in the electromagnet coil.

Figure 36 shows the simulation results of clutch engagement behaviors with different control strategies of current in an electromagnetic actuator for demonstration. As different shapes of the current feed the electromagnet coil for the same time, the clutch engaging time changes, significantly.

Figure 37 presents the construction of the clutch-brake system for different applications, which consist of a clutch-brake, dual-clutch, and brake. Based on the aforesaid concept, a clutch with an improved actuation solution is designed (see Figure 38). An additional bearing is employed to keep the actuation plate working in a non-rotary condition. 2 sets of solenoids are used as electromagnetic actuators. This design is dedicated to the dual-clutch application. However, with small adjustments, this design could also be used for clutch-brake or only brake applications. Compared with the rotary actuation solution, this design features lower abrasion and more stability.

Figure 39 presents the other construction of the clutch for a non-rotary actuation solution with stepper motor (Yn, Iseng, & Lin, 2010). Compared with the stepper motor actuation solution, the above-mentioned solenoid actuation solution can achieve better dynamic performance and smaller radial dimensions.

During vehicle driving in a hybrid or only pure internal combustion engine (ICE) mode, accelerating changes of gear speeds are necessary. The question how to avoid ICE over-speed during gear shifting appears.

In Figure 40, the feedback signals and torque-speed control signals are indicated, based on the next option of a compact hybrid planetary transmission power train.

Plug-In Hybrid Power Train Engineering, Modeling, and Simulation

During shifting of a 2-speed or 4-speed gearbox, the operating of the clutch in the gearbox needs about 0.3~2 seconds, including the synchronization of input and output shafts angular velocities. This means that in a maximum 2 seconds, the controlled clutch's plates slip process is over. During this period of time, the external load on the output shaft of a planetary gearbox is disconnected from the traction wheels by the clutch. This same time motor torque regulation signals a steered inverter output voltage, limiting its value up to zero, and the internal combustion engine's (ICE) speed signals cause decreasing engine speed (similar as in a conventional car's acceleration). When the vehicle's regenerative braking engine is switched off (its clutch is disengaged and shaft brake active) the motor-generator's torque signal controls inverter voltage and gear speeds shift in the opposite direction, as when accelerating. It has a significant influence on a vehicle's kinetic energy recuperation efficiency, as will be proved further.

3.2. Modeling of the Plug-In Hybrid Power Train

To evaluate the feasibility of the design and to optimize the parameters of the system, a dynamic model of the plug-in hybrid power train was built in a MATLAB - SIMULINK program environment, by using the mathematical or digital models of all components (Szumanowski, Liu, & Chang, 2010).

3.2.1. Planetary Transmission

Modeling of two degrees of freedom planetary gear was assessed in Chapter 8. Anyway, in this place, it is useful to remind ourselves of some basic information. A planetary two degrees of freedom gearbox combines power, torque and angular velocities of the internal combustion engine's (ICE) output shaft, the electric motor's output shaft, and the vehicle's speed and resistance torque, reduced on a third moving planetary transmission shaft. In this application, the ICE and electric motor are connected to the sun wheel and crown wheel of planetary gear respectively, and by yoke shaft, the vehicle is propelled (see Figure 39). The following equations describe the relations of torque and angular velocity. In these equations, k_p is the basic ratio of the planetary gear, whose adjustment has a big influence on the power distribution of the hybrid power train.

Plug-In Hybrid Power Train Engineering, Modeling, and Simulation

$$\begin{cases} J_1\dot{\omega}_1(t) = \eta_1 M_1(t) - \dfrac{1}{k_p}\eta_2 M_2(t) \\ J_3\dot{\omega}_3(t) = M_3(t) + \dfrac{k_p+1}{k_p}\eta_3 M_2(t) \\ \omega_1(t) + k_p\omega_2(t) - (1+k_p)\omega_3(t) = 0 \end{cases} \qquad (7)$$

where:

- $\omega_1, \omega_2, \omega_3$: Angular velocities of sun, ring and yoke wheels, respectively,
- k_p: The basic ratio of planetary gear,
- J_1: Total moment of inertia torque of the sun wheel and connecting elements reduced to the sun shaft,
- J_2: Total moment of inertia torque, obtained from a reduction of the vehicle mass, road wheels, and gears reducer inertial torques to the carrier shaft,
- η_1, η_2: Substitute coefficients of internal power losses,
- M_1, M_2: External torque acting on the sun shaft and ring shaft,
- M_3: External torque acting on the carrier and corresponding to the vehicle's motion resistance reduced to the appropriate shaft.

3.2.2 Internal Combustion Engine (ICE)

The ICE model is based on (see Chapter 7) the engine's map. By inputting torque and rotary speeds, the engine's fuel consumption rate is obtained as the output of the engine's map (see Figure 46).

3.2.3 Permanent Magnet PM Electric Motor

The electric motor model (see Chapters 3, 4) is based on the efficiency map of an electric motor with controller (see Figure 27). Power efficiency is obtained by inputting the torque and motor rotary speed into a look-up table. The efficiency map is obtained from a Unique Mobility 32kW PM motor with electronic controller, by reducing the torque and speed linearly.

Plug-In Hybrid Power Train Engineering, Modeling, and Simulation

Figure 33. The configurations of the new CHPTD with additional gearboxes: four-speed and clutch-brake system and a double two-speed and clutch-brake system (see construction of clutch-brake systems in Figures 36 and 37)

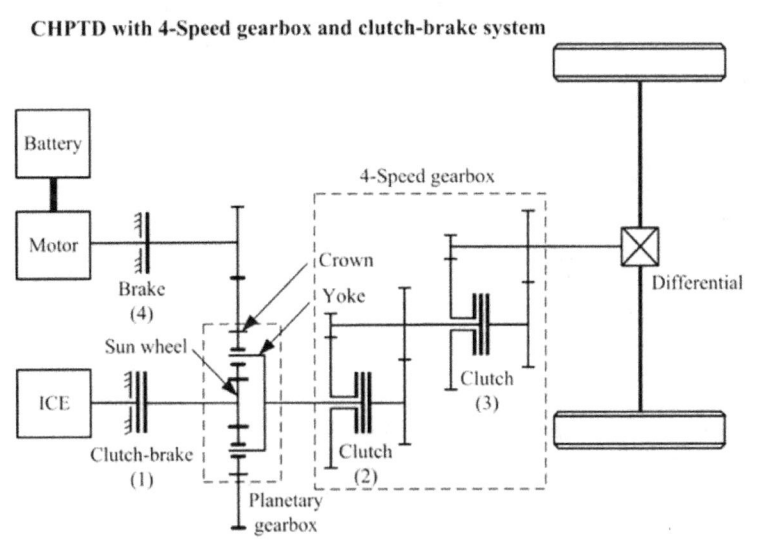

CHPTD with 4-Speed gearbox and clutch-brake system

CHPTD with 2x2-speed gearbox and clutch-brake system

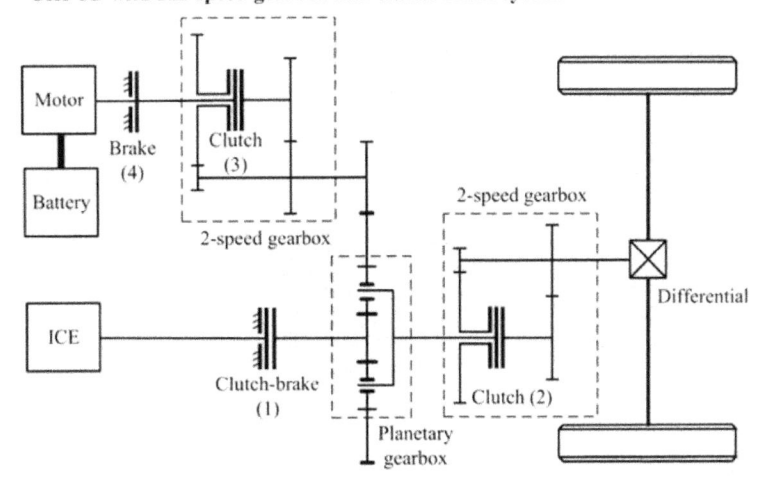

3.2.4. Lithium Ion Battery

A nonlinear dynamic battery model (see Chapters 5 and 6) is used in battery modeling. In this method, the electromotive force E and internal resistance R are resolved in a 6-order algebraic expression of the battery's state-of-charge factor SOC (k).

Figure 34. The illustration of the construction of the innovative dry clutch-brake device, during which its steady-state work, its electric energy delivery to the clutch's operation, controlled by electromagnetic coils is equal to zero: (a) the configuration of the innovative zero steady-states electrical energy consumption clutch-brake system; (b) operating position of the clutch, according to point B, C, and D in Figure 35c, respectively (point C is a transient state)

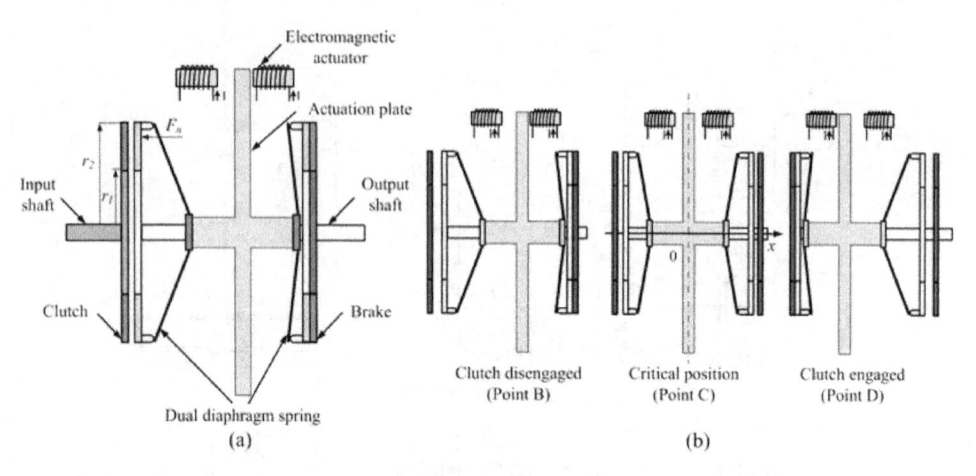

Figure 35. The characteristics of a dual diaphragm spring: (a) diaphragm spring; (b) the characteristic of a single diaphragm spring; (c) The characteristic of a dual diaphragm spring

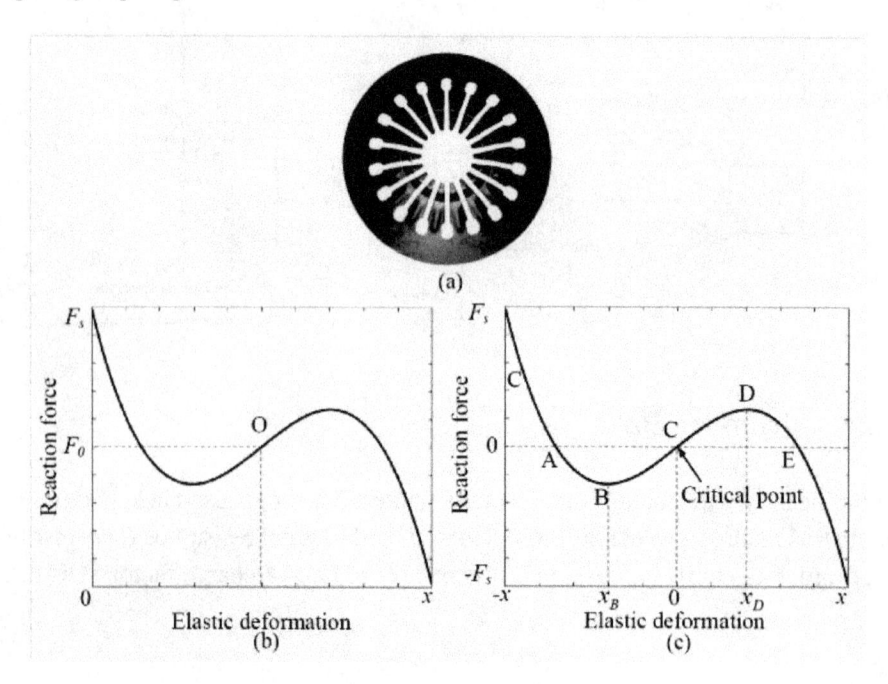

Plug-In Hybrid Power Train Engineering, Modeling, and Simulation

Figure 36. The different strategies of the clutch-brake device control: (a) current in an electromagnetic actuator with different engagement control strategies; (b) differential angular velocities of input and output shafts, with different engagement control strategies (ωin -ωout). When differential angular velocity between input and output shafts is zero, the clutch is fully engaged.

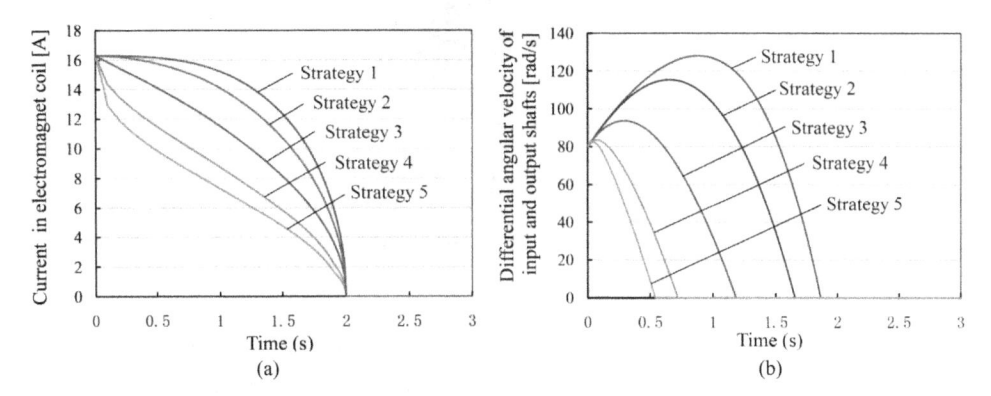

Figure 37. Zero steady-states, electrical energy consumption, clutch-brake systems for different applications

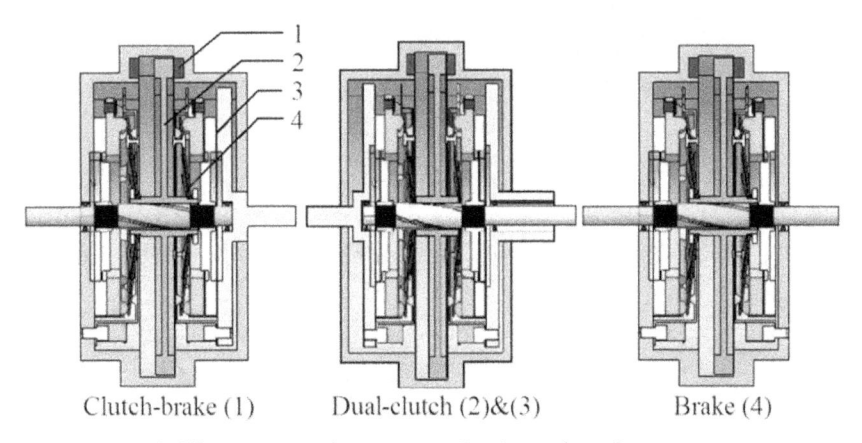

Clutch-brake (1)   Dual-clutch (2)&(3)   Brake (4)

1- Electromagnetic actuator   2 - Actuation plate
3 - Friction plate   4 - Diaphragm spring

$$\begin{cases} E(k) = A_e k^6 + B_e k^5 + C_e k^4 + D_e k^3 + E_e k^2 + F_e k + G_e \\ R(k) = A_r k^6 + B_r k^5 + C_r k^4 + D_r k^3 + E_r k^2 + F_r k + G_r \end{cases} \quad (8)$$

Figure 38. Assembly of zero steady-states, electrical energy consumption clutch with solenoid actuation for dual clutch application

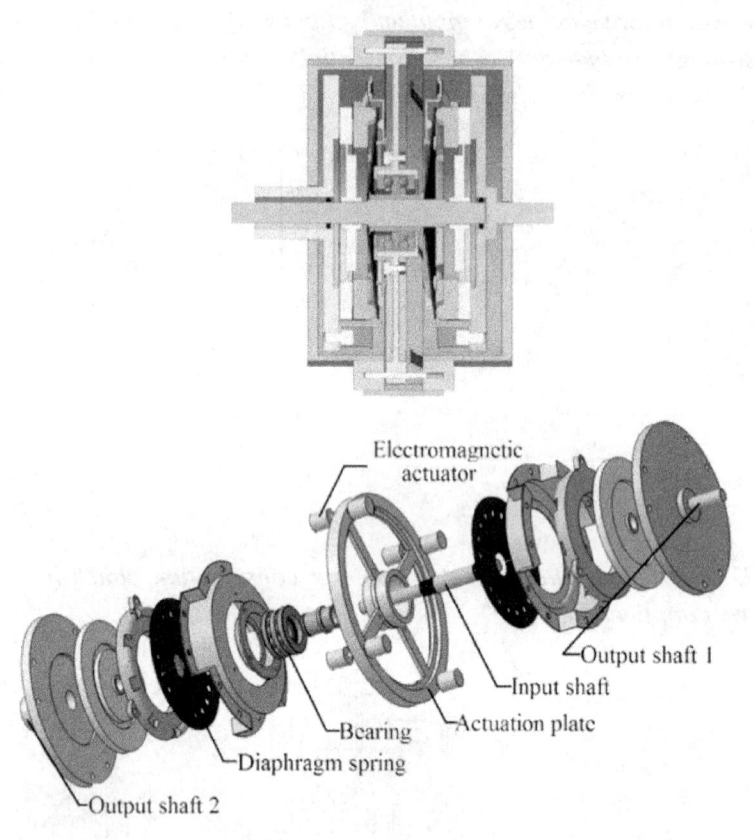

Figure 39. Clutch with a stepper motor actuation solution (Yn et al., 2010)

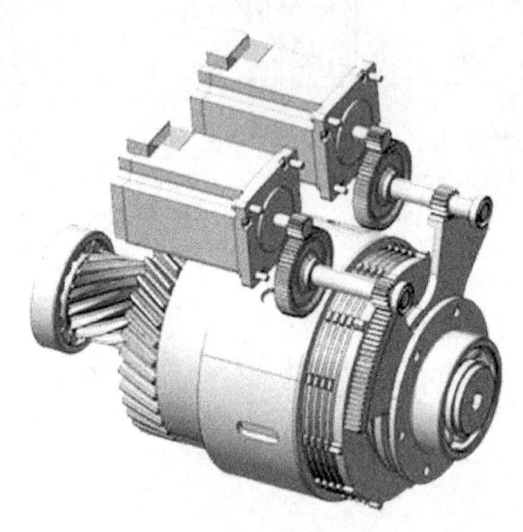

Figure 40. The next option of a compact hybrid planetary transmission drive, co-operating with an additional 4-speed mechanical transmission layout and its torque-speed control signals

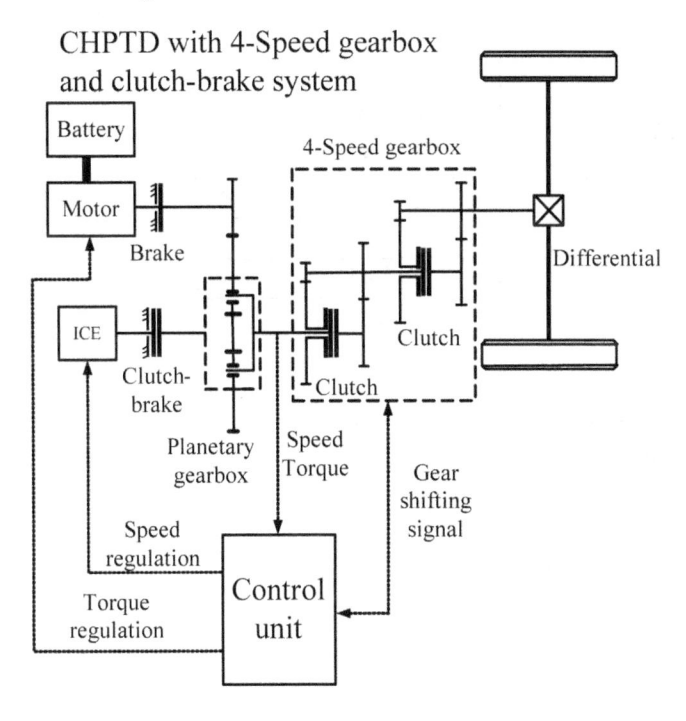

Table 2 includes the coefficient value of Equation 8 for a 30Ah - 43V Li-ion module from the SAFT company, which is used in the simulation. The approximated equation and other factors are based on battery discharging characteristics obtained from experiments.

3.2.5. Aggregate Simulation Model of the Considered Hybrid Power Train

By combining the specific models of the main components, a complete simulation model of a plug-in hybrid power train is built in a MATLAB – SIMULINK program environment, according to the configuration of a plug-in hybrid power train in Figure 40.

Table 1. The coefficients of Equation 8 for a 30Ah Li-ion battery module from the SAFT model

Factors	Internal resistance during discharging R(k)	Electromotive force E(k)
A	0.71806	-28.091
B	-2.6569	157.05
C	3.7472	-296.92
D	-2.5575	265.34
E	0.8889	-119.29
F	-0.14693	30.476
G	0.023413	38.757

Table 2. Simulation parameters of a planetary plug-in hybrid power train and its main components

Vehicle	
Vehicle mass (kg)	750
Rolling resistance coefficient	0.008
Aerodynamic drag coefficient	0.33
Front surface square (m^2)	1.6
Dynamic radius of the wheel (m)	0.257
Driving cycle	NEDC
Main reducer ratio	3.62

3.3. Simulation and Parameter Optimization

3.3.1. Simulation Parameters

In order to analyze the influence of different control strategies and parameters, different comparison simulations were conducted under the NEDC (New European Driving Cycle). An ultra-light basket-tube framed vehicle is considered as the vehicle model in simulation (see its sketch Figure 42). Table 3 shows the parameters of a planetary plug-in hybrid power train. The selected parameters are properly adjusted by analyzing the requirements and verified by simulation.

Plug-In Hybrid Power Train Engineering, Modeling, and Simulation

Figure 41. The dynamic model schematic of a plug-in hybrid power train.

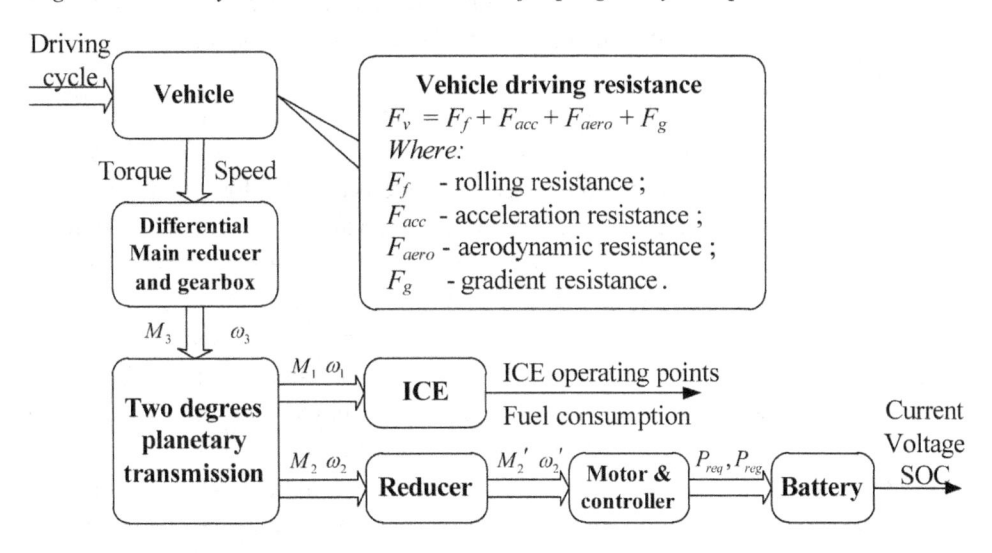

Table 3. Control signal of clutch-brake systems for different operating modes of a plug-in hybrid power train (see Figure 33)

Operating mode of a plug-in hybrid power train	Control signal of clutch-brake systems	
	Clutch-brake (1)*	Brake (4)**
Pure electric and regenerative braking mode	Off	Off
Pure engine mode	On	On
Hybrid mode	On	Off
Engine charge battery (when vehicle stops)	Off	Off

* 'On': clutch engaged and brake disabled; 'off': clutch disengaged and brake enabled.
** 'On': brake enabled; 'off': brake disabled.

3.3.2. Control Strategy

The control of the power train is connected with the clutch-brake operation. The relation between the control signal of a clutch-brake system and the operating modes of a power train is determined in Table 3.

The full control strategy is divided into basic parts and additional parts (see Figure 43). In basic control strategy, vehicle speed and the battery's SOC are used as feedback signals for changing the operating mode of a hybrid power train. To achieve lower fuel consumption and the functioning of a plug-in hybrid, two ad-

Plug-In Hybrid Power Train Engineering, Modeling, and Simulation

ditional control strategies are designed for demonstration. Separately, the torque on transmission shaft and demanded power of the vehicle influences the changing operating modes for a low-speed drive.

Furthermore, an additional threshold of the battery's state-of-charge factor (SOC) is set to determine the behaviors of a hybrid power train. When the battery's SOC is higher than the threshold, the pure electric mode is enabled for low and middle-speed drive. It means that more electric energy is consumed to limit the emissions. When the battery's SOC is lower than the threshold, the pure electric mode is only enabled for starting, which means the power train works just like a full hybrid HEV.

To investigate influences of different control strategies, a comparative simulation was made. Figure 43 and Table 4 show simulation results with different control strategies for the same conditions. For the same driving range, fuel consumption with Strategy I is 2% less than that with Strategy II. With Strategy II, the battery's SOC is limited to the proper set value 0.5 at the end of the simulation. While in Strategy I, the battery's SOC is out of control and decreases to 0.18. It means that the power train could work in hybrid mode for long distance driving, to achieve better emission performance without damaging the battery. Considering the requirements of a plug-in hybrid and similar fuel economy performance, Strategy II is better than Strategy I. The simulation in the following pages is based on Strategy II.

3.3.3. Simulation for Pure Electric Drive

Battery capacity influences the driving range of a pure electric drive of a plug-in hybrid power train. To fulfill the functionality of a plug-in hybrid, battery capacity

Figure 42. The exemplary construction of an ultra-light basket-tube framed vehicle

Figure 43. Control strategy of a planetary plug-in hybrid power train

Basic control strategy
Starting (0-15km/h)
  Pure electric mode
Low speed and middle speed (15-70km/h)
  Pure electric mode: SOC ∈ <0.6, 1>
  Hybrid mode: SOC ∈ <0.3, 0.6>
High speed (70-120km/h)
  Hybrid mode: SOC ∈ <0.6, 1>
  Pure engine mode: SOC ∈ <0.3, 0.6>

Additional control strategies

Strategy I
Speed-torque control
Feedback signal for changing operation mode:
- vehicle speed
- torque on transmission shaft

or

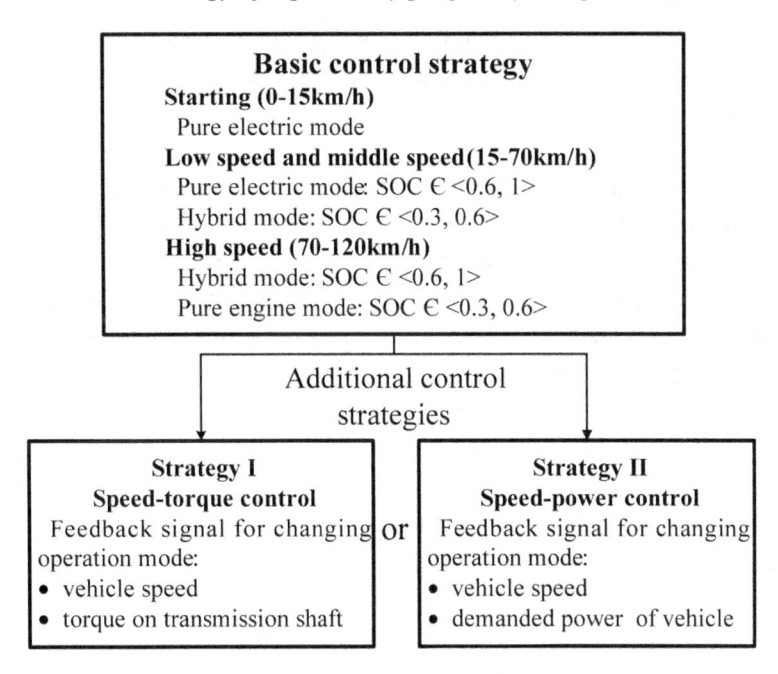

Strategy II
Speed-power control
Feedback signal for changing operation mode:
- vehicle speed
- demanded power of vehicle

Table 4. Simulation results of a plug-in hybrid analyzed power train for two different control strategies (50 x NEDC cycle repetition).

	Strategy I	Strategy II
Total driving range (km)	540	540
Total fuel consumption (l)	13.55	13.82
Average fuel consumption (l/100km)	2.51	2.56
SOC at the end of simulation	0.18	0.52

is adjusted to 3.9kWh. The NEDC driving cycle with a limited top speed of 65km/h is used in the simulation. According to the simulation results, it achieves a 55km driving range for a pure electric operating mode with the battery's SOC alteration changing from 0.95 to 0.4, by applying optimized shifting schedule of the gearbox.

3.3.4. Gear Ratio Optimization

According to the configuration of the power train (Figure 40), several sets of gears are equipped with transmission, which are as below:

- Planetary gear.
- Additional reducer between the ICE and planetary gear.
- Additional reducer between the electric motor and planetary gear.
- 4-speed gearbox.

The ratio of all these gears has an influence on the performance of power distribution and the operating points of the ICE and electric motor.

The target of gear ratio optimization minimizes internal losses and the energy consumption rate, which consists of fuel consumption and electric motor efficiency.

The gear ratio optimization is based on the digital computer simulation method. Other optimization methods are investigated and considered during the simulation's approach.

3.3.5. Basic Ratio of Planetary Gear

The basic ratio of planetary gear influences the power distribution of the hybrid power train. Simulation results in Table 5 demonstrate that a lower basic ratio of planetary gear could achieve better fuel economy performance. The basic ratio of a planetary gear may not have a specific number for it, as it is limited by manufacturing, dimensions, and other practical conditions. The ratio in Table 5 is properly selected by fulfilling these requirements.

The ICE's reducer ratio and electric motor reducer ratio are optimized with similar approaches.

3.3.6. Gear Ratio of 4-Speed Gearbox

The 4-speed gearbox is an important element for the plug-in hybrid power train. With properly adjusted gear ratios and a gear shifting schedule, it could increase

Table 5. Simulation results for a different basic ratio of planetary gear (Other gear ratio is assumed not to be changing*)

Basic ratio of planetary gear	Average fuel consumption (l/100km)	Average efficiency of motor (%)
1.80	2.186	76.95
1.875	2.201	76.32
1.99	2.218	75.97
2.25	2.260	75.09
2.99	2.383	73.56

*1st gear ratio: 2.00; 2nd gear ratio: 1.50; 3rd gear ratio: 0.95; 4th gear ratio: 0.83; ICE reducer ratio: 3.22; Electric motor reducer ratio: 1.98

** Simulation time: 30,000s; driving range: 270km; battery SOC alteration: 0.9 to 0.5.

Table 6. Simulation results for a different gear ratio of a 4-speed gearbox (the other gear ratio is assumed to be not changing) *

No.	1st gear	2nd gear	3rd gear	4th gear	Average fuel consumption (l/100km)	Average efficiency of motor (%)
1	2.00	1.50	1.10	0.90	2.301	75.40
2	2.00	1.50	1.00	0.90	2.259	75.95
3	2.00	1.50	1.00	0.90	2.251	76.25
4	2.00	1.50	1.10	0.83	2.227	76.21
5	2.00	1.60	0.95	0.83	2.213	75.99
6	2.00	1.50	1.00	0.83	2.206	76.36
7	2.00	1.50	0.95	0.85	2.199	76.73
8	2.50	1.50	0.95	0.83	2.192	76.42
9	2.20	1.50	0.95	0.83	2.189	76.58
10	2.00	1.50	0.95	0.83	2.186	76.69
11	2.00	1.45	0.95	0.83	2.176	76.95
12	1.20	1.20	1.20	1.20	2.567	74.90

The number 12 in Table 6 indicates a one-gear ratio, which means the gear transmission consists only of a constant ratio reducer.

* Basic ratio of planetary gear: 1.80; ICE reducer ratio: 3.22; Electric motor reducer ratio: 1.98.

** Simulation time: 30,000s; driving range: 270km; battery's state-of-charge factor SOC alteration: 0.9 to 0.5.

the energy efficiency for different driving conditions, relating to city traffic and suburban areas. As the focus is on dynamic performance and fuel economy, the gear shifting schedule is as follows:

- **1st-Gear:** 0~15km/h
- **2nd-Gear:** 15~40km/h
- **3rd-Gear:** 40~70km/h
- **4th-Gear:** 70~120km/h (or higher speed)

Tables 5 and 6 display the trend that gear ratios influence the fuel consumption and efficiency of the motor. In a selected range, the best ratios are indicated by a highlight in Tables 5 and 6. However, the real gear ratio optimization is a more complicated approach, because changing each gear ratio is connected with others. The adjustment of gear ratio should co-operate with observation of the operating points of the internal combustion engine (ICE) and the permanent magnet, PM electric motor, because gear ratio optimization is also limited by practical performance,

Plug-In Hybrid Power Train Engineering, Modeling, and Simulation

and other binding conditions. For example, the operating points of the ICE and PM electric motor should be located in a limited range; the proper power of the electric motor should be reserved for acceleration and grade ability.

Simulation results in Table 6 also show the necessity of a 4-speed gearbox. By equipping the analyzed plug-in hybrid power train with a 4-speed automatic gearbox, the average fuel consumption decreases 16.3% compared with a power train without a gearbox (see data - Nos. 11 and 12 in table 8). Figures 45, 46, 47, and 48 present the operating points of an internal combustion engine (ICE) and permanent magnet, PM electric motor installed in analyzing the plug-in hybrid power train equipped with (Figures 45 and 46), and without (Figures 47 and 48), a 4-speed gearbox transmission.

Figure 44. The results of a simulation test analyzing a planetary plug-in hybrid power train for two different control strategies (fifty times repetition of the European driving cycle, NEDC)

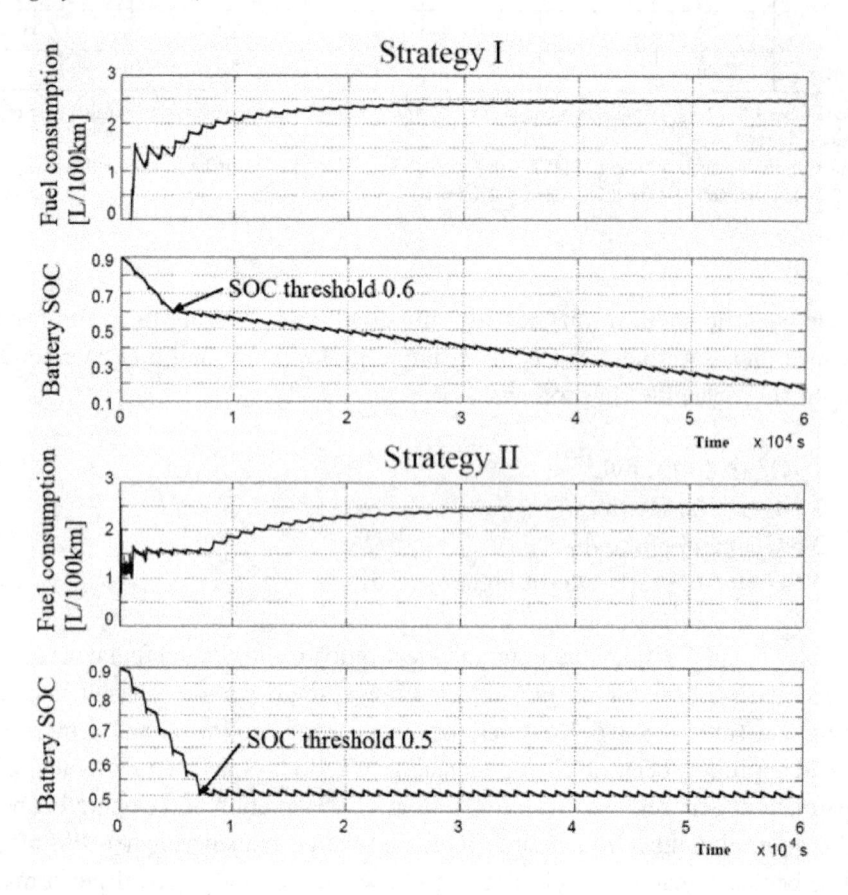

Plug-In Hybrid Power Train Engineering, Modeling, and Simulation

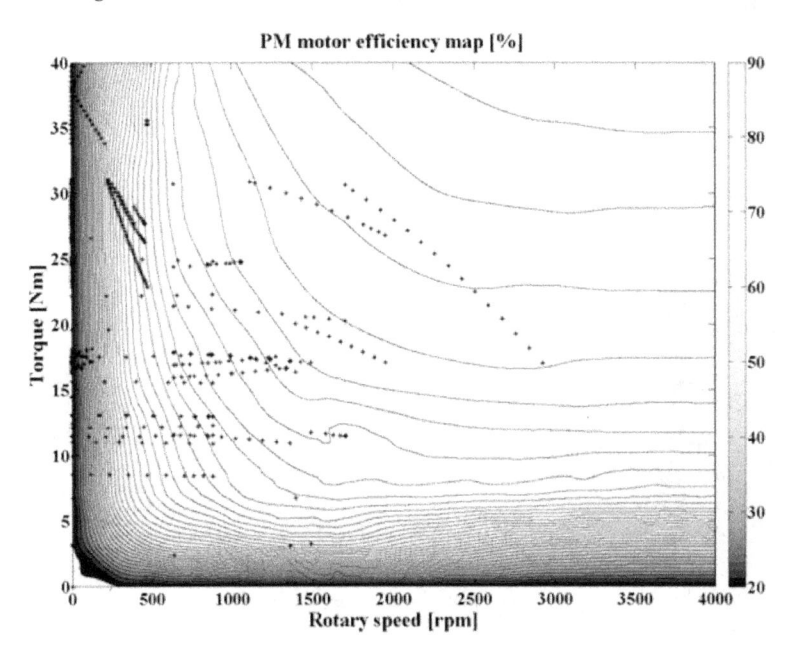

Figure 45. The operating points of an internal combustion engine (ICE) applied in analyzing a plug-in hybrid power train, equipped with a 4-speed gearbox transmission (according to data No. 11 in Table 6)

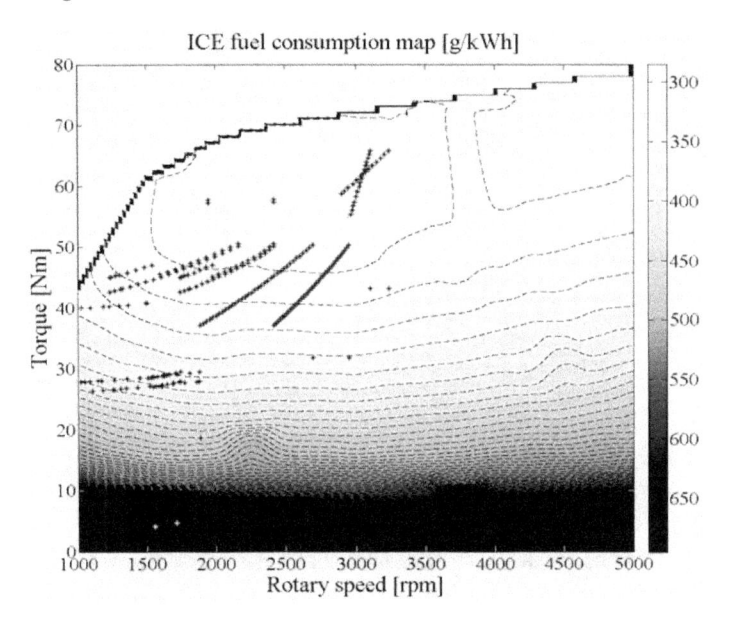

Figure 46. The operating points of a permanent, magnet motor (PM) applied in analyzing a plug-in hybrid power train, equipped with a 4-speed gearbox transmission (according to data No. 11 in Table 6)

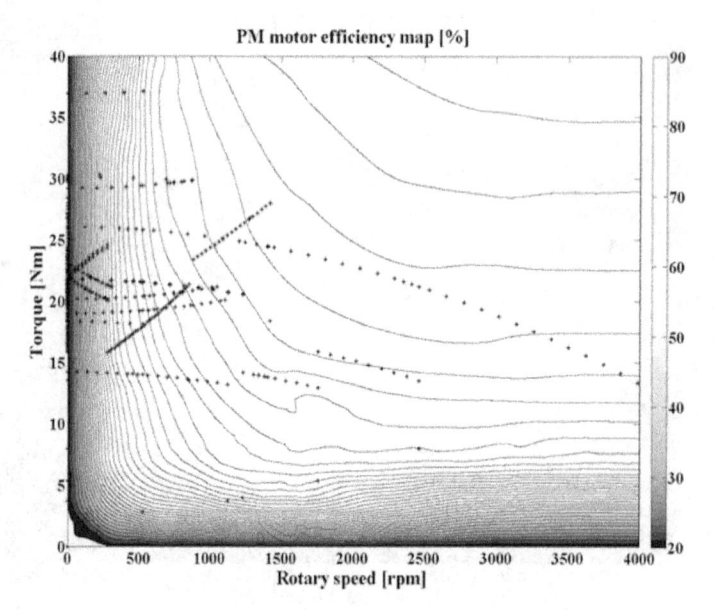

Figure 47. The operating points of an internal combustion engine (ICE) applied in analyzing a plug-in hybrid power train, equipped without a 4-speed gearbox transmission (according to data No. 12 in Table 6)

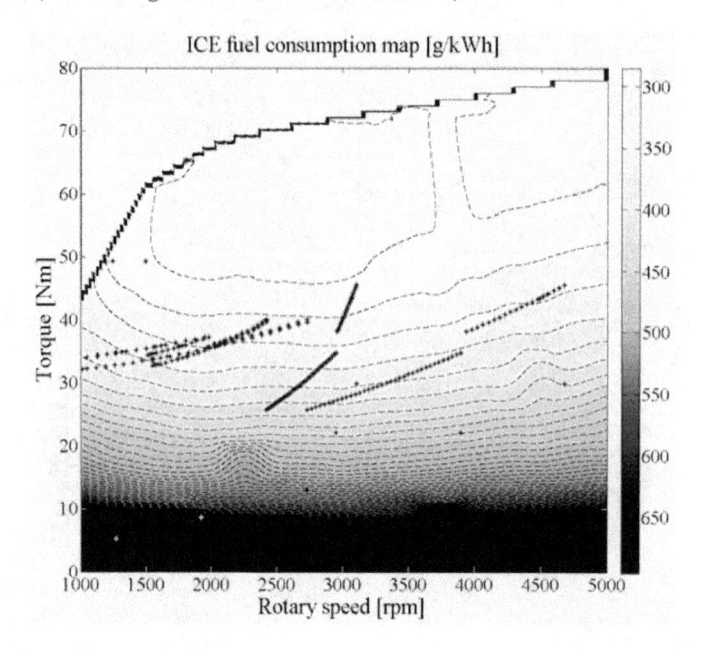

Figure 48. The operating points of a permanent, magnet motor (PM) applied in analyzing a plug-in hybrid power train, equipped without a 4-speed gearbox transmission (according to data No. 12 in Table 6)

Plug-In Hybrid Power Train Engineering, Modeling, and Simulation

Figure 49. The simulation of an analyzing plug-in hybrid power train's selected results: a) battery state-of-charge factor's value alteration SOC, according to data No. 11 in Table 6; b) simulation results in one driving cycle, with a high battery SOC factor value; according to Figure 49 a - point (b); c) simulation results in one driving cycle, with a low battery SOC factor value; according to Figure 49b

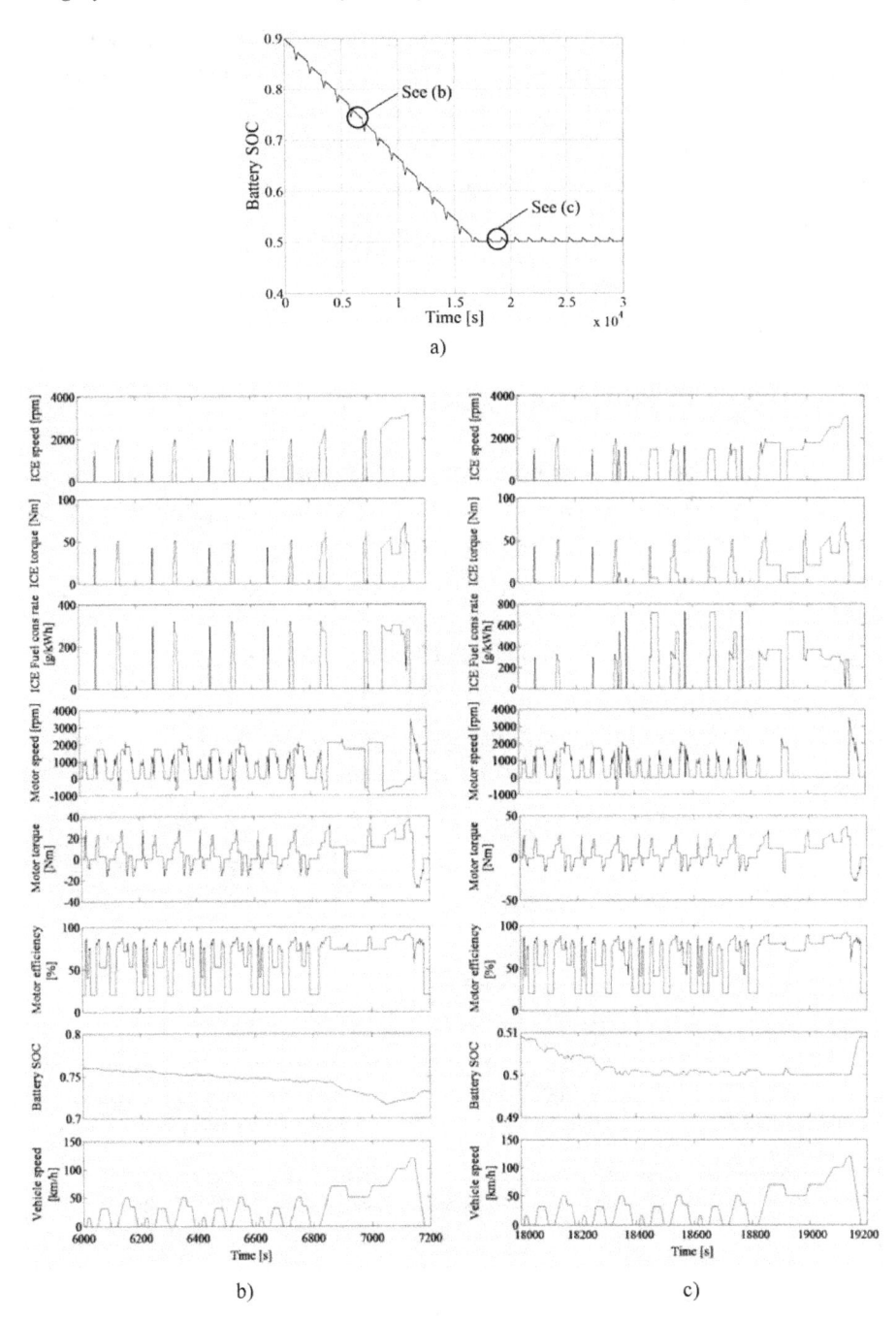

401

According to the operating points of the internal combustion engine indicated on its map (see in Figure 45), the analyzing IC engine has more power than is demanded. This means a smaller ICE should be selected for such a light vehicle. It also promotes the idea that simulation is an effective method to correct and verify the design.

To prove the practicability of the design, an additional simulation was carried out for a vehicle model with a normal mass of 1,200kg. For such a vehicle's parameters, it is not possible to keep the operating points of the ICE and motor in a limited range for the whole driving cycle, without equipping the 4-speed gearbox. The simulation results in Table 7, demonstrate that the 4-speed gearbox increases energy efficiency, significantly. However, the average efficiency of the motor is lower than that in the case of a 750kg vehicle mass. By observing operating points of the motor, the power of the motor is too small for this vehicle mass, which means the motor should be adjusted.

3.3.7. Energy Efficiency of Regenerative Braking

When the plug-in hybrid power train works in the regenerative braking mode, the equivalent torque on the electric motor shaft influences energy efficiency. With proper control of a 4-speed gearbox, it could change the operating points of the electric motor to increase the energy efficiency during regenerative braking. The energy efficiency of regenerative braking relates to the kinetic energy of the vehicle and electric energy received into the battery (see Figure 50).

Table 7. Simulation results for different gear ratios of the 4-speed gearbox with a 1,200kg vehicle mass (the other gear ratio is assumed to be not changing)*

No.	1st gear	2nd gear	3rd gear	4th gear	Average fuel consumption (L/100km)	Average efficiency of motor (%)
1	2.90	2.10	1.50	1.10	3.228	71.27
2	2.90	2.10	1.60	1.00	3.163	71.40
3	2.90	2.10	1.50	1.00	3.101	72.03
4	2.80	2.10	1.45	1.00	3.070	72.37
5	1.50	1.50	1.50	1.50	3.686	63.78

Number 5 in Table 9, indicates one gear ratio, meaning the gear transmission consists only of a constant ratio reducer.

* Basic ratio of planetary gear: 1.80; ICE reducer ratio: 3.22; Electric motor reducer ratio: 1.85.

** Simulation time: 30,000s; driving range: 270km; battery's state-of-charge factor SOC alteration: 0.9 to 0.5.

Plug-In Hybrid Power Train Engineering, Modeling, and Simulation

Figure 50. The results of analyzing the vehicle's regenerative braking efficiency: a) the operating points of the electric motor with a 4-speed gearbox transmission during regenerative braking (simulation data, according to Table 6, No. 11); b) The operating points of the electric motor without gearbox transmission during regenerative braking (simulation data, according to Table 6, No. 12)

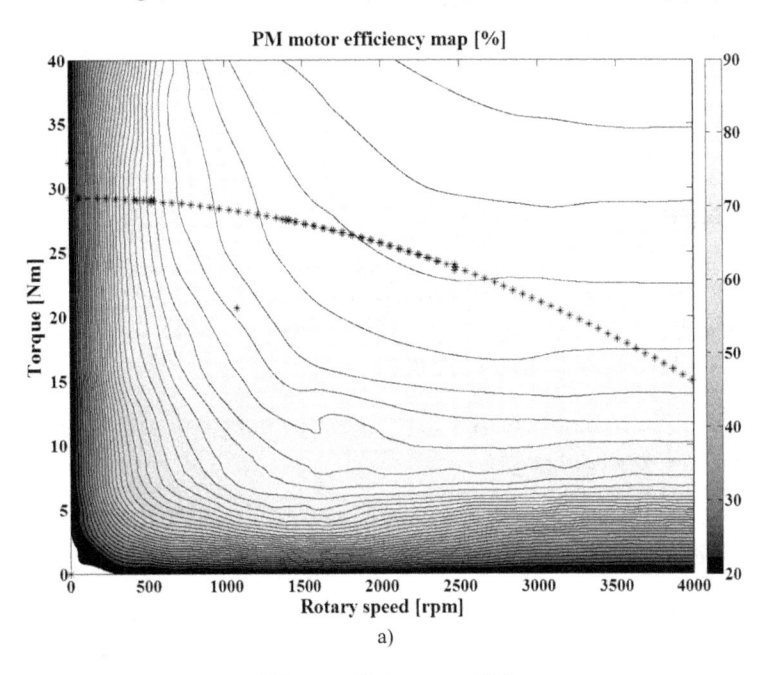

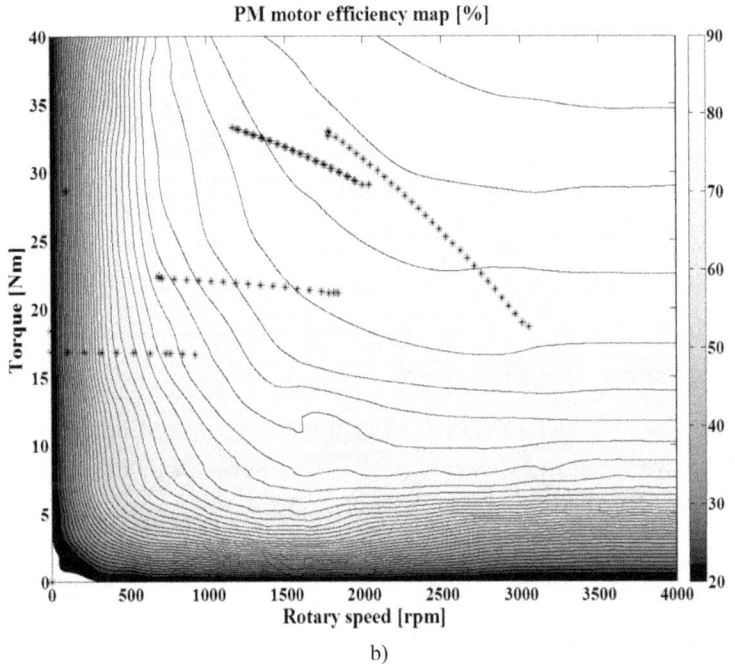

A comparative exemplary simulation was carried out to analyze the energy efficiency of regenerative braking with, and without, a gearbox for a 750kg vehicle mass. According to simulation results, the average efficiency of regenerative braking increases from 67.16% to 76.01% in the NEDC driving cycle. This means the application of a 4-speed gearbox transmission causing an increase of about 9 percent efficiency of the vehicle's regenerative braking.

Figure 50 demonstrates the operating points of the electric motor during regenerative braking from 120km/h to 0km/h within 35s.

REFERENCES

Chen, K., Bouscayrol, A., Berthon, A., Delarue, P., Hissel, D., & Trigui, R. (2009). Global modeling of different vehicles. *IEEE Vehicular Technology Magazine*, 4(2), 80–89. doi:10.1109/MVT.2009.932540

Debal, P., Faid, S., & Bervoets, S. (2010). Parallel hybrid (booster) range extender power train. In *Proceedings of EVS 25*. EVS.

Duvall, M. S. (2005). Battery evaluation for plug-in hybrid electric vehicles. In *Proceedings of Vehicle Power and Propulsion*. IEEE. doi:10.1109/VPPC.2005.1554580

Hellenbroich, G., & Huth, T. (2010). New planetary-based hybrid automatic transmission with electric torque converter and on-demand actuation. In *Proceedings of 20th Aachen Colloquium Automobile and Engine Technology*, (pp. 92-105). Aachen Colloquium.

Hofman, T., Hoekstra, D., van Druten, R. M., & Steinbuch, M. (2005). Optimal design of energy storage systems for hybrid vehicle drivetrains. In *Proceedings of Vehicle Power and Propulsion*. IEEE. doi:10.1109/VPPC.2005.1554535

Keith, M. (2007). Understanding power flows in HEV CVT's with ultra capacitor boosting plug-in hybrid electric vehicles R&D plan. In *Proceedings of EET*. EET.

Kim, J., Loe, S., & Yeo, T. (2010). Regenerative analysis of braking patterns for HEV. In *Proceedings of EVS 25*. EVS.

Kleimaier, A., & Schroder, D. (2002). An approach for the online optimized control of a hybrid powertrain. In *Proceedings of 7th International Workshop on Advanced Motion Control Proceedings*, (pp. 215-220). AMC.

Krawczyk, P. (2010). *Determining of chosen parameters of an urban electric vehicle equipped with a continuously variable transmission by simulation studies.* (Thesis). Warsaw University of Technology. Warsaw, Poland.

Krawczyk, P. (2011). *Comparative analysis of a CHPTD drive in various configurations.* (Master Thesis). Warsaw University of Technology. Warsaw, Poland.

Miller, J., Eschani, M., & Gao, J. (2005). Understanding power flows in HEV CVT's with ultra capacitor boosting. In *Proceedings of VPPC*. IEEE.

Mitsutani, N. Yamamoto, & Takaoka, T. (2010). Development of the plug-in hybrid system THS II plug-in. In *Proceedings of EVS 25*. EVS.

Sekrecki, M. (2010). *Mechanical transmission in hybrid and electric vehicle driving systems: Project assumption data for automatic gear transmission controlled by electric stepper motor.* (Thesis). Warsaw University of Technology. Warsaw, Poland.

Sekrecki, M. (2011). *Analysis and modeling of a hybrid propulsion system with automatic input gear changes and a planetary gear with two degrees of freedom.* (Master Thesis). Warsaw University of Technology. Warsaw, Poland.

Smaling, R., & Comits, H. (2010). A plug-in hybrid electric power train for commercial vehicles. In *Proceedings of EVS 25*. EVS.

Szumanowski, A., & Chang, Y. (2008). Battery management system based on battery nonlinear dynamics modeling. *IEEE Transactions on Vehicular Technology*, *57*(3), 1425–1432. doi:10.1109/TVT.2007.912176

Szumanowski, A., Chang, Y., Hajduga, A., & Piorkowski, P. (2007). Hybrid drive for ultalight city cars. In *Proceedings of EET*. EET.

Szumanowski, A., Chang, Y., & Piorkowski, P. (2005). Analysis of different control strategies and operating modes of compact hybrid planetary transmission drive. In *Proceedings of Vehicle Power and Propulsion*. IEEE. doi:10.1109/VPPC.2005.1554631

Szumanowski, A., Liu, Z., & Chang, Y. (2010). Design of planetary plug-in hybrid and its control strategy. In *Proceedings of EVS 25*. EVS.

Szumanowski, A., Liu, Z., & Hajduga, A. (2010). Zero steady-states electrical energy consumption clutch system. *High Technology Letters*, *16*(1), 58–62.

Takaoka, T., & Komatsu, H. (2010). Newly-developed toyota plug-in hybrid system and its vehicle perfomance. In *Proceedings of 20th Aachen Colloquium Automobile and Engine Technology*. Aachen Colloquium.

Yn, C., Iseng, C., & Lin, S. (2010). Development of clutchless AMT system for EV. In *Proceedings of EVS 25*. EVS.

Appendix

INDEX OF ABBREVIATIONS

- **AC:** Alternating Current
- **ACR:** Automatic Current Regulator
- **AMT:** Automatic Manual Transmission
- **ASR:** Automatic Speed Regulator
- **BLDC:** Permanent Magnet Brushless DC Motor
- **BMS:** Battery Management System
- **CHPTD:** Compact Hybrid Planetary Transmission Drive
- **CVT:** Continuously Variable Transmission
- **DC:** Direct Current
- **ECE:** Basic European City Driving Cycle
- **EM:** Electric Machine
- **EMF:** Electromotive Force
- **ESR:** Equivalent Series Resistance
- **EV:** Battery Powered Electric Vehicle
- **FC:** Fuel Cell
- **FUD:** Federal Urban Driving Cycle
- **HE:** High Energy Battery
- **HEV:** Hybrid Electric Vehicle
- **HP:** High Power Battery
- **HSSD:** Hybrid Split Sectional Drive
- **HV:** Full Hybrid Vehicle
- **ICE:** Internal Combustion Engine
- **IGBT:** Insulated Gate Bipolar Transistor
- **NEDC:** New Extended European Driving Cycle
- **PHEV:** Plug-in Hybrid Electric Vehicle
- **PM:** Permanent Magnet
- **PMSM:** Permanent Magnet Synchronous Motor
- **PPG:** Planetary Transmission
- **PS:** Primary Source
- **PWM:** Pulse Width Modulation

- **RMS:** Root-mean-square
- **SCR:** Silicon Controlled Rectifier
- **SOC:** Battery State of Charge
- **SS:** Secondary Source
- **UC:** Ultra Capacitor

Compilation of References

Abdulaziz, M., & Jufer, M. (1974). Magnetic and electric model of synchronous permanent – Magnet machines. *Bull. SEV, 74*(23), 1339–1340.

Anderson, W. M., & Cambier, C. (1990). An advanced electric drivetrain for EVs. In *Proceedings of EVS 10*. Hong Kong: EVS.

Antoniou, A. Komythy, Brench, J., & Emadi, A. (2005). Modeling and simulation of various hybrid electric configurations of the HMMWV. In *Proceedings of VPPC*. IEEE.

Ashihaga, T., Mizuno, T., Shimizu, H., Natori, K., Fujiwara, N., & Kaya, Y. (1992). Development of motors and controllers for electric vehicle. In *Proceedings of EVS 11*. Florence, Italy: EVS.

Ayad, M. Y., Rael, S., & Davat, B. (2003). Hybrid power source using supercapacitors and batteries. In *Proceedings of 10th European Conference on Power Electronics and Applications (EPE2003)*. EPE.

Barsaq, F., Blanchard, P., Broussely, M., & Sarre, G. (2004). Application of li-ion battery technology to hybrid vehicles. In *Proceedings of ELE European Drive Transportation Conference*. Estorial, Portugal: ELE.

Bartley, T. (2005). Ultra capacitors and batteries for energy storage in heavy-duty hybrid-electric vehicles. In *Proceedings of 22nd International Battery Seminar and Exhibit*. Fort Lauderdale, FL: IEEE.

Baucher, J. P. (2007). Online efficiency diagnostic method of three phases asynchronous motor. In *Proceedings of Powereng IEEE*. IEEE.

Berctta, J. (1998). New classification on electric–thermal hybrid vehicle. In *Proceedings of EVS 15*. Brussels: EVS.

Bin, W. (n.d.). Brushless DC motor speed control. *Ryerson Polytechnic University.*.

Blanchard, P., Gaignerot, L., Hemeyer, S., & Rigobert, G. (2002). Progress in SAFT li-ion cells and batteries for automotive application. In *Proceedings of EVS19*. EVS.

Braga, G., Farini, A., Fuga, F., & Manigrasso, R. (1991). Synchronous drive for motorized wheels without a gearbox for light rail systems and electric cars. In *Proceedings of EPE'91 European Conference on Power Electronics*. EPE.

Brusaglino, G., & Tenconi, A. (1992). System engineering with new technology for electrically propelled vehicles. In *Proceedings of EVS 11*. EVS.

Compilation of References

Bullock, K. J., & Hollis, P. G. (1998). Energy storage elements in hybrid bus applications. In *Proceedings of EVS15*. EVS.

Burke, A. (2002). Cost-effective combinations of ultra capacitors and batteries for vehicle application. In *Proceedings of AABC*. AABC.

Burke, A. F., & Heitner, K. L. (1992). Test procedures for hybrid/electric vehicles using different control strategies. In *Proceedings of EVS 11*. Florence, Italy.

Burke, A. F., & McDowell, R. D. (1992). The performance of advanced electric vans – Test and simulation. In *Proceedings of EVS 11*. EVS.

Burke, A., & Miller, M. (2003). Ultra capacitor and fuel cell applications. In *Proceedings of EVS20*. EVS.

Burke, A. F. (1992). *Development of test procedures for hybrid electric vehicles*. INEL US Department of Energy INEL Field Office.

Butler, K. L., Ehsani, M., & Kamath, P. (1999, Nov.). A matlab-based modeling and simulation package for electric and hybrid electric vehicle design. *IEEE Transactions on Vehicular Technology*, *48*(6), 1770–1778. doi:10.1109/25.806769.

Cackette, T., & Evaoshenk, T. (1995). *A new look at HEV in meeting California's clean air goals*. Paper presented at EPRI North American EV & Infrastructure Conference. Atlanta, GA.

Caumont, O., Moigne, P. L., Rombaut, C., Muneret, X., & Lenain, P. (2000). Energy gauge for lead acid batteries in electric vehicles. *IEEE Transactions Energy Conservation*, *15*(3), 354–360. doi:10.1109/60.875503.

Cegnar, E. J., Hess, H. L., & Johnson, B. K. (2004). A purely ultra capacitor energy storage system hybrid electric vehicles utilizing a based DC-DC boost converter. In *Proceedings of IEEE Applied Power Electronics Conference APEC'04*, (vol. 2, pp. 1160 – 1164). IEEE.

Ceraol, M., & Pede, G. (2001). Techniques for estimating the residual range of an electric vehicle. *IEEE Transactions on Vehicular Technology*, *50*(1), 109–111. doi:10.1109/25.917893.

Chan, C. C. (1994). The development of an advanced electric vehicle. In *Proceedings of EVS 12*. EVS.

Chan, C. C. (1994). The development of an advanced electric vehicle. In *Proceedings of EVS 12*. Los Angeles, CA: EVS.

Chan, C. C., & Lueng, W. S. (1990). A new permanent magnet motor drive for mini electric vehicles. In *Proceedings of EVS 10*. EVS.

Chan, C. C., Jiang, G. H., Chen, X. Y., & Wong, K. T. (1992). A novel high power density PM motor drive for electric vehicle. In *Proceedings of EVS 11*. EVS.

Chan, C. C. (2002). The state-of-the-art of electric and hybrid vehicles. *Proceedings of the IEEE*, *90*(2), 247–270. doi:10.1109/5.989873.

Chan, C. C., & Chau, K. T. (2001). *Modern electric vehicle technology*. Oxford, UK: Oxford University Press.

Chang, Y. (2005). *Battery modeling for HEV and battery parameters adjustment for series hybrid bus by simulation*. (MSc thesis). Warsaw University of Technology. Warsaw, Poland.

Compilation of References

Chen, K., Bouscayrol, A., Berthon, A., Delarue, P., Hissel, D., & Trigui, R. (2009). Global modeling of different vehicles. *IEEE Vehicular Technology Magazine*, *4*(2), 80–89. doi:10.1109/MVT.2009.932540.

Chris, C., & Luo, L. (2005). Analitical design of PM traction motors. In *Proceedings Vehicle Power and Propulsion Conference VPPC*. Chicago, IL: IEEE.

Chu, A. (2007). Nanophosphate li-ion technology for transportation application. In *Proceedings of EVS 23*. EVS.

Chu, A., & Braatz, P. (2002). Comparison of commercial supercapacitors and high-power lithium-ion batteries for power-assist applications in hybrid electric vehicles. *Journal of Power Sources*, *112*, 236–240. doi:10.1016/S0378-7753(02)00364-6.

Datla, M., & High, A. (2007). Performance decoupling control of an induction motor with efficient flux estimator. In *Proceedings of Powereng*. IEEE.

Debal, P., Faid, S., & Bervoets, S. (2010). Parallel hybrid (booster) range extender power train. In *Proceedings of EVS 25*. EVS.

Dietrich, P., Ender, M., & Wittmer, C. (1996). *Hybrid III power train update*. Electric & Hybrid Vehicle Technology.

Duvall, M. S. (2005). Battery evaluation for plug-in hybrid electric vehicles. In *Proceedings of Vehicle Power and Propulsion*. IEEE. doi:10.1109/VPPC.2005.1554580.

Ehsani, M., Gao, Y., Gay, L., & Emadi, A. (2004). *Modern electronic, hybrid electric and fuel cell vehicles – Fundamentals, theory and design*. Boca Raton, FL: CRC Press. doi:10.1201/9781420037739.

Eifert, M. (2005). Alternator control algorithm to minimize fuel consumption. In *Proceedings of VPPC*. IEEE.

Eiraku, A., Abe, T., & Yamacha, M. (1998). An application of hardware in the loop simulation to HEV. In *Proceedings of EVS 15*. EVS.

Ferraris, P., Tenconi, A., Brusaglino, G., & Ravello, V. (1996). Development of a new high-performance induction motor drive train. In *Proceedings of EVS 13*. EVS.

Fleckner, M., Gohring, M., & Spiegel, L. (2009). New strategies for an efficiency optimized layout of an operating control for hybrid vehicles. In *Proceedings of Aachen Colloquium*. Aachen Colloquium.

Fletcher, R. (1974). Minimization of a quadratic function of many variables subject only to lower and upper hounds. *J. Inst. Maths. Applies.*

Fletcher, R., & Powell, M. J. D. (1974). On the modification of LDLT factorization. *Mathematics of Computation*, 28.

Fujioka, N., Ikona, M., Kiruna, T., & Konomaro, K. (1998). Nickel metal-hydride batteries for hybrid vehicle. In *Proceedings of EVS 15*. EVS.

Gear, C. W. (1971). Simulations: Numerical solution of differential algebraic equations. *IEEE Transactions, 18*.

Gear, C. W. (1972). DIFSUB for the solution of ordinary differential equations. *Communications of the ACM*, 14.

Giglioli, R., Salutori, R., & Zini, G. (1992). Experience on a battery state of charge observer. In *Proceedings of EVS 11*. EVS.

Compilation of References

Gill, P. E., & Murray, W. (1972). Quasi–Newton methods for unconstrained optimization. *J. Inst. Applies.*

Gosden, D. F. (1992). Wide speed range operation of an AC PM EV drive. In *Proceedings of EVS 11*. EVS.

Gu, W., & Wang, C. (2000). Thermal-electrochemical modeling of battery systems. *Journal of the Electrochemical Society*, *147*(8), 2910–2922. doi:10.1149/1.1393625.

Hajduga, A. (2005). *Electrical-mechanical parameters adjustment of ICE hybrid drive using simulation method.* (PhD thesis). Warsaw University of Technology. Warsaw, Poland.

Haltori, N., Aoyama, S., Kitada, S., Matsuo, I., & Hamai, K. (1998). *Configuration and operation of a newly-developed parallel hybrid propulsion system.* Nissan Motor Co..

Haltori, N., Aoyama, S., Kitada, S., Matsuo, I., & Hamai, K. (2011). *Configuration and operation of a newly-developed parallel hybrid propulsion system.* Nissan Motor Co Technical Papers.

Hayasaki, K., Kiyota, S., & Abe, T. (2009). The potential of parallel hybrid system and nissan's approach. In *Proceedings of Aachen Colloquium*. Aachen Colloquium.

He, X., Parten, M., & Maxwell, T. (2005). Energy management strategies for HEV. In *Proceedings of IEEE Vehicle Power and Propulsion Conference, Illinois Institute of Technology*. Chicago, IL: IEEE.

He, Z., Zhang, C., & Sun, F. (2002). Design of EV BMS. In *Proceedings of EVS19*. EVS.

Hecke, R., & Plouman, S. (2009). Increase of recuperation energy in hybrid vehicles. In *Proceedings of Aachen Colloquium*. Aachen Colloquium.

Hellenbraich, G., & Rosenburg, V. (2009). FEV's new parallel hybrid transmission with single dry clutch and electric support. In *Proceedings of Aachen Colloquium*. Aachen Colloquium.

Hellenbroich, G., & Huth, T. (2010). New planetary-based hybrid automatic transmission with electric torque converter and on-demand actuation. In *Proceedings of 20th Aachen Colloquium Automobile and Engine Technology*, (pp. 92-105). Aachen Colloquium.

Hellenbrouch, G., Lefgen, W., Janssen, P., & Rosenburg, V. (2010). New planetary based hybrid automatic transmission with electric torque converter on demand actuation. In *Proceedings of Aachen Colloquium*. Aachen Colloquium.

Henneberger, G., & Lutter, T. (1991). Brushless DC–Motor with digital state controller. In *Proceedings of EPE '91 European Conference on Power Electronics and Application*. EPE.

Heywood, J. B. (1989). *Internal combustion engine fundamentals*. London: McGraw-Hill Company.

Hirschmann, D., Tissen, D., Schroder, S., & De Donecker, R. W. (2005). Inverter design for HEV considering mission profile. In *Proceedings of Vehicle Power and Propulsion Conference VPPC*. IEEE.

Hofman, T., & Van Druten, R. (2004). Research overview – Design specification for HV. In *Proceedings of ELE European Drive Transportation Conference*. Estorial, Portugal: ELE.

Hofman, T., Hoekstra, D., van Druten, R. M., & Steinbuch, M. (2005). Optimal design of energy storage systems for hybrid vehicle drivetrains. In *Proceedings of Vehicle Power and Propulsion*. IEEE. doi:10.1109/VPPC.2005.1554535.

Huang, H., Cambier, C., & Geddes, R. (1992). High constant power density wide speed range PM motor for EV application. In *Proceedings of EVS 11*. EVS.

Ippolito, L., & Rovera, G. (1996). Potential of robotized gearbox to improve fuel economy. In *Proceedings of International Symposium Power Train Technologies for a 3-Litre-Car*. Academic Press.

Jezernik, K. R. (1994). Induction motor control for electric vehicle. In *Proceedings of EVS 12*. EVS.

Johnson, V. H., & Pesaran, A. A. (2000). Temperature-dependent battery models for high-power lithium-ion batteries. In *Proceedings of International Electric Vehicle Symposium*. EVS.

Jozefowitz, W., & Kohle, S. (1992). Volkswagen golf hybrid – Vehicle result and test result. In *Proceedings of EVS 11*. Florence, Italy: EVS.

Jyunichi, L., & Hiroya, T. (1996). Battery state-of-charge indicator for electric vehicle. In *Proceedings of International Electric Vehicle Symposium*. EVS.

Kalman, P. G. (2002). Filter SOC estimation for Li PB HEV cells. In *Proceedings of EVS19*. EVS.

Karden, E., Buller, S., & De Doncker, R. W. (2002). A frequency-domain approach to dynamical modeling of electrochemical power sources. *Electrochimica Acta*, *47*(13–14), 2347–2356. doi:10.1016/S0013-4686(02)00091-9.

Keith, H. (2007). Doe plug in hybrid electric vehicles R&D plan. In *Proceedings of European ELE – Drive Transportation Conference*. Brussels: ELE.

Keith, M. (2007). Understanding power flows in HEV CVT's with ultra capacitor boosting plug-in hybrid electric vehicles R&D plan. In *Proceedings of EET*. EET.

Keith, H. (2007). Doe plug in hybrid electric vehicles R&D plan. In *Proceedings of European ELE – Drive Transportation Conference*. Brussels: ELE.

Kelly, K. (2007). Li-ion batteries in EV/HEV application. In *Proceedings of EVS 23*. EVS.

Kenjo, T., & Nagamori, S. (1985). *Permanent magnet and brushless DC motors*. Oxford, UK: Claderon Press.

Killman, G. (2009). The hybrid power train of the new Toyota Prius. In *Proceedings of Aachen Colloquium*. Aachen Colloquium.

Kim, J., Lee, S., & Cho, B. H. (2010). SOH prediction of li-ion battery based on hamming network using two patterns recognition. In *Proceedings of EVS 25*. EVS.

Kim, J., Loe, S., & Yeo, T. (2010). Regenerative analysis of braking patterns for HEV. In *Proceedings of EVS 25*. EVS.

King, R. D., & Konrad, C. E. (1992). Advanced on–board EV AC drive – Concept to reality. In *Proceedings of EVS 11*. EVS.

Kitada, S., Aoyama, S., Haltori, N., Maeda, H., & Matsuo, I. (1998). Development of parallel hybrid vehicles system with CVT. In *Proceedings of EVS15*. EVS.

Kleimaier, A., & Schroder, D. (2002). An approach for the online optimized control of a hybrid powertrain. In *Proceedings of 7th International Workshop on Advanced Motion Control Proceedings*, (pp. 215-220). AMC.

Compilation of References

Krawczyk, P. (2010). *Determining of chosen parameters of an urban electric vehicle equipped with a continuously variable transmission by simulation studies*. (Thesis). Warsaw University of Technology. Warsaw, Poland.

Krawczyk, P. (2011). *Comparative analysis of a CHPTD drive in various configurations*. (Master Thesis). Warsaw University of Technology. Warsaw, Poland.

Kruger, M., Cornetti, G., Greis, A., Weidmann, U., Schumacher, H., Gerhard, J., & Leonhard, R. (2010). Operational strategy of a diesel HEV with focus on the combustion engine. In *Proceedings of Aachen Colloquium*. Aachen Colloquium.

Kuhn, B., Pitel, G., & Krein, P. (2005). Electrical properties and equalization of li-ion cells in automotive application. In *Proceedings of VPPC*. IEEE.

Lecout, B., & Liska, I. (2004). NiMH advanced technologies batteries for hybrid public transportation system. In *Proceedings of ELE European Drive Transportation Conference*. ELE.

Ledowskij, A. N. (1985). *Electrical machines with high coercive force permanent magnets*. Moscow: Energoatimizdat.

Lukic, S. M., Wirasingha, S. G., Rodriguez, F., Cao, J., & Emadi, A. (2006). Power management of an ultra capacitor/battery hybrid energy storage system in an HEV. In *Proceedings of IEEE Power and Propulsion Conference*. IEEE.

Malkhandi, S., Sinha, S. K., & Muthukumar, K. (2001). Estimation of state-of-charge of lead-acid battery using radial basis function. In *Proceedings of Industrial Electronics Conference*, (vol. 1, pp. 131–136). IEC.

Marcos, J., Lago, A., Penalver, C. M., Doval, J., Nogueira, A., Castro, C., & Chamadoira, J. (2001). An approach to real behavior modeling for traction lead-acid batteries. In *Proceedings of Power Electronics Specialists Conference*, (vol. 2, pp. 620–624). PES.

Martines, J. E., Pires, V. F., & Gomes, L. (2009). Plug-in electric vehicles integration with renewable energy building facilities – Building vehicle interface. In *Proceedings of Powereng IEEE 2nd International Conference on Power Engineering, Energy and Electrical Drives*. Lisbon: IEEE.

Matsusa, K., & Katsuta, S. (1996). Fast rotor flux control of vector contolled induction motor operating at maximum efficiency for EV. In *Proceedings of EVS 13*. EVS.

McCann, R., & Domagatla, S. (2005). Analyses of MEMS-based rotor flux sensing in a hybrid reluctance motor. In *Proceedings of Vehicle Power and Propulsion Conference VPPC*. IEEE.

Miller, J. M., McCleer, P. J., & Everett, M. (2005). Comparative assessment of ultracapacitors and advanced battery energy storage systems in power split electronic-CVT vehicle power trains. In *Proceedings of IEEE International Electric Machines and Drives Conference IEMDC2005*. IEEE.

Miller, J. M., McCleer, P. J., Everett, M., & Strangas, E. (2005). Ultra capacitor plus battery energy storage system sizing methodology for HEV power split electronic CVT's. In *Proceedings of IEEE International Symposium on Industrial Electronics*. IEEE.

Miller, J., Eschani, M., & Gao, J. (2005). Understanding power flows in HEV CVT's with ultra capacitor boosting. In *Proceedings of VPPC*. IEEE.

Mitsutani, N. Yamamoto, & Takaoka, T. (2010). Development of the plug-in hybrid system THS II plug-in. In *Proceedings of EVS 25*. EVS.

Moore, T. (1996). Ultralight hybrid vehicle principles and design. In *Proceedings of EVS 13*. Osaka, Japan: EVS.

Morano, T. (2007). Pro-post hybrid parallel hybrid drive system. In *Proceedings of EVS 23*. EVS.

Morio, K., Kazuhiro, H., & Anil, P. (1997). Battery SOC and distance to empty meter of the Honda EV plus. In *Proceedings of International Electric Vehicle Symposium*, (pp. 1–10). EVS.

Moseley, P., & Cooper, A. (1998). Lead acid electric vehicle batteries – Improved performance of the affordable option. In *Proceedings of EVS 15*. EVS.

Nelson, R. F. (2000). Power requirements for battery in HEVs. *Journal of Power Sources*, *91*, 2–26. doi:10.1016/S0378-7753(00)00483-3.

Neuman, A. (2004). Hybrid electric power train. In *Proceedings of ELE European Drive Transportation Conference*. Estorial, Portugal: ELE.

Noil, M. (2007). Simulation and optimization of a full HEV. In *Proceedings of EVS 23*. EVS.

Ortmeyer, T. (2005). Variable voltage variable frequency options for series HV. In *Proceedings of Vehicle Power and Propulsion Conference VPPC*. Chicago, IL: IEEE.

Overman, B. (1993). Environmental legislation may initiate the EV and HV industry will economics sustain. In *Proceedings of ISATA*. Aachen, Germany: ISATA.

Ovshinski, S. R., Dhar, S. K., Venkatesan, S., Fetchenko, M. A., Gifford, P. R., & Corrigan, D. A. (1992). Performance advances in ovonic NiMH batteries for electric vehicles. In *Proceedings of EVS 11*. EVS.

Padmaraja, Y. (2003). *(BLDC) motor fundamentals*. Brushless, DC: Microchip Technology Inc..

Pang, S., Farrell, J., Du, J., & Barth, M. (2001). Battery state-of-charge estimation. In *Proceedings of American Control Conference*, (vol. 2, pp. 1644–1649). ACC.

Piller, S., Perrin, M., & Jossen, A. (2001). Methods for state–of–charge determination and their applications. *Journal of Power Sources*, *96*, 113–120. doi:10.1016/S0378-7753(01)00560-2.

Piórkowski, P. (2004). *Study of energy's accumulation efficiency in hybrid drives of vehicles*. (Ph.D. thesis). Warsaw University of Technology. Warsaw, Poland.

Plett, G. (2003). LiPB dynamic cell models for kalman-filter SOC estimation. In *Proceedings of EVS-20*. EVS.

Portmann, D., & Guist, A. (2010). Electric and hybrid drive developed by Mercedes – Benz vans and technical challenges to achieve a successful market position. In *Proceedings of Aachen Colloquium*. Aachen Colloquium.

Rodrigues, S., Munichandraiah, N., & Shukla, A. (2000). A review of state-of-charge indication of batteries by means of A.C. impedance measurements. *Journal of Power Sources*, *87*(1-2), 12–20. doi:10.1016/S0378-7753(99)00351-1.

Rsekranz, C. (2007). Modern battery systems for HEV. In *Proceedings of EVS 23*. EVS.

Compilation of References

Ruschmayer, R. Shussier, & Biermann, J.W. (2006). Detailed aspects of HV. In *Proceedings of Aachen Colloquium*. Aachen Colloquium.

Rutquist, P. (2002). Optimal control for the energy storage in a HEV. In *Proceedings of EVS19*. EVS.

Saenger, F., Zetina, S., & Neiss, K. (2008). Control approach for comfortable power shifting in hybrid transmission – ML 450 hybrid. In *Proceedings of Aachen Colloquium*. Aachen Colloquium.

Salkind, A., Atwater, T., Singh, P., Nelatury, S., Damodar, S., Fennie, C., & Reisner, D. (2001). Dynamic characterization of small lead-acid cells. *Journal of Power Sources*, *96*(1), 151–159. doi:10.1016/S0378-7753(01)00561-4.

Samper, Z. S., & Neiss, K. (2008). Control approach for comfortable powershifting in hybrid transmission. In *Proceedings of Aachen Colloquium*. Aachen Colloquium.

Sato, S., & Kawamura, A. (2002). A new estimation method of state of charge using terminal voltage and internal resistance for lead acid battery. *Processing Power*, *2*, 565–570.

Schofield, M., Mellor, P. H., & Howe, D. (1002). Field weakening of brushless PM motors for application in a hybrid electric vehicle. In *Proceedings of EVS 11*. EVS.

Schofield, N. (2006). Hybrid PM generators for EV application. In *Proceedings of Vehicle Power and Propulsion Conference VPPC*. IEEE.

Schupbach, R. M., & Balda, J. C. (2003). The role of ultra capacitors in an energy storage unit for vehicle power management. In *Proceedings of IEEE 58th Vehicular Technology Conference*. IEEE.

Schupbach, R. M., Balda, J. C., Zolot, M., & Kramer, B. (2003). Design methodology of a combined battery-ultra capacitor energy storage unit for vehicle power management. In *Proceedings of IEEE Power Electronics Specialists Conference*. IEEE.

Schussler, M. (2007). Predictive control for HEV – Development optimization and evaluation. In *Proceedings of ELE European Conference*. ELE.

Sekrecki, M. (2010). *Mechanical transmission in hybrid and electric vehicle driving systems: Project assumption data for automatic gear transmission controlled by electric stepper motor*. (Thesis). Warsaw University of Technology. Warsaw, Poland.

Sekrecki, M. (2011). *Analysis and modeling of a hybrid propulsion system with automatic input gear changes and a planetary gear with two degrees of freedom*. (Master Thesis). Warsaw University of Technology. Warsaw, Poland.

Shen, W. X., Chan, C. C., Lo, E. W. C., & Chau, K. T. (2002). Estimation of battery available capacity under variable discharge currents. *Journal of Power Sources*, *103*(2), 180–187. doi:10.1016/S0378-7753(01)00840-0.

Shen, W. X., Chau, K. T., Chan, C. C., & Lo, E. W. C. (2005). Neural network-based residual capacity indicator for nickel-metal hydride batteries in electric vehicles. *IEEE Transactions on Vehicular Technology*, *54*(5), 1705–1712. doi:10.1109/TVT.2005.853448.

Shimizu, K., & Semya, S. (2002). Fuel consumption test procedure for HEV. In *Proceedings of EVS19*. EVS.

Smaling, R., & Comits, H. (2010). A plug-in hybrid electric power train for commercial vehicles. In *Proceedings of EVS 25*. EVS.

Sporckman, B. (1992). Comparison of emissions from combustion engines and 'European' EV. In *Proceedings of EVS 11*. Florence, Italy: EVS.

Sporckman, B. (1995). *Electricity supply for electric vehicles in Germany*. Paper presented at EPRI North American EV & Infrastructure Conference. Atlanta, GA.

Stempel, R. C., Ovshinsky, S. R., Gifford, P. R., & Corrigan, D. A. (1998). Nickel-metal hydride: Ready to serve. *IEEE Spectrum*. doi:10.1109/6.730517.

Stienecker, A. W. (2005). A combined ultra capacitor – Lead acid battery energy storage system for mild hybrid electric vehicles. In *Proceedings of IEEE VPPC*. IEEE.

Strabnick, R., Naunin, D., & Freger, D. (2004). Online SOC determination and forecast for EV by use of different battery models. In *Proceedings of ELE European Drive Transportation Conference*. ELE.

Sweet, L.H., & Anhalt, D.A. (1978). Optimal control of flywheel hybrid transmission. *Journal of Dynamic System, Measurement and Control, 100*.

Szumanowski, A. (1993). Regenerative braking for one and two source EV drives. In *Proceedings of ISATA 26*. Aachen, Germany: ISATA.

Szumanowski, A. (1994). Simulation study of two and three-source hybrid drives. In *Proceedings of EVS12*. EVS.

Szumanowski, A. (1996). Generic method of comparative energetic analysis of HEV drive. In *Proceedings of EVS13*. Osaka, Japan: EVS.

Szumanowski, A. (1996). Simulation testing of the travel range of vehicles powered from battery under pulse load conditions. In *Proceedings of EVS13*. Osaka, Japan: EVS.

Szumanowski, A. (1996). Generalized method of comparative energetic analyses of HEV drives. In *Proceedings of EVS13*. Osaka, Japan: EVS.

Szumanowski, A. (1997). Advanced more efficient compact hybrid drive. In *Proceedings of EVS14*. EVS.

Szumanowski, A. (1999). *Evolution of two steps of freedom planetary transmission in hybrid vehicle application*. Global Power Train Congress.

Szumanowski, A. (2000). *Fundamentals of hybrid vehicle drives*. ISBN83-7204-114-8

Szumanowski, A., & Bramson, E. (1992). Electric vehicle drive control in constant power mode. In *Proceedings of ISATA 25*. Florence, Italy: ISATA.

Szumanowski, A., & Brusaglino, G. (1992). Analyses of the hybrid drive consisting of electrochemical battery and flywheel. In *Proceedings of EVS 11*. EVS.

Szumanowski, A., & Brusaglino, G. (1999). Approach for proper battery adjustment for HEV application. In *Proceedings of EVS 16*. EVS.

Szumanowski, A., & Hajduga, A. (1998). Energy management in HV drive advanced propulsion systems. In *Proceeding of GPS*. Detroit, MI: GPS.

Szumanowski, A., & Hajduga, A. (2006). Optimization series HEV drive using modeling and simulation methods. In *Proceedings of VPPC*. IEEE.

Compilation of References

Szumanowski, A., & Jaworowski, B. (1992). The control of the hybrid drive. In *Proceedings of EVS11*. EVS.

Szumanowski, A., & Krasucki, J. (1993). *Simulation study of battery engine hybrid drive*. Paper presented at the 2nd Polish-Italian Seminar Politecnico di Torino. Turin, Italy.

Szumanowski, A., & Nguyen, V. K. (1999). *Comparison of energetic properties of different two-source hybrid drive architectures*. Global Power Train Congress.

Szumanowski, A., & Piórkowski, P. (2004). Ultralight small hybrid vehicles: Why not? In *Proceedings of ELE European Drive Transportation Conference*. Estorial, Portugal: ELE.

Szumanowski, A., Chang, Y., & Piórkowski, P. (2005). Method of battery adjustment for hybrid drive by modeling and simulation. In *Proceedings of IEEE Vehicle Power and Propulsion (VPP) Conference*. IEEE.

Szumanowski, A., Chang, Y., & Piórkowski, P. (2005). Analysis of different control strategies and operating modes of a compact hybrid planetary transmission drive. In *Proceedings of IEEE Vehicle Power and Propulsion (VPP) Conference*. IEEE.

Szumanowski, A., Chang, Y., Hajduga, A., & Piorkowski, P. (2007). Hybrid drive for ultalight city cars. In *Proceedings of EET*. EET.

Szumanowski, A., Chang, Y., Piórkowski, P., Jankowska, E., & Kopczyk, M. (2005). Performance of city bus hybrid drive equipped with li-ion battery. In *Proceedings of EVS 21*. EVS.

Szumanowski, A., Dębicki, J., Hajduga, A., Piórkowski, P., & Chang, Y. (2003). Li-ion battery modeling and monitoring approach for hybrid electric vehicle applications. In *Proceedings of EVS-20*. EVS.

Szumanowski, A., Hajduga, A., & Piórkowski, P. (1998). Evaluation of efficiency alterations in hybrid and electric vehicles drives. In *Proceedings of Advanced Propulsion Systems*. GPC.

Szumanowski, A., Hajduga, A., & Piórkowski, P. (1998). Proper adjustment of combustion engine and induction motor in hybrid vehicles drive. In *Proceedings of EVS15*. EVS.

Szumanowski, A., Hajduga, A., Chang, Y., & Piórkowski, P. (2007). Hybrid drive for ultralight city cars. In *Proceedings of ELE European Conference*. ELE.

Szumanowski, A., Hajduga, A., Piórkowski, P., & Brusaglino, G. (2002). Dynamic torque speed distribution modeling for hybrid drives design. In *Proceedings of EVS19*. EVS.

Szumanowski, A., Hajduga, A., Piórkowski, P., Stefanakos, E., Moore, G., & Buckle, K. (1999). Hybrid drive structure and power train analysis for florida shuttle buses. In *Proceedings of EVS 16*. EVS.

Szumanowski, A., Liu, Z., & Chang, Y. (2010). Design of planetary plug-in hybrid and its control strategy. In *Proceedings of EVS 25*. EVS.

Szumanowski, A., Nguyen, K. V., & Piórkowski, P. (2000). Analysis of charging-discharging of nickel metal hydride (NiMH) battery and its influence on the fuel consumption of advanced hybrid drives. In *Proceedings of GPC*. GPC.

Szumanowski, A., Piórkowski, P., Hajduga, A., & Ngueyen, K. (2000). The approach to proper control of hybrid drive. In *Proceedings of EVS 17*. Montreal, Canada: EVS.

Szumanowski, A., Piórkowski, P., Sun, F., & Chang, Y. (2005). Control strategy choice: Influences on effectiveness of a HEV drive. In *Proceedings of EVS 21*. EVS.

Szumanowski, A. (2000). *Fundamentals of hybrid vehicle drives*. Warsaw: ITE Press.

Szumanowski, A. (2006). *Hybrid electric vehicle drive design based on urban buses*. Warsaw: ITE Press.

Szumanowski, A. (2010). *Nonlinear dynamics traction battery modeling* (pp. 199–220). INTECH.

Szumanowski, A., & Chang, Y. (2008). Battery management system based on battery nonlinear dynamics modeling. *IEEE Transactions on Vehicular Technology, 57*(3), 1425–1432. doi:10.1109/TVT.2007.912176.

Szumanowski, A., Chang, Y., & Piorkowski, P. (2005). Analysis of different control strategies and operating modes of compact hybrid planetary transmission drive. In *Proceedings of Vehicle Power and Propulsion*. IEEE. doi:10.1109/VPPC.2005.1554631.

Szumanowski, A., Chang, Y., & Piórkowski, P. (2006). Battery parameters adjustment for series hybrid bus by simulation. *Electrotechnical Review, 2*, 139.

Szumanowski, A., & Hajduga, A. (1998). Energy management in HV drive advanced propulsion systems. In *Proceedings of Advanced Propulsion Systems*. GPS.

Szumanowski, A., Hajduga, A., & Piórkowski, P. (1998). Evaluation of efficiency alterations in hybrid and electric vehicles drives. In *Proceedings of Advanced Propulsion Systems*. GPC.

Szumanowski, A., Liu, Z., & Hajduga, A. (2010). Zero steady-states electrical energy consumption clutch system. *High Technology Letters, 16*(1), 58–62.

Szumanowski, A., Piorkowski, P., & Chang, Y. (2007). Batteries and ultra capacitors set in hybrid propulsion systems. power engineering, energy and electrical drives. In *Proceedings of Powereng*. IEEE.

Takaoka, T., & Komatsu, H. (2010). Newly-developed toyota plug-in hybrid system and its vehicle perfomance. In *Proceedings of 20th Aachen Colloquium Automobile and Engine Technology*. Aachen Colloquium.

Tamburro, A., Mesiti, D., Ravello, V., Pesch, M., Schenk, R., & Glauning, J. (1997). *An intergrated motor-generator development for an effective drive train re-engineering*. CRF Technical Papers.

Timmermans, J.-M., Zadora, P., Cheng, Y., Van Mierlo, J., & Lataire, P. (2005). Modeling and design of super capacitors as a peak power unit for hybrid electric vehicles. In *Proceedings of IEEE 2005 VPPC*. IEEE. doi:10.1109/VPPC.2005.1554635.

Tojura, K., & Sekimori, T. (1998). Development of battery system for hybrid vehicle. In *Proceedings of EVS 15*. EVS.

Trackenbrodt, A., & Nitz, L. (2006). Two-mode hybrids = adoption power of intelligent system. In *Proceedings of Aachen Colloquium*. Aachen Colloquium.

Vaccaro, A., & Villaci, D. (2004). Prototyping a fussy based energy manager for parallel HEV. In *Proceedings of ELE European Drive Transportation Conference*. Estorial, Portugal: ELE.

Compilation of References

Van Mierlo, J., Maggetto, G., & Van den Bossche, P. (2003). Models of energy sources for EV and HEV: Fuel cells, batteries, ultracapacitors, flywheels and engine-generators. *Journal of Power Sources*, *128*(1), 76–89. doi:10.1016/j.jpowsour.2003.09.048.

Veinger, A. H. (1985). *Adjustable synchronous drive system*. Moscow: Energoatimizdat.

Wyczalek, F. A., & Wang, T. C. (1992). Regenerative braking for electric vehicles. In *Proceedings of ISATA 2J*. Florence, Italy: ISATA.

Yamagouchi, K. (1996). *Advancing the hybrid system*. Electric & Hybrid Vehicle Technology.

Yamaguci, K., Mityaishi, Y., & Kawamoto, M. (1996). Dual system – Newly-developed hybrid system. In *Proceedings of EVS 13*. EVS.

Yamamura, H. (1992). Development of power train system for Nissan. In *Proceedings of EVS 11*. Florence, Italy: EVS.

Yn, C., Iseng, C., & Lin, S. (2010). Development of clutchless AMT system for EV. In *Proceedings of EVS 25*. EVS.

Zhong, J. (2007). Regenerative braking system for a series hybrid electric city bus. In *Proceedings of EVS 23*. EVS.

Zolot, M. D., & Kramer, B. (2002). *Hybrid energy storage studies using batteries and ultra capacitors for advanced vehicles*. Paper presented at the 12th International Seminar on Double Layer Capacitors and Similar Energy Storage Devices. Deerfield Beach, FL.

About the Author

Antoni Szumanowski is a Professor at Warsaw University of Technology and the Head of the Department of the Multisource Propulsion System. He has been involved in Hybrid Electric Vehicles (HEVs) research since 1976. He earned both Ph.D. and D.Sc. degrees from Warsaw University of Technology, where he is the leader of the Scientific Division, whose activity is supported by government and university grants. He is a member of the Committee of EVS (Electric Vehicle Symposium), a board member of the European Association for Battery, Hybrid, and Fuel Cell Electric Vehicles (AVERE), and President of the Polish Society for Environmentally Friendly Vehicles. He is the reviewer of a few important journals such as *IEEE Transactions on Vehicular Technology* and the *Journal of Automobile Engineering*. He has been involved in EU FP6 (HyHEELS) and FP7 (AVTR, WIDE-MOB, ENIAC, INTRASME, SAGE) in the fields of electromobility and green transportation because of his rich experience and knowledge in the field of Hybrid & Electric Vehicles. He has published five monographs and books in Polish in the field of E&HEVs, alternative energy, and energy storage systems. Two monographs have been written in English, *Fundamentals of Hybrid Vehicle Drives* (2000) and *Hybrid Electric Vehicle Drives Design Edition Based on Urban Buses* (2006). The two English books were translated into Chinese and published in the Press Department of Beijing Institute of Technology. He has about 30 patents on the solutions of components and propulsion systems for E&HEVs.

Index

A

AC voltage 96-97
Alternative Current (AC) 52-53, 82, 167
armature windings 82-84, 92
Automatic Current Regulator (ACR) 106
Automatic Manual Transmission (AMT) 229, 253, 364
Automatic Speed Regulator (ASR) 106

B

Ball gearbox 364
Battery Electric Vehicle (BEV) 4
battery management 177, 193-195, 201-203, 221-222, 342, 361, 405
Battery Management System (BMS) 177, 201, 361
boost converter 226, 363-365
Brushless Direct Current (BLDC) 52, 127

C

cage-tube frame construction 372
carbon-metal fiber 208
Charging shuttling techniques 224
compact hybrid planetary transmission drive (CHPTD) 6, 14, 272, 283, 298, 380
Continuous Variable Transmission (CVT) 271, 373-375, 377
control shaft 364, 366
Coulomb capacity (Ah) 41, 173, 183, 195

D

diaphragm spring 381-382, 384, 388
differential gear box modeling 271
differential ratio 327
direct current (DC) 166, 168
double layer capacitor 207
Dual Clutch 229, 253, 390
dynamic engine modeling 231

E

Electric Motor (EM) 14, 276
Electric Vehicle (EV) 2
Electro-chemical capacitors 194, 206
electromagnetic actuator 293-294, 373, 381, 384, 389
electromagnetic torque 53, 56, 58-59, 75-76, 87, 100, 104-105, 126, 141, 240, 296, 298-299, 311-313, 315, 318
electromotive force (EMF) 151-152, 175, 177, 242
Euler-Lagrange equation 87
European Driving Cycle (ECE) 44
extended ECE (NEDC) 301

F

field weakening 92, 94-95, 107-108, 132, 134, 137-140, 150
Fuel Cell (FC) 7, 17

Index

G

Gear's method 40
general Nernst equation 158

H

high-dielectric materials 208
high-energy HE battery 201
high power (HP) 13, 153, 177
hybrid belt 366
Hybrid Electric Vehicle (HEV) 1-2, 4
hybrid mode acceleration 290
hybrid power train (HV) 13
hybrid split sectional drive (HSSD) 229, 260, 264

L

lithium ion (Li-Ion) 151, 156-157, 173

M

Maxwell Lab 209
metal hydride (MH) 155
motor rotation 366

N

New European Driving Cycle (NEDC) 7, 373, 376
nickel metal hydride (NiMH) 13, 24, 151, 155-156, 173, 183-185, 196, 203-204, 206, 228, 248, 326
Nonlinear Programming Problem (NPP) 43

P

Park-Goriev Equation 87

Permanent Machine (PM) 82
permanent magnet (PM) 52, 99, 166, 168, 241, 246, 300, 310, 377-378
permanent magnet synchronous (PMS) 120
Peukert formula 159, 162
planetary gear shaft 289
planetary transmission (PPG) 340
Plug-In Hybrid Electric Vehicle (PHEV) 4
power control strategy 323
power solid-state switches 81
Primary Source (PS) 23-24, 46
Pulse Width Modulation (PWM) 81-82, 91, 106

R

reducer ratio 343, 353, 396-397, 402

S

simulation analysis 254, 325
state of charge (SOC) 3, 9, 43-44, 151-152, 165, 167, 173, 175-176, 194
stator flux linkages 64, 66-68
steady state 47, 136, 293-294, 297, 370, 373, 381-382

T

thermal management 222
torque transmission 27
Total Transient Time (TTT) 254
transformation matrix 85

U

ultra capacitors 194-195, 208, 210-213, 223, 226-228